Functional dairy products

Related titles from Woodhead's food science, technology and nutrition list:

Dairy processing: improving quality (ISBN 978-1-85573-676-4)
With its distinguished international team of contributors, *Dairy processing* summarises key developments in the field and how they enhance dairy product safety and quality. The first part of the book discusses raw milk composition, production and quality. Part II reviews developments in processing from hygiene and HACCP systems to automation, high pressure processing and modified atmosphere packaging. The final part of the book considers developments for particular products such as fermented dairy products and cheeses.

Chilled foods Second edition (ISBN 978-1-85573-499-9)
Edited by Michael Stringer and Colin Dennis
The first edition of this book rapidly established itself as the standard work on the key quality issues in one of the most dynamic sectors in the food industry. This second edition has been substantially revised and expanded, and includes three new chapters on raw material selection for chilled foods.

'This book lives up to its title in reviewing a major section of the food industry.'
International Food Hygiene

Yoghurt: science and technology Third edition (ISBN 978-1-84569-213-1)
In its first edition, this book quickly established itself as the essential reference tool and only comprehensive book available in its field for both industry professionals and those involved in related fields of research. This completely revised and updated third edition incorporates the latest developments in scientific research underpinning the production of yoghurt of consistently high quality. Further scientific details on health-promoting yoghurts have been included, covering for example, the results of clinical studies and nutritional values of these products.

'No technical library should be without a copy and anyone in the business of manufacturing fermented milk or developing new milk products should obtain a copy.'
International Journal of Dairy Technology

Details of these books and a complete list of Woodhead's food science, technology and nutrition titles can be obtained by:

- visiting our web site at www.woodheadpublishing.com
- contacting Customer Services (e-mail: sales@woodheadpublishing.com; fax: +44 (0) 1223 893694; tel.: +44 (0) 1223 891358 ext. 130; address: Woodhead Publishing Limited, Abington Hall, Granta Park, Great Abington, Cambridge CB21 6AH, UK)

If you would like to receive information on forthcoming titles in this area, please send your address details to: Francis Dodds (address, tel. and fax as above; e-mail: francis.dodds@woodheadpublishing.com). Please confirm which subject areas you are interested in.

Woodhead Publishing Series in Food Science, Technology and Nutrition:
Number 79

Functional dairy products

Volume 1

Edited by
Tiina Mattila-Sandholm and Maria Saarela

CRC Press
Boca Raton Boston New York Washington, DC

WOODHEAD PUBLISHING LIMITED
Oxford   Cambridge   New Delhi

Published by Woodhead Publishing Limited, Abington Hall, Granta Park,
Great Abington, Cambridge CB21 6AH, UK
www.woodheadpublishing.com

Woodhead Publishing India Private Limited, G-2, Vardaan House, 7/28 Ansari Road,
Daryaganj, New Delhi – 110002, India

Published in North America by CRC Press LLC, 6000 Broken Sound Parkway, NW,
Suite 300, Boca Raton, FL 33487, USA

First published 2003, Woodhead Publishing Limited and CRC Press LLC
Reprinted 2007, 2009, 2010
© Woodhead Publishing Limited, 2003
The authors have asserted their moral rights.

British Library Cataloguing in Publication Data
A catalogue record for this book is available from the British Library.

Library of Congress Cataloging in Publication Data
A catalog record for this book is available from the Library of Congress.

Woodhead Publishing ISBN 978-1-85573-584-2 (print)
Woodhead Publishing ISBN 978-1-85573- 691-7 (online)
ISSN 2042-8049 Woodhead Publishing Series in Food Science, Technology and Nutrition (print)
ISSN 2042-8057 Woodhead Publishing Series in Food Science, Technology and Nutrition (online)
CRC Press ISBN 978-0-8493-1743-9
CRC Press order number: WP1743

Printed in the United Kingdom by Lightning Source UK Ltd

Contents

Contributors

Chapter 1

M. Saxelin*, R. Korpela and A.
  Mäyrä-Mäkinen
Valio Ltd, R&D
Meijeritie 4 A
PO Box 30
00039 Helsinki
Finland

Tel: +358 10381 3111
Fax: +358 10381 3019
Email: maija.saxelin@valio.fi

Chapter 2

C. Gill* and I. Rowland
Northern Ireland Centre for Diet and
  Health (NICHE)
University of Ulster
Coleraine Campus
Cromore Road
Coleraine
Co. Londonderry
BT52 1SA
UK

Tel: +44 (0) 28 7032 4675
Email: c.gill@ulst.ac.uk

Chapter 3

J. Lovegrove* and K. Jackson
School of Food Biosciences
The University of Reading
PO Box 226
Whiteknights
Reading
RG6 6AP
UK

Tel: +44 (0) 118 378 8700
Fax: +44 (0) 118 378 0080
Email: food@afnovell.reading.ac.uk

Chapter 4

R. Wood
Mineral Bioavailability Lab
USDA HNRCA at Tufts University
711 Washington St
Boston
MA 02111
USA

Email: richard.wood@tufts.edu

*Indicates main point of contact

Chapter 5

P.V. Kirjavainen
Department of Biochemistry and
    Food Chemistry and Functional
    Foods Forum
University of Turku
FIN-20014
Finland

Tel: +358 2 333 6861
Fax: +358 2 333 6860
Email: pirkka.kirjavainen@utu.fi

Chapter 6

H. Gill
Institute of Food, Nutrition and
    Human Health
Massey University
Palmerston North
New Zealand

Email: h.s.gill@massey.ac.nz

Chapter 7

F. Shanahan
University College Cork
Clinical Sciences Building
Cork University Hospital
Cork
Ireland

Tel: +353 21 490 1226
Fax: +353 21 434 5300
Email: f.shanahan@ucc.ie

Chapter 8

R.J. FitzGerald*
Life Sciences Department,
University of Limerick
Limerick
Ireland

Tel: +353 61 202 598
Fax: +353 61 331 490
Email: dick.fitzgerald@ul.ie

H. Meisel
Institut für Chemie und Technologie
    der Milch
PO Box 60 69
D-24121 Kiel
Germany

Email: meisel@bafm.de

Chapter 9

G. Boehm* and B. Stahl
Infant Nutrition Research
Numico Research Germany
Milupa GmbH & Co. KG
Bahnstrasse 14–30
61381 Friedrichsdorf
Germany

Tel: +49 6172 991320
Fax: +49 6172 991862
Email: guenther.boehm@milupa.de

Chapter 10

R. Fondén*
Arla Foods ICS
SE 10546 Stockholm
Sweden

Email: rangne.fonden@arlafoods.com

M. Saarela, J. Mättö and T. Mattila-
    Sandholm
VTT Biotechnology
Tietotie 2, Espoo
P.O.Box 1500
FIN-02044 VTT
Finland

Tel: +358-9-456 4466
Fax: +358-9-455 2103

Chapter 11

S. Gnädig, Y. Xue, O. Berdeaux, J.M.
    Chardigny and J-L. Sebedio*
Institut National de la Recherche
    Agronomique
Unité de Nutrition Lipidique
BP 86510 -17 rue Sully
21065 Dijon Cedex - France

Tel: +33 (0) 3 80 69 31 23
Fax: +33 (0) 3 80 69 32 23
Email: sebedio@dijon.inra.fr

Chapter 12

R. Rastall
School of Food Biosciences
The University of Reading
PO Box 226
Whiteknights
Reading
RG6 6AP
UK

Tel: +44 (0) 118 9316726
Fax: +44 (0) 118 9310080
Email: r.a.rastall@reading.ac.uk

Chapter 13

A.C. Ouwehand* and S. Salminen
University of Turku
Department of Biochemistry and
    Food Chemistry
FIN 20014 Turku
Finland

Tel: +358 2 333 6894
Fax: +358 2 333 6884
Email: arthur.ouwehand@utu.fi

Chapter 14

P. Marteau
European Hospital Georges
    Pompidou and Paris V University

20 Rue Leblanc
75908 Paris
Cedex 15
France

Email: philippe.marteau@egp.ap-hop-
    paris.fr

Chapter 15

L. Lähteenmäki
Sensory Quality and Food Choice
Group Manager
VTT Biotechnology
PO Box 1500
FIN 02044 VTT
Finland

Tel. +358 9 456 5965
Fax +358 9 455 2103
Email: liisa.lahteenmaki@vtt.fi

Chapter 16

T. Mattila-Sandholm,* L.
    Lähteenmäki and M. Saarela
VTT Biotechnology
PO Box 1500
FIN 02044 VTT
Finland

Tel: +358-50-5527243
Email: tiina.mattila-sandholm@vtt.fi

Chapter 17

L. Hoolihan
Nutrition Research Specialist
Dairy Council of California
222 Martin # 155
Irvine, CA 92612
USA

Tel: 0001 949 756 7892
Fax: 001 949 756 7896
Email: hoolihan@dairycouncilofca.org

1

Introduction: classifying functional dairy products

M. Saxelin, R. Korpela and A. Mäyrä-Mäkinen, Valio Ltd, Finland

1.1 Introduction

Dairy products form the major part of functional foods. To understand their success it is important to know that milk is a natural and highly nutritive part of a balanced daily diet. Designing and developing functionality in dairy-based products simply means modifying and/or enriching the healthy nature of the original base. This chapter is a brief introduction to the composition of milk and the nature of fermented milk products. It also gives a few definitions and introduces some of the functional dairy products on the market. The purpose of this chapter is not to evaluate the quality and depth of the science behind each product: some of these products are tested in their final state, while the functionality of others is based on accepted knowledge of a particular compound. At the same time, this chapter offers some 'good guesses' as to the potential development of functional dairy foods in the future.

1.2 Composition of milk

The milks of various mammalian species differ in the amount and type of their components. This review focuses on cows' milk and those products of which cows' milk forms a prominent ingredient. Cows' milk is mainly composed of water, with approximately 4.8% lactose, 3.2% protein, 3.7% fat, 0.19% non-protein nitrogen and 0.7% ash. The principal families of proteins in milk are caseins, whey proteins and immunoglobulins. About 80% of proteins are caseins (Banks and Dalgleish, 1990).

Caseins (α_{s1}-, α_{s2}-, β- and κ-) and whey proteins differ in their physiological and biological properties. Caseins form complexes called micelles with calcium. Globular α-lactalbumin and β-lactoglobulin are the main whey proteins. They constitute 70–80% of the total whey proteins, the remainder being immunoglobulins, glycomacropeptide, serum albumin, lactoferrin and numerous enzymes. Some of the biological properties of milk proteins are shown in Table 1.1. Milk proteins are a rich source of precursors of biologically active peptides. Bioactive peptides are formed by the enzymatic hydrolysis of proteins or by the proteolytic activity of lactic acid bacteria in microbial fermentations. Many of the peptides survive through the intestinal tract. Bioactive peptides are also formed *in vivo* by the enzymatic hydrolysis of the digestive enzymes. Table 1.2 shows some bioactive peptides derived from milk proteins, and also their functions.

Milk fat is a complex of lipids, and exists in microscopic globules in an oil-in-water emulsion in milk. The majority of milk lipids are triglycerides or the esters of fatty acids combined with glycerol (97–98%), and the minority are phospholipids (0.2–1%), free sterols (0.2–0.4%) and traces of free fatty acids. About 62% of milk fat is saturated, 30% monounsaturated (oleic acid), 4% polyunsaturated and 4% of minor types of fatty acids (Miller *et al.*, 2000).

Lactose is the principal carbohydrate in milk. It is a disaccharide formed from galactose and glucose. Lactose forms about 54% of the total non-fat milk solids. It also provides 30% of the energy of milk. In addition to high-value protein, milk also provides vital minerals and vitamins. It is an important source of minerals, in particular of calcium, phosphorus, magnesium, potassium and trace elements such as zinc. In many countries, especially in Europe, milk is the principal source of calcium, providing about 60–80% of the total calcium intake. Calcium forms soluble complexes with milk protein, casein, and phosphorus, and is easily absorbed. Milk contains all the vitamins known to be essential to humans. Vitamins A, D, E and K are associated with the fat component of milk. In northern countries where there is a shortage of sunshine in winter, milk and milk fat has traditionally been the major source of vitamin D. Milk also provides water-soluble vitamins (ascorbic acid, thiamin, riboflavin, niacin, pantothenic acid, vitamin B6, folate and vitamin B12) in variable quantities (Miller *et al.*, 2000).

1.3   Fermented milk products

The Scandinavian countries have a long tradition of using fermented dairy products. In the old days, the seasonal variation in milk production led the farms to preserve milk for the cold winter in the forms of butter and its by-product, buttermilk, as well as other traditional fermented milk products (Leporanta, 2001). Later, the industrial production of these products began, and selected product-specific starter cultures became commercially available. The consumption of milk and fermented milks in selected countries in Europe and some other countries is shown in Fig. 1.1. Cultured buttermilks, or fermented milk products as they are also called, are primarily consumed plain, but flavoured varieties are available,

Table 1.1 Biological activities of major cows' milk proteins (Korhonen *et al.*, 1998)

Protein	Suggested functions	Concentration (g/l)
Caseins (α, β and κ)	Iron carrier (Ca, Fe, Zn, Cu)	28
	Precursors of bioactive peptides	
α-Lactalbumin	Lactose synthesis in mammary gland, Ca carrier, immunomodulation, anticarcinogenic	1.2
β-Lactoglobulin	Retinol carrier, fatty acids binding, possible antioxidant	1.3
Immunoglobulins A, M and G	Immune protection	0.7
Glycomacropeptide	Antiviral, antibacterial, bifidogenic	1.2
	Releases protein to cause satiety?	
Lactoferrin	Toxin binding	0.1
	Antimicrobial, antiviral	
	Immunomodulation	
	Anticarcinogenic	
	Antioxidative	
	Iron absorption	
Lactoperoxidase	Antimicrobial	0.03
Lysozyme	Antimicrobial, synergistic with immunoglobulins and lactoferrin	0.0004

Table 1.2 Bioactive peptides derived from cows' milk proteins (Korhonen *et al.*, 1998; Clare and Swaisgood, 2000)

Bioactive peptides	Protein precursor	Bioactivity
Casomorphins	α- and β-Casein	Opioid agonists
α-Lactorphin	α-Lactalbumin	Opioid agonists
β-Lactorphin	β-Lactoglobulin	Opioid agonists
Lactoferroxins	Lactoferrin	Opioid antagonists
Casoxins	κ-Casein	Opioid antagonists
Casokinins	α- and β-Casein	Antihypertensive
Casoplatelins	κ-Casein, transferrin	Antithrombotic
Casecidin	α- and β-Casein	Antimicrobial
Isracidin	α-Casein	Antimicrobial
Immunopeptides	α- and β-Casein	Immunostimulants
Phosphopeptides	α- and β-Casein	Mineral carriers
Lactoferricin	Lactoferrin	Antimicrobial
Glycomacropeptide	Caseins	Anti-stress

too. Mesophilic *Lactococcus lactis* subsp. *lactis/cremoris/diacetylactis* and *Leuconoctoc cremoris* strains are used for fermentation at 20–30 °C for 16–20 h. Starter cultures other than mesophilic lactococci/leuconostoc can also be used for the fermentation of milk drinks. There are products on the market which are fermented with a special strain of lactobacilli (e.g. *L. casei*) or a mixture of several lactobacilli, lactococci and other genera/species. For example kefir, a traditional

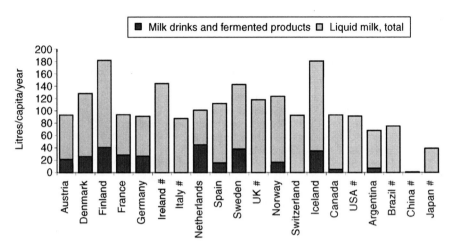

Fig. 1.1   The consumption of milk drinks and fermented products and total consumption of liquid milk in selected countries. #Data not available for fermented products.

fermented milk drink originating from the Balkans, is produced by a starter culture containing various species of *Lactococcus, Leuconostoc, Lactobacillus, Acetobacter* and yeasts, giving the product its special flavour and aroma.

The health effects of fermented milk products became known through the works of Professor Elie Metchnikoff (Pasteur Institute, Paris), who about a hundred years ago discovered that the secret of the long life of Bulgarian peasants lay in their high consumption of a fermented milk product, yoghurt. Since the 1950s, the flavouring of yoghurt with fruits has increased consumption radically. Today yoghurt is of ever-increasing popularity and there are various types of yoghurt on the market. All yoghurts have this in common: that the milk is fermented with *Streptococcus thermophilus* and *Lactobacillus delbrueckii* subsp. *bulgaricus*, which grow in synergy in milk. The fermentation is carried out at 30–43 °C for 2.5–20 h. The selection of the starter culture strains defines the fermentation time and thus the structure and flavour of the final product. Fruit preparations may then be added to the fermented milk base before packaging.

Quark-based products (fresh cheeses, etc.) are also made with microbial fermentations of milk, but the whey is separated after milk coagulation. The production processes vary, but many products contain live lactic acid bacteria. Matured cheeses are formed if coagulated milk protein and milk fat are further processed by pressing, salting and maturing in a cool temperature for various periods of time.

All fermented milk products contain live lactic acid bacteria, unless they are pasteurised after fermentation. In 2000 the total consumption of fermented milks and yoghurts in the EU was about 6.35 million tonnes (*Bulletin of the International*

Dairy Federation **368**, 2001). That means a total consumption of more than 10^{20} colony-forming units (cfu) of lactic acid bacteria. Consumption varies considerably according to country, the highest being in the Nordic countries and the Netherlands. Since Metchnikoff's time, fermented milks have been thought to offer health benefits. The addition of selected, well-documented health-effective strains (probiotics) to the fermentations is an easy and natural way of enhancing the functionality of these products. When one considers the healthy nature of milk, consumed on a daily basis, it is hardly surprising that the major part of functional foods is dairy based.

1.4   What do we mean by functional dairy products?

Functional foods are not defined in the EU directives. Some countries (e.g. the UK, Sweden, Finland) have national rules (guidelines on health claims) for the interpretation of the current legislation (Directives 65/65/EEC and 2000/13/EC) in relation to health claims, but as more products are advertised and marketed across borders, harmonisation at the EU level is needed (Smith, 2001). A draft proposal (working document Sanco/1832/2002) is under discussion. In Finland new guidelines were launched in June 2002. The European Functional Food Science Programme, funded by the European Union and led by the International Life Sciences Institute (ILSI), defines functional foods as follows (Diplock, 1999):

> A food can be regarded as 'functional' if it is satisfactorily demonstrated to affect beneficially one or more target functions in the body, beyond adequate nutritional effects in a way that is relevant to either an improved state of health and well-being and/or reduction of risk of disease.

What is actually meant by 'satisfactorily demonstrated'? One of the interpretations is that a food product can be called functional only if its health benefit has been shown in the consumption of a normal daily dose of the final product, or an effective dose of the ingredient is used and the impact of the food matrix is known. There is a general consensus that, in order to be 'satisfactorily demonstrated', at least two high-quality human intervention studies must have been completed.

Dairy foods can be divided into three groups:

* Basic products (milk, fermented milks, cheeses, ice cream, etc.).
* Added-value products, in which the milk composition has been changed, e.g. low-lactose or lactose-free products, hypoallergenic formulae with hydrolysed protein for milk-hypersensitive infants, milk products enriched with Ca, vitamins, etc. Primarily, these products are targeted at specific consumer groups, and, depending on individual opinions, are included or not in the functional food category.
* Functional dairy products with a proven health benefit. Products are based on milk that is enriched with a functional component, or the product is based on ingredients originating from milk. The most common functional dairy products

are those with probiotic bacteria, quite frequently enriched with prebiotic carbohydrates.

1.5   Examples of functional dairy products: gastrointestinal health and general well-being

1.5.1   Probiotic products

Probiotic bacteria are live microbial strains that, when applied in adequate doses, beneficially affect the host animal by improving its intestinal microbial balance. **Probiotic foods** are food products that contain a living probiotic ingredient in an adequate matrix and in sufficient concentration, so that after their ingestion, the postulated effect is obtained, and is beyond that of usual nutrient suppliers (de Vrese and Schrezenmeir, 2001).

It is clear, then, that the tradition of fermented dairy products is long, and to make these products 'functional' is a natural and fairly simple concept (Lourens-Hattingh and Viljoen, 2001). The probiotic strains used in dairy products most commonly belong to *Lactobacillus* and *Bifidobacterium* genera (see Table 1.3). The characteristics of probiotic strains vary, and each strain has to be studied individually. The primary requirement of a probiotic strain is that it should be adequately identified with methods based on genetics, and that the strain should be defined in the text of the product package. This makes it possible to analyse the scientific data behind any claims made.

Some probiotic strains are sufficiently proteolytic to grow excellently in milk, but others need growth stimulants. Those that do not ferment lactose need monosaccharides (Saxelin *et al.*, 1999; de Vrese and Schrezenmeir, 2001). Sometimes the texture or the taste of a milk product fermented with a probiotic does not meet with consumer approval or is technologically impractical. For this reason it is common to use probiotic strains together with standard starter cultures (yoghurt, mesophilic, etc). Probiotics can be added before the fermentation of the milk, or part of the milk can be fermented separately with the probiotic strain and the two parts mixed after the fermentations. Alternatively, a probiotic strain can be added to the fermented product after fermentation. Sometimes the milk is not fermented at all.

Table 1.3 The most common species of bacteria used in probiotic dairy foods

• *Lactobacillus acidophilus* group: *L. acidophilus, L. johnsonii, L. gasseri, L. crispatus*	• *Bifidobacterium lactis*
	• *B. bifidum*
	• *B. infantis*
• *L. casei/paracasei*	• *B. breve*
• *L. rhamnosus*	• *B. animalis*
• *L. reuteri*	• *B. adolescentis*
• *L. plantarum*	

The level of a probiotic strain has to be stable and viable during the shelf-life of the product. There are reports showing that this is not always the case (Shah, 2000). However, research on the subject has changed the situation and will further improve the quality of probiotic products. Today most of the defined probiotic strains used in dairy products have good storage stability. As to the testing of functionality, the easiest method is to develop one type of product and to test its health benefits. Multinational companies often operate in several countries with the same product image marketed under the same trade mark. The small bottle – the 'daily dose' concept – is a good example of this. Identical bottles of Yakult (with the *Lactobacillus casei* Shirota strain) or those of Danone Actimel (with the *L. casei* Imunitass strain DN 114 001) are marketed with the same product concept and the same marketing message all over the world.

However, to meet consumers' demands for probiotic foods in different countries, different types of products are also needed. One way to meet this challenge is to try to define an effective daily dose to be used in various types of products. For example, *Lactobacillus rhamnosus* GG is used in Finland in cultured buttermilks, 'sweet' milk, yoghurts, fermented whey-based drinks, set-type fermented milks ('*viili*'), cheeses, juices, and mixtures of milk and juice. It is not reasonable or scientifically interesting to repeat clinical studies with all the different types, especially when the overall claims to be used in marketing are the same general level. Milk is a protective food matrix for probiotics and improves the survival of the strain in the intestine. As can be seen in Fig. 1.2, if one wishes to re-isolate the

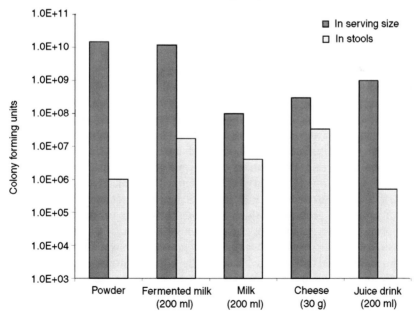

Fig. 1.2 The recovery of *Lactobacillus* GG in faecal samples during daily consumption of different product forms. The daily dose of the probiotic strain (log cfu) per serving and the level in stool samples (log cfu/g wet mass) are indicated in the vertical axis.

strain in stool samples during daily consumption, much lower doses of *Lactobacillus* GG can be used in milk or cheese than in capsules or in powders.

The most common probiotic dairy products worldwide are various types of yoghurt, other fermented dairy products (e.g. cultured buttermilks in Finland), various lactic acid bacteria drinks ('Yakult-type') and mixtures of probiotic (fermented) milks and fruit juice. Probiotic cheeses, both fresh and ripened, have also been launched recently. From January 2000 to May 2002, 25 functional cheeses were launched in Europe, 19 of which, it was claimed, contained an active culture or a probiotic strain (Mintel's Global New Products Database; www.gnpd.com). In addition to everyday products, probiotics are also used in indulgence products, e.g. ice creams.

Probiotic dairy foods (with certain specific strains) are known to relieve intestinal discomfort, prevent diarrhoea and improve recovery. However, no country will accept this claim, as it is too medical for use in the marketing of food. The most common health claim used for probiotic dairy foods may be 'improves natural defence systems', but as far as we know, the science behind that statement is not officially evaluated in any country for any product. In Japan, where functional food legislation is organised best, the package claims for the accepted Food for Specified Health Users (FOSHU) regulation lactobacilli products are that they balance gastrointestinal functions. Recently a claim that a yoghurt product enriched with a strain of *L. gasseri* suppressed *Helicobacter pylori* (one cause of peptic ulcers) was also accepted. There are other products that supposedly suppress the growth and activity of *H. pylori*, both in Europe and in the Korean Republic.

1.5.2   Prebiotic and synbiotic dairy products

Prebiotics are non-digestible food ingredients that beneficially affect the host by selectively stimulating the growth and/or activity of one or a limited number of bacteria in the colon. **Prebiotic foods** are food products that contain a prebiotic ingredient in an adequate matrix and in sufficient concentration, so that after their ingestion, the postulated effect is obtained, and is beyond that of usual nutrient suppliers. **Synbiotics** are mixtures of pro- and prebiotics that beneficially affect the host by improving the survival and implantation of selected live microbial strains in the gastrointestinal tract (de Vrese and Schrezenmeir, 2001).

In contrast to probiotics, which introduce exogenous bacteria into the human intestine, prebiotics stimulate the preferential growth of a limited number of bacteria already existing in a healthy, indigenous microbiota. The clue to prebiotic compounds is that they are not digested in the upper gastrointestinal tract, because of the inability of the digestive enzymes to hydrolyse the bond between the monosaccharide units. They act as soluble fibres and are digested in the colon, enhancing microbial activity and stimulating the growth mainly of bifidobacteria and lactobacilli. Consumption of higher doses may encourage the formation of gas, flatulence and intestinal discomfort. The end-products in the gut fermentation are mainly short chain fatty acids (acetic, propionic and butyric acid), lactic acid, hydrogen, methane and carbon dioxide. Short chain fatty acids, especially butyric acid, are known to act as an energy source for enterocytes (Wollowski *et al.*, 2001). The main dairy products enriched with prebiotics are yoghurts and yoghurt drinks, but spreads, fresh cheeses and milks are also on the market.

Galactooligosaccharide, a milk-based prebiotic, is derived from lactose by the β-galactosidase enzyme. It is a natural prebiotic of human breast milk, and facilitates the growth of bifidobacteria and lactobacilli in breast-fed infants. Galactooligosaccharides are commercially used principally in Japan and other parts of Asia.

In Europe inulin and fructooligosaccharides are widely used in various functional foods, including dairy-based products. Inulin is a group of fructose polymers linked by β(2–1) bonds that limit their digestion by enzymes in the upper intestine. Their chain lengths range from 2 to 60. Oligofructose is any fructose oligosaccharide containing two to ten monosaccharide units linked with glycosidic linkage. Both inulin and fructooligosaccharides (oligofructoses) are extracted from plant material (e.g. chicory) or synthesised from sucrose. The role of inulin and the oligofructoses in a food matrix is bi-functional. They do not increase the viscosity of a milk product but give a richer texture to liquid products and spreads.

1.5.3   Low-lactose and lactose-free milk products

In the human intestine lactose is hydrolysed by a lactase enzyme developed in the brush border of the small intestine. When a person has a lactase deficiency and lactose causes intestinal discomfort and other symptoms, this is called lactose intolerance, and is quite common in most parts of the world. The incidence of lactose intolerance is low only in the Nordic countries, the British Isles, Australia and New Zealand. Most people can drink one glass of milk (~10 g lactose) in a single dose taken with a meal, without suffering symptoms, but not a 50 g dose ingested on an the empty stomach, the dose used in lactose tolerance tests.

There is a general consensus of opinion that probiotic dairy products alleviate lactose intolerance. This is true of all fermented dairy products, especially yoghurt, owing to the β-galactosidase activity of the yoghurt culture and the higher consistency of fermented milks compared with ordinary milk. However, a much more sophisticated and efficient way of reducing symptoms caused by lactose is to hydrolyse it in the milk enzymatically. In long-life milks the enzyme is generally added to the milk after sterilisation, and the product is released for sale after a certain period, when the level of lactose has decreased. In fermented milks the enzyme is added before fermentation or at the same time as the culture. If added with the culture, the enzyme must be active in acidic conditions. In Finland, Valio Ltd has a large range of lactose-hydrolysed (HYLA®) milk products, altogether around 80 varieties.

The hydrolysis of lactose changes the taste of the milk, making it sweeter, because glucose and galactose are sweeter than lactose. This is an accepted fact in fermented milk products, especially if they are additionally sweetened. However, this sweetness is not popular in milk for drinking, and thus milk consumption drops. Recently, this problem, too, has been solved. In 2001 Valio Ltd launched a lactose-free milk in which the lactose has been completely removed physically. The sweetness has been restored to its normal level and the taste is that of normal fresh milk.

1.5.4   Others
Sphingolipids contain compounds such as ceramides, sphingomyelin, cerebrosides, sulphatides and gangliocides. Sphingolipids are found in millk, butter and cheese – approximately 2 mg/100 g milk. Because they exist in cell membranes rather than in fat droplets, they are found in fat-free, low-fat as well as in full-fat dairy products. *In vitro* and experimental studies indicate that sphingolipids influence cell regulation, and thus carcinogenesis and tumour formation (Miller *et al.*, 2000). In 2000, a yoghurt brand called 'Inpulse' was launched in Belgium (Büllenger Butterei). The low-fat product was said to be rich in natural milk lecithin (45 mg/100 g) and sphingolipids (phospholipids 144 mg/100 g). A variety launched since then contains 0.6 g fat, 115 mg phospholipids, 36 mg phosphatidylcholine and 18.4 mg sphingolipids. The information on the product declares that 'lecitin and sphingolipids are biomembranes, which re-establish the biological equilibrium of the cells, protect against bacterial infections and help digestion'.

1.6   Examples of functional dairy products: cardiovascular health

Coronary heart disease (CHD) is a serious form of cardiovascular disease and the most common – the leading cause of death in developed industrialised countries. Many risk factors, both genetic and environmental, contribute to the development of coronary heart disease. The three most important modifiable risk factors for this are cigarette smoking, high blood pressure and high blood cholesterol levels, particularly of low-density lipoprotein (LDL) cholesterol. Other risk factors likely to contribute to the risk of CHD are diabetes, physical inactivity, low high-density lipoprotein (HDL) cholesterol, high blood triglyceride levels, and obesity. Oxidative stress, homocysteine, lipoprotein and psychosocial factors may also increase the risk. To choose a healthy, low-fat diet with high levels of fruits and vegetables, an active lifestyle and no smoking seems to reduce the risk of heart diseases. The inclusion of semi-skimmed or non-fat milk products in an otherwise healthy diet adds many essential vitamins, not to mention milk calcium, which has a vital role in controlling blood pressure (Miller *et al.*, 2000). Milk products specifically developed to reduce dietary risk factors are already on the market.

1.6.1   Products for controlling hypertension
There are a few products on the market for lowering blood pressure. Several milk peptides are known to have an inhibitory effect on the angiotensin converting

enzyme (ACE inhibition). ACE is needed for converting angiotensin I to angiotensin II, increasing blood pressure and aldersterone, and inactivating the depressor action of bradykinin. ACE inhibitors derived from caseins are called casokinins, and they are derived from the tryptic digestion of bovine β- and κ-caseins. In two commercial products, these peptides are isoleucine–proline–proline and valine–proline–proline, which are formed from β-casein by the fermentation of milk with *Lactobacillus helveticus*. The *L. helveticus* bacterium is generally used in cheese-making and the fermentation is a normal dairy process. The Calpis Amiel drink (Japan) is a sterile product, without living bacterial cells. The fermented milk drink Evolus, more recently developed by Valio Ltd (Finland), contains, in addition to the active tripeptides, living bacterial cells and an improved composition of minerals (Ca, K, Mg). Both products have been tested in animal studies with spontaneously hypertensive rats (Sipola, 2002) and in clinical human trials (Hata *et al.*, 1996; Seppo *et al.*, 2002). The Japanese product has official FOSHU status.

In Finland there is a cheese on the market that has been shown to have ACE inhibitory activity (Festivo cheese, Agricultural Research Centre, Jokioinen, Finland). The bioactive peptides are shown to be αs_1-casein N-terminal peptides but the researchers thought that they might be too long to be absorbed intact in the intestine. The quantity also varied during the maturation and age of the cheese, and the effect of the cheese on human blood pressure remains to be tested (Ryhänen *et al.*, 2001). Another idea, not yet commercially launched in dairy products, is based on whey proteins that are hydrolysed so that the whey protein isolate has an ACE inhibitory activity (Davisco Foods International Inc., USA). The effect of this product seems to be much faster than those based on the tripeptides, but the mechanism is not yet known (Pins and Keenan, 2002).

1.6.2 Products for controlling cholesterol

Natural cows' milk fat contains high levels of saturated fatty acids. Replacing the consumption of full-cream milk with semi-skimmed or non-fat milk will reduce the intake of saturated fatty acids. Sometimes it is not enough just to reduce the intake of saturated fats and cholesterol, since most cholesterol is synthesised within our own bodies. On the other hand, plant sterols and stanols have long been known to reduce the assimilation of dietary cholesterol. Since the mid-1990s there have been products enriched with plant stanols specially targeted at those people with (moderately) high cholesterol levels. A few years later plant sterols were also accepted as food ingredients by the EU Novel Foods legislation, and now the Food and Drug Administration in the USA has also accepted plant sterols and stanols. Sterols are building blocks of the cell membranes in both plant and animal cells. Isolated plant stanols, hydrated forms of sterols, are crystallised particles. They effectively bind cholesterol and are not absorbed by the human body. Esterified plant stanols are fat-soluble and easy to use as a food ingredient. Intestinal enzymes hydrolyse the ester bond and the insoluble stanol is free to bind cholesterol and to be secreted. Basically, the effect of plant sterols is based on the same mechanism.

Several milk-based functional foods including plant sterols or stanols are commercially available. They all are semi-skimmed or non-fat products. Products containing Benecol (Raisio Benecol Ltd, Finland), the only **plant stanol** ester ingredient, are on the market in several countries. In some products the 'effective daily dose' has to be collected from several servings (e.g. Benecol milk, yoghurt, various spreads in the UK), in some other countries the dose is contained in one serving (e.g. Valio Benecol yoghurt in Finland). **Plant sterols** are also added to functional milk products, especially to milk (e.g. Mastellone Hnos SA, Argentina). In March 2001, Marks & Spencer launched a range of 20 products, including yoghurt, enriched with **soy proteins** (& More brand, UK). The daily consumption of 25 g soy proteins has been shown to lower cholesterol by 10%.

The safety risk of overdosing with plant sterols and stanols has been the subject of discussion by the scientific committee on food of the European Commission. The consumption of this kind of product requires a fairly good knowledge on the part of the consumer, as she or he has to be familiar with the products with the compound and also to know the quantity of the active ingredient in various products. For that reason the labelling must be informative enough.

Matured cheeses contain quite high levels of milk fats. Replacing milk fat with vegetable oil can reduce the intake of saturated fatty acids. In Finland there are cheeses on the market in which milk fat has been replaced by rapeseed oil (Julia and Julius with 17% and 25% rapeseed oil, respectively; Kyrönmaan Osuusmeijeri, Finland). The products, when included daily in a low-fat diet, reduced blood cholesterol statistically significantly (Karvonen *et al.*, 2002).

1.6.3 Omega-3 fatty acids

There are two major classes of polyunsaturated fatty acids: omega-3 fatty acids found in fish oils and as a minor constituent of some vegetable oils, and omega-6 fatty acids, which include the essential fatty acid linoleic acid, found in vegetable oils such as corn, sunflower and soybean. Omega-3 fatty acids are said to contribute to the good functioning of the cardiovascular system, on the basis of various physiological effects. Before omega-3 fatty acids could be added to milk products, the fishy taste and odour had to be disguised and the easy oxidation of the oil overcome. It took several years before these problems were solved, but nowadays there are a few suppliers selling good-quality fish-oils to be added to milk. The pioneer in launching an omega-3-enriched milk was the Italian dairy company Parmalat. Its 'Plus Omega 3' milk was launched in 1998 and is a semi-skimmed milk enriched with 80 mg omega-3. It is recommended for use by all health-conscious consumers in a dose of half a litre per day (Mellentin and Heasman, 1999). Since then other producers all over the world have followed with their own omega-3-enriched milks. Milk is often also enriched with the antioxidative vitamins A, C and E.

1.7   Examples of functional dairy products: osteoporosis and other conditions

The cause of osteoporosis, as with other chronic diseases, is multifactorial, involving both genetic and environmental factors. An accumulation of scientific evidence indicates that a sufficient intake of calcium throughout one's life offers protection against osteoporosis. The bone mass reaches its peak when a person is 30 years of age and then the density decreases with age, especially after the menopause. The fortification of semi-skimmed and non-fat milk with vitamin D is important, as this vitamin is essential to improve calcium absorption and is also removed when fat is removed. Milk is the richest source of calcium. There are several milks and milk products enriched with calcium, and both inorganic and milk-based calcium (e.g. TruCal, Glanbia Ingredients Inc.) are used. The absorption of calcium may be enhanced with bioactive milk proteins. Caseino-phosphopeptides (CPPs) are known to increase the solubility of calcium, but controversy exists as to whether CPPs enhance calcium absorption in the body. The authors do not know of any commercial applications of CPPs in dairy products.

1.7.1   Products for enhancing immune functions

Some of the probiotic dairy products have been shown to enhance immune functions and thus to reduce the risk of infection. Milk contains natural immunoglobulins, which can be isolated and concentrated, either from normal milk or from colostrum, which contains a high proportion of them. There are milk-based products on the market in which the product is further enriched with immunoglobulins. In the USA and Australia, Lifeway Foods is marketing kefir under the brand name Basic Plus. The product is said to be probiotic, although the probiotic strains are not specified. The active ingredient, an extract of colostrum, has been developed by GalaGen Inc. and is targeted at maintaining intestinal health and the natural microbiota. Basic Plus was launched in 1998 and is the first dairy-based food supplement sold in the USA in the refrigerated sections of health food and grocery stores.

Milk immunoglobulins are used in new drinks in the USA under the brand name of 'NuVim'. The production of immunoglobulins is boosted in a selected herd in New Zealand by an immune stimulant, and isolated under carefully controlled conditions in order to preserve the micronutrients. The product is said to be lactose-free and fat-free, to have beneficial effects on the immune system and to improve the health of muscles and joints (Heasman and Mellentin, 2002).

1.7.2   Milks to help with sleeping problems

Melatonin is a hormone that controls the body's day and night rhythm. The secretion of melatonin is high in early childhood and decreases rapidly with ageing. Stress conditions and age cause a lowering of the level of melatonin. It is secreted at nights in both humans and bovines. The concentration at night in cows' milk is about four times higher than in milk collected during the day. The first product based on a standardised milking system at night was launched in Finland in 2000 (Yömaito, Ingman Foods Ltd). Since no human trials have been published so far, the company does not make any health claims. In spring 2002 an organic milk, 'Slumbering Bedtime Milk' (Red Kite Farm, UK), was launched in the UK. It is said to contain higher levels of melatonin than ordinary milk. The company says that the level of melatonin in the milk complements that of the human body and the drink will not induce drowsiness if drunk during the day, or the following morning if drunk at night/late in the evening.

1.8   Future trends

Research and discussion on pro- and prebiotics have encouraged basic research in the field of the intestinal microbial flora and its metabolism. This has also led to improved research funding from public resources, both nationally and from the European Union. Not enough is known of the composition and metabolism of the bacteria in the intestines in health and disease. Also the knowledge on the role of the microbiota in the development and function of immune response needs more investigation. Development and improvements in research methods, and *in vitro*, *ex vivo* and *in vivo* models, have provided important information on the mechanisms behind the effects, and new biomarkers to be followed in human studies. The more we know about the composition and function of the intestinal microbiota, the greater the potential to develop functional foods for targeted consumer groups. Considering the healthy population there may be potential to develop targeted products for different age groups. In the reduction of risk and treatment of various diseases, pro- and prebiotics have resulted in promising benefits. However, it is important to understand the mechanisms behind the effects. When the mechanisms are known, it will be also possible to control the activity or the dose of the effective compounds. We also need official definitions of functional foods, and relevant regulation of physiological claims and health claims. The production of functional foods that have to follow the rules of production of medicines is hardly in the interest of normal dairy companies. It may be unrealistic to apply the same rules to medicines as to everyday foods with a short shelf-life.

Milk is a rich source of nutritive compounds which can be enriched and/or further modified. Milk fat does not consist merely of saturated fatty acids, but also of monounsaturated and polyunsaturated fatty acids. The role of conjugated linoleic acid (CLA) in preventing the risk of certain diseases, and in particular, the problem of how to increase its quantity in milk has evoked wide interest among several research groups. Milk proteins and bioactive peptides may supply new products to help protect against several common health risk factors. There are bioactive peptides potentially to be used to give satiety or to better tolerate stress. Lactose derivatives can be used as soluble fibre to relieve constipation and to modulate the intestinal flora. Milk minerals can be used to replace sodium in salt, supporting a healthy diet for avoiding hypertension. Milk components are natural, and applications for novel foods are seldom needed. There is also a huge selection of lactic acid bacteria used for milk fermentations, which have a long tradition of safe use. Genetically modified strains may be needed for special purposes, though perhaps not in products for the general public.

In developing functional dairy products, various groups of experts are needed. The basis must be in the scientific research of effects, requiring medical experts, nutritionists and microbiologists. Food technologists are needed for product development, process technologists and biotechnologists for processing the compounds, chemists to analyse the compounds and, finally, experts for marketing the products. Marketing is a big challenge, as it has to tell the public about the health benefits in such a simple way that every layperson understands. Medical and nutritional messages need to be simplified. It is important to remember that functional dairy products are mainly for supplying nutritive foods for everyday consumption. Nutrimarketing is also needed to explain research results to health-care professionals and to convince them of the benefits of functional foods.

1.9   Sources of further information and advice: links

www.gnpd.com
www.new-nutrition.com
www.scirus.com
www.just-food.com
www.ifis.org
http://www.foodlineweb.co.uk
www.fst.ohio-state.edu/People/HARPER/Functional-foods/Functional-Foods.htm
www.valio.com
www.benecol.com
www.daviscofood.com
www.kefir.com
www.ific.org
www.effca.com
www.usprobiotics.org
www.elintarvikevirasto.fi/english

1.10   References

BANKS W and DALGLEISH D G (1990), 'Milk and milk processing' in Robinson R K, *Dairy Mircobiology,* Volume 1, *The Microbiology of Milk,* second edition, London, Elsevier Science Publishers Ltd, 1–35.
CLARE D A and SWAISGOOD H E (2000), 'Bioactive milk peptides: a prospectus', *J Dairy Sci,* **83**, 1187–1195.
DE VRESE M and SCHREZENMEIR J (2001), 'Pro and prebiotics', *Innov Food Technol,* **May/ June**, 49–55.
DIPLOCK A T (1999), 'Scientific concepts of functional foods in Europe: Consensus document', *Br J Nutr,* **81**(Suppl 1), S1–S27.

HATA Y, YAMAMOTO M, OHNI M, NAKAJIMA K, NAKAMURA Y and TAKANO T (1996), 'A placebo-controlled study of the effect of sour milk on blood pressure in hypertensive sugjects', *Am J Clin Nutr*, **64**, 767–771.

HEASMAN M and MELLENTIN J (2002), 'New NuVim prepares to be swallowed up', *NNB*, **7**(8), 29–30.

KARVONEN H M, TAPOLA N S, UUSITUPA M I and SARKKINEN E S (2002), 'The effect of vegetable oil-based cheese on serum total and lipoprotein lipids', *Eur J Clin Nutr*, **56**, 1094–1101.

KORHONEN H , PIHLANTO-LEPPÄLÄ A, RANTAMÄKI P and TUPASELA T (1998), 'Impact of processing on bioactive proteins and peptides', *Trends Food Sci Technol*, 8, 307–319.

LEPORANTA K (2001), 'Developing fermented milks into functional foods', *Innov Food Technol*, **10**, 46–47.

LOURENS-HATTINGH A and VILJOEN B C (2001), 'Yoghurt as probiotic carrier food', *Int Dairy J*, **11**, 1–17.

MELLENTIN J and HEASMAN M (1999), 'Functional foods are dead. Long live functional foods', *NNB*, **4**(7), 16–19.

MILLER G D, JARVIS J K and MCBEAN L D (2000), *Handbook of Dairy Foods and Nutrition*, second edition, Boca Raton, London, New York, Washington DC, CRC Press.

PINS J and KEENAN J M (2002), 'The antihypertensive effects of a hydrolysed whey protein isolate supplement (BioZate1®)', *Cardiovasc Drugs Ther*, **16** (Suppl 1), 68.

RYHÄNEN E-L, PIHLANTO-LEPPÄLÄ A and PAHKALA E (2001), 'A new type of ripened, low-fat cheese with bioactive properties', *Int Dairy J*, **11**, 441–447.

SAXELIN M, GRENOW B, SVENSSON U, FONDEN R, RENIERO R and MATTILA-SANDHOLM T (1999), 'The technology of probiotics', *Trends Food Sci Technol*, **10**, 387–392.

SEPPO L, JAUHIAINEN T, POUSSA T and KORPELA R (2002), 'A fermented milk, high in bioactive peptides, has a blood pressure lowering effect in hypertensive subjects', *Am J Clin Nutr*, in press.

SHAH N P (2000), 'Probiotic bacteria: selective enumeration and survival in dairy foods', *J Dairy Sci*, **83**(4), 894–907.

SIPOLA M (2002), 'Effects of milk products and milk protein-derived peptides on blood pressure and arterial function in rats', PhD Thesis, Institute of Biomedicine/ Pharmacology, University of Helsinki; electronic PDF version: http://ethesis.helsinki.fi/julkaisut/laa/biola/vk/sipola/.

SMITH J (2001), 'Defining health claims for Europe', *Funct Foods Nutraceut*, **November/December**, 12.

WOLLOWSKI I, RECHKEMMER G and POOL-ZOBEL B L (2001), 'Protective role of probiotics and prebiotics in colon cancer', *Am J Clin Nutr*, **73**(2 Suppl), 451S–455S.

Part I

The health benefits of functional dairy products

2

Cancer

C. Gill and I. Rowland, University of Ulster, UK

2.1 Introduction

Colorectal cancer (CRC) is the fourth most common cause of cancer-related mortality in the world. It is estimated that up to 15% of all colorectal cancer is due to genetic predisposition, with a further 60% due to sporadic tumours that appear to develop from adenomatous polyps. Adenomas and carcinomas develop through a stepwise accumulation of somatic mutations, termed the 'adenoma–carcinoma sequence'. Diet has major implications for the aetiology of the disease, with those rich in vegetables associated with protection.

Worldwide, milk and dairy products contribute approximately 5% of total energy. But among the traditionally pastoral people of China, India, Africa and Northern Europe, dairy and milk products supply approximately 10% total energy and 15–25% dietary protein and fat intake. Dairy products have been tentatively suggested to play a protective role in the prevention of CRC. This chapter will examine both experimental and epidemiological data for dairy products and their significant components (including calcium, casein and conjugated linoleic acid) to determine if a basis for the protective hypothesis exists. The emerging evidence for a protective role for fermented milks, probiotics, prebiotics and other functional dairy products is reviewed.

2.2 The relationship between diet and cancer

Cancer is a significant global public health problem. Yearly 10.1 million new cancer cases are diagnosed, with a further 6.2 million people losing their lives worldwide. This disease accounts for 25% of deaths in countries with a westernised lifestyle (IARC, 2000). Colorectal cancer is the fourth most frequent cause of cancer-related mortality in the world: approximately 944,000 new cases were diagnosed globally in 2000 and this accounts for 9.2% of all new cancer cases (IARC, 2000). Within New Zealand, Australia, North America and Europe it is the second most prevalent cancer after lung/breast (Boyle and Langman, 2000): 363,000 new cases were reported in Europe in 2000 and it affects 6% of men and women by age 75, in almost equal proportion. Generally the incidence and mortality of the disease are escalating (Cummings *et al.*, 1996; Boyle and Langman, 2000). The Modena colorectal registry (Italy) reported a 12.2% increase in incident rates from 1985 to 1997 and other European studies have reported a similar trend (Johansen *et al.*, 1993; Kemppainen *et al.*, 1997).

Worldwide incident rates show an approximate 20-fold variation, with the developed world suffering the highest rates and India one of the lowest (Ferlay *et al.*, 2001). Rates even within countries may vary, as in India where the strictly vegetarian Janists have a lower rate of colorectal cancer than the westernised Parsi population (ICS, 1985). These fluctuations are generally attributed to both genetic factors and environmental factors, especially diet. Migrant studies (Japan to the USA, Eastern Europe to North America) give additional support to the role of environmental factors in the aetiology of colorectal malignancies, with reported incident rates of migrants and their descendants reaching those of the host country, sometimes within one generation (WCRF, 1997). The highest rates of colorectal cancer are currently seen within Hawaiian Japanese men with an incidence of 53.5 per 100,000 (IARC, 1997).

Evidence suggests that diet plays an important role in the aetiology of colorectal cancer. However, identifying conclusively which constituents (e.g. vegetables, meat, fibre, fat, micronutrients) exert an effect on risk has been more problematic due to inconsistent data (for a detailed review of the epidemiological studies see Potter, 1999). The 1997 World Cancer Research Fund report concluded that the evidence (mainly from case-control studies) for diets rich in vegetables protecting against colorectal cancer was convincing, but that the effect of fruits could not be judged because the data are limited and contradictory. Data from prospective studies are less convincing than case-control studies (Bingham, 2000). Diets high in fibre were reported to possibly reduce the risk of colorectal cancer, with suggested protective mechanisms including toxin adsorption/dilution (WCRF, 1997; AGA, 2000). Furthermore several micronutrients including carotenoids, ascorbate and folate have been examined epidemiologically to account for the protective effect associated with vegetables, but the results have frequently been incongruous, and coupled with the paucity of data, no strong associations were observed (Giovannucci *et al.*, 1993; Slattery *et al.*, 1997). Studies examining the effect of meat consumption (especially red and processed meats) on colorectal cancer have collectively produced neither strong nor consistent findings, but it is believed that the weight of evidence points towards a slighty elevated risk (WCRF, 1997; Norat and Riboli, 2001), although the mechanisms by which meat affects colon carcinogenesis remains unclear. High saturated/animal fat intake may be related to elevated risk (Potter, 1999; Zock, 2001) but does not appear to contribute to the risk associated with meat consumption (Giovannucci and Goldin, 1997).

A slightly elevated risk in beer drinkers versus abstainers was first reported by Stocks (1957); since then alcohol has been suspected as a risk factor for colorectal neoplasms. Studies on the topic have provided contentious results as detailed in a review by Potter *et al.* (1993). Overall, however, raised alcohol consumption probably increases the risk of cancers of the colon and rectum and this association is related to total ethanol intake rather than the type of alcoholic drink (WCRF, 1997). Despite the weight of the epidemiological evidence for diet playing an important role in colorectal cancer risk, definitive evidence for causal association is lacking owing to the difficulties of conducting dietary intervention studies. It is necessary to look elsewhere for stronger criteria of such a link. The next section provides a brief summary of the processes of carcinogenesis in the colon.

2.3   Colon carcinogenesis

Almost 70% of colorectal malignancies appear to be restricted to the left large bowel (descending) between the lower rectum and the splenic fissure, though curiously this subsite distribution appears to be undergoing a proximal shift towards the right large bowel (ascending), for reasons unknown (Faivre *et al.*, 1989; Ponz de Leon and Roncucci, 2000). The colonic microarchitecture is characterised by crypts, which are approximately 50 cells in depth. The normal replicative dynamics and structure of these crypts ensure that both stem cells and immediate daughter cells replicate in the lowest region. When the immediate daughter cells divide and migrate they give rise to all the cells that line the crypt. Eventually these cells will reach the surface, by which stage they are fully differentiated columnar epithelial cells, covered with microvilli, intimately connected via numerous tight junctions and involved in water and electrolyte transport. The constant outward movement of cells from the crypts should ensure that no interaction occurs between the luminal environment and replicating cells; thus any mutagens should then only affect the differentiated colonocytes and effectively have no impact upon the integrity of the crypt cell population (Potter, 1999).

The nature of the microarchitecture was used by Potter to argue that the first mutagenic event occurring to a progenitor cell must be a blood-borne agent rather than luminal. Further it was suggested that, for luminal constituents to play any role in carcinogenesis, a pre-existing polyp must be in contact with the faecal stream. Alternatively, Shih *et al.* (2001) recently postulated that the development of an adenomatous polyp may proceed from a top-down mechanism, whereby the inter-cryptal cells are transformed presumably from a luminal agent and the altered mucosae spread laterally and downward to form new crypts, which connect to and eventually replace existing crypts. Other authors who follow the more classical mechanism for colorectal carcinogenesis, offer the assertion that a luminal agent could provide the 'first hit' (mutation) if a focal failure in the epithelial barrier occurred as a result of insult (physical/chemical) or of failure in terminal differentiation. An agent could then affect directly or indirectly cells in the crypt and lead to the formation of a polyp (Bruce *et al.*, 2000). Mutation of the *APC* gene (adenomatous polyposis coli) permits an adenomatous polyp to develop, and such a formation is considered an important predisposing risk factor for CRC. However, this does not mean that all polyps become malignancies (only about 5%) nor does it prohibit the possibility that *de novo* colorectal tumorigenesis may occur (Owen, 1996). Adenomas are well-demarcated clumps of epithelial dysplasia, classified into three histological types – tubulovillous, tubular and villous – which increase in prevalence with age, being present in 24–40% of people over 50 years old (Ponz de Leon and Roncucci, 2000).

Approximately 15% of all colorectal cancer is due to genetic predisposition, with a further 60% the result of sporadic tumours that appear to develop from adenomatous polyps. Adenomas and carcinomas develop through a stepwise accumulation of somatic mutations (Fig. 2.1). While the precise sequence of genetic events is not completely understood, it involves inactivation of various tumour-suppressing genes (e.g. *APC*, *p53*), activation mutations in proto-oncogenes (e.g. K-*ras*, c-*myc*), and loss of function in DNA repair genes (e.g. *hMLH1*, *hMSH2*). This archetypal multi-step model has been termed the 'adenoma–carcinoma sequence' (Vogelstein *et al.*, 1988; Fearon and Vogelstein, 1990). For detailed information on the genetic events and pathways to colorectal cancer the reader is referred to Potter (1999), Chung (2000), Ponz de Leon and Percesepe (2000) and Souza (2001).

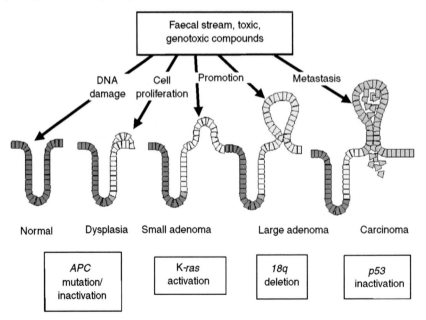

Fig. 2.1   The adenoma–carcinoma sequence. Inactivation of the *APC* gene facilitates the development of a hyperplastic epithelium. Activation of oncogenic *ras* occurs in early adenomas. Deletion of *18q* is found in dysplastic late adenomas. Ultimately, with the inactivation of *p53*, colorectal cancer develops. Components of the faecal stream are believed to modulate colorectal carcinogenesis at various stages.

2.4   Colorectal cancer and dairy products

Globally, milk and dairy products contribute approximately 5% of total energy, but consumption is higher in peoples from China, India, Africa and Northern Europe. Within these traditionally pastoral people, dairy and milk products supply approximately10% total energy and 15–25% dietary protein and fat intake (WCRF, 1997). Nutritionally, cows' milk (per 100 g) provides around 4 g of fat, 4.6 g lactose and 3 g protein. Saturated fatty acids account for almost 66% of milk fat, while polyunsaturated fatty acids make up less than 4%. Yoghurt, milk and cheese are good sources of calcium, riboflavin and vitamin B_{12}, whereas the high-fat dairy products, such as cream and butter, are important sources of vitamin A but provide little calcium.

The epidemiological data on the relationship between dairy products and colorectal cancer are inconsistent at best (see Table 2.1). Several studies have suggested that increased consumption of milk elevated the risk of colorectal cancer (Kune et al., 1987; Mettlin et al., 1990). In contrast a number of studies (including Bostick et al., 1993; Kearney et al., 1996; Boutron-Ruault et al., 1999) have reported no significant association between risk of colorectal cancer and dairy products. A few studies have described a significant inverse association of milk intake with colon cancer. In a US study yoghurt exhibited a positive association with decreased risk, but milk showed no protective effect (Peters et al., 1992). Cheese has both been reported to have no association with colorectal cancer (Tuyns et al., 1988) and to be associated with elevated risk (Bidoli et al., 1992). The conflicted data resulted in a recent WRCF report (1997) stating that:

> It seems plausible that, in relation to colorectal cancer, any increased risk associated with dairy products may be due to fat whereas any decreased risk may be a consequence of vitamin D and calcium content and possibly, for some dairy products, conjugated linoleic acid. As it stands, however, the evidence on the relationship between colorectal cancer and dairy products is inconsistent; no judgement is possible.

Even though the epidemiological data for dairy products and their effects on colorectal cancer have proved inconsistent, a somewhat stronger case may be presented for the role of individual constituents of dairy products (e.g. calcium, casein, whey, conjugated linoleic acid and sphingolipids) and functional dairy products (e.g. probiotics and prebiotics) in modulating the risks for CRC.

2.5   Calcium

A vast body of epidemiological literature has addressed the possibility that calcium might reduce colorectal cancer risk (for reviews see Bostick et al., 1993; Lipkin and Newmark, 1995; Martinez and Willett, 1998). The results are conflicting, with recent case-control studies providing further evidence that no strong association between dietary calcium and reduced CRC risk appears to exist. Boutron-Ruault et

Table 2.1   Colon cancer and dairy products

Study design/size	Location	Food	Association with colorectal cancer risk	Author
Cohort, 461,443, 146 colon cancer deaths	USA	Milk products	No significant association	Phillips and Snowdon (1985)
Case-control, 100ca/100 Hc	Greece	Milk products	No significant association	Manousos et al. (1983)
Cohort, 9959 M + F, 76 colorectal cancer	Sweden	Milk	No significant	Jarvinen et al. (2001)
Case-control, ca 171/pc 309	France	Dairy products	No significant	Boutron-Ruault et al. (1999)
Cohort, 9,159 M and 8,585 F, 331M + 350F colorectal polyp cases	USA	Cheese, whole milk, skim milk, fermented dairy	No significant association	Kampman et al. (1994a)
Cohort, 12,852, 215 colon cancer	Netherlands	Fermented milk, hard cheese, unfermented milkcalcium	No significant association	Kampman et al. (1994b)
Cohort, 35,216, 212 colon cancer	USA	Dairy products	No significant association	Bostick et al. (1993)
Case-control, 746 ca/746 pc	USA	Dairy products	No significant association	Peters et al. (1992)
Case-control, 132 ca/426 hc	Singapore	Milk products calcium	No significant association	Lee et al. (1989)
Case-control, 93 ca/93 hc	Japan	Dairy products	No significant association	Tajima and Tominaga (1985)
Cohort 47,935, 203 colon cancer	USE	Milk, fermented dairy	No significant association	Kearney et al. (1996)
Case-control, 286ca/295 pc and 203 hc	Spain	Dairy, calcium	No significant association	Benito et al. (1990)
Case-control, 558 ca/1032 hc	Italy	Milk, cheese, calcium	No significant association	Negri et al. (1990)
Case-control, 3334 cancer, 504 colon ca/ 1300 pc	USA	Whole milk, 2% milk Skim milk	Significant positive association No significant association	Mettlin et al. (1990)
Case-control, 392 ca/727 hc and pc	Australia	Milk	Significant positive (1987)	Kune et al.
Cohort, 27,111, 185 colon cancer	Finland	Calcium	Significant inverse association	Pietinen et al. (1999)
Case-control,424 ca/414 hc and pc	USA	Dairy products	Significant inverse association	Shannon et al. (1996)
Case-control, 210 distal and 152 proximal colon ca/618 pc	USA	Cheese	Significant inverse association (distal colon only)	Young and Wolf (1988)

Table 2.1 cont'd

Study design/size	Location	Food	Association with colorectal cancer risk	Author
Case-control, 231 ca/391 pc	USA	Dairy	Significant inverse association	Slattery *et al.* (1988)
Case-control, 232 ca, 259 pc	Nether-lands	Milk	Significant inverse association	Kampman *et al.* (1994c)
		Skim milk, fermented dairy, cheese	No significant, association	
Case control, 399 ca/399 hc	France	Milk	Significant inverse trend	Macquart-Moulin *et al.* (1986)
		cheese, butter	No significant association	

ca cases, pc population controls, hc hospital controls, M male, F female.

al. (1999) reported no association between the factors in a French case-control study, as did a Swedish case-control study, although it suggested that dietary vitamin D might reduce the risk of colorectal cancer (Pritchard *et al.*, 1996). In Finnish men consuming a diet high in fat, meat and fibre and low in vegetables, high calcium intake was associated with lowered risk of colorectal cancer (Pietinen *et al.*, 1999). Further case-control studies on fermented dairy products and milk indicated that total dietary calcium was positively but non-significantly associated with colon cancer risk after adjustment for confounding factors (Kampman *et al.*, 1994c; Ma *et al.*, 2001). Similarly a prospective study examining the influence of dairy foods reported a non-significant inverse association between the intake of calcium from food/supplements and colon cancer risk (Kearney *et al.*, 1996).

Notwithstanding the burgeoning press on the apparent lack of effect of calcium, sufficient studies have reported a positive association to suggest that calcium may in some manner protect against CRC (for reviews see Holt, 1999; Mobarhan, 1999; Kampman *et al.*, 2000). Recently, a case-control study by Marcus and Neucomb (1998) reported a protective effect of higher levels of calcium intake and CRC risk in women. Calcium supplementation and consumption of total low-fat dairy products were inversely associated with colon cancer risk in both men and women (Kampman *et al.*, 2000).

In contrast to conflicting epidemiological studies on calcium and colon cancer, experimental studies in animals and humans more regularly report a beneficial role. Two recent studies on calcium supplementation and colorectal adenoma recurrence have shown significant, though moderate, reduction in the risk, with a third reporting a modest but non-significant reduction (Baron *et al.*, 1999; Bonithon-Kopp *et al.*, 2000; Wu *et al.*, 2002). Wu *et al.* (2002) reported reduced risk only for distal colon cancer and suggested that the observed risk pattern was consistent with a threshold effect, indicating that calcium intake beyond moderate level

(700 mg/d) may not be associated with a further reduction in risk. High calcium intake may offer a protective effect from fat-induced promotion of carcinogenesis by binding cytotoxic bile and fatty acids (Kleibeuker *et al.*, 1996a) or by reducing proliferation in the upper part of the colonic crypt (Bostick *et al.*, 1995). Proliferative zone expansion in the colonic crypts and an increased rate of epithelial proliferation are often viewed as an early step in carcinogenesis. Stimulation of proliferative activity in colonic epithelial may in part be mediated by chemical or physical cytotoxic mechanisms, such that epithelial damage induced by these sources would increase cell loss at the epithelial surface. This would result in a compensatory increase in the mitotic activity of the crypts, thus elevating the risk for colorectal cancer. Such considerations led to the development of assays to assess cytotoxic activity in faecal water (aqueous phase faeces) towards colon cells *in vitro* (Rafter *et al.*, 1987). It is thought that bile acids, especially secondary bile acids, make a major contribution to faecal water cytotoxicity (Rafter *et al.*, 1987). Dietary calcium reduces the cytotoxicity of faecal water, presumably by precipitating soluble bile acids (Van-der-Meer *et al.*, 1991, 1997; Govers *et al.*, 1993; Lapre *et al.*, 1993; Sesink *et al.*, 2001). Interestingly a recent study showed that a shift from a dairy-rich diet (high fat, high risk) to a dairy-free diet (low fat, low risk) showed an increase in cytotoxicity of the faecal water, possibly as a result of, among other things, decreased calcium (Glinghammar *et al.*, 1997).

The formation of aberrant crypt foci (ACF) represent putative preneoplastic lesions that are induced in the colon of carcinogen-treated rodents and are present in humans with a high risk for developing colon cancer (Pretlow *et al.*, 1992; Bird, 1995). ACFs are characterised by an increase in the size of the crypts, the epithelial lining and the pericryptal zone. The prevention of azoxymethane (AOM)-induced ACF is used as a bioassay in animal studies, and has demonstrated the positive chemoprotective effects of dietary calcium supplementation (various concentrations) and the efficacy of different calcium salts (lactate, phosphate, glucurate) (Wargovich *et al.*, 1990, 2000; Pereira *et al.*, 1994; Li *et al.*, 1998). An ACF study using 1,2-dimethylhydrazine (DMH) reported that low dietary calcium lactate supplementation inhibited colorectal carcinogenesis, and changed tumour location towards the distal colon (Vinas-Salas *et al.*, 1998). Calcium also appears to induce apoptosis in normal mouse distal colonic epithelium without affecting cell proliferation. Such a mechanism may contribute to calcium's putative chemopreventive role in colorectal carcinogenesis (Penman *et al.*, 2000). Considering these data as a whole, it appears likely that calcium has a promising role to play in protecting against colorectal cancer, but the strength of that role is as yet undetermined.

2.6   Casein

The casein content of milk represents approximately 80% of the milk proteins, in the form of α-, β- and κ-casein. Caseins are mostly phosphate-conjugated proteins with low solubility with the degree of calcium binding by the protein being proportional to the phosphate content. No epidemiological data currently exist for casein: several animal studies indicated that dietary supplementation with casein reduced the risk of colon cancer. Casein diets provided a greater protective effect against the development of DMH-induced intestinal tumours in rats than did red meat or soybean diets, as evidenced by a reduced tumour incidence (McIntosh *et al.*, 1995). Further, a casein–wheat diet resulted in a decreased tumour mass index but not tumour burden in DMH-induced rats when compared with a chickpea–wheat diet (McIntosh *et al.*, 1998). *In vitro* casein appears to exert an antimutagenic effect (Yoshida and Ye-Xiuyun, 1992; van-Boekel *et al.*, 1993), which may be enhanced by enzymatic digestion (Goëptar *et al.*, 1997). The converse, however, is true if the protein is thermolysed. Casein cooked at 180 °C and fed as a supplement promotes the growth of ACFs in AOM-induced animals and appears to do so in a dose-dependent and thermolysis time-dependent manner (Zhang *et al.*, 1992). The thermolysis of casein reduces its digestibility, so more protein reaches the colon and consequently raises colonic protein fermentation, and therefore the concentration of the potential tumour promoter ammonia. However, the effect of thermolysed casein on colorectal carcinogenesis does not appear to be mediated by the protein's fermentation products (Corpet *et al.*, 1995), neither is the promoting effect associated with the formation of heterocyclic amines (Corpet and Cassand, 1995). Molecules are created during the cooking of protein-rich foods, which when activated become mutagens and carcinogens capable of producing tumours in various tissues both in rodents and non-human primates (Nagao and Sugimura, 1993). The promotion effect of cooked casein has been suggested as the result of mucosal abrasion due to the physical properties of the protein, as coarse cooked casein resulted in significantly more ACFs than fine ground (Corpet and Chatelin-Pirot, 1997).

Although the direct effects of casein on colorectal carcinogenesis appear negative, data suggest that the protein and its degradation products may help to decrease the risk of colorectal cancer through indirectly affecting calcium. Several animal supplementation studies have reported that casein phosphopeptides (CPPs) (created by proteolytic degradation of α- and β-casein) significantly increase calcium and zinc absorption (Lee *et al.*, 1983; Saito *et al.*, 1998). CPPs have also been reported to help overcome the inhibitory effect of phytate on zinc and calcium absorption (Hansen *et al.*, 1996) and to improve the absorption of these metals from rice-based cereals but not wholegrain cereal in humans (Hansen *et al.*, 1997). CPPs appear to alter calcium absorption not by influencing membrane-bound receptors or ion channels, but rather by acting as calcium carriers (ionophores) across the membrane (Ferraretto *et al.*, 2001). However, one recent study has indicated that CPPs had no effect, whereas casein enhanced calcium absorption (Bennett *et al.*, 2000).

Evidence for casein's anticancer activity is accumulating, as are the data for an effect on calcium absorption. High levels of calcium may be protective for colorectal cancer. If calcium's effect on colonic mucosal proliferation is in part a direct mechanism as suggested by Nobre-Leitao *et al.* (1995), rather than the effect being wholly mediated through binding of cytotoxic bile acids, then the ability of casein and/or its phosphopeptides to enhance the metal's absorption may be important, especially in conjunction with the protein's apparent antimutagenic activity.

2.7   Whey

Whey proteins constitute a further 18% of the milk proteins, principally α-lactoglobulin, α-lactalbumin and bovine serum albumin. Whey proteins are globular in nature, more water-soluble than caseins and subject to heat denaturation. In comparison to casein, whey protein appears to exhibit a more direct effect on colorectal carcinogenesis. Animal studies examining DMH- and AOM-induced tumour formation have reported that whey protein supplementation inhibits incidence and growth of chemically induced tumours and as a result may reduce the risk of developing colorectal cancer (Bounous et al., 1988; Papenburg et al., 1990; Hakkak et al., 2001). Current studies are attempting to attribute the anticarcinogenic effects of whey protein to specific group members, i.e. bovine lactoferrin inhibits carcinogenesis in AOM-treated rats with no overt toxicity (Tsuda et al., 2000), while bovine α-lactalbumin exerts an antiproliferative effect on human colon adenocarcinoma cell lines (Caco2, Ht-29) (Sternhagen and Allen, 2001).

2.8   Conjugated linoleic acid

Conjugated linoleic acid (CLA) is a collective term for positional and geometric isomers of linoleic acid that are formed during the process of biohydrogenation of polyunsaturated fatty acids in ruminant animals. This reaction is catalysed by enzymes present in anaerobic microorganisms in the rumen, and therefore milk, dairy products and ruminant meat fat represent the richest sources of this fatty acid. The concentrations of CLA in dairy products range from 2.9 to 8.92 mg/g fat with the most abundant isomer 9, trans-11 CLA, contributing approximately 73–93% of the total amount (MacDonald, 2000). CLA has been shown unequivocally to inhibit carcinogenesis in animal models (NRC, 1996), protecting against 7,12-dimethylbenz[a]anthracene (DMBA)-induced mouse epidermal tumours (Ha et al., 1987), benzo[a]pyrene-induced mouse forestomach neoplasia (Ha et al., 1990), DMBA and methylnitroso urea-induced rat mammary tumours (Ip et al., 1994; Banni et al., 1999). While a lot of data confirm the beneficial role of CLA, only a few studies exist to support a role in protecting against colorectal carcinogenesis. Dietary supplementation with CLA protected against 2-amino-3-methylimidazo[4,5-f] quinoline (IQ) adduct formation in the rat colon, and also significantly inhibited ACF formation in AOM tumour-induced animals (Liew et al., 1995). A later study reported that under certain conditions, dietary CLA lowered IQ and 2-amino-1-methyl-6-phenylimidazo[4,5-6]pyridine (PhIP) adduct formation, in various rat tissues including the colon (Josyula and Schut, 1998). Physiological concentrations of CLA inhibited proliferation of colorectal cancer cells *in vitro* (Shultz *et al.*, 1992). Although anti-carcinogenic effects of CLA are dramatic, little information exists on the mechanisms of action. CLA anticancer activity may result from its potential function as an antioxidant or by modifying synthesis of eicosanoids, which are associated with tumour promotion (Banni *et al.*, 1995; Sebedio *et al.*, 1997). The activation of PPARgamma (peroxisome proliferator-activated receptors) by CLA is also suggested to account for a degree of CLA's broad anticarcinogenic activity, as agonists of PPARgamma are reported to promote apoptosis and inhibit clonal expansion both *in vitro* and *in vivo* (McCarty, 2000). For further information on CLA the reader is referred to reviews by Parodi (1997), McGuire and McGuire (1999) and Lawson *et al.* (2001). Although currently no direct evidence exists that CLA protects against colon cancer in humans, the wealth of data suggests that CLA may contribute to the potential anticancer effect of dairy food and therefore cannot be ignored.

2.9   Sphingolipids

Sphingolipids are situated in plasma membranes, lipoproteins and lipid-rich structures. They are components of most food types, but are most prevalent in eggs, soybeans and dairy products with concentrations ranging from 1 to 2 g/kg (Vesper *et al.*, 1999). The molecules are characterised by the presence of a sphingoid base backbone which has numerous configurations depending upon many factors, including degree of saturation, alkyl chain length and branching of specific chains (for review see Schmelz *et al.* (1996). In addition to the structural function of sphingolipids in membranes, these molecules and their active metabolites (ceramide and sphinogosine) play important roles in cell regulation and transmembrane signalling (for reviews see Riboni *et al.*, 1997; Kolesnick and Kronke, 1998). Ceramide and sphingosine affect apopotosis, differentiation and cell growth in a number of cell types (Hannun and Obeid, 1995; Sweeney *et al.*, 1998). Dietary sphingomyelin (a major sphingolipid) decreases the incidence of ACF formation (70%) and the ratio of adenoma to adenocarcinoma in colons of DMH-treated mice (Dillehay *et al.*, 1994; Schmelz *et al.*, 1996). Other major sphingolipids, e.g. glucosylceramide and lactosylamide, have anti-cancer effects comparable to those of sphingomyelin (Schmelz *et al.*, 2000). The activity of these molecules may be mediated through the active metabolites, as sphingolipids are hydrolysed throughout the gastrointestinal tract. To that end, recent work has centred on functionalising a ceramide analogue by attaching D-glucuronic acid, so that the compound becomes a substrate for *Escherichia coli* β-glucuronidase and colonic digestion, thus facilitating a more colon-specific delivery for the effects of the anticancer metabolite (Schmelz *et al.*, 1999). Clearly there is evidence to support a protective dietary effects of sphingolipids against colorectal carcinogenesis, but as these associations are based on only a few studies, the need for more investigations on both animals and humans is evident.

2.10   Prebiotics and probiotics

The relationship between colorectal cancer, diet and gut microflora is complex and intimate. The substances entering the colon from the ileum and the resident microflora represent key determinants of colon physiology. These, together with the innate biology of the colon (e.g. epithelium, motility), are pertinent to colorectal carcinogenesis. The concentration of bacteria resident in the colon increases distally with an estimated 300–400 different cultivatable species (belonging to 190 genera) resident within a healthy adult colon (Holzapfel *et al.*, 1998). About a hundred of these species are resident in the human colon at concentrations around 10^{11} bacteria per gram. The anaerobes, bifidobacteria, eubacteria and bacteroides, and clostridia account for more than 99% of those species present in the colon. Once microflora is established, little qualitative variation in the composition occurs over time (Kleibeuker *et al.*, 1996b), although there is extensive evidence that the metabolic activity of the microflora can be modulated by diet, especially non-digestible carbohydrates (fibre, oligosaccharides) (Rowland *et al.*, 1985; Rowland, 1988). With the capacity for the microflora to modulate colonic conditions established, it becomes obvious why analysing their dynamic interaction with the colonic environment and mucosa is of such importance in terms of colorectal cancer and why there is currently intense interest in dietary modulation of microflora with food ingredients such as pro- and prebiotics (reviewed in Burns and Rowland, 2000).

The original definition of a probiotic was 'a live microbial feed supplement which beneficially affects the host animal by improving its intestinal microbial balance' (Fuller, 1989). Recent definitions are more general, omitting the aspect of intestinal balance to encompass more directly effects on the immune system. Salminen *et al.* (1998) define a probiotic as 'a live microbial food ingredient that is beneficial to health'. For the most part probiotics are members of two genera of lactic acid-producing bacteria (LAB), *Bifidobacterium* and *Lactobacillus*, but *Enterococcus* and *Saccharomyces* are also used. Many of the bacteria used for probiotic preparations have been isolated from human faecal samples to maximise the chance of compatibility with the human gut and thereby enhance their probability of survival. The poor survival of ingested probiotics in the inimical environment of the stomach and small intestine has led to interest in the use of prebiotics.

A prebiotic is a 'non-digestible food ingredient that beneficially affects the host by selectively stimulating the growth and/or activity of one or a limited number of bacteria in the colon that have the potential to improve host health' (Gibson and Roberfroid, 1995). Various poorly digested carbohydrates fall into the category of prebiotics including certain resistant starches and fibres (Silvi *et al.*, 1999), but the most extensively described prebiotics are the non-digestible oligosaccharides (NDOs). These are low molecular weight carbohydrates with two to ten degrees of polymerisation, which are poorly digested in the human intestine, thus reaching the colon largely unaltered and can act as a substrate for the colonic microflora. NDOs appear to specifically stimulate certain bifidobacteria and lactobacilli, often at the expense of other microflora components such as clostridia, *E. coli* and bacteroides

Table 2.2   Effect of probiotics and prebiotics on cancer in humans – epidemiological studies

Study design/size	Type of cancer	Probiotic/ prebiotic	Outcome	Author
Case control, 209 M and 62 F ca/182 M and 245 F pc	Small and large colon adenoma, and colon cancer	Yoghurt consumption	Inverse relationship yoghurt consumption (0.5–1 pots per day) with risk of large adenomas in men and women	Boutron et al. (1996)
Case-control, 210 distal and 152 proximal colon ca/618 pc	Colon cancer	Fermented milk consumption	Inverse association of colon cancer with the consumption of fermented milk	Young and Wolf (1988)
Case-control, 746 ca/746 pc	Colon cancer	Yoghurt	Protective effect against colon cancer	Peters et al. (1992)
Cohort, 9159 M and 8585 F, 331 M + 350 F colorectal polyp ca	Colorectal adenomas	Fermented dairy products	Inverse association (non-significant) between yoghurt consumption and adenomas in men and women	Kampman et al. (1994a)
Case-control, 232 ca, 259 pc	Adeno-carcinoma of colon	Fermented dairy products	Positive, significant association (OR 1.52) in men; negative, non-significant association in women	Kampman et al. (1994c)

ca cases, pc population controls, hc hospital controls, M male, F female.

(Rowland and Tanaka, 1993; Gibson and Roberfroid, 1995). The combination of probiotics and prebiotics can result in synergistic or additive effects on gastrointestinal function. Such combinations have been referred to as synbiotics. The definition of a synbiotic is:

> a mixture of probiotics and prebiotics that beneficially affects the host by improving the survival and implantation of live microbial dietary supplements in the gastrointestinal tract, by selectively stimulating the growth and/or activating the metabolism of one or a limited number of health-promoting bacteria, and thus improving host welfare (Gibson and Roberfroid, 1995).

Although limited epidemiological data are available to assess the effects of LAB on colorectal carcinogenesis, the data are indicative of a protective role for probiotics (Table 2.2). However, at the epidemiological level, the studies are related to normal yoghurt rather than varieties enhanced with specific probiotics. Yoghurt is made by inoculating certain bacteria (starter culture), usually *Streptococcus thermophilus* and *Lactobacillus bulgaricus*, into milk. Whilst neither of these strains is classically considered a probiotic, a number of studies have demonstrated that intake of yoghurt enhances lactose digestion in individuals with low intestinal levels of lactase, the enzyme necessary to digest lactose or milk's

Table 2.3   Effects of probiotics and prebiotics on bacterial enzyme activity and in animal and human studies

Subjects/species	End-point	Pro-/prebiotic	Result	Author
Lister hooded rat (HFA)	β-glucuronidase and β-glucosidase activity	L. acidophilus (NCFM) or Bifidobacterium adolescentis (2204) (10⁹cells/day for 3 days)	A significant decrease in enzyme activity for L. acidophilus only.	Cole et al. (1989)
F344 rat	Faecal levels of enzyme reaction products after administration of test substances	L. acidophilus NCFM, 1×10⁹ CFU/day (for 21 days)	Animals given L. acidophilus had significantly lower free amines in faeces and 50% fewer conjugates.	Goldin and Gorbach (1984a)
Elderly patients	Faecal p-cresol	Neosugar (8 g per day for 14 days)	A 10-fold increase in Bifidobacteria.	Hidaka et al. (1986)
Rat	Faecal short chain fatty acid (SCFA) levels	Neosugar (10–20% dietary for 6–8 weeks)	Significantly increased SCFA concentration in faeces.	Tokunga et al. (1986)
Germ-free Lister hooded rat	Various caecal enzymes	Transgalactosylated oligo-saccharides (TOS) (5% w/w in diet for 4 weeks) or TOS + Bifidobacterium breve (Yakult strain) 1×10⁹ CFU/day for 4 weeks	β-glucuronidase and nitrate reductase activities, pH and the conversion of IQ to 7-OHIQ significantly reduced in caecal contents of the TOS-fed rats. Bacterial β-glucosidase activity was increased in TOS-fed rats.	Rowland and Tanaka (1993)
Male Sprague–Dawley rats	Faecal enzymes and ammonia	Bifidobacterium longum 25 6×10⁹ CFU/day and inulin (5%) for 12 weeks	Significant decrease in β-glucuronidase and ammonia. Probiotic plus prebiotic was more effective.	Rowland et al. (1998)
21 healthy subjects	Faecal enzyme activities	Milk supplemented with L. acidophilus (1×10⁹ CFU/day for 4 weeks)	Faecal β-glucuronidase activity was reduced from 1.7–2.1 units to 1.1 units in all subjects.	Goldin and Gorbach (1984b)

Subjects	Measure	Treatment	Result	Reference
14 colon cancer patients	Faecal β-glucuronidase activity	L. acidophilus (given as a fermented product, between 1.5×10^{11} and 6×10^{11} CFU/day for 6 weeks)	A 14% decrease in mean β-glucuronidase activity after 2 weeks.	Lidbeck et al. (1991)
20 healthy male subjects (40–65 years old)	Faecal β-glucuronidase and β-glucosidase activity	Lactobacillus casei Shirota (3×10^{11}CFU/ day for 4 weeks)	Significant decrease in β-glucuronidase and β-glucosidase activity ($P < 0.05$).	Spanhaak et al. (1998)
9 healthy adults	Faecal β-galactosidase β-glucosidase and β-glucuronidase	Ofilus™ (a commercial fermented milk containing L. acidophilus (3×10^{9} CFU/ day) strain A1, B. bifidum B1 (3×10^{10} CFU/ day), Lactoccus lactis and Lactococcus cremoris (3×10^{10} CFU/day for 3 weeks)	No change in faecal β-galactosidase and b-glucuronidase. Significant increase in β-glucosidase activity.	Marteau et al. (1990b)
3 male and 9 female healthy subjects	Faecal β-glucosidase and β-glucuronidase activity	Digest™ containing viable L. acidophilus (strain DDS1) (three 8 oz cups of milk containing 2×10^{6} cfu/ml per day for 4 weeks)	A decrease in β-glucuronidase and β-glucosidase activity.	Ayebo et al. (1980)
21 young women aged 21–35 years with severe pre-menstrual syndrome	19 faecal enzyme activities	L. acidophilus and B. bifidum (1×10^{9} of each type of bacteria/capsule – three capsules per day for 2 months	Reduction in β-glucosidase activity. A decrease in β-glucuronidase activity.	Bertazzoni-Minelli et al. (1996)
Human volunteers	Viable bacterial count	Crossover study, 2 treatments 1) Soybean oligosaccharides (SOE) 10g/day for 3 weeks. 2) SOB = SOE 10g/day + B. longum 105 (6×10^{9}CFU)	No significant difference in levels of p-cresol, indole or phenol between various dietary periods.	Hayakawa et al. (1990)

sugar (i.e., a condition called lactase nonpersistence) (Marteau, 1990a). The beneficial effect may be explained by the ability of the starter cultures used in the manufacture of yoghurt with live, active cultures to produce the enzyme lactase, which digests the lactose in yoghurt. Yoghurt without live cultures, unfermented or sweet acidophilus milk, and cultured buttermilk is tolerated about the same as milk by individuals with lactase nonpersistence. To that end these active cultures in the yoghurt beneficially affect the host, and it has been reported that the bacteria used to produce yoghurt can inactivate carcinogens and prevent DNA damage in the colon of rats (Wollowski, 1999). For these reasons we shall consider the above bacteria as probiotic. A case-control study by Boutron *et al.* (1996) showed a significant inverse relationship between consumption of moderate amounts of yoghurt (half to one pot per day) and the risk of large colonic adenomas in both women and men. Further case-control studies have provided additional evidence of inverse associations of colorectal cancer risk and consumption of fermented dairy products (Young and Wolf, 1988) and yoghurt (Peters *et al.*, 1992). Kampman *et al.* (1994a) reported a non-significant inverse relationship between colonic adenomas and yoghurt consumption. This finding, however, was not corroborated in a further case-control study in the Netherlands of colorectal cancer risk and fermented dairy products, which revealed a small significant positive association in men (odds ratio, OR 1.52) and a small, non-significant inverse association in women (OR 0.77) (Kampman *et al.*, 1994c).

The data generated from probiotic and prebiotic human intervention prospective and animal studies suggest a beneficial role for probiotics and serve to strengthen the information provided by the population-based trials. The intervention studies used to evaluate the ability of probiotics to prevent cancer in humans have been based largely upon the use of biomarkers for assessing cancer risk (reviewed in Gill and Rowland, 2002). Owing to their non-invasive nature, markers in faeces and urine have been mostly used. Faecal markers are composed of two main categories: those examining the activity of bacterial enzymes or bacterial metabolites (e.g. IQ activation, β-glucuronidase and ammonia production) and those based on bioassays on faecal water and metabolites. The aqueous phase of human faeces (faecal water) is considered to be an important source of modulators and inducers for colorectal carcinogenesis and methods exist for assessing biological activities related to cancer risk in such samples.

Copious toxic, carcinogenic and endogenously produced compounds, e.g. steroids, are metabolised in the liver and conjugated to glucuronic acid by phase II enzymes, before being excreted via the bile into the small intestine. In the colon, the bacterial enzyme β-glucuronidase can hydrolyse these conjugates, thereby releasing the parent compound or its hepatic metabolite. In the case of carcinogens and mutagens, the activity of β-glucuronidase in the colon may enhance the probability of tumour induction. For example, the colon carcinogen DMH is metabolised in the liver, small amounts of the procarcinogenic conjugate of the activated metabolite methylazoxymethanol (MAM) are excreted in the bile. Hydrolysis of the conjugate by colonic bacteria releases MAM in to the colon. Germ-free animals treated with DMH or MAM result in fewer colon tumours than do conventional animals (Reddy *et al.*, 1974). Both animal and human studies have frequently shown that consumption of prebiotics and probiotics can decrease β-glucuronidase activity. These studies are summarised in Table 2.3 and some examples are discussed in more detail below. In a conventional rat study (F344), *Lactobacillus acidophilus* (10^9–10^{10} CFU/day) supplementation (21 days) of a high meat diet (72% beef) significantly reduced the activity of faecal β-glucuronidase and nitroreductase (by 40–50%) (Goldin and Gorbach, 1976). The modulating effect of the *Lactobacillus* strain was dependent on the type of basal diet fed – no significant effects were seen when the rats were fed a grain-based diet. In a similar study in human flora associated (HFA) rats, Cole *et al.* (1989) demonstrated a significant reduction in β-glucuronidase and β-glucosidase activities when *L. acidophilus* was fed for three days, with the effect evident for seven days post-dose.

Lactobacillus acidophilus supplementation (fermented milk) of the diets of 14 colon cancer patients for a period of six weeks resulted in a 14% decrease in β-glucuronidase activity and a reduction in total bile acids and deoxycholic acid of 15% and 18%, respectively. Coincident with these effects were alterations in the faecal microflora, as *E. coli* numbers were depressed and lactobacilli augmented. Similar results were obtained by Spanhaak *et al.* (1998) who reported a significant decrease in the activity of faecal β-glucuronidase and β-glucosidase activity in a group of 20 healthy male subjects given *L. casei* Shirota (approximately 10^{11} CFU/day for a four-week test period).

More direct evidence for protective properties of probiotics and prebiotics against cancer has been obtained by assessing the ability of cultures to prevent mutations and DNA damage in cell cultures or in animals (Table 2.4). The Ames *Salmonella* assay has been used to study the effect of lactic acid bacteria (LAB) on the induction of mutations by a wide variety of model carcinogens *in vitro*. The carcinogens used include *N*-methylnitrosourea (MNU), *N*-nitrosocompounds, *N*-methyl-*N*-nitro-*N*-nitrosoguanidine (MNNG), heterocyclic amines (e.g. IQ and related compounds) and aflatoxin B1. In general the results indicate that the various LAB can inhibit genotoxicity of dietary carcinogens *in vitro*. The degree of inhibition was strongly species-dependent, e.g. Pool-Zobel *et al.* (1993) demonstrated that *L. confusus* and *L. sake* had no effect on the mutagenic activity of nitrosated beef, whereas *L. casei* and *L. lactis* diminished the effect by over 85%.

It would appear probably these results, in conjunction with similar results by other workers, are a consequence of binding of the mutagens by the LAB (Zhang and Ohta, 1991; Bolognani *et al.*, 1997). Whether such a mechanism operates *in vivo* is questionable, since binding appears not to affect uptake of carcinogens from the gut, neither does it have any apparent effect *in vivo* on mutagenicity in the liver, and further the process appears to be highly pH-dependent and easily reversed (Bolognani *et al.*, 1997).

Using the comet assay (single cell microgel electrophoresis technique), Pool-Zobel *et al.* (1996) investigated the ability of a range of species of LAB to inhibit DNA damage in the colon mucosa of rats treated with the carcinogens MNNG or DMH. All the strains of lactobacilli and bifidobacteria tested – *L. acidophilus* (isolated from a yoghurt), *L. confusus* DSM 20196, *L. gasseri* P76, *B. longum* and

Table 2.4 Antigenotoxicity of LAB and prebiotics *in vitro* and *in vivo*

Species	Mutagen	Probiotic/prebiotic	End-point	Result	Author
Salmonella enterica sv. *Typhimurium* TA 98	Trp-P-1, Trp-P-2 and Glu-P-1	Lactic acid bacteria isolated from a traditional Chinese cheese	*In vitro* mutagenicity (Ames)	Lyophilised cells of all strains inhibited Trp-P-1 and Trp-P-2 mutagenicity. Some strains inhibited Glu-P-1.	Zhang *et al.* (1990)
Salmonella enterica sv. *Typhimurium* TA 1538	Nitrosated beef extract	Ten isolated *Lactobacillus* strains	*In vitro* mutagenicity (Ames)	*L. casei* and *L. lactis* inhibited mutation by > 85% and *L. sake* and *L. confusus* had no effect.	Pool-Zobel *et al.* (1993)
Salmonella enterica sv. *Typhimurium* TA 98	Glu-P-1, Glu-P-2, IQ, MeIQ, MeIQx, Trp-P-1, Trp-P-2	22 strains of intestinal bacteria	*In vitro* mutagenicity (Ames)	The majority of strains inhibited mutagenicity.	Morotomi and Mutai (1986)
Salmonella enterica sv. *Typhimurium* TA100 and TA97	Nitrovin and 2-aminofluorene	Nine strains of LAB	*In vitro* mutagenicity (Ames)	Significant anti-genotoxic activity exerted by six of the nine strains tested (*S. thermophilus, S. carnosus, En. faecalis, L. bulgaricus, L. rhamnosus, En. faecium*).	Ebringer *et al.* (1995)
Female, 4-week-old BALB/c mice	B(a)P, AFB1, IQ, MeIQ, MeIQx, PhIP and Trp-P-2	*L. acidophilus* and *Bifidobacterium longum*, single dose 10×10^{12} CFU/kg body weight	*In vitro* binding and *in vivo* mutagenicity in liver	Bacterial strains tested were able to bind carcinogens *in vitro*. No effect on *in vivo* mutagenicity or absorption.	Bolognani *et al.* (1997)
Male Sprague–Dawley rats	MNNG and DMH	*L. acidophilus, L. gasseri* P79, *L. confusus* DSM20196, *B. longum, B. breve, S. thermophilus* NCIM 50083, *L. acidophilus*. Single dose 1×10^{10} CFU/kg body weight p.o.	*In vivo* DNA damage in colon (comet assay)	Most LAB tested strongly inhibited genotoxicity in the colon. *Strep. thermophilus* had no effect. Heat treatment abolished LAB effect.	Pool-Zobel *et al.* (1996)
F344 rats (human flora associated)	DMH	Lactulose (3% in diet) for four weeks	*In vivo* DNA damage in colon (comet assay)	Lactulose significantly decreased extent of DNA damage ($P < 0.05$).	Rowland *et al.* (1996)

B. breve – prevented MNNG-induced DNA damage when given at a dose of 10^{10} CFU/kg body mass, 8 h before the carcinogen. In the majority of cases the DNA damage was reduced to a level similar to that in control animals. *Streptococcus thermophilus* NCIM 50083 was less effective than the other LAB strains. The protective effect was dose-dependent: diminished doses of *L. acidophilus* (50% and 10% of the original dose) were less effective in inhibiting MNNG-induced DNA damage. The importance of viable cells was also emphasised as heat-treatment of *L. acidophilus* abolished its antigenotoxic potential. Similar results were obtained when the LAB strains were tested in rats given DMH as the DNA damaging agent. Again, all the lactobacilli and bifidobacteria strongly inhibited DNA damage in the colon mucosa, whereas *S. thermophilus* was much less effective with only one of three strains tested exhibiting any protection against DNA damage.

A range of animal studies have been carried out using DMH or AOM to determine the effects of specific probiotics and prebiotics on ACF formation and tumour incidence (Table 2.5). In the most recent work by Tavan *et al.* (2002) it was reported that rats treated with various heterocyclic aromatic amines (HAAs) and supplemented with a range of dairy products (including non-fermented milk (30%), *Bifidobacterium animalis* DN-173010 fermented milk (30%) or *Streptococcus thermophilus* DN-001158 fermented milk (30%)) showed a 66%, 96% and 93% decrease in ACF incidence respectively. Further, intermediate biomarkers showed that there was a reduction in HAA metabolism, faecal mutagenicity and colon DNA lesions. Kulkarni and Reddy (1994) reported 50% inhibition of ACF formation in AOM tumour-induced male F344 rats fed *Bifidobacterium longum* in the diet (1.5% and 3% of a lyophilised culture containing 2×10^{10} CFU/g) for five weeks. As dietary treatments were started five weeks prior to administration of the carcinogen dose results do not allow deductions to be made about the stage of carcinogenesis affected.

An analogous study carried out by Challa *et al.* (1997) observed a 23% decrease in total colonic ACF and a 28% reduction in total AC in rats given a diet containing 0.5% *B. longum* BB536 (1×10^{8} CFU/g of feed). Animals were fed the experimental diet before treatment with AOM and throughout the experiment. Abdelali *et al.* (1995) compared the effects of *Bifidobacterium* sp. Bio Danone strain 173010 administered in diet and also fed as a fermented milk product. The amounts of organisms consumed were similar (6×10^{9} CFU/day) in both experimental variants. The LAB was given four weeks before DMH and continued for a further four weeks before ACF assessment. The bifidobacteria-fermented milk appeared to be slightly less effective in reducing ACF than the dietary bifidobacteria (49% and 61% reduction, respectively). Interestingly, however, skim milk alone reduced ACF numbers by 51%. In AOM-treated Sprague–Dawley rats, supplementation of *B. longum* 25 (6×10^{9} cfu/day) for 12 weeks, caused a significant reduction (26%) in total ACF by comparison to control animals (Rowland *et al.*, 1998). The changes were reported in only small ACF (one to three AC per focus). Given that the probiotic treatment began one week after the carcinogen exposure, the results indicate an effect on the early promotional phase of carcinogenesis.

Table 2.5   Effect of probiotics/prebiotics and synbiotics on colonic aberrant crypt foci and tumour incidence in laboratory animals

Species	Carcinogen	Pre/probiotic exposure stage*	Probiotic/prebiotic	Result	Author
Male Sprague–Dawley rats	DMH (s.c.)	Initiation and promotion	L.acidophilus (Delvo Pro LA-1) L. acidophilus Delvo Pro LA-1 + B. animalis CSCC1941; L. rhamnosus; S. thermophilus DD145 (10^{10} CFU/g, 1% dietary intake for 36 weeks)	Significant decrease in tumour no. (70%) and tumour mass index for L. acidophilus only. Tumour incidence decreased in large intestine L.acidophilus (25%); L. acidophilus + B. animalis (20%); L. rhamnosus (20%); S. thermophilus (10%).	McIntosh et al. (1999)
Male F344 rats	AOM (s.c.)	Initiation and promotion	Lyophilised B. longum (1.5% and 3% dietary for 13 weeks (suppl. from 2 10^{10} CFU/g stock)	Significant inhibition of total ACF ($P < 0.01$). Significant reduction in total AC per colon ($P < 0.001$).	Kulkarni and Reddy (1994)
Male F344 rats	HAA mixture IQ, MeIQ, phIP(1:1:1 ratio)	Initiation and promotion	B. animalis DN-173-010 or S. thermophilus DN-001-158 fermented milk (30% dietary for 15 weeks)	B. animalis DN-173-010 reduced ACF incidence 96%. S. thermophilus DN-001-158 reduced ACF incidence 93%.	Tavan et al. (2002)
Male F344 rats	AOM (s.c.)	Initiation and promotion	B. longum (1×10⁸ CFU/g of feed, rats fed ad libitum) lactulose (2.5%) or both for 13 weeks	Significant reduction in ACF in rats consuming BI, L, BI + L (P, 0.05). Rats fed BI + L had significantly fewer ACF than rats consuming BI or L alone.	Challa et al. (1997)
Male F344 rats	DMH (i.p.)	Initiation and promotion	Bifidobacterium sp. Bio Danone strain 173010 (6×10⁹ CFU/animal/day) in cell suspension or fermented milk. Skim milk powder for 4 weeks).	Significant ($P < 0.05$) inhibition of AC: 61% (Bifidobacterium) 51% (skim milk) 49% (fermented milk).	Abdelali et al. (1995)

Animal model	Carcinogen	Protocol	Treatment	Results	Reference
Male Sprague–Dawley rats	AOM (s.c.)	Promotion	B. longum (4×10^8 viable CFU/g of diet) or inulin (5% w/w dietary) for 12 weeks.	Total ACF decreased by 74% in rats treated with probiotic + prebiotic (synbiotic effect). Numbers of the large ACF (>4 AC per focus) were significantly decreased ($P < 0.05$) by 59% in rats fed probiotic + prebiotic.	Rowland et al. (1998)
Male wistar rats	DMH (gavage)	Promotion	Skim milk, skim milk + bifidobacteria (10^8 CFU/day) skim milk + fructooligosaccharide and skim milk + bifidobacteria (10^8 CFU/day) + fructooligosaccharide; L. acidophilus (10^6 CFU/g) for 4 weeks	Inconsistent results	Gallaher et al. (1996)
Weanling male F344 rat	AOM (s.c.)	Initiation and promotion	Inulin (10%) in diet	No significant effect on ACF but reduced the number of AC/cm^2.	Rao et al. (1998)
F344 rat	AOM (s.c.)	Initiation and promotion	L. acidophilus NCFMTM (lyophilised) in diet	Significant suppression of colonic ACF.	Rao et al. (1999)
Male F344 rat	AOM (s.c.)	Initiation and promotion	Oligofructose (10%) or inulin (10%) in diet for 12 weeks	Significant inhibition of ACF/colon – more pronounced for oligofructose ($P < 0.0006$) than for inulin ($P < 0.02$). Crypt multiplicity also inhibited in animals fed inulin ($P < 0.02$) or oligofructose ($P < 0.04$).	Reddy et al. (1997)
F344 rats	DMH	N/A	L. acidophilus NCFM 1×10^{10} CFU/day, 20 weeks	Colon tumour incidence lower in probiotic fed animals (40% vs 77% in controls).	Goldin & Gorbach (1980)
C57BL/C J-Min/+mice	N/A	N/A	Short chain fructo-oligosaccharides (5.8%) for 6 weeks	Significant reduction in colon tumours ($P < 0.01$).	Pierre et al. (1997)
Male F344 rats	DMH (s.c.)	N/A	Lactobacillus GG (2–4×10^{10} CFU/day for 27 weeks	Lower incidence of tumours ($P < 0.012$) and tumour multiplicity ($P < 0.001$) when rats given LGG throughout experiment. No effect when LGG given after DMH.	Goldin et al. (1996)
Sprague–Dawley rats	AOM (s.c.)	N/A	Lactobacillus casei strain Shirota, 5% dietary for 18, 12 and 25 weeks (suppl. from 2 1×10^{10} CFU/g stock)	Significant reduction in colon tumours ($P < 0.01$).	Yamazaki et al. (2000)

The 'initiation and promotion' protocol involves feeding the probiotic for about a week, followed by dosing with the carcinogen and then continued probiotic administration until animals are sacrificed prior to ACF assessment. In the 'promotion' protocol, the rats are dosed with carcinogen prior to probiotic treatment. s.c. = subcutaneous. i.p. = interperitoneal.

Alone, prebiotics appear to give inconsistent results on carcinogen-induced ACF induction, which may be partly a consequence of differences in treatment regimes and carcinogens used. For example Rao *et al.* (1998) reported that 10% dietary inulin had no significant effect on total ACF in colon, or their multiplicity, in F344 rats, even though curiously a significant reduction in ACF/cm^2 of colon was reported. A study on *Bifidobacterium* spp. and fructo-oligosaccharides (FOS) (2% in diet) gave inconsistent results with only one out of three experiments showing a reduction in DMH-induced ACF (Gallaher *et al.*, 1996). Reddy *et al.* (1997) compared long (inulin) and short chain (FOS) oligosaccharides incorporated at a level of 10% in diet on AOM-induced ACF in rats. The NDOs were fed prior to carcinogen treatment and, throughout the experiment, significant decreases in total ACF were reported at approximately 35% and 25%, respectively. The decreases observed were almost exclusively in the smaller ACF (< 3 AC per focus) and inhibition by inulin appeared to be more pronounced than that of FOS. Rowland *et al.* (1998) reported similar results, when inulin (5% in diet) was given a week after AOM dose. A decrease of 41% in small ACF was observed, but inulin appeared to have no effect on large ACF.

An investigation of the effect of *L. acidophilus* NCFM on colon tumour incidence in DMH-treated rats, showed a reduction in colon cancer incidence (40% *vs* 77% in controls) in animals receiving *L. acidophilus* 1×10^{10} CFU/day after 20 weeks. No difference was apparent at 36 weeks, suggesting that the lactobacilli had increased the latency period, or induction time, for tumours (Goldin and Gorbach, 1980). Administration of dietary *B. longum* (0.5% lyophilised *B. longum* in diet, 1×10^{10} CFU/day) to male rats significantly decreased the formation of IQ-induced liver and colon tumours and the multiplicity of tumours in liver, colon and small intestine (Reddy and Rivenson, 1993; Rumney *et al.*, 1993). The percentage incidence of tumours was also decreased by 100% in colon and 80% in liver. In female rats, dietary supplementation with *Bifidobacterium* cultures also decreased the IQ induced mammary carcinogenesis to 50% and liver carcinogenesis to 27% of that on the control diet, but the differences were not significant. There were, however, significant changes in tumour multiplicity in the mammary gland. McIntosh *et al.* (1999) reported that freeze-dried *L. acidophilius* (10^{10} CFU/g, 1% of dietary intake) fed daily was protective, as indicated by a 70% decrease in tumour numbers and a large decrease in tumour mass in the large intestine of DMH-treated Sprague–Dawley rats. A number of other LAB strains tested also decreased non-significantly tumour burden relative to control animals including *L. rhamnosus* (GG) (20% reduction), *B. animalis* CSSC 1941 + *L. acidophilius* Delvo Pro LA-1 (20% reduction) and *S. thermophilius* DD145 (10% reduction). Bolognani *et al.* (2001) report that background diet fed to rats influenced the protective effects of probiotics, such that probiotics reduced ACF incidence only when rats were fed a high-fat (20%) diet.

The *Min* mouse model recently developed provides animals, which are heterozygous for a nonsense mutation of the *Apc* gene, the murine homologue of APC. These mice develop spontaneous adenomas throughout the small intestine and colon within a few weeks and have been used for testing of chemopreventive agents targeted against cancerous lesions. In one such study *Min* mice were fed various diets containing resistant starch, wheat bran or FOS (5.8% in diet) for six weeks. Tumour numbers remained unchanged from the control (fibre-free diet) in the mice fed either resistant starch or wheat bran, but a significant decrease in colon tumours was observed in rats receiving the diet supplemented with FOS. Furthermore 40% of the FOS-fed animals ($n = 10$) were totally free of colon tumours (Pierre *et al.*, 1997).

2.11   Mechanisms of anticarcinogenicity and antigenotoxicity for probiotics and prebiotics

2.11.1   Binding of carcinogens

There are numerous reports describing the binding or adsorption *in vitro* by LAB and other intestinal bacteria, of a range of food-borne carcinogens including the heterocyclic amines, the fungal toxin aflatoxin B1, benzo(a)pyrene and the food contaminant AF2 (Morotomi and Mutai, 1986; Zhang *et al.*, 1990; Zhang and Ohta, 1991; Orrhage *et al.*, 1994; Bolognani *et al.*, 1997). Several of these studies reported a concomitant decrease in mutagenicity. The extent of the binding was dependent on the mutagen and bacterial strain used. In general the least binding was seen with aflatoxin B1 and AF2 and the most with the heterocyclic amines. The adsorption appeared to be a physical phenomenon, chiefly due to a cation exchange mechanism.

Whilst binding represents a credible mechanism for the inhibition of mutagenicity and genotoxicity by LAB *in vitro*, it does not appear to have any influence *in vivo*. Bolognani *et al.* (1997) reported that simultaneous administration to mice of LAB with various carcinogens did not affect absorption of the compounds from the gastrointestinal tract, nor did it affect the *in vivo* mutagenicity of the carcinogens in the liver. These results are contradictory to those of Zhang and Ohta (1993), who found that absorption from the rat small intestine of Trp-P-1 was significantly inhibited by co-administration of freeze-dried LAB. The latter study was confounded, however, by the use of rats that had been starved for four days, which would induce severe physiological and nutritional stresses on the animals.

2.11.2   Effects on bacterial enzymes, metabolite production

The studies listed in Table 2.3 demonstrate that the increase in concentration of LAB as a consequence of consumption of LAB and/or prebiotics leads to reductions in certain bacterial enzymes ostensibly involved in activation or synthesis of carcinogens, genotoxins and tumour promoters. Such effects would appear to be due to the low specific activity of these enzymes in LAB (Saito *et al.*, 1992). These changes in enzyme activity or metabolite concentration have been suggested to be responsible for the reduced level of tumours or preneoplastic lesions seen in carcinogen-treated rats administered pro- and prebiotics (Rumney *et al.*, 1993;

Rowland *et al.*, 1998). Even though a causal link has not been demonstrated, this remains a likely hypothesis.

2.11.3 Stimulation of protective enzymes

Numerous food-borne carcinogens such as heterocyclic amines and polycyclic aromatic hydrocarbons, are known to be conjugated to glutathione (GSH), which appears to result in inactivation. The enzyme involved, glutathione-*s*-transferase (GST), is found in the liver and in other tissues including the gut. Challa *et al.* (1997) in a study of the effect of lactulose and *B. longum* BB536 on AOM-induced ACF in the colon, showed that GST activity in the colonic mucosa was inversely related to the ACF numbers. The prebiotic-resistant starch has also been shown to increase colonic GST when incorporated into diets of rats (Treptow van Lishaut *et al.*, 1999). Such a mechanism of protection would be effective against a wide range of dietary carcinogens.

2.11.4 Increase in immune response

An alternative mechanism suggested by Perdigon *et al.* (1998) to account for LAB anti-tumour activity is that of reducing the inflammatory immune response. In a study of tumour growth in DMH-treated mice, yoghurt was observed to suppress the inflammatory immune response with an increase in CD4[+] T lymphocytes and in IgA-secreting cells. In those animals, a distinct reduction in tumours was observed. The study is consistent with the work of Schiffrin *et al.* (1996) and Marteau *et al.* (1997) who have provided evidence of immune system modulation in human subjects consuming probiotics. The observed changes were increases in levels of antibody-secreting cells and elevated phagocytic activity of monocytes and granulocytes.

Taken as a whole, studies in *in vitro* systems and in a wide range of animal models provide considerable evidence that LAB, and to a lesser extent prebiotics, have the potential to decrease colon cancer risk. The evidence from humans is less persuasive, but nonetheless is suggestive of a cancer-preventing effect of fermented foods.

2.12   Future trends

No doubt exists that in the last few decades we have greatly expanded our knowledge about the causes and risk factors associated with colorectal cancer. However, it is painfully obvious from the conflicting epidemiological information provided on dairy products that population-based studies must be more selective in the factors they seek to examine, if a clearer and more consistent picture is to be created. In addition, greater regard must be given to the means of dietary assessment applied to studies as recent trials have indicated that some methods of dietary assessment are associated with very large measurement error, which obscures relationships between disease and diet within single homogeneous populations. The introduction of biomarkers to calibrate measurement error is a critical development in the struggle to improve accuracy of large-scale prospective studies investigating the interactions of diet and disease (S.A. Bingham, pers. comm.). To establish causal relationships between diet and colorectal cancer risk and to identify more precisely the dietary components involved, human intervention trials are required. The problem with human intervention studies is that cancer is an impractical end-point in terms of numbers, cost, study duration and ethical considerations. The long lag phase (up to 20 years) between exposure to a carcinogenic event and appearance of tumours is a particular problem. Another strategy is the use of intermediate end-point biomarkers of cancer, which may be cellular, molecular, biochemical or rooted in pathologic change (e.g. faecal water, recurrence of polyps, epithelial markers). Biomarkers have been developed from an understanding of the sequence of events leading to colonic cancer, the biology of normal mucosa and the factors associated with changes symptomatic of progression towards cancer and the manifestation of cancer itself. The specific advantages of biomarkers are that they represent short/intermediate term end-points, which facilitate intervention on a reasonable time scale and that ethical approval is more readily obtainable for biomarker studies as they are minimally invasive, with measurements occurring on accessible material (e.g. faeces and small biopsies). Preferably, biomarkers should be reproducible, sensitive and rigorously validated, although this is not the case with all biomarkers in the cancer area. Biomarkers ought to be correlated or causally linked with cancer and consequently of biological significance. Accordingly, validation of a biomarker is critical to its application as a research tool, as such an apposite response from the marker is required when assayed in healthy individuals on low-risk and high-risk diets for colorectal cancer or in cancer patients (e.g. familial adenomatous polyposis).

At present a wide range of potential biomarkers exists to augment investigations into the activity of specific foods, compounds and metabolites in colorectal cancer. The majority of these assays still require further validation before they can be used with complete confidence and compromises must be made in terms of the strength of the causal information provided and study complexity. Clearly, we now need carefully controlled intervention studies in human subjects using biomarkers of cancer risk in order to strengthen the evidence for the role of functional dairy products. Such studies, if conducted correctly, should help to expose dietary interactions at vital stages in the carcinogenesis process and within a relatively short time begin to provide a more accurate and reliable picture.

2.13   Sources of further information and advice

For further information on the topic of colorectal cancer, the reader is directed to a number of sources within the IARC website including www-dep.iarc.fr/globocan/globocan.html (globocan 2000) and www-dep.iarc.fr/dataava/infodata.html. These sources provide excellent data on a wide variety of cancer statistics. To consider in greater detail the epidemiology related to colon cancer and diet, the reader is advised to examine a number of reviews quoted within the body of this chapter. These references represent comprehensive insight given by authors including Potter, Slattery and Bingham.

2.14   Acknowledgement

This chapter has been carried out with the financial support from the Commission of the European Communities, specific RTD programme 'Quality of Life Management of Living Resources', QLK-2000-00067 'Functional foods, gut microflora and healthy ageing'. It does not necessarily reflects its views and in no way anticipates the Commission's future policy in this area.

2.15   References

ABDELALI H, CASSAND P, SOUSSOTTE V, DAUBEZE M, BOULEY C and NARBONNE J F (1995) Effect of dairy products on initiation of precursor lesions of colon cancer in rats. *Nutrition and Cancer* **24**, 121–132.

AGA (2000) American Gastroenterologists Association Technical Review: Impact of dietary fiber on colon cancer occurence. *Gastroenterology* **118**, 1235–1257.

AYEBO AD, ANGELO I D and SHAHANI K M (1980) Effect of ingesting *Lactobacillus acidophilius* milk upon faecal flora and enzyme activity in humans. *Milchwissenschaft* **35**, 730–733.

BANNI S, DAY B W, EVANS R W, CORONGIU F P and LOMBARDI B (1995) Detection of conjugated diene isomers of linoleic acid in liver lipids of rats fed a choline-devoid diet indicated that the diet does not cause lipoperoxidation. *Journal of Nutritionant Biochemistry* **6**, 281–289.

BANNI S, ANGIONI E, CASU V, MELIS M P, CARTA G, CORONGIU F P, THOMPSON H and IP C (1999) Decrease in linoleic acid metabolites as a potential mechanism in cancer risk reduction by conjugated linoleic acid. *Carcinogenesis* **20**, 1019–1024.

BARON J A, BEACH M, MANDEL J S, VAN STOLK R U, HAILE R W, SANDLER R S, ROTHSTEIN R, SUMMERS R W, SNOVER D C, BECK G J, BOND J H and GREENBERG E R (1999) Calcium supplements for the prevention of colorectal adenomas. Calcium Polyp Prevention Study Group. *New England Journal of Medicine* **340**, 101–107.

BENITO E, OBRADOR A, STIGGELBOUT A, BOSCH F X, MULET M, MUNOZ N and KALDOR J (1990) A population-based case-control study of colorectal cancer in Majorca. I. Dietary factors. *International Journal of Cancer/Journal International Du Cancer* **45**, 69–76.

BENNETT T, DESMOND A, HARRINGTON M, MCDONAGH D, FITZGERALD R, FLYNN A and CASHMAN K D (2000) The effect of high intakes of casein and casein phosphopeptide on calcium absorption in the rat. *British Journal of Nutrition* **83**, 673–680.

BERTAZZONI-MINELLI E, BENINI A, VICENTINI L, ANDREOLI E, OSELLADORE M and CERUTTI R (1996) Effect of *Lactobacillius acidophilius* and *Bifidobacterium bifidum* administration on colonic microbiotia and its metabolic activity in premenstrual syndrome. *Microbiol Ecology in Health and Disease* **9**, 247–260.

BIDOLI E, FRANCESCHI S, TALAMINI R, BARRA S and LA VECCHIA C (1992) Food consumption and cancer of the colon and rectum in north-eastern Italy. *International Journal of Cancer/Journal International Du Cancer* **50**, 223–229.

BINGHAM S A (2000) Diet and colorectal cancer prevention. *Biochemical Society Transactions* **28**, 12–16.

BIRD R P (1995) Role of aberrant crypt foci in understanding the pathogenesis of colon cancer. *Cancer Letters* **93**, 55–71.

BOLOGNANI F, RUMNEY C J and ROWLAND I R (1997) Influence of carcinogen binding by lactic acid-producing bacteria on tissue distribution and *in vivo* mutagenicity of dietary carcinogens. *Food and Chemical Toxicology* **35**, 535–545.

BOLOGNANI F, RUMNEY C, POOL ZOBEL B L and ROWLAND I (2001) Effect of lactobacilli, bifidobacteria and inulin on the formation of aberrant crypt foci in rats. *European Journal of Nutrition* **40**, 293–300.

BONITHON-KOPP C, KRONBORG O, GIACOSA A, RATH U and FAIVRE J (2000) Calcium and fibre supplementation in prevention of colorectal adenoma recurrence: a randomised intervention trial. *The Lancet* **356**, 1300–1306.

BOSTICK R M, POTTER J D, SELLERS T A, MCKENZIE D R, KUSHI L H and FOLSOM A R (1993) Relation of calcium, vitamin D, and dairy food intake to incidence of colon cancer among older women. The Iowa Women's Health Study. *American Journal of Epidemiology* **137**, 1302–1317.

BOSTICK R M, FOSDICK L, WOOD J R, GRAMBSCH P, GRANDITS G A, LILLEMOE T J, LOUIS T A and POTTER J D (1995) Calcium and colorectal epithelial cell proliferation in sporadic adenoma patients: a randomized, double-blinded, placebo-controlled clinical trial. *Journal of the National Cancer Institute* **87**, 1307–1315.

BOUNOUS G, PAPENBURG R, KONGSHAVN P A, GOLD P and FLEISZER D (1988) Dietary whey protein inhibits the development of dimethylhydrazine induced malignancy. *Clinical and Investigative Medicine/Medecine Clinique et Experimentale* **11**, 213–217.

BOUTRON M C, FAIVRE J, MARTEAU P, COUILLAULT C, SENESSE P and QUIPOURT V (1996) Calcium, phosphorus, vitamin D, dairy products and colorectal carcinogenesis: a French case – control study. *British Journal of Cancer* **74**, 145–151.

BOUTRON RUAULT M C, SENESSE P, FAIVRE J, CHATELAIN N, BELGHITI C and MEANCE S (1999) Foods as risk factors for colorectal cancer: a case-control study in Burgundy (France). *European Journal of Cancer Prevention* **8**, 229–235.

BOYLE P and LANGMAN J S (2000) ABC of colorectal cancer – Epidemiology. *British Medical Journal* **321**, 805–808.

BRUCE W R, GIACCA A and MEDLINE A (2000) Possible mechanisms relating diet and risk of colon cancer. *Cancer Epidemiology Biomarkers & Prevention* **9**, 1271–1279.

BURNS A J and ROWLAND I R (2000) Anti-carcinogenicity of probiotics and prebiotics. *Current Issues* in *Intestinal Microbiology* **1**, 13–24.

CHALLA A, RAO D R, CHAWAN C B and SHACKELFORD L (1997) *Bifidobacterium longum* and lactulose suppress azoxymethane-induced colonic aberrant crypt foci in rats. *Carcinogenesis* **18**, 517–521.

CHUNG D C (2000) The genetic basis of colorectal cancer: insights into critical pathways of tumorigenesis. *Gastroenterology* **119**, 854–865.

COLE C B, FULLER R and CARTER S M (1989) Effect of probiotic supplements of *Lactobacillus acidophilus* and *Bifidobacterium adolescentis* 2204 on β-glucosidase and β-glucuronidase activity in the lower gut of rats associated with a human faecal flora. *Microbial Ecology in Health and Disease* **2**, 223–225.

CORPET D E and CASSAND P (1995) Lack of aberrant crypt promotion and of mutagenicity in extracts of cooked casein, a colon cancer-promoting food. *Nutrition and Cancer* **24**, 249–256.

CORPET D E and CHATELIN-PIROT V (1997) Cooked casein promotes colon cancer in rats, may be because of mucosal abrasion. *Cancer Letters* **114**, 89–90.

CORPET D E, YIN Y, ZHANG X M, REMESY C, STAMP D, MEDLINE A, THOMPSON L, BRUCE W R and ARCHER M C (1995) Colonic protein fermentation and promotion of colon carcinogenesis by thermolyzed casein. *Nutrition and Cancer* **23**, 271–281.

CUMMINGS J H, BEATTY E R, KINGMAN S M, BINGHAM S A and ENGLYST H N (1996) Digestion and physiological properties of resistant starch in the human large bowel. *British Journal of Nutrition* **75**, 733–747.

DILLEHAY D L, WEBB S K, SCHMELZ E M and MERRILL A H (1994) Dietary sphingomyelin inhibits 1,2-dimethylhydrazine-induced colon cancer in CF1 mice. *Journal of Nutrition* **124**, 615–620.

EBRINGER L, FERENCIK M, LAHITOVA N, KACANI L and MICHALKOVA D (1995) Antimutagenic and immuno-stimulatory properties of lactic acid bacteria. *World Journal of Microbiology and Biotechnology* **11**, 294–298.

FAIVRE J, BEDENNE L, BOUTRON M C, MILAN C, COLLONGES R and ARVEUX P (1989) Epidemiological evidence for distinguishing subsites of colorectal cancer. *Journal of Epidemiology and Community Health* **43**, 356–361.

FEARON E R and VOGELSTEIN B (1990) A genetic model for colorectal tumorigenesis. *Cell* **61**, 759–767.

FERLAY J, BRAY F, PISANI P and PARKIN D M (2001) *GLOBOCAN 2000: Cancer Incidence, Mortality and Prevalence Worldwide*, Lyon: IARC Press.

FERRARETTO A, SIGNORILE A, GRAVAGHI C, FIORILLI A and TETTAMANTI G (2001) Casein phosphopeptides influence calcium uptake by cultured human intestinal HT-29 tumor cells. *Journal of Nutrition* **131**, 1655–1661.

FULLER R (1989) Probiotics in man and animals. *Journal of Applied Bacteriology* **66**, 365–378.

GALLAHER D D, STALLINGS W H, BLESSING L L, BUSTA F F and BRADY L J (1996) Probiotics, cecal microflora, and aberrant crypts in the rat colon. *Journal of Nutrition* **126**, 1362–1371.

GIBSON G R and ROBERFROID M B (1995) Dietary modulation of the human colonic microbiota: introducing the concept of prebiotics. *Journal of Nutrition* **125**, 1401–1412.

GILL CIR and ROWLAND I (2002) Diet and cancer: assessing the risk. *British Journal of Nutrition* **88** (suppl. 1), 51–516.

GIOVANNUCCI E and GOLDIN B (1997) The role of fat, fatty acids, and total energy intake in the etiology of human colon cancer. *American Journal of Clinical Nutrition* **66**, 1564S–1571S.

GIOVANNUCCI E, STAMPFER M J, COLDITZ G A, RIMM E B, TRICHOPOULOS D, ROSNER B A, SPEIZER F E and WILLETT W C (1993) Folate, methionine, and alcohol intake and risk of colorectal adenoma. *Journal of the National Cancer Institute* **85**, 875–884.

GLINGHAMMAR B, VENTURI M, ROWLAND I R and RAFTER J J (1997) Shift from a dairy product-rich to a dairy product-free diet: influence on cytotoxicity and genotoxicity of fecal water – potential risk factors for colon cancer. *American Journal of Clinical Nutrition* **66**, 1277–1282.

GOEPTAR A R, KOEMAN J H, VAN BOEKEL M A J S and ALINK G M (1997) Impact of digestion on the antimutagenic activity of the milk protein casein. *Nutrition Research* **17**, 1363–1379.

GOLDIN B R and GORBACH S L (1976) The relationship between diet and rat fecal bacterial enzymes implicated in colon cancer. *Journal of the National Cancer Institute* **57**, 371–375.

GOLDIN B R and GORBACH S L (1980) Effect of *Lactobacillus acidophilus* dietary supplements on 1,2-dimethylhydrazine dihydrochloride-induced intestinal cancer in rats. *Journal of the National Cancer Institute* **64**, 263–265.

GOLDIN B R and GORBACH S L (1984a) Alterations of the intestinal microflora by diet, oral antibiotics, and lactobacillus: decreased production of free amines from aromatic nitro compounds, azo dyes, and glucuronides. *Journal of the National Cancer Institute* **73**, 689–695.

GOLDIN B R and GORBACH S L (1984b) The effect of milk and lactobacillus feeding on human intestinal bacterial enzyme activity. *American Journal of Clinical Nutrition* **39**, 756–761.

GOLDIN B R, GUALTIERI L J and MOORE R P (1996) The effect of *Lactobacillus* on the initiation and promotion of DMH-induced intestinal tumors in the rat. *Nutrition and Cancer* **25**, 197–204.

GOVERS M J, LAPRE J A, DE VRIES H T and VAN DER MEER R (1993) Dietary soybean protein compared with casein damages colonic epithelium and stimulates colonic epithelial proliferation in rats. *Journal of Nutrition* **123**, 1709–1713.

HA Y L, GRIMM N K and PARIZA M W (1987) Anticarcinogens from fried ground beef: heat-altered derivatives of linoleic acid. *Carcinogenesis* **8**, 1881–1887.

HA Y L, STORKSON J and PARIZA M W (1990) Inhibition of benzo(a)pyrene-induced mouse forestomach neoplasia by conjugated dienoic derivatives of linoleic acid. *Cancer Research* **50**, 1097–1101.

HAKKAK R, KOROURIAN S, RONIS M J, JOHNSTON J M and BADGER T M (2001) Dietary whey protein protects against azoxymethane-induced colon tumors in male rats. *Cancer Epidemiology Biomarkers & Prevention* **10**, 555–558.

HANNUN Y A and OBEID L M (1995) Ceramide: an intracellular signal for apoptosis. *Trends in Biochemical Sciences* **20**, 73–77.

HANSEN M, SANDSTROM B and LONNERDAL B (1996) The effect of casein phosphopeptides on zinc and calcium absorption from high phytate infant diets assessed in rat pups and Caco-2 cells. *Pediatric Research* **40**, 547–552.

HANSEN M, SANDSTROM B, JENSEN M and SORENSEN S S (1997) Casein phosphopeptides improve zinc and calcium absorption from rice-based but not from whole-grain infant cereal. *Journal of Pediatric Gastroenterology and Nutrition* **24**, 56–62.

HAYAKAWA K, MIZUTANI J, WADA K, MASAI T, YOSHIHARA I and MITSUOKA T (1990) Effects of soybean oligosaccharides on human faecal microflora. *Microbial Ecology in Health and Disease* **3**, 293–303.

HIDAKA H, EIDA T, TAKIZAWA T, TOKUNGA T and TASHIRO Y (1986) Effects of fructo-oligosaccharides on intestinal flora and human health. *Bifidobacteria Microflora* **5**, 37–50.

HOLT P R (1999) Dairy foods and prevention of colon cancer: human studies. *Journal of the American College of Nutrition* **18**, 379S–391S.

HOLZAPFEL W H, HABERER P, SNEL J, SCHILLINGER U and HUIS IN'T VELD J H (1998) Overview of gut flora and probiotics. *International Journal of Food Microbiology* **41**, 85–101.

IARC (1997) *Cancer Incidence in Five Continents*, Volume VII. France: IARC Scientific Publications, i–xxxiv, 1–1240.

ICS (1985) Cancer incidence in Greater Bombay, by religion and sex 1973–1978. Bombay: Indian Cancer Society.

IP C, SINGH M, THOMPSON H J and SCIMECA J A (1994) Conjugated linoleic acid suppresses mammary carcinogenesis and proliferative activity of the mammary gland in the rat. *Cancer Research* **54**, 1212–1215.

JARVINEN R, KNEKT P, HAKULINEN T and AROMAA A (2001) Prospective study on milk products, calcium and cancers of the colon and rectum. *European Journal of Clinical Nutrition* **55**, 1000–1007.

JOHANSEN C, MELLEMGAARD A, SKOV T, KJAERGAARD J and LYNGE E (1993) Colorectal cancer in Denmark 1943–1988. *International Journal of Colorectal Disease* **8**, 42–47.

JOSYULA S and SCHUT H A (1998) Effects of dietary conjugated linoleic acid on DNA adduct formation of PhIP and IQ after bolus administration to female F344 rats. *Nutrition and Cancer* **32**, 139–145.

KAMPMAN E, GIOVANNUCCI E, VAN'T VEER P, RIMM E, STAMPFER M J, COLDITZ G A, KOK F J and WILLETT W C (1994a) Calcium, vitamin D, dairy foods, and the occurrence of colorectal adenomas among men and women in two prospective studies. *American Journal of Epidemiology* **139**, 16–29.

KAMPMAN E, GOLDBOHM R A, VAN DEN BRANDT P A and VAN'T VEER P (1994b) Fermented dairy products, calcium, and colorectal cancer in The Netherlands Cohort Study. *Cancer Research* **54**, 3186–3190.

KAMPMAN E, VAN'T VEER P, HIDDINK F J, VAN AKEN-SCHNEIJDER P, KOK F J and HERMUS R J (1994c) Fermented dairy products, dietary calcium and colon cancer: a case-control study in The Netherlands. *International Journal of Cancer/Journal International Du Cancer* **59**, 170–176.

KAMPMAN E, SLATTERY M L, CAAN B and POTTER J D (2000) Calcium, vitamin D, sunshine exposure, dairy products and colon cancer risk (United States). *Cancer Causes & Control* **11**, 459–466.

KEARNEY J, GIOVANNUCCI E, RIMM E B, ASCHERIO A, STAMPFER M J, COLDITZ G A, WING A, KAMPMAN E and WILLETT W C (1996) Calcium, vitamin D, and dairy foods and the occurrence of colon cancer in men. *American Journal of Epidemiology* **143**, 907–917.

KEMPPAINEN M, RAIHA I and SOURANDER L (1997) A marked increase in the incidence of colorectal cancer over two decades in southwest Finland. *Journal of Clinical Epidemiology* **50**, 147–151.

KLEIBEUKER J H, MULDER N H, CATS A, VAN DER MEER R and DE VRIES E G (1996a) Calcium and colorectal epithelial cell proliferation. *Gut* **39**, 774–775.

KLEIBEUKER J H, NAGENGAST F M and VAN DER MEER R (1996b) Carcinogenesis in the colon. In *Prevention and Early Detection of Colorectal Cancer* (G P Young, P Rozen and B Levin, editors). London, Philadelphia, Toronto, Sydney, Tokyo: WB Saunders Company Ltd, 46–62.

KOLESNICK R N and KRONKE M (1998) Regulation of ceramide production and apoptosis. *Annual Review of Physiology* **60**, 643–665.

KULKARNI N and REDDY B S (1994) Inhibitory effect of *Bifidobacterium longum* cultures on the azoxymethane-induced aberrant crypt foci formation and fecal bacterial beta-glucuronidase. *Proceedings of the Society for Experimental Biology and Medicine* **207**, 278–283.

KUNE S, KUNE G A and WATSON L F (1987) Case-control study of dietary etiological factors: the Melbourne Colorectal Cancer Study. *Nutrition and Cancer* **9**, 21–42.

LAPRE J A, DE VRIES H T and VAN DER MEER R (1993) Cytotoxicity of fecal water is dependent on the type of dietary fat and is reduced by supplemental calcium phosphate in rats. *Journal of Nutrition* **123**, 578–585.

LAWSON R E, MOSS A R and GIVENS D I (2001) The role of dairy products in supplying conjugated linoleic acid to man's diet: a review. *Nutrition Research Reviews* **14**, 153–172.

LEE H P, GOURLY L, DUFFY S W, ESTEVE J, LEE J and DAY H E (1989) Colorectal cancer and diet in an Asian population: a case-control study among Singapore Chinese. *International Journal of Cancer* **43**, 1007–1016.

LEE Y S, NOGUCHI T and NAITO H (1983) Intestinal absorption of calcium in rats given diets containing casein or amino acid mixture: the role of casein phosphopeptides. *British Journal of Nutrition* **49**, 67–76.

LI H, SHIMURA H, AOKI Y, DATE K, MATSUMOTO K, NAKAMURA T and TANAKA M (1998) Hepatocyte growth factor stimulates the invasion of gallbladder carcinoma cell lines *in vitro*. *Clinical and Experimental Metastasis* **16**, 74–82.

LIDBECK A, GELTNER ALLINGER U, ORRHAGE K M, OTTOVA L, BRISMAR B, GUSTAFSSON J A, RAFTER J J and NORD C E (1991) Impact of *Lactobacillus acidophilus* supplements on the faecal microflora and soluble faecal bile acids in colon cancer patients. *Microbial Ecology in Health and Disease* **4**, 81–88.

LIEW C, SCHUT H A, CHIN S F, PARIZA M W and DASHWOOD R H (1995) Protection of conjugated linoleic acids against 2-amino-3-methylimidazo[4,5-f]quinoline-induced colon carcinogenesis in the F344 rat: a study of inhibitory mechanisms. *Carcinogenesis* **16**, 3037–3043.

LIPKIN M and NEWMARK H (1995) Calcium and the prevention of colon cancer. *Journal of Cellular Biochemistry* **Supplement 22**, 65–73.

MA J, GIOVANNUCCI I, POLLAK M, CHAN M J, GAZIANO M, WILLET W and STAMPFER M J (2001) Milk intake, circulating levels of insulin-like growth factor-I, and risk of colorectal cancer in men. *Journal of the National Cancer Institute* **93**, 1330–1336.

MACDONALD H B (2000) Conjugated linoleic acid and disease prevention: a review of current knowledge. *Journal of the American College of Nutrition* **19**, 111S–118S.

MACQUART-MOULIN G, RIBOLI E, CORNEE J, CHARNAY B, BERTHEZENE P and DAY N (1986) Case-control study on colorectal cancer and diet in Marseilles. *International Journal of Cance/Journal International Du Cancer* **38**, 183–191.

MANOUSOS O, DAY N E, TRICHOPOULOS D, GEROVASSILIS F, TZONOU A and POLYCHRONO-

POULOU A (1983) Diet and colorectal cancer: a case-control study in Greece. *International Journal of Cancer/Journal International Du Cancer* **32**, 1–5.

MARCUS P M and NEWCOMB P A (1998) The association of calcium and vitamin D, and colon and rectal cancer in Wisconsin women. *International Journal of Epidemiology* **27**, 788–793.

MARTEAU P, FLOURIE B, POCHART P, CHASTANG C, DESJEUX J F and RAMBAUD J C (1990a) Effect of the microbial lactase (EC 3, 2, 1, 23), activity in yoghurt on the intestinal absorption of lactose: an *in vivo* study in lactase-deficient humans, *The British Journal of Nutrition*, **64**, 71–79.

MARTEAU P, POCHART P, FLOURIE B, PELLIER P, SANTOS L, DESJEUX J F and RAMBAUD J C (1990b) Effect of chronic ingestion of a fermented dairy product containing *Lactobacillus acidophilus* and *Bifidobacterium bifidum* on metabolic activities of the colonic flora in humans. *American Journal of Clinical Nutrition* **52**, 685–688.

MARTEAU P, MINEKUS M, HAVENAAR R and HUIS IN'T VELD J H (1997) Survival of lactic acid bacteria in a dynamic model of the stomach and small intestine: validation and the effects of bile. *Journal of Dairy Science* **80**, 1031–1037.

MARTINEZ M E and WILLETT W C (1998) Calcium, vitamin D, and colorectal cancer: a review of the epidemiologic evidence. *Cancer Epidemiology, Biomarkers and Prevention* **7**, 163–168.

MCCARTY M F (2000) Activation of PPARgamma may mediate a portion of the anticancer activity of conjugated linoleic acid. *Medical Hypotheses* **55**, 187–188.

MCGUIRE M A and MCGUIRE M K (1999) Conjugated linoleic acid (CLA): a ruminant fatty acid with beneficial effects on human health. *Proceedings of the American Society of Animal Science* 1–8.

MCINTOSH G H, REGESTER G O, LE LEU R K, ROYLE P J and SMITHERS G W (1995) Dairy proteins protect against dimethylhydrazine-induced intestinal cancers in rats. *Journal of Nutrition* **125**, 809–816.

MCINTOSH G H, WANG Y H and ROYLE P J (1998) A diet containing chickpeas and wheat offers less protection against colon tumors than a casein and wheat diet in dimethylhydrazine-treated rats. *Journal of Nutrition* **128**, 804–809.

MCINTOSH G H, ROYLE P J and PLAYNE M J (1999) A probiotic strain of *L. acidophilus* reduces DMH-induced large intestinal tumors in male Sprague–Dawley rats. *Nutrition and Cancer* **35**, 153–159.

METTLIN C J, SCHOENFELD E R and NATARAJAN N (1990) Patterns of milk consumption and risk of cancer. *Nutrition and Cancer* **13**, 89–99.

MOBARHAN S (1999) Calcium and the colon: recent findings. *Nutrition Reviews* **57**, 124–126.

MOROTOMI M and MUTAI M (1986) *In vitro* binding of potent mutagenic pyrolysates to intestinal bacteria. *Journal of the National Cancer Institute* **77**, 195–201.

NAGAO M and SUGIMURA T (1993) Carcinogenic factors in food with relevance to colon cancer development. *Mutation Research* **290**, 43–51.

NEGRI E, LA VECCHIA C, D'AVANZO B and FRANCESCHI S (1990) Calcium, dairy products, and colorectal cancer. *Nutrition and Cancer* **13**, 255–262.

NOBRE-LEITAO C, CHAVES P, FIDALGO P, CRAVO M, GOUVEIA-OLIVEIRA A, FERRA M A and MIRA F C (1995) Calcium regulation of colonic crypt cell kinetics: evidence for a direct effect in mice. *Gastroenterology* **109**, 498–504.

NORAT T and RIBOLI E (2001) Meat consumption and colorectal cancer: a review of epidemiologic evidence. *Nutrition Reviews* **59**, 37–47.

NRC (1996) *Carcinogens and Anticarcinogens in the Human Diet*. The National Academy of Sciences. Washington DC: National Academy Press.

ORRHAGE K, SILLERSTROM E, GUSTAFSSON J A, NORD C E and RAFTER J (1994) Binding of mutagenic heterocyclic amines by intestinal and lactic acid bacteria. *Mutation Research* **311**, 239–248.

OWEN D A (1996) Flat adenoma, flat carcinoma, and de novo carcinoma of the colon. *Cancer* **77**, 3–6.

PAPENBURG R, BOUNOUS G, FLEISZER D and GOLD P (1990) Dietary milk proteins inhibit the development of dimethylhydrazine-induced malignancy. *Tumour Biology: the Journal of the International Society for Oncodevelopmental Biology and Medicine* 11, 129–136.

PARODI P W (1997) Cows' milk fat components as potential anticarcinogenic agents. *Journal of Nutrition* 127, 1055–1060.

PENMAN I D, LIANG Q L, BODE J, EASTWOOD M A and ARENDS M J (2000) Dietary calcium supplementation increases apoptosis in the distal murine colonic epithelium. *Journal of Clinical Pathology* 53, 302–307.

PERDIGON G, VALDEZ J C and RACHID M (1998) Antitumour activity of yoghurt: study of possible immune mechanisms. *Journal of Dairy Research* 65, 129–138.

PEREIRA M A, BARNES L H, RASSMAN V L, KELLOFF G V and STEELE V E (1994) Use of azoxymethane-induced foci of aberrant crypts in rat colon to identify potential cancer chemopreventive agents. *Carcinogenesis* 15, 1049–1054.

PETERS R K, PIKE M C, GARABRANT D H and MACK T M (1992) Diet and colon cancer in Los Angeles County, California. *Cancer Causes and Controls* 3, 457–473.

PHILLIPS R L and SNOWDON D A (1985) Dietary relationships with fatal colorectal cancer among Seventh-Day Adventists. *Journal of the National Cancer Institute* 74, 307–317.

PIERRE F, PERRIN P, CHAMP M, BORNET F, MEFLAH K and MENANTEAU J (1997) Short-chain fructo-oligosaccharides reduce the occurrence of colon tumors and develop gut-associated lymphoid tissue in *Min* mice. *Cancer Research* 57, 225–228.

PIETINEN P, MALILA N, VIRTANEN M, HARTMAN T J, TANGREA J A, ALBANES D and VIRTAMO J (1999) Diet and risk of colorectal cancer in a cohort of Finnish men. *Cancer Causes & Control* 10, 387–396.

PONZ DE LEON M and PERCESEPE A (2000) Pathogenesis of colorectal cancer. *Digestive and Liver Disease* 32, 807–821.

PONZ DE LEON M and RONCUCCI L (2000) The cause of colorectal cancer. *Digestive and Liver Disease* 32, 426–439.

POOL-ZOBEL B L, MUNZNER R and HOLZAPFEL W H (1993) Antigenotoxic properties of lactic acid bacteria in the *S. typhimurium* mutagenicity assay. *Nutrition and Cancer* 20, 261–270.

POOL-ZOBEL B L, NEUDECKER C, DOMIZLAFF I, JI S, SCHILLINGER U, RUMNEY C, MORETTI M, VILARINI I, SCASSELLATI-SFORZOLINI R and ROWLAND I (1996) Lactobacillus- and bifidobacterium-mediated antigenotoxicity in the colon of rats. *Nutrition and Cancer* 26, 365–380.

POTTER J D (1999) Colorectal cancer: molecules and populations. *Journal of the National Cancer Institute* 91, 916–932.

POTTER J D, SLATTERY M L, BOSTICK R M and GAPSTUR S M (1993) Colon cancer: a review of the epidemiology. *Epidemiologic Reviews* 15, 499–545.

PRETLOW T P, O'RIORDAN M A, PRETLOW T G and STELLATO T A (1992) Aberrant crypts in human colonic mucosa: putative preneoplastic lesions. *Journal of Cellular Biochemistry* **Supplement 16G**, 55–62.

PRITCHARD R S, BARON J A and GERHARDSSON DE VERDIER M (1996) Dietary calcium, vitamin D, and the risk of colorectal cancer in Stockholm, Sweden. *Cancer Epidemiology, Biomarkers and Prevention* 5, 897–900.

RAFTER J J, CHILD P, ANDERSON A M, ALDER R, ENG V and BRUCE W R (1987) Cellular toxicity of fecal water depends on diet. *American Journal of Clinical Nutrition* 45, 559–563.

RAO C V, CHOU D, SIMI V, KU H and REDDY B S (1998) Prevention of colonic aberrant crypt foci and modulation of large bowel microbial activity by dietary coffee fiber, inulin and pectin. *Carcinogenesis* 19, 1815–1819.

RAO C V, SANDERS M E, INDRANIE C, SIMI B and REDDY B S (1999) Prevention of colonic preneoplastic lesions by the probiotic *Lactobacillus acidophilus* NCFMTM in F344 rats. *International Journal of Oncology* 14, 939–944.

REDDY B S and RIVENSON A (1993) Inhibitory effect of *Bifidobacterium longum* on colon, mammary, and liver carcinogenesis induced by 2-amino-3-methylimidazo[4,5-f]quinoline, a food mutagen. *Cancer Research* 53, 3914–3918.

REDDY B S, WEISBURGER J H and WYNDER E L (1974) Fecal bacterial beta-glucuronidase: control by diet. *Science* **183**, 416–417.

REDDY B S, HAMID R and RAO C V (1997) Effect of dietary oligofructose and inulin on colonic preneoplastic aberrant crypt foci inhibition. *Carcinogenesis* **18**, 1371–1374.

RIBONI L, VIANI P, BASSI R, PRINETTI A and TETTAMANTI G (1997) The role of sphingolipids in the process of signal transduction. *Progress in Lipid Research* **36**, 153–195.

ROWLAND I R (1988) Factors affecting metabolic activity of the intestinal microflora. *Drug Metabolism Reviews* **19**, 243–261.

ROWLAND I R and TANAKA R (1993) The effects of transgalactosylated oligosaccharides on gut flora metabolism in rats associated with a human faecal microflora. *Journal of Applied Bacteriology* **74**, 667–674.

ROWLAND I R, MALLETT A K and WISE A (1985) The effect of diet on the mammalian gut flora and its metabolic activities. *Critical Reviews in Toxicology* **16**, 31–103.

ROWLAND I R, BEARNE C A, FISCHER R and POOL-ZOBEL B l (1996) The effect of lactulose on DNA damage induced by DMH in the colon of human flora-associated rats. *Nutrition and Cancer* **26**, 37–47.

ROWLAND I R, RUMNEY C J, COUTTS J T and LIEVENSE L C (1998) Effect of *Bifidobacterium longum* and inulin on gut bacterial metabolism and carcinogen-induced aberrant crypt foci in rats. *Carcinogenesis* **19**, 281–285.

RUMNEY C J, ROWLAND I R, COUTTS T M, RANDERATH K, REDDY R, SHAH A B, ELLUL A and O'NEILL I K (1993) Effects of risk-associated human dietary macrocomponents on processes related to carcinogenesis in human-flora-associated (HFA) rats. *Carcinogenesis* **14**, 79–84.

SAITO Y, TAKANO T and ROWLAND I R (1992) Effects of soybean oligosaccharides on the human gut microflora in *in vitro* culture. *Microbial Ecology in Health and Disease* **5**, 105–110.

SAITO Y, LEE Y S and KIMURA S (1998) Minimum effective dose of casein phosphopeptides (CPP) for enhancement of calcium absorption in growing rats. *International Journal for Vitamin and Nutrition Research* **68**, 335–340.

SALMINEN S, BOULEY C, BOUTRON-RUAULT M C, CUMMINGS J H, FRANCK A, GIBSON G R, ISOLAURI E, MOREAU M C, ROBERFROID M and ROWLAND I (1998) Functional food science and gastrointestinal physiology and function. *British Journal of Nutrition* **80** Suppl 1, S147–171.

SCHIFFRIN E, ROCHAT F, LINK-AMSTER H, AESCHLIMANN J and DONNET-HUGHES A (1996) Immunomodulation of blood cells following ingestion of lactic acid bacteria. *Journal of Dairy Science* **78**, 491–497.

SCHMELZ E M, DILLEHAY D L, WEBB S K, REITER A, ADAMS J and MERRILL A H (1996) Sphingomyelin consumption suppresses aberrant colonic crypt foci and increases the proportion of adenomas versus adenocarcinomas in CF1 mice treated with 1,2-dimethylhydrazine: implications for dietary sphingolipids and colon carcinogenesis. *Cancer Research* **56**, 4936–4941.

SCHMELZ E M, BUSHNEV A S, DILLEHAY D L, SULLARDS M C, LIOTTA D C and MERRILL A H (1999) Ceramide-beta-D-glucuronide: synthesis, digestion, and suppression of early markers of colon carcinogenesis. *Cancer Research* **59**, 5768–5772.

SCHMELZ E M, SULLARDS M C, DILLEHAY D L and MERRILL A H (2000) Colonic cell proliferation and aberrant crypt foci formation are inhibited by dairy glycosphingolipids in 1,2-dimethylhydrazine-treated CF1 mice. *Journal of Nutrition* **130**, 522–527.

SEBEDIO J L, JUANEDA P, DOBSON G, RAMILISON I, MARTIN J D and CHARDIGNY J M (1997) Metabolites of conjugated isomers of linoleic acid (CLA) in the rat. Biochimica et Biophysica Acta **1345**, 5–10.

SESINK A L, TERMONT D S, KLEIBEUKER J H and VAN DER MEER R (2001) Red meat and colon cancer: dietary haem-induced colonic cytotoxicity and epithelial hyperproliferation are inhibited by calcium. *Carcinogenesis* **22**, 1653–1659.

SHANNON J, WHITE E, SHATTUCK A L and POTTER J D (1996) Relationship of food groups and water intake to colon cancer risk. *Cancer Epidemiology, Biomarkers and Prevention* **5**, 495–502.

SHIH I M, WANG T L, TRAVERSO G, ROMANS K, HAMILTON S R, BEN-SASSON A, KINZLER K W and VOGELSTEIN B (2001) Top-down morphogenesis of colorectal tumors. *Proceedings of the National Academy of Sciences of the United States of America* **98**, 2640–2645.

SHULTZ T D, CHEW B P, SEAMAN W R and LUEDECKE L O (1992) Inhibitory effect of conjugated dienoic derivatives of linoleic acid and beta-carotene on the *in vitro* growth of human cancer cells. *Cancer Letters* **63**, 125–133.

SILVI S, RUMNEY C J, CRESCI A and ROWLAND I R (1999) Resistant starch modifies gut microflora and microbial metabolism in human flora-associated rats inoculated with faeces from Italian and UK donors. *Journal of Applied Microbiology* **86**, 521–530.

SLATTERY M L, SORENSON A W and FORD M H (1988) Dietary calcium intake as a mitigating factor in colon cancer. *American Journal of Epidemiology* **128**, 504–514.

SLATTERY M L, POTTER J D, COATES A, MA K N, BERRY T D, DUNCAN D M and CAAN B J (1997) Plant foods and colon cancer: an assessment of specific foods and their related nutrients (United States). *Cancer Causes and Control* **8**, 575–590.

SOUZA R F (2001) Review article: a molecular rationale for the how, when and why of colorectal cancer screening. *Alimentary Pharmacology & Therapeutics* **15**, 451–462.

SPANHAAK S, HAVENAAR R and SCHAAFSMA G (1998) The effect of consumption of milk fermented by *Lactobacillus casei* strain Shirota on the intestinal microflora and immune parameters in humans. *European Journal of Clinical Nutrition* **52**, 899–907.

STERNHAGEN L G and ALLEN J C (2001) Growth rates of a human colon adenocarcinoma cell line are regulated by the milk protein alpha-lactalbumin. *Advances in Experimental Medicine and Biology* **501**, 115–120.

STOCKS P (1957) *Cancer Incidence in North Wales and Liverpool Region in Relation to Habit and Environment*. British Imperial Cancer Campaign, 35th Annual Report, 1.

SWEENEY E A, INOKUCHI J and IGARASHI Y (1998) Inhibition of sphingolipid induced apoptosis by caspase inhibitors indicates that sphingosine acts in an earlier part of the apoptotic pathway than ceramide. *Federation of European Biochemical Societies Letters* **425**, 61–65.

TAJIMA K and TOMINAGA S (1985) Dietary habits and gastro-intestinal cancers: a comparative case-control study of stomach and large intestinal cancers in Nagoya, Japan. *Japanese Journal of Cancer Research: Gann* **76**, 705–716.

TAVAN E, CAYUELA C, ANTOINE J, TRUGMAN G, CHAUGIER C and CASSAND P (2002) Effects of dairy products on heterocyclic aromatic amine-induced rat colon carcinogenesis. *Carcinogenesis* **23**, 477–483.

TOKUNGA T, OKU T and HOYSOYA N (1986) Influence of chronic intake of new sweetener fructooligosaccharide (Neosugar) on growth and gastrointestinal function of the rat. *Journal of Nutritional Science and Vitaminology* **32**, 111–121.

TREPTOW VAN LISHAUT S, RECHKEMMER G, ROWLAND I, DOLARA P and POOL-ZOBEL B L (1999) The carbohydrate crystalean and colonic microflora modulate expression of glutathione S-transferase subunits in colon of rats. *European Journal of Nutrition* **38**, 76–83.

TSUDA H, SEKINE K, USHIDA Y, KUHARA T, TAKASUKA N, IIGO M, HAN B S and MOORE M A (2000) Milk and dairy products in cancer prevention: focus on bovine lactoferrin. *Mutation Research* **462**, 227–233.

TUYNS A J, KAAKS R and HAELTERMAN M (1988) Colorectal cancer and the consumption of foods: a case-control study in Belgium. *Nutrition and Cancer* **11**, 189–204.

VAN BOEKEL M A, WEERENS C N, HOLSTRA A, SCHEIDTWEILER C E and ALINK G M (1993) Antimutagenic effects of casein and its digestion products. *Food and Chemical Toxicology* **31**, 731–737.

VAN DER MEER R, LAPRE J A, GOVERS M J and KLEIBEUKER J H (1997) Mechanisms of the intestinal effects of dietary fats and milk products on colon carcinogenesis. *Cancer Letters* **114**, 75–83.

VAN DER MEER R, TERMONT D S and DE VRIES H T (1991) Differential effects of calcium ions and calcium phosphate on cytotoxicity of bile acids. *American Journal of Physiology* **260**, G142–147.

VESPER H, SCHMELZ E M, NIKOLOVA-KARAKASHIAN M N, DILLEHAY D L, LYNCH D V and MERRILL A H (1999) Sphingolipids in food and the emerging importance of sphingolipids to nutrition. *Journal of Nutrition* **129**, 1239–1250.

VINAS-SALAS J, BIENDICHO-PALAU P, PINOL-FELIS C, MIGUELSANZ-GARCIA S and PEREZ-HOLANDA S (1998) Calcium inhibits colon carcinogenesis in an experimental model in the rat. *European Journal of Cancer* **34**, 1941–1945.

VOGELSTEIN B, FEARON E R, HAMILTON S R, KERN S E, PREISINGER A C, LEPPERT M, NAKAMURA Y, WHITE R, SMITS A M and BOS J L (1988) Genetic alterations during colorectal-tumor development. *New England Journal of Medicine* **319**, 525–532.

WARGOVICH M J, ALLNUTT D, PALMER C, ANAYA P and STEPHENS L C (1990) Inhibition of the promotional phase of azoxymethane-induced colon carcinogenesis in the F344 rat by calcium lactate: effect of simulating two human nutrient density levels. *Cancer Letters* **53**, 17–25.

WARGOVICH M J, JIMENEZ A, MCKEE K, STEELE V E, VELASCO M, WOODS J, PRICE R, GRAY K and KELLOFF G J (2000) Efficacy of potential chemopreventive agents on rat colon aberrant crypt formation and progression. *Carcinogenesis* **21**, 1149–1155.

WCRF (1997) *Diet, Nutrition and the Prevention of Cancer: A Global Perspective.* Washington DC: World Cancer Research Fund/American Institute.

WOLLOWSKI I JI S T, BAKALINSKY A T, NEUDECKER C and POOL-ZOBEL B L (1999), Bacteria used for the production of yoghurt inactivate carcinogens and prevent DNA damage in the colon of rats. *J. Nutr.*, **129**, 77.

WU K, WILLETT W C, FUCHS C S, COLDITZ G A AND GIOVANNUCCI E L (2002) Calcium intake and risk of colon cancer in women and men. *Journal of the National Cancer Institute* **94**, 437–446.

YAMAZAKI K, TSUNODA A, SIBUSAWA M, TSUNODA Y, KUSANO M, FUKUCHI K, YAMANAKA M, KUSHIMA M, NOMOTO K and MOROTOMI M (2000) The effect of an oral administration of *Lactobacillus casei* strain Shirota on azoxymethane-induced colonic aberrant crypt foci and colon cancer in the rat. *Oncology Reports* **7**, 977–982.

YOSHIDA S and YE XIUYUN (1992) The binding ability of bovine milk caseins to mutagenic heterocyclic amines. *Journal of Dairy Science* **75**, 958–961.

YOUNG T B and WOLF D A (1988) Case-control study of proximal and distal colon cancer and diet in Wisconsin. *International Journal of Cancer/Journal International Du Cancer* **42**, 167–175.

ZHANG X B and OHTA Y (1991) *In vitro* binding of mutagenic pyrolyzates to lactic acid bacterial cells in human gastric juice. *Journal of Dairy Science* **74**, 752–757.

ZHANG X B and OHTA Y (1993) Microorganisms in the gastrointestinal tract of the rat prevent absorption of the mutagen-carcinogen 3-amino-1,4-dimethyl-5H-pyrido(4,3-b)indole. *Canadian Journal of Microbiology* **39**, 841–845.

ZHANG X B, OHTA Y and HOSONO A I (1990) Antimutagenicity and binding of lactic acid bacteria from a Chinese cheese to mutagenic pyrolyzates. *Journal of Dairy Science* **73**, 2702–2710.

ZHANG X M, STAMP D, MINKIN S, MEDLINE A, CORPET D E, BRUCE W R and ARCHER M C (1992) Promotion of aberrant crypt foci and cancer in rat colon by thermolyzed protein. *Journal of the National Cancer Institute* **84**, 1026–1030.

ZOCK P L (2001) Dietary fats and cancer. *Current Opinion in Lipidology* **12**, 5–10.

3

Coronary heart disease

J. Lovegrove and K. Jackson, The University of Reading, UK

3.1 Introduction

Coronary heart disease (CHD) is one of the major causes of death in adults in the Western world. Although there has been a trend over the past 15 years towards a reduction in the rates of CHD in the major industrialised nations (including the USA, Australia and the UK) due to better health screening, drug treatment, smoking and dietary advice, there has been an alarming increase in CHD in Eastern Europe.

CHD is a condition in which the main coronary arteries supplying the heart are no longer able to supply sufficient blood and oxygen to the heart muscle (myocardium). The main cause of the reduced flow is an accumulation of plaques, mainly in the intima of arteries, a disease known as atherosclerosis. This is a slow but progressive disease which usually begins in childhood, but does not usually manifest itself until later life. Depending on the rate of the narrowing of the arteries and its ultimate severity, four syndromes may occur during the progression of CHD. These include angina pectoris, unstable angina pectoris, myocardial infarction and sudden death. Angina pectoris literally means a 'strangling sensation in the chest' and is often provoked by physical activity or stress. This pain, which often radiates from the chest to the left arm and neck, is caused by reduced blood flow in the coronary artery but this pain fades quite quickly when the patient is rested. Unstable angina pectoris occurs as the condition worsens and pain is experienced not only during physical activity but also during rest. This type of angina is thought to involve rupture or fissuring of a fixed lesion which, when untreated, leads to an acute myocardial infarction. The condition is responsible for a large proportion of deaths from CHD and occurs as a result of prolonged occlusion of the coronary artery leading to death of some of the heart muscle. Disturbances to the contraction of the heart can lead to disruption in the electrical contraction of heart muscle and the heart may go into an uncoordinated rhythm (ventricular fibrillation) or may completely stop. Sudden death from a severe myocardial infarction occurs within one hour of the attack and is generally associated with advanced atherosclerosis.

A number of risk factors known to predispose an individual to CHD have been categorised into those that are not modifiable, such as age, sex, race and family history, and those that are modifiable, such as hyperlipidaemia (high levels of lipids (fat) in the blood), hypertension (high blood pressure), obesity, cigarette smoking and lack of exercise. Epidemiological studies examining CHD risks in different populations have observed a positive correlation between an individual's fasting cholesterol level (i.e. cholesterol level before eating a meal), especially elevated levels of low density lipoprotein (LDL) cholesterol and development of CHD.[1] Low levels of high density lipoprotein (HDL) cholesterol have been shown to be another risk factor for CHD since an inverse relationship exists with CHD development. This lipoprotein is involved in reverse cholesterol transport, carrying cholesterol from cells to the liver for removal from the body and a low level of HDL cholesterol reflects an impairment in this process.

A recently recognised risk factor for CHD is an elevated level of triacylglycerol (TAG; major fat in the blood) in both the fasted and fed (postprandial) state and the strength of this association has been demonstrated in a number of observation studies.[2] The atherogenic lipoprotein phenotype (ALP) is a newly recognised condition that describes a collection of abnormalities of both classical and newly defined risk factors which predisposes an apparently healthy individual to an increased CHD risk.[3] The lipid abnormalities of the ALP are discussed in more detail in section 3.3.3. The development of techniques by which to characterise the LDL subclass of individuals with CHD and those without CHD has enabled the prevalence of those with ALP to be identified in the population. Heritability studies have revealed that 50% of the variability in expression is due to genetic factors, with the remainder being associated with environmental influences such as diet, smoking and a sedentary lifestyle.

Individuals with CHD and those with a high risk of developing the condition are treated in a number of ways to help lower their LDL cholesterol and TAG concentrations while elevating their HDL cholesterol. Many lines of evidence suggest that adverse dietary habits are a contributory factor in CHD and so the first line of treatment for individuals with moderately raised cholesterol (5.5–7.0 mmol/l) and/or TAG (1.5–3.0 mmol/l) levels is to modify their diet. This is implemented by reducing the percentage of dietary energy derived from fat to approximately 30%, with a reduction of saturated fatty acids (SFAs) to 10% of the dietary energy derived from fat. Candidate fats for the replacement of a large proportion of SFAs include polyunsaturated fatty acids (PUFAs) of the n-6 (linoleic) series and monounsaturated fatty acids (MUFAs). Although both PUFAs (n-6 series) and MUFAs have been shown to decrease total cholesterol and LDL cholesterol levels, MUFAs have also been shown to maintain or increase HDL cholesterol concentrations when they are used to substitute SFAs compared with n-6 PUFAs. PUFAs of the *n*-3 series (eicosapentaneoic acid (EPA) and docosahexaneoic acid (DHA)) have been shown to reduce plasma TAG levels in both the fed and fasted states and reduce thrombosis when added to the diet, but they do not usually reduce total and LDL cholesterol levels.

Individuals with cholesterol levels above 8.0 mmol/l and/or TAG levels above 3.0 mmol/l require not only dietary modification but also lipid-lowering drugs to help reduce their disorder. Drugs that are available are effective in a number of ways, including the following:

- reduced synthesis of very low density lipoprotein (VLDL) and LDL (e.g. nicotinic acid)
- enhanced VLDL clearance (e.g. benzafibrate)
- enhanced LDL clearance (e.g. cholestyramine) and hydroxy-methyl-glutaryl CoA (HMG-CoA) reductase inhibition (e.g. simvasatin).

A recently available drug, fenofibrate, has been shown not only to reduce cholesterol and TAG levels, but it also increases HDL cholesterol concentrations and is able to shift the LDL class profile away from the more atherogenic small dense LDL. These drugs are more aggressive and, although they cause greater reductions in lipid levels compared with dietary modification, they are often associated with unpleasant side-effects. Therefore this supports the need for effective dietary strategies that can reduce circulating lipid levels in both the fed and fasted state and which offer long-term efficacy comparable with most effective drug treatments. One dietary strategy that has been proposed to benefit the lipid profile involves the supplementation of the diet with prebiotics, probiotics and synbiotics, which are mechanisms to improve the health of the host by supplementation and/or fortification of certain health-promoting gut bacteria.

A *prebiotic* is a non-viable component of the diet that reaches the colon in an intact form and which is selectively fermented by colon bacteria such as bifidobacteria and lactobacilli. The term 'prebiotic' refers to non-digestible oligosaccharides derived from plants and also synthetically produced oligosaccharides. Animal studies have shown that dietary supplementation with prebiotics markedly reduced circulating TAG and, to a lesser extent, cholesterol concentrations. The generation of short chain fatty acids (SCFAs) during fermentation of the prebiotic by the gut microflora has been proposed to be one of the mechanisms responsible for their lipid-lowering effects via inhibition of enzymes involved in *de novo* lipogenesis. Of the human studies conducted to date, there have been inconsistent findings with respect to changes in lipid levels, although on the whole there have been favourable outcomes.

A *probiotic* is a live microbial feed supplement, which beneficially affects the host animal by improving its intestinal microbial balance and is generally fermented milk products containing lactic acid bacteria such as bifidobacteria and/or lactobacilli. The putative health benefits of probiotics include improved resistance to gastrointestinal infections, reduction in total cholesterol and TAG levels and stimulation of the immune system.[4] A number of mechanisms have been proposed to explain their putative lipid-lowering capacity and these include a 'milk factor', which has been thought to inhibit HMG-CoA reductase and the assimilation of cholesterol by certain bacteria.

A combination of a prebiotic and a probiotic, termed a *synbiotic*, is receiving attention at present since this association is thought to improve the survival of the probiotic bacteria in the colon. However, further well-controlled nutrition trials are required to investigate the mechanisms of action and effects of prebiotics, probiotics and synbiotics.

Manufacturers in the USA and Europe are starting to explore the commercial opportunities for foods that contain health-promoting food ingredients (probiotics, prebiotics and synbiotics). Issues considered important to the continuing development of the growing market include safety, consumer education, price and appropriate health claims. Until recently, most of the food products marketed were probiotics incorporated into dairy products. However, with the increasing interest in prebiotics, a more diverse food market has been opened up since prebiotics can be incorporated into many long-life foods ranging from bread to ice-cream. Although there is increasing interest in the use of prebiotics, probiotics and synbiotics as supplements to the diet, there is a need to ensure that claims for these products are based on carefully conducted human trials, which exploit up-to-date methodologies.

3.2   Risk factors in coronary heart disease

Diseases of the circulatory system account for an appreciable proportion of total morbidity and mortality in adults throughout the world. In 1990, CHD accounted for 27% of all deaths in the UK and stroke for 12%. The rates of mortality due to CHD throughout the world vary. For example, in one study among men aged 40–59 years, initially free of CHD, the annual incidence rate (occurrence of new cases) varied from 15 per 100,000 in Japan to 198 per 100,000 in Finland.[5] In addition to different rates of disease of the circulatory system between countries, the conditions assume varying degrees of importance in developing and affluent countries. Rheumatic heart disease is common in developing countries, whereas CHD has assumed almost epidemic proportions in affluent societies.

Figure 3.1 shows trends in CHD standardised mortality rates for the USA, Australia and the UK. Rates in the majority of countries are reducing, but the decline in CHD mortality in the UK began later than in the USA and Australia.[6] However, in Eastern Europe, rates of CHD are increasing dramatically, which is in contrast to many other parts of the world. Improved surgical procedures, more extensive use of cholesterol-lowering drugs and other medication, a reduction in smoking and increased health screening are in part responsible for this downturn. In addition, changes in diet, and especially in the type of fat consumed, may also have had a beneficial impact on disease incidence.

3.2.1   Pathology of CHD development

The underlying basis for clinical cardiovascular disease is a combination of

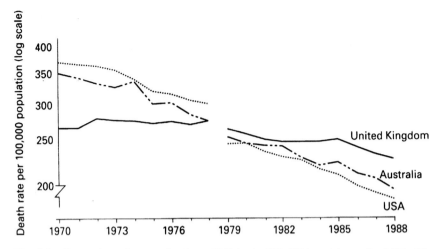

Fig. 3.1 Comparisons in mortality from CHD in the UK, USA and Australia (1970–88). Data between the years 1978 and 1979 were by a change in classification from ICD8 to ICD9 (adapted from WHO, 1989[6]).

atherosclerosis and thrombosis. Atherosclerosis is a condition in which the arterial lining is thickened in places by raised plaques as a result of excessive accumulation of modified lipids, and of the proliferation and migration of smooth muscle cells from deeper layers of the arterial wall. The atherosclerotic plaque usually develops at a point of minor injury in the arterial wall. Tissue macrophages (a type of white blood cell) are attracted to this point of damage and engulf and accumulate LDL particles from the blood. Recent studies have shown that LDL particles that have become oxidised are more likely to be taken up and cause cholesterol accumulation in macrophages. The cholesterol-loaded macrophages are transformed into lipid-laden foam cells, which remain in the arterial wall. At a later stage, the plaque becomes sclerosed and calcified (hence the term 'hardening of the arteries').

Formation of an atherosclerotic plaque can occlude one or more of the arteries, mainly the coronary and cerebral arteries, resulting in CHD or a stroke, respectively. In addition, a superimposed thrombus or clot may further occlude the artery. An example of an artery that has atherosclerotic lesions and has been completely occluded by a thrombus is shown in Fig. 3.2. Blood clot formation is in part determined by a release of eicosanoids from platelets and the vessel walls. PUFAs released from platelet membranes are metabolised into thromboxane (an eicosanoid), which stimulates platelet aggregation and vasoconstriction. Simultaneously, the vessel walls also release PUFAs, which are converted to prostacyclins (antagonistic eicosanoids), which inhibit platelet aggregation and cause vasodilation. The balance between production of thromboxane and prostacyclin, and the relative potencies of these two eicosanoids, will determine the extent of the blood clot formed.

Two major clinical conditions are associated with these processes: angina

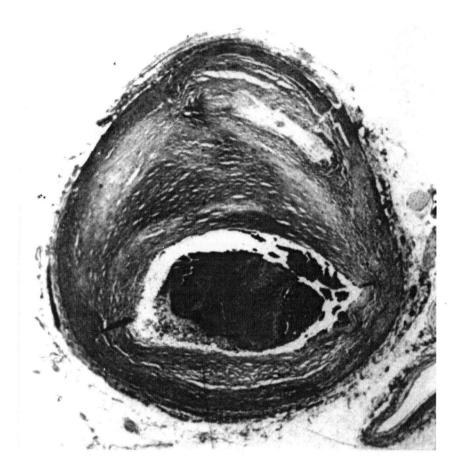

Fig. 3.2   Atherosclerotic plaque and thrombus completely occluding a coronary artery.

pectoris and coronary thrombosis or myocardial infarction. Angina pectoris is characterised by pain and discomfort in the chest, which is brought on by exertion and stress. It results from a reduction or temporary block in the blood flow through the coronary artery to the heart and seldom lasts for more than 15 minutes. A coronary thrombosis results from prolonged total occlusion of the artery, which causes infarct or death of some of the heart muscle and is associated with prolonged and usually excruciating central chest pain.

A variety of cells and lipids are involved in arterial thrombus, including lipoproteins, cholesterol, TAG, platelets, monocytes, endothelial cells, fibroblasts and smooth muscle cells. Nutrition may influence the development of CHD by modifying one or more of these factors and this will be discussed in more detail in Section 3.4. For the purposes of this chapter the disease of the circulatory system that will be addressed is CHD, with little discussion of other CVD such as strokes.

3.2.2   Risk factors for the development of CHD

CHD is a multifaceted condition that has no single cause. The term 'risk factor' is used extensively, and often very loosely, to describe features of lifestyle and behaviour, as well as physical activity and biochemical attributes, which predict the likelihood of developing disease. Potential risk factors are continually being refined as research into the aetiology of CHD progresses. The known risk factors for development of CHD can be categorised into those that cannot be modified, those that can be changed, those associated with disease states and those related to geographic distribution, as shown in Table 3.1.

Some risk factors have a greater influence on CHD development than others. It has been demonstrated that there is a strong and consistent relationship between total plasma cholesterol and CHD risk.[7] The positive association is largely confined to the LDL fraction, which transports about 70% of cholesterol in the blood. In a large prospective study published in 1986, a fivefold difference in CHD mortality, over a range of plasma cholesterol levels, was observed in the US population.[7] In a recent cholesterol-lowering drug trial in a healthy population, a reduction of 20% and 26% in total and LDL cholesterol respectively was associated with a 31% reduction in the five-year incidence of myocardial infarction and CHD death.[8] It is this relationship between the plasma cholesterol levels and its link with CHD that forms the basis of most dietary guidelines, which recommend reductions in total fat and SFA intakes.

If an individual presented with any one, or a combination, of risk factors, it is not inevitable that that person will suffer from CHD. The ability to predict the occurrence of a myocardial infarct in individuals is fraught with complications. An obese, middle-aged man who suffers from diabetes, consumes a high-fat diet, smokes 40 cigarettes a day and has a stressful job may never suffer from CHD;

Table 3.1   Risk factors for the development of coronary heart disease

Unmodifiable	Being male
	Increasing age
	Genetic traits (including lipid metabolism abnormalities)
	Body build
	Ethnic origin
Modifiable	Cigarette smoking
	Some hyperlipidaemias (increased plasma cholesterol and triacylglycerol)
	Low levels of high density lipoprotein (HDL)
	Obesity
	Hypertension
	Low physical activity
	Increased thrombosis (ability to clot)
	Stress
	Alcohol consumption
Diseases	Diabetes (glucose intolerance)
Geographic	Climate and season: cold weather
	Soft drinking water

Table 3.2 Relative risk and yield of cases in the top fifth of the ranked distribution of risk factors in the men from the British Regional Heart Study after five years of follow-up (adapted from Shaper et al., 1986[9])

Factor	Relative risk	Yield in top fifth (%)
Age	4.7	34
Total cholesterol	3.1	31
Systolic BP	3.0	36
Diastolic BP	3.1	34
Body Mass Index (BMI)	1.8	28
'Smoking years'	5.1	38

whereas a slim, physically active non-smoker who consumes a low-fat diet may die from a myocardial infarct prematurely. Despite this anomaly, individuals and populations are deemed at increasing risk of CHD according to the severity and number of identified risk factors. Table 3.2 shows the relative risk and yield of cases.[9]

The purpose of relative risk scores is in prediction, as it clearly contains items that cannot be modified. However, for the purpose of intervention, one must go beyond the items used to predict risk and consider issues such as diet, body weight, physical activity and stress (factors not used in the scoring system), as well as blood pressure and cigarette smoking, which are taken into account. General population dietary recommendations have been provided by a number of bodies, which are aimed at reducing the incidence and severity of CHD within the population. Those specifically related to dietary factors are discussed in detail in section 3.4.

3.3 Relevant lipid particles

3.3.1 Plasma lipoprotein metabolism

Plasma lipoproteins are macromolecules representing complexes of lipids such as TAG, cholesterol and phospholipids, as well as one or more specific proteins referred to as apoproteins. They are involved in the transport of water-insoluble nutrients throughout the blood stream from their site of absorption or synthesis, to peripheral tissue. For the correct targeting of lipoproteins to sites of metabolism, the lipoproteins rely heavily on apoproteins associated with their surface coat.

The liver and intestine are the primary sites of lipoprotein synthesis and the two major transported lipid components, TAG and cholesterol, follow two separate fates. TAGs are shuttled primarily to adipose tissue for storage, and to muscle, where the fatty acids are oxidised for energy. Cholesterol, in contrast, is continuously shuttled among the liver, intestine and extrahepatic tissues by HDL.[10] Human lipoproteins are divided into five major classes according to their flotation density (Table 3.3). The density of the particles is inversely related to their sizes, reflecting the relative amounts of low density, non-polar lipid and high density surface protein present. The two largest lipoproteins contain mainly TAG within their core structures. These are chylomicrons (CMs), secreted by the enterocytes (cells of the small intestine), and VLDL, secreted by the hepatocytes (liver cells). Intermediate density lipoprotein (IDL) contains appreciable amounts of both TAG and cholesterol esters in their core. The two smallest classes, LDL and HDL, contain cholesterol esters in their core structures and the mature forms of these particles are not secreted directly from the liver but are produced by metabolic processes within the circulation. LDLs are produced as end products of the metabolism of VLDL, whereas components of HDL are secreted with CMs and VLDL. The lipid metabolic pathways can be divided into the exogenous and endogenous cycles, which are responsible for the transport of dietary and hepatically derived lipid, respectively.[11]

Exogenous pathway
Following the digestion of dietary fat in the small intestine, long chain fatty acids are absorbed by the enterocyte. The nascent CM particle consists of a core of TAG (84–9% of the mass), cholesterol ester (3–5%) and surface free cholesterol (1–2%) and on the surface, phospholipids (7–9%) and apoproteins (1.5–2.5%).[12] Following secretion, the TAG component of the CM particle is hydrolysed by lipoprotein lipase (LPL) bound to the luminal surface of the endothelial cells in adipose tissue and muscle. Approximately 70–90% of the TAG is removed to produce a cholesterol ester-rich lipoprotein particle termed a CM remnant.[13] As the core of the CM remnant particle becomes smaller, surface materials, phospholipids, cholesterol and apolipoproteins are transferred to HDL to maintain stability of the particle. Uptake of these particles probably requires interaction with hepatic lipase (HL), which is situated on the surface of the liver. HL further hydrolyses the TAG and phospholipid components of the CM remnant before uptake by a receptor-mediated process in the liver.[14] The remnants are endocytosed and catabolised in lysosomes from which cholesterol can enter metabolic pathways in hepatocytes, including excretion into the bile (Fig. 3.3).

Endogenous pathway
VLDL provides a pathway for the transport of TAG from the liver to the peripheral circulation. Two subclasses of VLDL are released from the liver, $VLDL_1$ (large TAG-rich lipoprotein) and $VLDL_2$ (smaller, denser particles). LPL in the capillary bed extracts TAG from the secreted $VLDL_1$ but less efficiently when compared to CMs.[15] LDL receptors on the surface of the liver recognise the $VLDL_2$ particles and mediate the endocytosis of a fraction of these particles. Prolonged residency of some VLDL particles in the plasma results in further metabolism by LPL, and to some extent HL, to form the higher density IDL. HL, on the surface of hepatocytes, further hydrolyses IDL with eventual formation of LDL.[16] LDL is a heterogeneous population consisting of larger LDL species (LDL-I and LDL-II subclass) and smaller denser LDL particles (LDL-III). LDL can be taken up by LDL receptors present on the surface of hepatocytes and on extrahepatic cells.

Table 3.3   Structure and function of lipoproteins (adapted from Erkelens, 1989)

Lipoprotein	Structural apolipoproteins	Apolipoproteins attached	Function
CM	B-48	A-I, A-IV, C-I, C-II and C-III	Carries exogenous TAG from gut to adipose tissue, muscle and liver
CM remnant	B-48	C-I, C-II, C-III	Carries exogenous cholesterol to the liver and periphery
VLDL (VLDL$_1$ and VLDL$_2$)	B-100	C-I, C-II, C-III and E	Carries endogenous TAG to the periphery
IDL	B-100	E	Carries endogenous cholesterol to the periphery
LDL	B-100	–	Carries cholesterol to the liver and periphery
HDL$_2$ and HDL$_3$	A-I and A-II	C-I, C-II, C-III, E, LCAT and CETP	Reverse cholesterol transport

Note:
Abbreviations: CM, chylomicron; VLDL, very low density lipoprotein; IDL, intermediate density lipoprotein; LDL, low density lipoprotein; HDL, high density lipoprotein; LCAT, lecithin cholesterol acyltransferase; CETP, cholesterol ester transfer protein.

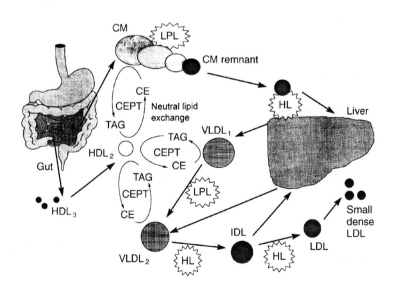

Fig. 3.3   Simplified overview of lipoprotein metabolism showing the inter-relationships between the exogenous and endogenous pathways.

Reverse cholesterol transport
A function of HDL is to trigger the flux of cholesterol from peripheral cells and from membranes undergoing turnover to the liver for excretion. The process of reverse cholesterol transport is mediated by an enzyme, lecithin cholesterol acyltransferase (LCAT), bound to species of HDL particles. It acts by trapping cholesterol into the core of the nascent HDL following interaction of this particle with a cell surface protein. The cholesterol is transferred to HDL_3, which in turn is converted to HDL_2. This particle can follow one of two pathways (direct or indirect) to deliver cholesterol to the liver.[17] In the direct pathway, the HDL_2 is removed via receptor-mediated endocytosis by the LDL receptor or via selective uptake of cholesterol ester from the HDL particle by the liver. The cholesterol ester transfer protein (CETP) is involved in the indirect pathway and transfers cholesterol ester to lower density lipoproteins (CMs and VLDL), in return for TAG, and is followed by their uptake by the liver (neutral lipid exchange) (Fig. 3.3).

The balance between the forward (exogenous and endogenous pathways) and the reverse pathway, which is tightly regulated by the secretion rates of the lipoproteins, determines the concentration of cholesterol in the plasma.

3.3.2   Classical lipid risk factors in coronary heart disease
Many epidemiological studies have shown a positive correlation between fasting total cholesterol levels, especially LDL cholesterol levels and CHD mortality.[1] Accumulation of LDL in the plasma leads to a deposition of cholesterol in the arterial wall, a process that involves oxidative modification of the LDL particles. The oxidised LDL is taken up by macrophages, which eventually become foam cells and forms the basis of the early atherosclerotic plaque. It has been estimated that every 1% increase in LDL cholesterol level leads to a 2–3% increase in CHD risk.[18] The role of cholesterol lowering as a public health strategy in the primary prevention of CHD was unequivocally supported by the findings of the West Scotland Heart Study.[8] This lipid-lowering trial involved 6,595 middle-aged men with moderately raised cholesterol levels (5.8–8.0 mmol/l) but without any history of CHD. The study showed a 22% reduction in overall mortality and a 33% reduction in cardiovascular mortality in men receiving the active drug (prevastatin) compared with a placebo, in a five-year follow-up. Total cholesterol levels were reduced by 20%, with LDL cholesterol levels being reduced by 26% on the active drug treatment.

HDL cholesterol levels may influence the relationship between total cholesterol levels and CHD risk. A strong inverse relationship between fasting plasma HDL levels and the risk of development of CHD has been reported.[18] On average, HDL cholesterol levels are higher in women than in men. Factors that may lead to reduced HDL cholesterol levels include smoking, low physical activity and diabetes mellitus; whereas those that increase levels include moderate alcohol consumption. The Münster Heart Study, carried out between 1979 and 1991, investigated cardiovascular risk factors, stroke and mortality in people at work. Examination of fasting lipid parameters at the beginning of the study and with a follow-up six to seven years later demonstrated that HDL cholesterol concentrations were significantly lower in the group with major coronary events compared to the group that was free of such coronary events. Low HDL levels reflect a compromised pathway for the excretion of cholesterol and have been associated with a fivefold increase in risk of CHD compared to normal HDL values.[19]

3.3.3   Newly recognised risk factors

Elevated fasting and post-prandial TAG levels
Almost all of the epidemiological evidence for CHD risk has been determined from fasting lipoprotein concentrations, obtained following a 12-hour overnight fast. However, individuals spend a large proportion of the day in the post-prandial state when the lipid transport system is challenged by fat-containing meals. The magnitude and duration of post-prandial TAG concentrations following a fat load have been correlated with the risk of development of CHD. This has been shown in patients with CHD, who show a more pronounced and prolonged TAG response following a meal compared with matched people without CHD, even though both groups showed similar fasting TAG concentrations.[2] The strength of the association between TAG levels and CHD has been demonstrated in a number of observational studies.[20] In the Framingham study, individuals were segregated according to their HDL cholesterol levels (classical risk factor) and it was observed that the group with the highest TAG and lowest HDL cholesterol concentrations had the greatest risk for CHD.[21] This also agreed with the observation by O'Meara *et al.*,[22] where the highest magnitude of lipaemia was found in individuals who had high fasting TAG compared with lower fasting TAG concentrations. Although univariate analysis has demonstrated the association between TAG concentrations and CHD risk, multivariate analysis, including other lipid parameters and especially HDL cholesterol levels, has abolished this statistical significance. However, this statistical correlation is a controversial issue since an inverse relationship exists between the levels of TAG and HDL cholesterol levels due to their metabolic interrelationships. Recent findings from the Münster Heart Study have demonstrated, using multivariate analysis, that total cholesterol, LDL cholesterol, HDL cholesterol and log-transformed values of TAG showed a significant age-adjusted association with the presence of major coronary events. This correlation with TAGs remained after adjustment for LDL and HDL cholesterol levels.[19] Therefore, a single measurement of these lipid parameters in an individual may provide insufficient information and so underestimate any association between these variables.

Elevated levels of chylomicrons and chylomicron remnants
The development of specific methods to differentiate between the exogenous and endogenous TAG-rich lipoproteins, CMs and VLDL, in post-prandial samples has enabled the atherogenicity of these lipoproteins to be investigated. A recent finding by Karpe *et al.*[23] has provided evidence that a delayed uptake of small CM

remnants is associated with the progression of atherosclerosis. The abnormal clearance of CMs and CM remnants after a fat load have been implicated directly and/or indirectly with the presence of CHD. This may reflect the ability of CM remnants to infiltrate the arterial wall directly. The mechanism whereby CM remnants provide the building blocks of arterial lesions are thought to occur by one of two processes. First, the CM remnants may bind and penetrate the arterial surfaces (in just the same way as plasma LDL), therefore the rate of atherogenesis should be proportional to the plasma remnant concentrations. Second, CMs may be absorbed and then degraded to remnants on the arterial surface.[24] In each instance, the reaction leading to the endocytosis of remnants by smooth muscle cells may take place at sites where local injury has removed the endothelium.

The magnitude of post-prandial lipaemia is dependent on a number of factors including rates of clearance by peripheral tissue and receptor-mediated uptake of the remnants by the liver. Defects in any of these processes will cause an accumulation of CMs and their remnants which in turn can influence endogenous lipoproteins known to be atherogenic. In particular the transfer of TAG from CMs to LDL and HDL and a reciprocal transfer of cholesterol esters to the CMs by CETP may increase the atherogenic lipid profile. This is known as neutral lipid exchange. The transfer of cholesterol ester to CMs and VLDL makes them resistant to lipase action which impedes the normal metabolism of these TAG-rich lipoproteins. The cholesterol ester-enriched CM remnants are then able to be taken up by the macrophages in the arterial lesion. LDL and HDL, on the other hand, become suitable substrates for LPL and HL, causing a reduction in the size of these particles. This results in the development of an atherogenic lipoprotein profile in which the TAG-rich lipoproteins become cholesterol ester-enriched, LDL size is reduced and HDL cholesterol levels are reduced. The small dense nature of the LDL makes it unrecognisable by the LDL receptors in the liver and so makes it a favourable candidate for uptake by scavenger receptors present on macrophages in the arterial lesion.[25]

Elevated levels of small dense LDL (LDL-III)
It has long been established that elevated circulating LDL cholesterol levels represent a major risk factor for the development of CHD. Recently, with the use of density gradient ultracentrifugation techniques, LDL has been separated into three major subclasses. These are subdivided as light large LDL (LDL-I), intermediate size LDL (LDL-II) and small dense LDL (LDL-III).[26] In healthy normolipidaemic males, a preponderance of LDL-II are seen, with only a small percentage of LDL-III being present.

A number of case control and cross-sectional studies has examined the relationship between LDL subclass and risk of CHD.[27] In studies in men with CHD, or those who had survived a myocardial infarction, it was demonstrated that small dense LDL-III was more common in the men with CHD than without. In 1988, Austin *et al.*[26] proposed that a preponderance of LDL-III in young men was associated with a threefold increase in CHD. However, more recently, a greater relative risk of CHD associated with raised LDL-III has been proposed.[28] The increased atherogenicity of small dense LDL-III is thought to be due to the increased residency of these particles in the circulation due to their slow uptake by the LDL receptor in the liver and peripheral tissues. This allows time for the small dense LDL to infiltrate into the intima of the arterial wall where it is thought that these particles are retained by extracellular matrix components before oxidation of the LDL particles occur. The modified small dense LDL-III is then taken up by the scavenger receptor on macrophages leading to the subsequent formation of foam cells.[29]

Atherogenic lipoprotein phenotype
The ALP is a collection of lipid abnormalities, which confers an increased risk of CHD upon normal, healthy individuals. It has been proposed that 30–35% of middle-aged men in the Western world may be affected.[3] In the fasting state, this phenotype is characterised by a moderately raised TAG concentration (1.5 to 2.3 mmol/l), low HDL concentration (less than 1 mmol/l) and a predominance of small dense LDL-III (greater than 35% of LDL mass) and HDL particles. It is important to note that total cholesterol levels are usually in the normal range or are only moderately raised. The atherogenic nature of the ALP may arise from the impairment of the removal of TAG-rich lipoproteins (CMs and VLDL) leading to the conversion of small, atherogenic cholesterol enriched remnants, LDL and HDL by neutral lipid exchange. It is generally considered that this collection of lipid abnormalities is closely associated with the insulin resistance syndrome.[30]

3.4   Diet and coronary heart disease

There is a substantial, diverse and generally consistent body of evidence linking diet with cardiovascular disease. The evidence is most extensive for the relationship between CHD and dietary fat. Diet is believed to influence the risk of CHD through its effects on certain risk factors described in Table 3.1, for example blood lipids, blood pressure and probably also through thrombogenic mechanisms. More recently, evidence suggests a protective role for dietary antioxidants such as vitamins E and C and carotenes, possibly through a mechanism that prevents the oxidation of LDL cholesterol particles.

3.4.1   Dietary lipids and CHD risk
Epidemiological and clinical evidence clearly shows that the likelihood of death from CHD is directly related to the circulating level of total cholesterol (and more specifically LDL cholesterol). More recent evidence suggests that an exaggerated post-prandial TAG response to fat-containing meals is also a significant risk factor for CHD.[2] Numerous studies have shown that the kinds and amounts of fat in the diet significantly influence plasma cholesterol levels. In the Seven Countries Study, mean concentrations of cholesterol of each group were highly correlated with percentage energy derived from SFAs, and even more strikingly related to a formula that also took into account the intake of PUFAs (see Fig. 3.4).[5]

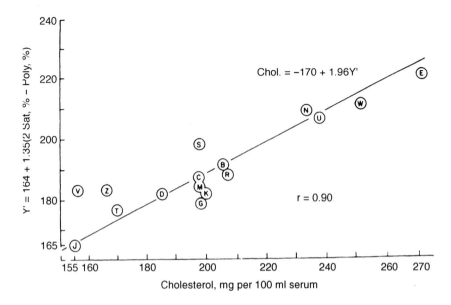

Fig. 3.4   Relationship of mean serum cholesterol concentration of the cohorts of Seven Countries Study to fat composition of the diet expressed in the multiple regress equation, including intakes of saturated and polyunsaturated fatty acids (B = Belgrade, C = Crevalcore, D = Dalmatia, E = East Finland, G = Corfu, J = Ushibuka, K = Crete, M = Montegiorgio, N = Zutphen, R = Roma, S = Slavonia, T = Tanushimara, U = USA, V = Velika Kisna, W = West Finland, Z = Zregnjamin). (Adapted from Keys, 1980.[5])

Since this research, numerous other studies have supported the relationship between dietary intake of saturates and the raised plasma cholesterol levels. However, not all SFAs are equally potent in raising plasma cholesterol. Palmitic acid (16:0), the principal SFA in most diets, and myristic acid (14:0) are the most effective at elevating cholesterol; whereas stearic acid (18:0), lauric acid (12:0) and medium chain fatty acids (8:0 and 10:0) have little effect on plasma cholesterol levels.[31] The cholesterol response to a particular SFA may depend in part on the TAG structure,[32] and in part on LDL receptor activity.[33] The mechanism that is responsible for the increase in plasma cholesterol due to SFAs is at present unclear, but a reduction in the LDL receptor activity is one possibility. It has been recommended that the average contribution of SFAs to dietary energy be reduced to no more than 10%.[34] The current UK dietary intake is shown in Table 3.4. A considerable reduction from current levels of intake of dietary SFAs (average 33%) would be required to meet this recommendation.[35]

The effect of the ingestion of different fatty acids on plasma cholesterol levels is varied and is summarised in Table 3.5. Substitution of SFAs by MUFAs or *n*-6 PUFAs significantly reduces LDL cholesterol levels, although *n*-6 PUFA are more effective in this respect. There has been doubt as to whether MUFAs are effective in cholesterol-lowering or whether the observed decrease in plasma cholesterol was simply due to a replacement of SFAs. Some studies have suggested that the

Table 3.4 Dietary intakes of selected nutrients for men and women aged 16–64 years (adapted from Gregory *et al.*, 1990[35])

Nutrients	Men	% total energy	Women	% total energy
Total energy (KJ)	10462		7174	
Fat (g)	102	37.6	74	39.2
SFA (g)	42	15.4	31	16.5
MUFA (g)	31	11.6	22	11.8
n-3 PUFA (g)	1.9	0.7	1.3	0.7
n-6 PUFA (g)	13.8	5.1	9.6	5.1
trans FA (g)	5.6	2.0	4.0	2.1
P:S ratio	0.40		0.38	
Cholesterol (mg)	390		280	
Carbohydrate (g)	272	41.6	193	43.0
Alcohol (g)	25	6.9		2.8
Fibre (g)	24.9	18.6		

effect of oleic acid (*cis* 18:1) (the major MUFA in the diet) and linoleic acid (18:2) (major *n*-6 PUFA in the diet) on LDL cholesterol are similar, and that the greater effect of linoleic acid on total cholesterol is through reduction of HDL cholesterol.[36] In addition to this it has been reported that PUFAs incorporated into lipoproteins can increase their susceptibility to oxidation if there is insufficient antioxidant protection. Despite the beneficial lowering of LDL cholesterol associated with increased dietary PUFA, it has been proposed that the decreased levels of beneficial HDL cholesterol and greater susceptibility of lipoproteins containing *n*-6 PUFA to oxidation, result in pro-atherogenic effects of diets in which *n*-6 PUFA provide greater than 10% energy.[37] Substitution of saturates by oleic acid would avoid this and therefore MUFAs have theoretical advantages over PUFAs. It has long been recognised that, in Mediterranean populations, there is a significantly lower risk of CHD.[5] Their diet traditionally contains high amounts of olive oil in addition to fruit and vegetables compared to the UK diet. It has been speculated that a higher intake of MUFAs could contribute to the lower rate of

Table 3.5 Effects of fatty acids on plasma lipoprotein concentrations

Fatty acid	Total cholesterol	LDL cholesterol	HDL cholesterol	Triacylglycerol
Saturated FA	Increase	Increase	Neutral	Neutral
n-6 PUFA	Decrease	Decrease	Decrease	Neutral
n-3 PUFA	Unchanged	Unchanged*	Increase	Decrease
Trans FA	Increase	Increase	Decrease	NA
MUFA	Decrease	Decrease	Neutral	Neutral
Cholesterol	Increase	Increase	NA	NA

Notes:
FA – fatty acid. PUFA – polyunsaturated fatty acid. MUFA – monounsaturated fatty acid. NA – not available. * May increase in hyperlipidaemics.

CHD within this population. However, it could also be due to the higher antioxidants found in the fruit, vegetables and within virgin olive oil itself, in addition to other dietary and lifestyle factors. Research into the reasons for this link between the Mediterranean lifestyle and CHD risk is necessary to explain this observation fully.

As regards CHD, *trans* fatty acids appear to act similarly to SFA in their effects on blood cholesterol except that *trans* FA decreased HDL whereas SFA have little effect on HDL. In a trial conducted to compare the effects of different fatty acids, elaidic (*trans* 18:1), the principal *trans* fatty acid found in the diet, was found to decrease HDL and increase LDL levels significantly.[38] Some epidemiological evidence also supports these findings. Hydrogenated fats are a major dietary source of *trans* FA and are abundant in vegetable margarines and processed foods. However, due to the reported link between CHD risk and *trans* fatty acids, the level of these fatty acids has been substantially reduced in many margarines and manufactured foods. The weight of evidence supports the view that raising cholesterol content of the diet increases plasma cholesterol, primarily LDL cholesterol, although there is considerable inter-individual response. Studies in humans over the past 25 years have indicated a threshold for an effect at an intake of about 95 mg/4300 kJ with a ceiling at about 450 mg/4300 kJ. Excess cholesterol is either not absorbed or suppresses endogenous production. As daily intake (Table 3.4) is at the lower end of this range, it is recommended that current dietary cholesterol intakes, measured per unit of dietary energy, should not rise.[34]

In contrast to dietary SFAs, MUFAs, *n*-6 PUFAs and *trans* fatty acids whose effects on cardiac health primarily reside in their ability to modify plasma LDL and HDL cholesterol levels, the benefits of an increased intake of *n*-3 PUFA lie in their ability to reduce thrombosis and decrease plasma TAG levels. The low incidence of CHD in Greenland Eskimos and Japanese fishermen, despite a high fat intake (fat providing 80% of dietary energy), has been attributed to their intake of marine foods high in EPA (20:5, *n*-3) and DHA (22:6, *n*-3).[39] In the DART trial, individuals who had suffered a previous myocardial infarction were advised to consume 1–2 portions of fatty fish per week. After a two-year follow-up period, a 29% reduction in mortality from CHD was reported in those advised to consume the fatty fish.[40]

Evidence has shown that fasting TAG levels and post-prandial (following a meal) TAG levels are a significant risk factor for CHD.[2] The mechanism is not fully understood but is believed to be associated with the TAG-rich lipoprotein particle (CM and VLDL) remnants which are potentially atherogenic. High levels of these remnants, can contribute towards an increased risk of neutral lipid exchange (see Section 3.3) which will in turn lead to an accumulation of small dense, more easily oxidised LDL-III, and a decreased circulation of beneficial HDL particles. Supplementation studies with *n*-3 PUFA have observed a significant reduction in blood TAG levels.[41] The proposed mechanism for this decrease in TAG is a reduction in the level of VLDL synthesis by the liver and an increased clearance of TAG-rich particles from the blood by the enzyme LPL (the enzyme that hydrolyses TAG within VLDL and CMs).

Fatty acids from the n-3 PUFA series also have a beneficial effect on thrombogenesis (blood clot formation). The utilisation of n-3 PUFAs instead of n-6 PUFAs in the production of eicosanoids (substances involved with the formation of blood clotting) can significantly reduce the rate of blood clot formation and thus reduce the risk of myocardial infarction. In addition, n-3 PUFA ingestion may result in the reduction in cardiac arrhythmias.[42] Sudden cardiac death is a serious problem in Western countries and no drug treatment, to date, has had any significant effect on incidence. Epidemiological studies have shown a reduced incidence of sudden death with n-3 PUFA intake, and animal and limited human evidence also suggests an anti-arrhythmic effect.[43]

At the present time most of the national and international guidelines for population intakes of dietary fat are based on the known adverse effects of SFAs in raising blood cholesterol levels with some consideration given to possible benefits of n-3 PUFAs on thrombosis. Table 3.6 shows the dietary guidelines of the European Union[44] and those of the UK Committee of Medical Aspects of Food Policy.[34] These recommendations are aimed at both the food industry and the general population. Due to high SFA intake compared with those recommended, which reflect consumer resistance to these recommendations, future recommendations may take greater account of the benefits of substituting SFAs with MUFAs, and the overall significance to human health of the absolute and relative amounts of n-6: n-3 PUFAs.

3.4.2   Carbohydrate intake and CHD risk

Recommendations for reductions in total fat in the diet have important implications for dietary carbohydrate intake. There is little variation in the proportion of energy derived from dietary protein; therefore there is a reciprocal relationship between

Table 3.6   Recommended daily intakes of dietary fat and fatty acids of the EEC (EU) and UK and the average daily consumption of these nutrients (% dietary energy)

	Total fat	SFA	PUFA	MUFA	Trans FA
Recommendations					
European	20–30[1]	10[1]	0.5[2] (n-3)	NR	NR
(EEC, 1992)			2.0[2] (n-6)		
UK	<35	<10	0.2g/day (n-3)	NR	2
(DH, 1994)			5.0 (n-6)		
Current intake					
UK (1990)	39	16	5 (n-6)	12	2
USA (1985)	37	13	7 (n-6)	13	–
France (1985–7)	36	15	6 (n-6)	13	–
Germany (1988)	40	15	6 (n-6)	12	–
Netherlands (1988)	40	16	7 (n-6)	15	–

Notes:
1. Ultimate goal. 2. Population reference intake. NR – no recommendations.

the contributions of dietary fat and carbohydrate to energy. The metabolic effects of exchanging carbohydrate for fat depend mainly on the degree of substitution. Diets where about 60% of food energy is derived from carbohydrate are associated with lower HDL levels and higher TAG levels, and despite lower LDL levels have been suggested to be associated with a higher risk of CHD.[45, 46] However, a smaller increase in dietary carbohydrate levels to accommodate a reduction in dietary fat to 30% of energy has been reported to result in a small rise in TAG levels and no fall in HDL levels,[46] resulting in an overall positive benefit in CHD risk.

3.4.3   Non-starch polysaccharides and CHD risk

Four prospective studies have shown an inverse relationship between dietary 'fibre' intake and CHD. The studies have varied in the source of 'fibre' that was found to be effective. Morris *et al.*[47] found that cereal fibre was inversely related to cases of CHD, whereas others found that vegetable sources of fibre were associated with decreased risk.[48] On the contrary, a two-year intervention study in men who had suffered a previous heart attack, found no effect of increasing cereal 'fibre' intake on subsequent risk of mortality from cardiovascular disease.[40] The effect of a number of soluble NSPs which are selectively digested by the colonic gut microflora (classed as prebiotics) on the blood levels of cholesterol and especially LDL cholesterol will be discussed in detail in section 3.6.

3.4.4   Antioxidant nutrients and CHD risk

The cells of the body are under constant attack by activated oxygen species, which are produced naturally in the body. The protection of cells from the detrimental effects of these species is due to defence mechanisms, component parts of which are, or are derived from, the micronutrients called 'antioxidants'. These include the essential trace elements selenium, zinc and magnesium, and vitamin C and the various forms of vitamin E (the tocopherols and the tocotrienols). In addition, betacarotene, and other carotenoids, such as lutein and lycopene (present in large amounts in tomatoes) and flavonoids, present in red wine, tea and onions, probably also play a role.

There is a large body of evidence supporting a protective effect of the antioxidant vitamins E and C.[49] There are, however, inconsistencies in the amount of these antioxidants associated with reduced risk of CHD; this is especially the case for vitamin E. One explanation for these discrepancies is that vitamin E may be a marker for a diet that contains other dietary constituents (such as carotenoids and flavonoids) which have antioxidant potential. When vitamin E is taken in isolation as a supplement, higher levels might be necessary to achieve the same effect as diets that contain combinations of antioxidants. No specific recommendations on the levels of antioxidant intake have been given but a diet rich in vegetable and fruit and containing nuts and seeds is recommended.[50] The beneficial effects of red wine and olive oil on CHD risk could in part be attributed to antioxidants such as flavonoids and polyphenols contained within these foods.

3.4.5   Sodium and potassium and CHD risk

Sodium intake appears to be an important determinant of blood pressure in the population as a whole, at least in part by influencing the rise in blood pressure with age.[51, 52] A diet lower in common salt and higher in potassium would be expected to result in lower blood pressure and a smaller rise in blood pressure with age. Salt is the predominant source of sodium in the diet, with manufactured foods contributing to 65–85% of the total salt ingested. Blood pressure is an important risk factor for the development of CHD and strokes. It has been recommended that the population should reduce its salt intake by 3 g/day by reducing salt at the table and also the consumption of processed foods. Moreover, higher potassium levels within the diet are believed to reduce the blood pressure and foods such as fruit and vegetables, which contain this mineral, should be taken in higher amounts.

3.4.6   Alcohol and CHD risk

The debate surrounding the benefits of alcohol consumption and, more specifically, red wine consumption and the risk of CHD is one that has been running for many years. There is evidence that high consumption of alcohol is related to increased mortality, especially from CHD. However, alcohol consumption appears to be associated with relatively low risk of CHD across a variety of study populations. The benefit that is associated with alcohol consumption is almost entirely due to the increased levels of HDL cholesterol which is associated with reduced CHD risk.[53] However, other factors such as lower platelet activity, reducing the risk of thrombosis, and antioxidant properties of some drinks such as red wine have also been suggested. The proposed benefit of red wine over other drinks would only explain the potential increase in antioxidant levels as the effect of alcohol on HDL and haemostatic factors have been attributed to alcohol itself. A consumption of two units (one unit is equivalent to 8 g alcohol) a day of alcohol is believed to be beneficial in the risk of CHD, but the debate continues.[54]

3.4.7   Coffee and CHD risk

For over 20 years it has been suspected that caffeine or coffee consumption may contribute to the development of CHD, but the evidence remains inconsistent. There is little evidence that caffeine itself has any relation to CHD risk but there are other components of coffee, which might in part account for some observed associations. The Scandinavian practice of boiling coffee during its preparation appears to generate a hypercholesterolaemic fraction. Significant levels of these compounds have also been found in cafeteria coffee which have been found to increase plasma LDL cholesterol levels.[55] The relevance of this to the UK population, whose consumption of boiled coffee is very low, is unclear.

3.4.8   Diet and CHD risk

The diet is one of the modifiable risk factors associated with CHD risk.

Recommending reducing total fat (especially saturated fat), increasing NSP intake and consumption of fruit and vegetables is advice that is likely to be associated with overall benefits on health. However, there is great inertia for dietary change, and many new products that claim to reduce the risk of CHD and other chronic diseases, without altering lifestyle factors such as diet, clearly attract a great deal of attention.

3.5   The effects of probiotics on coronary heart disease

Current dietary strategies for the prevention of CHD advocate adherence to low fat/low saturated fat diets.[34] Although there is no doubt that, under experimental conditions, low fat diets offer an effective means of reducing blood cholesterol concentrations, on a population basis they appear to be less effective, largely due to poor compliance attributed to low palatability and acceptability of these diets to the consumer. Due to the low consumer compliance, attempts have been made to identify other dietary components that can reduce blood cholesterol levels. These have included investigations into the possible hypocholesterolaemic properties of milk products, usually in a fermented form. The role of fermented milk products as hypocholesterolaemic agents in humans is still equivocal, as the studies performed have been of varying quality, design and statistical analysis with incomplete documentation being the major limitation of most studies. However, since 1974 when Mann and Spoerry[56] showed an 18% fall in plasma cholesterol after feeding 4–5 litres of fermented milk per day for three weeks to Masai warriors, there has been considerable interest in the effect of probiotics on human lipid metabolism.[56]

3.5.1   Evidence from the Masai

As epidemiological evidence suggested that an environmental agent that contributed to hypocholesterolaemia had been introduced into the Western world around 1900, Mann and Spoerry intended to investigate the effect of an exogenous surfactant material, known to be hypercholesterolaemic in animals.[56] The authors' original intention was to feed 4–5 litres of pasteurised European milk fermented with a wild culture of *Lactobacillus* to 24 male Masai warriors, for six days out of every week. However, demand rose to over 8.3 litres per day (23,000 kJ/day) and an exercise programme was started to prevent weight gain. However, eight of the subjects gained considerable weight (> 5 lb) after feigning injury to avoid this exercise. The eight subjects who gained weight surprisingly had a significant fall (18.2%) in serum cholesterol of 0.73 mmol/l. The total group's mean cholesterol level also significantly decreased (9.8%), as shown in Table 3.7. Although the marked reductions in serum cholesterol were striking, considering the weight increase (usually associated with a raised cholesterol concentration), this study is now considered simply as a curiosity. Due to the introduction of an exercise programme and the inability to control food intake, the work cannot lend much to the causality of the findings.

Table 3.7 Human studies to evaluate the hypocholesterolaemic properties of fermented milk products (adapted from Taylor and Williams[112])

Author	Subjects (n)	Product (vol/type)	Duration	Total cholesterol	LDL cholesterol
Mann and Spoerry [56]	24M	8.3 l lacto/yog	3 weeks	−9.6% (P<0.001)	NA
Mann[57]	3 M 1F	4 l WMY	12 days	−16.8% (P<0.05)	NA
	3M 2F	2 l SMY	12 days	−23.2% (P<0.05)	NA
Howard and Marks[58]	10	3 l	3 weeks	5.5% (P<0.05)	NA
Hepner et al.[59]	6M 4F	720 ml (A)	4 weeks	−5.4% (P<0.01)	NA
	5M 3F	720 ml (B)	4 weeks	8.9% (P<0.01)	NA
Rossouw et al.[68]	11M	2 l yoghurt	3 weeks	+16% (P<0.01)	+12% (P<0.001)
Thompson et al.[61]	13	1 l UPY	3 weeks	NS	NS
Bazzare et al.[111]	5M 16F	550g yoghurt	1 week	−8.7% NA	NA
Massey[62]	30F	480 ml yoghurt	4 weeks	NS	NS
Jaspers et al.[64]	10M	681 g yoghurt	2 weeks	−11.6% transient (P<0.05)	NS
McNamara et al.[63]	18M	16 oz LFY	4 weeks	NS	NS
Agerbaek et al.[65]	58M	200 ml UPY	6 weeks	−6.1% (P<0.001)	−9.8% (P<0.001)
Richelsen et al.[66]	47M 43F	200 ml UPY	21 weeks	NA	−9% transient (P<0.05)
Sessions et al.[60]	78M 76F	200 ml UPY	12 weeks	NS	NS
Bertolami et al.[67]	11M 21F	200 ml UPY	8 weeks	−5.3% (P<0.004)	6.2% (P<0.001)

Notes:
F = female; M = male; WMY = whole milk yoghurt; UPY = unpasteurised yoghurt; SMY = skimmed milk yoghurt; PY = pasteurised yoghurt; LFY = low fat yoghurt;
NA = data not available; NS = not significant.

3.5.2   Evidence for the 'milk factor'

As a follow-up to the Masai trial, Mann fed a small group of US volunteers ($n = 4$) 4 litres per day of yoghurt (microbiological activity unspecified) over a 12-day period and reported a significant fall of 37% in serum cholesterol values (however, the tabulated data indicated only a 16.8% fall).[57] When intake of the yoghurt was reduced to 2 litres, the hypocholesterolaemic effect was maintained, although an intake of 1 litre per day resulted in a return to baseline cholesterol levels. The rate of cholesterol biosynthesis was monitored by measuring the specific activity of plasma digitonin-precipitated sterols, two hours after a pulse of [^{14}C] acetate. A 28% fall in acetate incorporation was reported by 16 days after a 12-day ingestion of the high dose of yoghurt (4 litres per day). Mann proposed the presence of a 'milk factor' to explain the fall in serum cholesterol, such as a 3HMG-CoA reductase inhibitor.[57] Investigating possible candidates for the 'milk factor', Howard and Marks fed lactose ± Ca/Mg, cheese whey or yoghurt to volunteers over a two-week period.[58] The yoghurt, but not the lactose ± Ca/Mg or cheese whey, significantly reduced plasma cholesterol by 5.5%. However, this trial was subject to the same problems of lack of dietary control with substantial changes from the volunteers' habitual diet resulting in a number of confounding factors.

Most of the early studies introduced confounding factors due to the lack of control of the subjects' diet. Hepner et al.[59] performed a study that attempted to control for these. This was a cross-over study in which 720 ml of yoghurt and 750 ml milk were given to the subjects for a four-week period. Significant reductions in plasma cholesterol were observed after the first week of both supplementation periods.[59] The observation that cholesterol levels can significantly fall after acceptance onto a study has been well documented. This is probably due to a conscious or even unconscious modification of the diet by the volunteer due to an awareness of dietary assessment. In an attempt to reduce this, baseline run in periods are essential.[60] Of the early negative studies that have been published, those of Thompson et al.,[61] Massey[62] and McNamara et al.[63] incorporated a run in period. The study performed by McNamara et al.[63] was one of the more carefully designed studies. They investigated the effects of the ingestion of 480 ml unspecified yoghurt and reported no significant cholesterol reduction. This was a well-controlled study, which included a three-week run-in period and four-week intervention period, and investigated the effect of 16 oz of a low fat yoghurt (unspecified microbiological nature) and a non-fermented milk concentrate (as a control). Dietary intake and body weight remained constant and there was no change in serum cholesterol, LDL, HDL or TAG levels.[63] From the studies mentioned above, it can be concluded that there is little evidence that fermented milk products affect serum lipid parameters per se.

3.5.3   Probiotic effect on lipid parameters

Hepner et al.[59] were the first to attempt to discern whether the presence of live bacteria was important for the reported affects of yoghurt on lipid parameters. The aim of the study was to compare the effects of 750 ml pasteurised and unpasteurised yoghurt, using milk as a placebo. After a 12-week intervention period, all treatments significantly reduced plasma cholesterol levels, with milk resulting in a lesser reduction. Unfortunately, the nutritional and microbiological content of the products used was not reported, which severely hampers comparison with other study data. Thompson et al.[61] assessed a wide range of milk-based products including milk laced with *Lactobacillus acidophilus* (titre 1.3×10^7 counts/ml), buttermilk (a milk product fermented with *Streptococcus cremoris* and *Streptococcus lactis* – titre 6.4×10^8 counts/ml) and a yoghurt (fermented with *Lactobacillus bulgaricus* and *Streptococcus thermophilus* – titre 1.2×10^9 counts/ml). One litre of supplement was fed for a three-week period, but no significant change was reported in serum total cholesterol, LDL or HDL. The possible importance of variation in yoghurt cultures stimulated Jaspers et al.[64] to assess the effect of 681 g/day of three strains of a yoghurt fermented with a 1:1 ratio of *Lactobacillus bulgaricus* and *Streptococcus thermophilus*. Two strains (CH-I and CH-II) were taken separately over a 14-day period and two batches of a third strain (SH-IIIA and SH-IIIB) were taken separately over 14 days and 7 days, which ran consecutively, with a 21-day 'washout period' between each of the intervention periods. Body weight remained constant in the subjects and there were only differences in the dietary intakes in minerals and vitamins. Significant falls in serum total and LDL cholesterol levels occurred after one week with one strain (CH-II) and two weeks with SH-IIIA strain. These transient changes could be explained by the effect of commencing a study as discussed previously, or this could be a true difference between the efficacious properties of different strains and indeed different types of probiotics.

Agerbaek et al.[65] tested the effect of 200 ml per day of a yoghurt that contained *Enterococcus faecium* which was shown to have hypocholesterolaemic properties when tested on animals. The study was a parallel design, and the active yoghurt was tested against identical yoghurt that had been chemically fermented with an organic acid (delta-gluco-lactone). The intervention period was for a six-week period in 58 middle-aged men with moderately raised cholesterol levels (5.0–6.5 mmol/l). They observed a 9.8% reduction in LDL cholesterol levels ($P < 0.001$) for the live yoghurt group, which was sustained over the intervention period (Table 3.7). This was a well-controlled study, which excluded many variables such as age, sex and body weight. However, an unforeseen skew in the randomisation resulted in a significantly different baseline total and LDL cholesterol levels in the two groups. The fall in these parameters observed in the live yoghurt group could be ascribed to a regression towards the mean. Another study performed using the same yoghurt and a similar design for a longer period (six months) was carried out in 87 men and women aged 50–70 years.[66] It was reported that at 12 weeks there was a significant drop in LDL cholesterol levels in the group taking the active yoghurt. These reductions were not sustained and this was partly explained by a reduction in the titre of the yoghurt at 12 weeks. At the end of the study, a non-significant reduction in LDL or total cholesterol levels was observed between the two groups. A recent publication investigating the same product used a 200 g/day ingestion of the yoghurt, for an eight-week period in a randomised double-blind placebo controlled trial, with 32 patients who had mild to moderate hypercholesterolaemia. The patients were asked to follow a lipid-lowering diet for eight weeks and were then given the test or control product for two eight-week periods. The results showed a significant reduction of 5.3% (P = 0.004) for total cholesterol and a 6.2% (P = 0.01) reduction in LDL cholesterol levels after the active product. However, the authors did question whether the average reduction of approximately 5% for total and LDL cholesterol was clinically important.[67]

A similar trial of the same product containing *E. faecium* was conducted in 160 middle-aged men and women with moderately raised cholesterol.[60] The study was a randomised, double-blind, multi-centre, placebo-controlled parallel study. Volunteers consumed 200 ml per day of either the active or chemically fermented yoghurt for a 12-week period. Stratified randomisation was used to ensure that the groups were comparable for age, sex, body mass index (BMI) and baseline fasting cholesterol levels. The importance of not changing their dietary habits and lifestyle during the study was emphasised and adherence to the protocol confirmed by dietary assessment. Due to the importance of the titre of the bacterial content of the yoghurt, this was monitored throughout the study. The levels were found to be no lower than 1×10^6 counts/ml at any time tested. During the two-week run-in period, both groups showed significant reductions in blood cholesterol levels ($P<0.05$), but thereafter there was no further change in either of the groups or between the groups at any of the time points. These data are consistent with the conclusions drawn by Rossouw *et al.*,[68] which indicated that apparent effects of some probiotics on blood cholesterol levels may be attributed to reductions in blood lipids observed in subjects who commence an intervention trial. While these reductions are well recognised but are difficult to prevent, they highlight the importance of the inclusion of a run-in period within such studies.

3.5.4   Possible mechanisms of action

Before the possible mechanisms are considered, it is important to highlight that, since viable and biological active micro-organisms are usually required at the target site in the host, it is essential that the probiotics not only have the characteristics that are necessary to produce the desired biological effects, but also have the required viability and are able to withstand the host's natural barriers against ingested bacteria. The classic yoghurt bacteria, *Streptococcus thermophilus* and *Lactobacillus bulgaricus*, are technologically effective, but they do not reach the lower intestinal tract in a viable form. Therefore, intrinsic microbiological properties, such as tolerance to gastric acid, bile and pancreatic juice are important factors when probiotic organisms are considered.[69]

The mechanism of action of probiotics on cholesterol reduction is unclear, but there are a number of proposed possibilities. These include physiological actions of the end products of fermentation SCFAs, cholesterol assimilation, deconjugation of bile acids and cholesterol binding to bacterial cell walls. The SCFAs that are produced by the bacterial anaerobic breakdown of carbohydrate are acetic, propionic and butyric. The physiological effects of these are discussed in more detail in section 3.6.3.

It has been well documented that microbial bile acid metabolism is a peculiar probiotic effect involved in the therapeutic role of some bacteria. The deconjugation reaction is catalysed by conjugated bile acid hydrolase enzyme, which is produced exclusively by bacteria. Deconjugation ability is widely found in many intestinal bacteria including genera *Enterococcus, Peptostreptococcus, Bifidobacterium, Fusobacterium, Clostridium, Bacteroides* and *Lactobacillus*.[70] This reaction liberates the amino acid moiety and the deconjugated bile acid, thereby reducing cholesterol reabsorption, by increasing faecal excretion of the deconjugated bile acids. Many *in vitro* studies have investigated the ability of various bacteria to deconjugate a variety of different bile acids. Grill *et al.*[71] reported *Bifidobacterium longum* as the most efficient bacterium when tested against six different bile salts. Another study reported that *Lactobacillus* species had varying abilities to deconjugate glycocholate and taurocholate.[72] Studies performed on *in vitro* responses are useful but *in vivo* studies in animals and humans are required to determine the full contribution of bile acid deconjugation to cholesterol reduction. Intervention studies on animals and ileostomy patients have shown that oral administration of certain bacterial species led to an increased excretion of free and secondary bile salts.[73, 74]

There is also some *in vitro* evidence to support the hypothesis that certain bacteria can assimilate (take up) cholesterol. It was reported that *L. acidophilus*[75] and *Bif. bifidum*[76] had the ability to assimilate cholesterol in *in vitro* studies, but only in the presence of bile and under anaerobic conditions. However, despite these reports there is uncertainty whether the bacteria are assimilating cholesterol or whether the cholesterol is co-precipitating with the bile salts. Studies have been performed to address this question. Klaver and Meer[77] concluded that the removal of cholesterol from the growth medium in which *L. acidophilus* and a *Bifidobacterium sp.* were growing was not due to assimilation, but due to bacterial bile salt deconjugase activity. The same question was addressed by Tahri *et al.*,[78] with conflicting results, and they concluded that part of the removed cholesterol was found in the cell extracts and that cholesterol assimilation and bile acid deconjugase activity could occur simultaneously.

The mechanism of cholesterol binding to bacterial cell walls has also been suggested as a possible explanation for hypocholesterolaemic effects of probiotics. Hosona and Tono-oka[79] reported *Lactococcus lactis* subsp. *biovar* had the highest binding capacity for cholesterol of bacteria tested in the study. It was speculated that the binding differences were due to chemical and structural properties of the cell walls, and that even killed cells may have the ability to bind cholesterol in the intestine. The mechanism of action of probiotics on cholesterol reduction could be one or all of the above mechanisms with the ability of different bacterial species to have varying effects on cholesterol lowering. However, more research is required to elucidate fully the effect and mechanism of probiotics and their possible hypocholesterolaemic action.

3.6   The effects of prebiotics on coronary heart disease

3.6.1   Prebiotics

In recent years, there has been increasing interest in the important nutritional role of prebiotics as functional food ingredients. This interest has been derived from animal studies that showed markedly reduced TAG and total cholesterol levels when diets containing significant amounts of a prebiotic (oligofructose (OFS)) were fed. A prebiotic is defined as 'a non-digestible food ingredient that beneficially affects the host by selectively stimulating the growth and/or the activity of one or a number of bacteria in the colon, that has the potential to improve health'.[80] Prebiotics, most often referred to as non-digestible oligosaccharides, are extracted from natural sources (e.g. inulin and OFS) or synthesised from disaccharides (e.g. transgalacto-oligosaccharides). The most commonly studied of the prebiotics include inulin and OFS which are found in many vegetables, including onion, asparagus, Jerusalem artichoke and chicory root (Dysseler and Hoffem).[81] These consist of between 2 and 60 fructose molecules joined by $\beta2$–1 osidic linkages, which, due to the nature of this type of linkage, escape digestion in the upper gastrointestinal tract and remain intact but are selectively fermented by colonic microflora. Inulin is currently found as a food ingredient of bread, baked goods, yoghurt and ice-cream because it displays gelling and thickening properties and helps to improve the mouth feel and appearance of lower energy products.[81] In Europe, the estimated intake of inulin and OFS is between 2 g and 12 g per day.[82]

3.6.2   The effect of prebiotics on lipid metabolism in humans

Several studies that have investigated the effects of prebiotics on fasting plasma lipids have generated inconsistent findings (Table 3.8). In studies with individuals with raised blood lipids, three studies showed significant decreases in fasting total and LDL cholesterol, with no significant changes in TAG levels,[83–85] whereas one recent study in type II diabetics did not observe any change in cholesterol levels.[86] In normolipidaemic volunteers, only one study has demonstrated significant changes in both fasting TAG (–27%) and total cholesterol (–5%) levels with inulin[87] with another study showing a significant decrease only in TAG levels.[88] However, Luo et al.[89] and Pedersen et al.[90] have reported no effect of OFS or inulin treatment on plasma lipids levels in young healthy subjects. In a group of middle-aged men and women, lower plasma TAG levels were observed at eight weeks compared with a placebo.[91] It was suggested that the lack of lipid lowering noted in the studies of Pedersen et al.[90] and Luo et al.[89] may be due to insufficient duration of supplementation. In the study by Jackson et al.,[91] follow-up blood samples were taken four weeks after completion of the inulin supplementation period by which time the concentrations of TAG had returned to baseline values, supporting the conclusion that inulin feeding may have been responsible. These findings are in line with observed effects of inulin on lipid levels in animals in which the predominant effect is on TAG rather than cholesterol concentrations.

Raised post-prandial TAG concentrations have also been recognised as a risk

Table 3.8 Summary of human studies to examine the effects of fructan supplementation on blood lipids

Author	Subjects	Fructan	Dose	Study design	Duration	Vehicle	Significant changes observed in blood lipids glucose	
Yamashita et al.[83]	8M and 10F NIDDM	OFS	8 g	DB, parallel	2 wks	Packed coffee drink Canned coffee jelly	↓ TC ↓ LDL-C	↓ glucose
Hidaka et al.[84]	37 (M and F) hyperlipidaemic	OFS	8 g	DB, parallel	5 wks	Confectionery	↓ TC	NS
Canzi et al.[87]	12 M normolipidaemic	Inulin	9 g	Sequential	4 wks	Breakfast cereal	↓TAG ↓ TC	N/A
Luo et al.[89]	12 M normolipidaemic	OFS	20 g	DB cross-over	4 wks	100 g biscuits	NS	NS
Pedersen et al.[90]	66 F normolipidaemic	Inulin	14 g	DB cross-over	4 wks	40 g margarine	NS	N/A
Causey et al.[88]	9 M normolipidaemic	Inulin	20 g	DB cross-over	3 wks	Low fat ice-cream	↓TAG	NS
Davidson et al.[85]	21 M and F hyperlipidaemia	Inulin	18 g	DB cross-over	6 wks	Chocolate bar/paste or coffee sweetener	↓ LDL-C ↓ TC	N/A
Alles et al.[86]	9 M and 11 F Type II diabetes	OFS	15 g	SB cross-over	3 wks	Supplement not specified	NS	NS
Jackson et al.[91]	54 M and F normolipidaemic	Inulin	10 g	DB, parallel	8 wks	Powder added to food and drinks	TAG	↓ insulin

Notes:
M, male; F, female; DB, double blind; N/A, not measured; NS, not significant; SB, single blind; TC, total cholesterol; LDL-C, LDL cholesterol; OFS, oligofructose; TAG, triacylglycerol.

factor for CHD.[2] Data from studies in rats have shown a 40% reduction in postprandial TAG concentrations when diets containing 10% OFS (w/v) were fed.[92] However, very little information regarding the effect of prebiotics on postprandial lipaemia in human subjects are available, although one recent study in middle-aged subjects has shown no effect of inulin treatment on post-prandial TAG levels.[93]

The marked reduction in fasting lipid levels, notably TAGs, observed in animal studies have not been consistently reproduced in human subjects. Only two studies in normolipidaemic subjects have shown significant reductions in fasting TAG levels with inulin,[87, 88] with one study showing a significant effect of inulin treatment over time on fasting TAG levels compared with the placebo group.[91] The amount of fructans used in the human studies in Table 3.8 varies between 9 g and 20 g and this amount is small compared to that which is used in animal studies (50–200 g per kg of rat chow of OFS),[94] which is equivalent to a dose in humans of approximately 50–80 g of OFS/inulin per day. The prebiotic nature of OFS and inulin restricts its dosage in humans to 15–20 g per day since doses greater than this can cause gastrointestinal symptoms such as stomach cramps, flatulence and diarrhoea.[90] It is not known whether, at the levels used in human studies, significant effects would be observed in animals.[95]

The types of food vehicles used to increase the amount of OFS/inulin in the diet differ. In the case of Luo *et al.*,[89] 100 g of biscuits were eaten every day and for Pedersen *et al.*,[90] 40 g of margarine was consumed which may have contributed to the negative findings in blood lipids. In the case of Davidson *et al.*,[85] significant changes in total and LDL cholesterol levels were observed over six weeks with inulin in comparison with the placebo (sugar). The percentage change in each of the lipid parameters was calculated over each of the six-week treatment periods and, unexpectedly, there was an increase in total cholesterol, LDL cholesterol and TAG during the placebo phase. Non-significant falls in these variables were observed during inulin treatment and so when the net changes in the variables were calculated (change during inulin minus change during control treatment) there were significant differences in total and LDL cholesterol between the two treatments. The authors attributed the increase in total and LDL cholesterol levels in the placebo phase to be due to the increased intake of SFAs in the chocolate products that were used as two of the vehicles in the study.[85] In later studies, the use of inulin in its powder form enabled it to be added to many of the foods eaten in the subjects' normal diet without any need for dietary advice, thus avoiding changes in body weight. Since inulin has water binding properties, in its powder form it could be added to orange juice, tea, coffee, yoghurt and soup.[91]

The significant relationship between subjects' initial TAG concentration and percentage change in TAG levels over the eight-week study demonstrated by Jackson *et al.*[91] lends support to the hypothesis that initial TAG levels could be important in determining the degree of the TAG response to inulin. The lack of response in some individuals may be as a result of them being less responsive to inulin, variations in their background diet, or non-compliance with the study protocol. Speculation as to possible reasons for variability in response would be aided by a better understanding of the mechanism of action of inulin on plasma TAG levels.

The length of the supplementation period used in the studies in Table 3.8 may be another factor for inconsistent findings in changes in TAG levels in human subjects with inulin. The studies were conducted over two to eight weeks, with significant effects occurring in only two studies conducted over three to four weeks.[87, 88] The lack of lipid lowering noted in the studies of Alles et al.,[86] Pedersen et al.[90] and Luo et al.[89] may be due to insufficient duration of supplementation. In the study of Jackson et al.,[91] a trend for TAG lowering on inulin treatment was seen some time between four and eight weeks and this reflects the time needed for the composition of the gut microflora to be modified. A four-week wash-out seemed to be sufficient for the TAG concentrations to return to baseline values. This may provide an explanation for the significant findings in TAG levels in the studies of Canzi et al.[87] and Causey et al.[88] who used 3–4 week sequential and cross-over designs with very short wash-out periods.

While some of the studies, to date, support beneficial effects of inulin on plasma TAG, the findings are by no means consistent and more work is required to provide convincing evidence of the lipid-lowering consequences of prebiotic ingestion.

3.6.3   Mechanism of lipid lowering by prebiotics

Prebiotics have been shown to be an ideal substrate for the health-promoting bacteria in the colon, notably bifidobacteria and lactobacilli.[96] During the fermentation process a number of byproducts are produced including gases (H_2S, CO_2, H_2, CH_4), lactate and SCFAs (acetate, butyrate and propionate). The SCFAs, acetate and propionate enter the portal blood stream, where they are utilised by the liver. Acetate is converted to acetyl CoA in the liver and acts as a lipogenic substrate for de novo lipogenesis, whereas propionate has been reported to inhibit lipid synthesis.[97, 98] Butyrate, on the other hand, is taken up by the large intestinal cells (colonocytes) and has been shown to protect against tumour formation in the gut.[99] The type of SCFAs that are produced during the fermentation process is dependent on the microflora which can be stimulated by the prebiotic. Inulin has been shown to increase both acetate and butyrate levels, whereas synthetically produced prebiotics, for example galacto-oligosaccharides, increase the production of acetate and propionate and xylo-oligosaccharides increase acetate only.[99]

Inulin and OFS have been extensively studied to determine the mechanism of action of prebiotics in animals. Early in vitro studies using isolated rat hepatocytes suggested that the hypolipaemic action of OFS was associated with the inhibition of de novo cholesterol synthesis by the SCFA propionate following impairment of acetate utilisation by the liver for de novo lipogenesis.[98] This is in agreement with human studies in which rectal infusions of acetate and propionate resulted in propionate inhibiting the incorporation of acetate into TAGs released from the liver.[100] Fiordaliso et al.[101] demonstrated significant reductions in plasma TAGs, phospholipids and cholesterol in normolipidaemic rats fed a rat chow diet containing 10% (w/v) OFS. The TAG-lowering effect was demonstrated after only one week of OFS and was associated with a reduction in VLDL secretion. TAG and phospholipids are synthesised in the liver by esterification of fatty acids and glycerol-3-phosphate before being made available for assembly into VLDL, suggesting that the hypolipidaemic effect of OFS may be occurring in the liver. The reduction observed in cholesterol levels in the rats was only demonstrated after long-term feeding (16 weeks) of OFS. Recent evidence has suggested that the TAG lowering effect of OFS occurs via reduction in VLDL TAG secretion from the liver due to the reduction in activity of all lipogenic enzymes (acetyl-CoA carboxylase, fatty acid synthase, malic enzyme, ATP citrate lyase and glucose- 6-phosphate dehydrogenase), and in the case of fatty acid synthase, via modification of lipogenic gene expression (see Fig. 3.5).[102]

3.6.4   The effect of prebiotics on glucose and insulin levels

Very little is known about the effects of prebiotics on fasting insulin and glucose levels in humans. Of the supplementation studies conducted in humans, a significant reduction in glucose was observed in NIDDM subjects[83] and a trend for a reduction in glucose was observed in hyperlipidaemic subjects[84] with OFS. However, a recent study has reported no effect of OFS on blood glucose levels in type II diabetics.[86] A significant reduction in insulin levels was observed in healthy middle-aged subjects with inulin, although this was not accompanied by changes in plasma glucose levels.[91] The effect of the ingestion of acute test meals containing OFS on blood glucose, insulin and C-peptide levels in healthy adults showed a trend for a lower glycaemic response and peak insulin levels following the OFS enriched meals.[103]

The mechanism of action of prebiotics on lowering glucose and insulin levels has been proposed to be associated with the SCFAs, especially propionate. A significant reduction in post-prandial glucose concentrations was observed following both acute and chronic intakes of propionate-enriched bread.[104] The effect of propionate intake on post-prandial insulin levels was not investigated. A recent animal study has shown an attenuation of both post-prandial insulin and glucose levels following four weeks of feeding with OFS. These effects were attributed to the actions of OFS on the secretion of the gut hormones, glucose-dependent insulinotropic polypeptide (GIP) and glucagon-like peptide 1 (GLP-1).[92] These hormones are secreted from the small intestine (GIP) and the terminal ileum and colon (GLP-1) and contribute to the secretion of insulin following a meal in the presence of raised glucose levels.[105]

In summary, the mechanisms of action of prebiotics, especially inulin and OFS, have been determined largely from animal studies. Present data suggest inhibition of *de novo* lipogenesis as the primary mode of action of prebiotics in mediating their lipid-lowering effects via down-regulation of the enzymes involved. If this is the case, more modest or inconsistent effects might be expected in humans, in whom *de novo* lipogenesis is extremely low, or variable depending on their background diet. In animal studies, rats are fed a diet that is low in fat and high in carbohydrate and so *de novo* lipogenesis is an up-regulated pathway in these

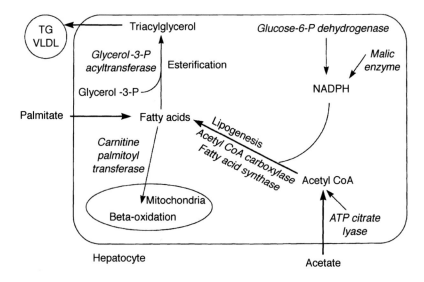

Fig. 3.5   Hepatic fatty acid metabolism.

animals for the synthesis of fatty acids. It is interesting to note that, when rats are fed OFS along with a high fat diet typical of the Western-style diet, TAG levels are thought to be decreased by a different mechanism involving the enhanced clearance of TAG-rich lipoproteins. An increased GIP secretion in OFS-treated rats was observed by Kok et al.[92] and this gut hormone has been shown to enhance the activity of LPL, the principal enzyme involved in the clearance of TAG-rich lipoproteins following the ingestion of fat.[106] The release of GIP and GLP-1 in the intestine and colon may act as mediators of the systemic effect of prebiotics such as inulin and OFS, on blood lipid, insulin and glucose levels. However, further work is required to determine the metabolic pathways that are influenced by prebiotics. Their effect on gastrointestinal kinetics such as gastric emptying and its modification of the levels of TAG-rich lipoproteins (CMs) and glucose in the circulation has recently been proposed as a potential modulator of systemic effects.[107] Therefore the design of future studies to investigate the effect of prebiotics in humans should consider the choice of subjects, length of supplementation period and type of vehicle used to increase the intake of prebiotics in the diet, as these variables may influence the outcome of the study.

3.7   The effects of synbiotics on coronary heart disease

A synbiotic is defined as 'a mixture of a prebiotic and a probiotic that beneficially affects the host by improving the survival and the implantation of live microbial dietary supplements in the gastrointestinal tract, by selectively stimulating the growth and/or by activating the metabolism of one or a limited number of health promoting bacteria'.[108] The use of synbiotics as functional food ingredients is a new and developing area and very few human studies have been performed looking at their effect on risk factors for CHD. Research conducted so far with synbiotics have looked at their effect on the composition of the gut microflora. In one study in healthy subjects, a fermented milk product containing a *Bifidobacterium spp.* with or without 18 g of inulin was given daily for 12 days.[109] The authors concluded that the administration of the fermented milk product (probiotic) substantially increased the proportion of bifidobacteria in the gut, but that this increase was not enhanced by the addition of 18 g of inulin. The composition of the gut microflora was then assessed two weeks after completing the supplementation period and it was found that subjects who received the fermented milk product and inulin maintained their bifidobacterial population in the gut compared with the subjects receiving the fermented milk product only. Although a synergistic effect on bifidobacteria in the gut was not observed with the synbiotic, these results suggest that either there was better implantation of the probiotic or a prebiotic effect on indigenous bifidobacteria.[109] Maintenance of high numbers of bifidobacteria in the gut may be beneficial in terms of healthy gut function; however, its effect on the lowering of blood lipid levels remains to be investigated. A more recent study has shown that a lower dose of prebiotic (2.75 g) added to a lactobacillus fermented milk was able to increase significantly the number of bifidobacteria when fed over a seven-week period in healthy human subjects.[108] If this effect was a result of the synbiotic product used in this study, the use of lower doses of prebiotics that can be used in synbiotic preparations should help to reduce gastrointestinal complaints observed when a prebiotic is used alone and improve the acceptability of these types of products by the general public.

3.8 Future trends

In recent years, a number of food manufacturers in the USA and Europe have been interested in the commercial opportunities for foodstuffs containing health-promoting probiotics and prebiotics. These food ingredients have received attention for their beneficial effects on the gut microflora and links to their systemic effects on the lowering of lipids known to be risk factors for CHD, notably cholesterol and TAG. Early attention was given to the incorporation of probiotics into dairy products such as fermented milk products (Yakult and Actimel Orange milk drink) and yoghurts, whose market is currently estimated at US$2 billion.[110] Prebiotics are currently gaining interest and this has opened up the market away from the dairy industry to other areas of the food industry since prebiotics can be baked into bread, cereals, cakes, biscuits and even added to soups. Synbiotics have also generated interest with some food manufacturers who are exploiting the effects of combining a prebiotic with a probiotic (Symbalance and PROBIOTIC plus oligofructose).

While consumers are interested in the concept of improving their health and well-being through diet, this is not quite so straightforward as originally thought. Recent bad press regarding food safety and the introduction of GM foods has made the public sceptical about new ingredients in foods. For progress to be made, the consumers need to be educated about the various health benefits and how they will be able to use these products in their own diet without adverse consequences. Although the introduction of probiotics onto the supermarket shelves are slowly being accepted in the UK population, carefully controlled nutrition studies need to be carried out to determine the beneficial effects of prebiotics, probiotics and synbiotics before substantial health claims can be made.

To make these foods attractive to the consumer, the products need to be priced in such a way that they are accessible to the general public. The low doses of prebiotics and probiotics needed to help maintain a healthy gut microflora should be made available to the general public, whereas products that contain higher amounts of prebiotics in order to help reduce blood lipids will need to be restricted in order for the appropriate population group to be targeted.

3.9   Sources of further information and advice

3.9.1   General biochemistry and metabolic regulation text books
DEVLIN T M *Textbook of Biochemistry with Clinical Correlations*, 3rd edn, New York, John Wiley, 1992.
FRAYN K N *Metabolic Regulation: A Human Perspective*, London, Portland Press, 1996.

3.9.2   Published reports on dietary intake and coronary heart disease
Department of Health, *Nutritional Aspects of Cardiovascular Disease, Report on Health and Social Subjects*, no. 46, London, HMSO.

3.9.3   Prebiotic, probiotic and synbiotic information
ADAMS C A 'Nutricines', *Food Components in Health and Nutrition*, Nottingham, Nottingham University Press, 1999.
HASLER C M and WILLIAMS C M 'Prebiotics and probiotics: where are we today?', *BJN*, **80**, supp. 2.
LEEDS A R and ROWLAND I R *Gut Flora and Health: Past, Present and Future*, International Congress and Symposium Series 219, Royal Society of Medicine, 1996.
SADLER M J and SALTMARSH M *Functional Foods: The Consumer, the Products and the Evidence*, Bath, Bookcraft, 1997.

3.10   References

1 KEYS A 'Coronary artery disease in seven countries', *Circulation* (1970), **41**, 11–1211.
2 PATSCH J R, MEISENBÖCK G, HOPFERWEISER T, MULHBERGER V, KNAPP E, DUNN J K, GOTTO A M and PATSCH W. 'Relation of triglyceride metabolism and coronary artery disease: studies in the postprandial state', *Arterioscler Thromb* (1992), **12**, 1336–1345.

88 Functional dairy products

3 GRIFFIN B A and ZAMPELAS A 'Influence of dietary fatty acids on the atherogenic lipoprotein phenotype', *Nut Res Rev* (1995), **8**, 1–26.

4 GIBSON G R and BEAUMONT A 'An overview of human colonic bacteriology in health and disease'. In A R Leeds and I R Rowland (eds), pp. 3–11, *Gut Flora and Health: Past, Present and Future*, International Congress and Symposium Series 219, London, Royal Society of Medicine (1996).

5 KEYS A *Seven Countries: A Multivariate Analysis of Death and Coronary Heart Disease*, Harvard University Press, Cambridge, MA (1980).

6 WORLD HEALTH ORGANISATION, 'World Health Statistics Annual 1989', Geneva, WHO (1989).

7 MARTIN M J, BROWNER W S, WENTWORTH J, HULLEY S B. and KULER LH. 'Serum cholesterol, blood pressure and mortality: implications from a cohort of 361,662 men', *Lancet* (1986) **2**, 933–936.

8 SHEPHERD J, COBBE S M, FORD I, ISLES C G, LORIMER A R, MACFARLANE P W, MCKILLOP J H. and PACKARD C J 'Prevention of coronary heart disease with pravastatin in men with hypercholesterolaemia', *New Engl J Med* (1995) **333**(20), 1301–1307.

9 SHAPER A G, POCOCK S J, PHILIPS A N and WALKER M 'Identifying men at high risk of heart attacks – strategy for use in general practice', *BMJ* (1986) **293** (6545), 474–479.

10 FIELDING P E and FIELDING C J 'Dynamics of lipoprotein transport in the circulatory system'. In D E Vance and J Vance (eds), pp. 427–459, *Biochemistry of Lipids, Lipoproteins and Membranes*, Amsterdam, Elsevier (1991).

11 SCHNEIDER W J 'Removal of lipoproteins from plasma'. In D E Vance and J Vance (eds), pp. 461–487, *Biochemistry of Lipids, Lipoproteins and Membranes*, Amsterdam, Elsevier (1991).

12 DEVLIN T M *Textbook of Biochemistry with Clinical Correlations*, 3rd edn, New York, John Wiley (1992).

13 REDGRAVE T G *Gastrointestinal Physiology IV*, Baltimore, University Park Press (1983).

14 BRASAEMLE D L, CORNLEY-MOSS K and BENSADOUN A 'Hepatic lipase treatment of chylomicron remnants increases exposure of apolipoprotein E', *J Lipid Res* (1993) **34**, 455–463.

15 KARPE F and HULTIN M 'Endogenous triglyceride-rich lipoproteins accumulate in rat plasma when competing with a triglyceride emulsion for a common lipolytic pathway', *J Lipid Res* (1995) **36**, 1557–1566.

16 GRIFFIN B A and PACKARD C J 'Metabolism of VLDL and LDL subclasses', *Curr Opin Lipidol* (1994) **5**, 200–206.

17 TALL A R 'Plasma cholesterol ester transfer protein', *J Lipid Res* (1993) **27**, 361–367.

18 GENSINI G F, COMEGLIO M and COLELLA A 'Classical risk factors and emerging elements in the risk profile for coronary artery disease', *Eur Heart J* (1998) **19** (supp. A), A52–A61.

19 ASSMANN G, CULLEN P and SCHULTE H 'The Münster Heart Study (PROCAM): Results of follow-up at 8 years', *Eur Heart J* (1998) **19** (supp. A), A2–A11.

20 AUSTIN M A 'Plasma triglyceride and coronary artery disease', *Arterioscl Thromb* (1991) **11**, 2–14.

21 WILSON P W F 'Established risk factors and coronary artery disease: the Framingham study', *Am J Hypertens* (1994) **7**, 7S–12S.

22 O'MEARA N M, LEWIS G F, CABANA V G, IVERIUS P H, GETZ G S and POLONSKY K S 'Role of basal triglyceride and high density lipoprotein in determination of postprandial lipid and lipoprotein responses', *J Clin Endocrin Metab* (1991) **75**, 465–471.

23 KARPE F, STEINER G, UFFELMAN K, OLIVECRONA T and HAMSTEN A 'Postprandial lipoproteins and progression of coronary heart disease', *Atherosclerosis* (1994) **106**, 83–97.

24 ZILVERSMIT D 'Atherogenesis: a postprandial phenomenon', *Circulation* (1979) **60**, 473–485.

25 MEISENBÖCK G and PATSCH J R 'Coronary artery disease: synergy of triglyceride-rich lipoproteins and HDL', *Cardiovascular Risk Factors* (1991) **1**, 293–299.
26 AUSTIN M A, BRESLOW J L, HENNEKENS C H, BURLING J E, WILLETT W C and KRAUSS R M 'Low density lipoprotein subclass patterns and risk of myocardial infarction', *JAMA* (1988) **260**, 1917–1921.
27 AUSTIN M A, HOKANSON J E and BRUNZELL J D 'Characterisation of lowdensity lipoprotein subclasses: methodological approaches and clinical relevance', *Curr Opin Lipidol* (1994) **5**, 395–403.
28 GRIFFIN B A 'Low density lipoprotein heterogeneity', *Ballière's Clinical Endocrinology and Metabolism* (1995) **9**, 687–703.
29 CHAPMAN M J, GUÉRIN M and BRUCKERT E 'Atherogenic, dense lowdensity lipoproteins: pathophysiology and new therapeutic approaches', *Eur Heart J* (1998) **19** (supp. A), A24–A30.
30 REAVEN G M, CHEN Y-D I, JEPPENSEN J, MAHEUX P and KRAUSS R M 'Insulin resistance and hyperinsulinaemia in individuals with small, dense, low density lipoprotein particles', *J Clin Invest* (1993) **92**, 141–146.
31 MENSINK R P 'Effects of individual saturated fatty acids on serum lipids and lipoprotein concentrations', *Am J Clin Nutr* (1993) **57**(s), 711S–714S.
32 HEGSTED D M, ANSMAN L M, JOHNSON J A and DALLAL G E 'Dietary fat and serum lipids: an evaluation of the experimental data', *Am J Clin Nutr* (1993) **57** 875–883.
33 HAYES K C and KHOSLA P 'Dietary fatty acids thresholds and cholesterolaemia', *FASEB J* (1992) **6**, 2600–2607.
34 DEPARTMENT OF HEALTH, *Nutritional Aspects of Cardiovascular Disease: Report of the Cardiovascular Review Group Committee on Medical Aspects of Food Policy*, London, HMSO (1994). Report on Health and Social Subjects 46.
35 GREGORY J, FOSTER K, TYLER H and WISEMAN M *The Dietary and Nutritional Survey of British Adults: A Survey of the Dietary Behaviour, Nutritional Status and Blood Pressure of Adults aged 16 to 64 living in Great Britain'*, Office of Population Census and Surveys, Social Survey Division, London, HMSO (1990).
36 GRUNDY S M, NIX D, WHELAN M F and FRANKLIN L 'Comparison of three cholesterol lowering diets in normolipidaemic men', *JAMA* (1986) **256**, 2351.
37 BERRY E M 'The effects of nutrients on lipoprotein susceptibility to oxidation', *Current Opinions in Lipidology* (1992) **3**, 5–11.
38 MENSINK R P and KATAN M B 'Effect of dietary *trans* fatty acids in high density and low density lipoprotein cholesterol levels in healthy subjects', *N Engl J Med* (1990) **323**, 439–445.
39 DYBERG J and BANG H O 'Haemostatic function and platelet polyunsaturated fatty acids in Eskimos', *Lancet* (1979) **2**, 433–435.
40 BURR M L, FEHILY A M, GILBERT J F *et al.* 'Effects of changes of fat, fish and fibre intakes on frequency of myocardial infarctions: DART Study', *Lancet* (1989) **2**, 757–762.
41 HARRIS W S '*n*-3 Fatty acids and serum lipoproteins: human studies', *Am J Clin Nutr* (1997) **65** (S), 1645S–1654S.
42 SISCOVICK D S, RAGHUNATHAN T E, KING I and WEINMAN S 'Dietary intake and cell membrane levels of long chain *n*-3 polyunsaturated fatty acids and the risk of primary cardiac arrest', *JAMA* (1996) **274**, 1363–1367.
43 DE DECKERE E A M, KORVER O, VERSCHUREN P M and KATAN M B 'Health aspects of fish and n-3 polyunsaturated fatty acids from plant and marine origin', *Euro J Clin Nutr* (1998) **52**, 749–753.
44 EEC SCIENTIFIC COMMITTEE FOR FOOD, *Reference Nutrient Intakes for the European Community*, Brussels, EC (1992).
45 MENSINK R P and KATAN M B 'Effects of monosaturated fatty acids versus complex carbohydrates on high density lipoproteins in healthy men and women', *Lancet* (1987) **1**, 122–125.

46 GRUNDY S M 'Comparison of monounsaturated fatty acids and carbohydrates for lowering plasma cholesterol', *N Engl J Med* (1986) **314**, 745–748.

47 MORRIS J N, MARR J W and CLAYTON D G 'Diet and heart: a postscript'. *BMJ* (1977) **2**, 1307–1314.

48 KUSHI L H, LEW R A, STARE F J *et al.*, 'Diet and 20-year mortality from coronary heart disease: the Ireland–Boston Diet–Heart Study', *N Engl J Med* (1985) **312**, 811–818.

49 RIEMERSMA R A, WOOD D A, MACINTYRE C C, ELTON R A, GEY K F and OLIVER M F 'Risk of angina pectoris and plasma concentrations of vitamin A, C and E and carotene', *Lancet* (1991) **337**, 1–5.

50 STEPHENS N G, PARSONS A, SCHOFIELD P M, KELLY F and CHEESEMAN K 'Randomised controlled trial of vitamin E in patients with coronary disease: Cambridge Heart Antioxidant Study (CHAOS)' (1996) *Lancet*, **348**, 781–786.

51 ELLIOT P, DYER A and STAMLER J 'Correcting for regression dilution in INTERSALT', *Lancet* (1993) **342**, 1123.

52 LAW M R, FROST C D and WALD N J 'By how much does dietary salt reduction lower blood pressure? I-III- Analysis of observational data among populations', *BMJ* (1991) **302**, 811–824.

53 GAZIANO J M, BURING J E, BRESLOW J L, GOLDHABER S Z, ROSNER B , VANDENBURGH M, WILLIET W C and HENNEKENS C H 'Moderate alcohol intake, increased levels of high density lipoprotein and its subfractions, and decreased risk of myocardial infarction', *New Eng J Med* (1993) **329**, 1829–1834.

54 LAW M and WALD N 'Why heart disease mortality is low in France: the time lag explanation', *BMJ* (1999) **318**, 1471–1480.

55 URGERT R, MEYBOOM S, KUILMAN M, REXWINKEL H, VISSERS M N, KLERK M and KATAN M B 'Comparison of effect of cafetiere and filter coffee on serum concentrations of liver aminotransferases and lipids: six month randomised controlled trial', *BMJ* (1996) **313**, 1362–1366.

56 MANN G V and SPOERRY A 'Studies of a surfactant and cholesteremia in the Maasai', *Am J Clin Nutr* (1974) **27**, 464–469.

57 MANN G V 'A factor in yoghurt which lowers cholesteremia in man', *Atherosclerosis* (1977) **26**, 335–340.

58 HOWARD A N and MARKS J 'Effect of milk products on serumcholesterol', *Lancet* (1979) **II**, 957.

59 HEPNER G, FRIED R, ST JEOR S, FUSETTI L and MORIN R 'Hypocholesterolemic effect of yogurt and milk', *Am J Clin Nutr* (1979) **32**, 19–24.

60 SESSIONS V A, LOVEGROVE J A, TAYLOR G R J, DEAN T S, WILLIAMS C M, SANDERS T AB, MACDONALD I and SALTER A 'The effects of a new fermented milk product on total plasma cholesterol: LDL cholesterol and apolipoprotein B concentrations in middle aged men and women', *Proceedings of the Nutrition Society* (1997) **56**, 120A.

61 THOMPSON L U, JENKINS D J A, AMER V, REICHERT R, JENKINS A and KAMULSKY J 'The effect of fermented and unfermented milks on serum cholesterol', *Am J Clin Nutr* (1982) **36**, 1106–1111.

62 MASSEY L K 'Effect of changing milk and yogurt consumption on human nutrient intake and serum lipoproteins', *J Dairy Sci* (1984) **67**, 255–262.

63 MCNAMARA D J, LOWELL A E and SABB J E 'Effect of yogurt intake on plasma lipid and lipoprotein levels in normolipidemic males', *Atherosclerosis* (1989) **79**, 167–171.

64 JASPERS D A, MASSEY L K and LUEDECKE L O 'Effect of consuming yogurts prepared with three culture strains on human serum lipoproteins', *J Food Sci* (1984) **49**, 1178–1181.

65 AGERBAEK M, GERDES L U and RICHELSEN B 'Hypocholesterolaemic effects of a new product in healthy middle-aged men', *Eur J Clin Nutr* (1995) **49**, 346–352.

66 RICHELSEN B, KRISTENSEN K and PEDERSEN S B 'Long-term (6 months) effect of a new fermented milk product on the level of plasma lipoproteins: a placebo-controlled and double blind study', *Eur J Clin Nutr* (1996) **50**, 811–815.

67 BERTOLAMI M C 'Evaluation of the effects of a new fermented milk product (Gaio) on primary hypercholesterolemia', *Eur J Clin Nutr* (1999) **53**, 97–101.

68 ROSSOUW J E, BURGER E-M, VAN DER VYVER P and FERREIRA J J 'The effect of skim milk yoghurt, and full cream milk on human serum lipids', *Am J Clin Nutr* (1981) **34**, 351–356.

69 HUIS IN'T VELD J H J and SHORTT C 'Selection criteria for probiotic microorganisms'. In A.R. Leeds and I.R. Rowland (eds), pp. 27–36, *Gut Flora and Health: Past, Present and Future*, International Congress and Symposium Series No. 219, London, New York, Royal Society of Medicine Press (1996).

70 HYLEMOND P B 'Metabolism of bile acids in intestinal microflora'. In H. Danielson and J. Sjovall (eds), pp. 331–343, *Sterols and Bile Acids*, New York, Elsevier Science (1985).

71 GRILL J P, MANGINOT-DURR C, SCHNEIDER F and BALLONGUE J 'Bifidobacteria and probiotic effects: action of *Bifidobacterium* species on conjugated bile salts', *Curr Microbiol* (1995) **31**, 23–27.

72 GILLILand S E, NELSON C R and MAXWELL C 'Assimilation of cholesterol by *Lactobacillus acidophilus*', *Appl Environ Microbiol* (1985) **49**(2), 377–381.

73 DE SMET I, DE BOEVER P and VERSTRAETE W 'Cholesterol lowering in pigs through enhanced bacterial bile salt hydrolase activity', *BJN* (1998) **79**, 185–194.

74 MARTEAU P, GERHARDT M F, MYARA A, BOUVIER E, TRIVIN F and RAMBAUD J C 'Metabolism of bile salts by alimentary bacteria during transit in human small intestine', *Microbiol Ecol in Health and Disease* (1995) **8**, 151–157.

75 GILLILand S E, NELSON C R and MAXWELL C 'Assimilation of cholesterol by *Lactobacillus acidophilus*', *Appl and Environ Microbiol* (1985) **49** (2), 377–381.

76 RASIC J L, VUJICIC I F, SKRINJAR M and VULIC M 'Assimilation of cholesterol by some cultures of lactic acid bacteria and bifidobacteria', *Biotech Lett* (1992) **14** (1), 39–44.

77 KLAVER F A M and VAN DER MEER R 'The assumed assimilation of cholesterol by lactobacilli and *Bifidobacterium bifidum* is due to their bile salt-deconjugating activity', *Appl Environ Microbiol* (1993) **59**(4), 1120–1124.

78 TAHRI K, CROCIANI J, BALLONGUE J and SCHNEIDER F 'Effects of three strains of bifidobacteria on cholesterol', *Lett Appl Microbiol* (1995) **21**, 149–151.

79 HOSONO A and TONO-OKA T 'Binding of cholesterol with lactic acid bacterial cells', *Milchwissenschaft* (1995) **50**(20), 556–560.

80 GIBSON G R and ROBERFROID M B 'Dietary modulation of the human colonic microbiota: introducing the concept of prebiotics', *J Nutr* (1995) **125**, 1401–1412.

81 DYSSELER P and HOFFEM D 'Inulin, an alternative dietary fibre: properties and quantitative analysis', *Eur J Clin Nutr* (1995) **49**, S145–S152.

82 VAN LOO J, COUSSEMENT P, DE LEENHEER L, HOEBREGS H and SMITS G 'On the presence of inulin and oligofructose as natural ingredients in the Western diet', *Crit Rev Food Sci and Nutr* (1995) **35**, 525–552.

83 YAMASHITA K, KAWAI K and ITAKURA M 'Effects of fructo-oligosaccharides on blood glucose and serum lipids in diabetic subjects', *Nutr Res* (1984) **4**, 961–966.

84 HIDAKA H, TASHIRO Y and EIDA T 'Proliferation of bifidobacteria by oligosaccharides and their useful effect on human health', *Bifidobacteria Microflora* (1991) **10**, 65–79.

85 DAVIDSON M H, SYNECKI C, MAKI K C and DRENNEN K B 'Effects of dietary inulin in serum lipids in men and women with hypercholesterolaemia', *Nutr Res* (1998) **3**, 503–517.

86 ALLES M S, DE ROOS N M, BAKX J C, VAN DE LISDONK E, ZOCK P L and HAUTVAST J G A J 'Consumption of fructooligosaccharides does not favourably affect blood glucose and serum lipid concentrations in patients with type 2 diabetes', *Am J Clin Nutr* (1999) **69**, 64–69.

87 CANZI E, BRIGHENTI F, CASIRAGHI M C, DEL PUPPO E and FERRARI A 'Prolonged consumption of inulin in ready to eat breakfast cereals: effects on intestinal ecosystem, bowel habits and lipid metabolism', *Cost 92. Workshop on Dietary Fibre and Fermentation in the Colon*, Helsinki (1995).

88 CAUSEY J L, GALLAHER D D and SLAVIN J L 'Effect of inulin consumption on lipid and glucose metabolism in healthy men with moderately elevated cholesterol', *FASEB* (1998) **12**, 4737.

89 LUO J, RIZKALLA S W, ALAMOWITCH C, BOUSSAIRI A, BLAYO A, BARRY J-L, LAFFITTE A, GUYON F, BORNET F R J and SLAMA G 'Chronic consumption of short-chain fructooligosaccharides by healthy subjects decreased basal hepatic glucose production but no effect on insulin stimulated glucose metabolism', *Am J Clin Nutr* (1996) **63**, 939–945.

90 PEDERSEN A, SANDSTRÖM B and VAN AMELSVOORT J M M 'The effect of ingestion of inulin on blood lipids and gastrointestinal symptoms in healthy females', *BJN* (1997) **78**, 215–222.

91 JACKSON K G, TAYLOR G R J, CLOHESSY A M and WILLIAMS C M 'The effect of the daily intake of inulin on fasting lipid, insulin and glucose concentrations in middle-aged men and women', *BJN* (1999) **82**, 23–30.

92 KOK N N, MORGAN L M, WILLIAMS C M, ROBERFROID M B, THISSEN J-P and DELZENNE N M 'Insulin, glucagon-like peptide 1, glucose dependent insulinotropic polypeptide and insulin-like growth factor 1 as putative mediators of the hypolipidemic effect of oligofructose in rats', *J Nutr* (1998) **128**, 1099–1103.

93 WILLIAMS C M 'Effects of inulin on blood lipids in humans', *J Nutr* (1998) **7**, S1471–S1473.

94 ROBERFROID M 'Dietary fiber, inulin and oligofructose: a review comparing their physiological effects', *Crit Rev Food Sci and Nutr* (1993) **33**, 102–148.

95 DELZENNE N M, KOK N, FIORDALISO M-F, DEBOYSER D M, GOETHALS F M and ROBERFROID R M 'Dietary fructo-oligosaccharides modify lipid metabolism', *Am J Clin Nutr* (1993) **57**, 820S.

96 GIBSON G R and MCCARTNEY A L 'Modification of gut flora by dietary means', *Biochem Soc Trans* (1998) **26**, 222–228.

97 WOLEVER T M S, BRIGHENTI F, ROYALL D, JENKINS A L and JENKINS D J A 'Effect of rectal infusion of short chain fatty acids in human subjects', *Am J Gastroenterol* (1989) **84**, 1027–1033.

98 DEMIGNÉ C, MORAND C, LEVRAT M-A, BESSON C, MOUNDRAS C and RÉMÉSEY C 'Effect of propionate on fatty acid and cholesterol synthesis and on acetate metabolism in isolated rat hepatocytes', *BJN* (1995) **74**, 209–219.

99 VAN LOO J, CUMMINGS J, DELZENNE N, ENGLYST H, FRANCK A, HOPKINS M, KOK N, MACFARLANE G, NEWTON D, QUIGLEY M, ROBERFROID M, VAN VLIET T and VAN DEN HEUVEL E 'Functional food properties of nondigestible oligosaccharides: a consensus report from the ENDO project (DGXII AIRII-CT94-1095)', *BJN* (1999) **81**, 121–132.

100 WOLEVER T M S, SPADAFORA P J, CUNNANE S C and PENCHARZ P B 'Propionate inhibits incorporation of colonic [1,2-13C]acetate into plasma lipids in humans', *Am J Clin Nutr* (1995) **61**, 1241–1247.

101 FIORDALISO M-F, KOK N, DESAGER J-P, GOETHALS F, DEBOYSER D, ROBERFROID R M and DELZENNE N 'Dietary oligofructose lowers triglycerides, phospholipids and cholesterol in serum and very low density lipoproteins in rats', *Lipids* (1995) **30**, 163–167.

102 DELZENNE N M and KOK N 'Effect of non-digestible fermentable carbohydrates on hepatic fatty acid metabolism', *Biochem Soc Trans* (1998) **26**, 228–230.

103 RUMESSEN J J, BODE S, HAMBERG O and GUDMAND-HOYER E 'Fructans of the Jerusalem artichokes: intestinal transport, absorption, fermentation, and influence on blood glucose, insulin, and C-peptide responses in healthy subjects', *Am J Clin Nutr* (1990) **52**, 675–681.

104 TODESCO T, RAO A V, BOSELLO O and JENKINS D J A 'Propionate lowers blood glucose and alters lipid metabolism in healthy subjects', *Am J Clin Nutr* (1991) **54**, 860–865.

105 MORGAN L M 'The role of gastrointestinal hormones in carbohydrate and lipid metabolism and homeostasis: effects of gastric inhibitory polypeptide and glucagon-like peptide-1', *Biochem Soc Trans* (1998) **26**, 216–222.

106 KNAPPER J M, PUDDICOMBE S M, MORGAN L M and FLETCHER J M 'Investigations into the actions of glucose dependent insulinotrophic polypeptide and glucagon-like peptide-1 (7-36) amide on lipoprotein lipase activity in explants of rat adipose tissue', *J Nutr* (1995) **125**, 183–188.

107 DELZENNE N 'The hypolipidaemic effect of inulin: when animal studies help to approach the human problem', *BJN* (1999) **82**, 3–4.

108 ROBERFROID M B 'Prebiotics and synbiotics: concepts and nutritional properties', *BJN* (1998) **80** (supp. 2), S197–S202.

109 BOUHNIK Y, FLOURIE B, RIOTTOT M, BISETTI N, GAILING M, GUIBERT A, BORNET F and RAMBAUD J 'Effect of fructo-oligosaccharides ingestion on fecal bifidobacteria and selected metabolic indexes of colon carcinogenesis in healthy humans', *Nutrition and Cancer* (1996) **26**, 21–29.

110 YOUNG J 'European market developments in prebiotic- and probioticcontaining foodstuffs', *BJN* (1998) **80** (supp. 2), S231–S233.

111 BAZARRE T L, WU S L and YUHAS J A 'Total and HDL-cholesterol concentration following yoghurt and calcium supplementation', *Nutritional Reports International* (1983) **28**, 1225–1232.

112 TAYLOR G B J and WILLIAMS C M 'Effect of probiotics and prebiotics on blood lipids', *BJN* (1998) **80** (supp. 5), S225–S230.

4

Osteoporosis

R. Wood, Tufts University, USA

4.1   Introduction

Osteoporosis is a metabolic bone disease characterized by a decrease in bone mass (osteopenia) with decreased bone mineral density (BMD) and enlargement of bone spaces producing porosity and fragility, which leads to an increased risk of bone fracture. There are many different etiologies of osteoporosis, but the most common is due to a universal age-dependent loss of bone in adult populations.

Proposed risk factors for osteoporosis include genetic, nutritional, lifestyle and endocrine factors. The role of genetic factors is illustrated by differences in BMD between races and sexes, and familial similarities in BMD. Nutritional factors known to influence the risk of osteoporosis include low calcium intake, high alcohol intake, high caffeine and high sodium intake. Lifestyle factors that can contribute to osteoporosis include cigarette smoking and low physical activity. An important endocrine factor is estrogen deficiency (postmenopausal and oophorectomized women).

4.2   The epidemiology of osteoporosis

Osteoporosis is the most prevalent metabolic bone disease in the United States and in other developed countries. Osteoporosis is a cross-racial and emerging global public health problem that especially affects older women, but also needs to be recognized as very common in older men. Osteoporosis-related skeletal fractures, especially of the hip, are a significant cause of morbidity and mortality in the elderly in both the USA and elsewhere. The prevalence of a vertebral fracture in women aged 65 years in the USA has been estimated to be 27% (Cooper *et al.*,

1993). Among elderly persons in the USA who sustain a hip fracture, 27% enter a nursing home for the first time within a year of the event (Cooper, 1997). Beyond the evident personal pain and suffering as well as family disruption brought about by these bone fractures, the added financial burden to the national health-care bill is estimated to be 13 billion dollars per year in the USA (Ray *et al.*, 1997). Given the rising number of elderly around the globe, development of an effective osteoporosis prevention program is a key factor in the management of this growing public health problem.

There are many environmental and genetic factors that have been shown to be associated with an increased risk of osteoporotic fracture. A very important determinant of the risk of fracture is low BMD. The prevalence of low femoral BMD of a representative sample of the elderly (50 years and older) in the USA has been reported from the NHANES III survey (Looker *et al.*, 1995, 1997). Based on this sample, the number of older women in the USA with osteoporosis (defined as a bone mineral density <2.5 SD below the mean BMD of young adult non-Hispanic white women) was 6 million, while 1 million men could be classified with osteoporosis using the same young female BMD cut-off value. Thus, although osteoporosis is clearly more common in older women than older men, there is still a substantial number of older men affected by this disease. Moreover, osteoporosis is a universal phenomenon of aging affecting all races, although currently more white women than blacks or Hispanics in the USA have osteoporosis (Looker *et al.*, 1995). In the next 50 years, the proportion of elderly minority members in the US population will double, leading to a proportionate increase in the osteoporotic burden in these groups. Likewise, by the year 2050, there will be an estimated 6.3 million osteoporotic fractures worldwide and 50% of these will occur in Asia. Racial differences in hip fracture incidence are substantial. Differences in bone mass between the races, such as between whites and blacks, probably have an important effect on the risk of bone fracture. However, other more subtle genetic factors may also play a significant role in determining the risk of osteoporotic fracture at any given level of BMD.

4.2.1   Osteoporosis in African–Americans

It is generally appreciated that black populations have high bone mineral density and a low prevalence of osteoporotic fracture, despite often low dietary calcium intakes. This observation clearly indicates that there is more to the osteoporosis equation than calcium intake alone. There is very little systematic information on bone mineral density and fracture risk in African populations, but some information is available on African–American populations, which will be highlighted here to indicate the potential impact of racial differences on bone health. On the whole, African–American (black) women have more dense bones than white women (Looker *et al.*, 1995). As a consequence, the risk of osteoporosis is lower in black women. Differences in bone mass between black and white men are also evident. For example, whole body bone mineral content is greater in black than in white men, even after controlling for lean body mass and height (Barondess *et al.*, 1997).

Because of these racial differences resulting in higher average BMD, it is often not appreciated that osteoporosis is also an issue in black populations, albeit to less of an extent than in white populations. For example, according to data from the NHANES III survey, there are approximately 1.7 million black women in the USA with low femoral bone mineral density indicative of either osteopenia or a more severe stage of bone loss, osteoporosis (Looker *et al.*, 1995).

The question arises whether racial differences in bone mineral density reflect differences in the attainment of peak bone mass or differences in the rates of bone loss with aging. Although this question has not been fully settled, it appears that differences in peak bone mass may be particularly important. In a relatively small study of bone loss (Luckey *et al.*, 1996) in African–American and white women conducted in 122 white and 121 African–American women, BMD was measured at six-month intervals in the forearm and spine over a three- to four-year period. These investigators found that, in early menopause, bone loss in the forearm was faster in white than in black women. A similar trend was also evident in the spine. This apparent difference in the loss of bone density in the spine of –2.2% per year in whites compared with –1.3% per year in blacks was not, however, statistically significant. In women more than five years postmenopausal, bone loss did not differ by ethnic group. These findings are consistent with the idea that the higher bone mineral density evident in blacks may largely reflect the attainment of a greater peak bone mass ending in early adulthood, although there could be some contribution of slower rates of estrogen-dependent bone loss in the early post-menopausal years in black women. In another small study, Perry and colleagues (1996) also concurred in the finding of generally similar rates of bone loss in black and white women.

African–American populations have lower rates of osteoporotic fracture than age-matched white populations. A study in the Washington DC area of the USA (Farmer *et al.*, 1984) used the National Hospital Discharge Survey for the years 1974 to 1979 and found that the incidence of hip fracture for white women was 1.5 times greater than for black females after age 40. Likewise, a retrospective study from 1987 to 1989 among Tennessee Medicaid enrollees in the USA, age 65 years and more, indicated that the incidence of non-vertebral fractures in blacks was only half of that in whites, and differences in the fracture rates were evident in 13 different fracture sites examined (Griffin *et al.*, 1992). Overall, findings in blacks support the notion that high BMD is associated with a low rate of bone fracture.

4.2.2 Osteoporosis in Asian populations

The complexity of predicting osteoporotic bone fracture risk from calcium intake or BMD measures alone is illustrated by comparing bone fracture risk in Asian compared with white populations. Calcium intake is low in Asian populations, and many Asians have low BMD. In a study of BMD of Hong Kong Chinese it was observed that women aged 60 and above have mean BMD 30% lower than young women (Ho *et al.*, 1999). Moreover, the age-dependent increase in prevalence of spinal osteoporosis increased dramatically from 10% in the 50–59 age group to 45% in the 60–69 group, and over half of the women over 70 had osteoporosis of the hip. Comparison of trabecular BMD, measured by quantitative computed tomography, between Japanese women and American women 20–85 years old found that BMD was lower in the Asian women and that this difference became greater with age (Ito *et al.*, 1997). In premenopausal women, these investigators found that BMD began to decrease at the age of 20, whereas in American women peak bone mass was sustained until 35. In late postmenopausal women, the rate of bone loss was increased in the Japanese women, –2.2% *vs* –1.4%.

However, despite generally lower BMD, hip fracture rates are substantially lower in Asian countries than in the USA (Ho *et al.*, 1993). These differences probably represent real racial differences in propensity to fracture rather than environmental factors or unintended data collection bias because rates of hip fracture are also lower in Chinese–Americans, Japanese–Americans and Korean–Americans compared with USA whites (Lauderdale *et al.*, 1997).

Why is the general notion that lower BMD is associated with higher rates of osteoporotic bone fracture not evident in the case of Asian populations? One possibility is racial (genetic) differences in bone architecture that influence the propensity to fracture. It has been shown for example that Asian women have shorter hip axis lengths than non-Asian women (Cummings *et al.*, 1994). However, it needs to be pointed out that, despite a substantially lower risk of sustaining a hip fracture in Asian populations, because of its huge population it has been projected that 50% of all hip fractures in the world will occur in Asia during the next century (Lau and Cooper, 1996).

4.2.3   Osteoporosis in Hispanic populations

Owing to the changing demographic profiles in various parts of the globe, steep increases in hip fracture incidences are expected in the future in both Asia and Latin America (Cooper *et al.*, 1992). BMD of Hispanics is similar to non-Hispanic whites. For example, there is no difference in BMD in Puerto Rican women in the proximal femur compared with a standard mixed US population (Haddock *et al.*, 1996). In the USA, many of the studies of BMD and bone fracture have focused on Mexican–Americans, the largest Hispanic group. Bone mineral content of Mexican–American children aged 3–18 is not different from that of non-Hispanic whites (Ellis *et al.*, 1997).

However, despite similarities in BMD, available data on skeletal fracture rates in Mexican–Americans indicate that they are less likely to sustain a hip fracture. Hip fracture rates and vertebral fracture rates in Mexican–Americans are about one-half those seen in non-Hispanic whites (Bauer and Deyo, 1987; Bauer, 1988). The factors influencing lower rates of skeletal fracture in Hispanic populations are poorly defined. Nevertheless, observations of the discordance between BMD and fracture risk in Hispanic and Asian populations compared with white populations suggest that a clear understanding of the factors that influence bone fracture risk will be complex.

4.3   Dairy products, calcium intake and calcium absorption

Among the nutritional factors known to influence the risk of low BMD, low dietary calcium intakes have received considerable research and public attention. In general, low BMD in older persons may result from a lower rate of net bone accrual (peak bone mass) achieved during childhood and young adulthood and/or a relatively increased rate of bone loss during aging. Dietary calcium intake has been shown to influence both of these determinants of bone mass. Therefore, it is generally believed that a high calcium intake throughout the entire lifespan will achieve the greatest benefit in ameliorating the risks of osteoporosis. The Food and Nutrition Board of the Institute of Medicine in the United States reflected this notion in their recommendations for the need of a high 'Adequate Intake' level for dietary Ca in all age groups. The current 'adequate intake' levels for dietary calcium in the USA are shown in Table 4.1. On the other hand, actual intakes of calcium in the general population are relatively low when considered in the context of recommended calcium intakes to maintain optimal bone health. As shown in Table 4.2, mean calcium intakes of various age–sex groups adjusted for day-to-day variation as determined by the 1994 USDA Continuing Survey of Food Intakes by Individuals fall considerably below recommended calcium intake levels. Since dairy products are the main source of calcium intake in the US population, low calcium intakes largely reflect relatively low dairy food intakes. Despite high dietary calcium needs in older people, consumption of milk-based foods in the USA decreases as people age. Clearly, there is a great need for nutrition education to convince large segments of the adult population that they need to alter their eating patterns to provide additional calcium intake. In addition, efforts by the food industry to provide more alternative calcium-fortified foods for groups who prefer to minimize or not to consume milk-based dairy products are needed to fill the calcium intake gap.

As a rule of thumb, about 20–30% of dietary calcium is absorbed. At low dietary calcium intakes, the majority of calcium absorbed is through a vitamin D-dependent pathway. At high dietary calcium intakes, most of the calcium that is absorbed is via a vitamin D-independent concentration-dependent pathway (Sheikh et al., 1988). The determinants of intestinal calcium absorption, especially any genetic components, are poorly understood (Wood, 2000). The bioavailability of calcium can vary considerably depending on the dietary source of the calcium. High dietary fiber intake reduces the bioavailability of dietary calcium (Allen, 1982; Knox et al., 1991). In general, a plant-based diet provides insufficient bioavailable dietary calcium to meet current recommendations (Weaver et al., 1999). It should be noted that, despite the fact that calcium bioavailability from some vegetables can be high (Heaney and Weaver, 1990), when one considers the generally low calcium density (calcium per serving) of many plant foods, it becomes impractical in the absence of calcium supplementation to fully meet one's calcium needs from non-dairy food sources alone.

It has been known for a long time that the absorption efficiency of a low dose of calcium decreases as a function of age (Bullamore et al., 1970). This probably

Table 4.1   Dietary reference intakes for calcium

Age	Adequate intake (mg/d)
0–6 months	210
7–12 months	270
1–3 years	500
4–8 years	800
9–13 years	1300
14–18 years	1300
19–30 years	1000
31–50 years	1000
50–70 years	1200
>70 years	1200

No added calcium intake is recommended for pregnant or lactating women beyond the level recommended for their age. Source: *Dietary Reference Intakes for Calcium, Phosphorus, Magnesium, Vitamin D, and Fluoride.* Standing Committee on the Scientific Evaluation of Dietary Reference Intakes, Institute of Medicine, Food and Nutrition Board. Washington DC, National Academy Press (1997).

Table 4.2   Mean calcium intakes as a percentage of the current recommended intakes from the 1994–1996 USDA Continuing Survey of Food Intakes by Individuals

Age group (years)		Adequate intake level (%)
Males		
	9–13	78
	14–18	86
	19–30	96
	31–50	88
	51–70	65
	≥71	61
Females		
	9–13	65
	14–18	54
	19–30	67
	31–50	63
	51–70	49
	≥71	49

represents a decrease in the vitamin D-dependent calcium absorption pathway. Moreover, there is increasing evidence that ageing causes intestinal resistance to the action of vitamin D (Wood *et al.*, 1998; Pattanaungkul *et al.*, 2000). Thus, high calcium intakes, which are predominantly absorbed through the vitamin D-independent transport pathway, become all the more important in older populations. There are a variety of ways that higher calcium intakes can be achieved, including the use of calcium supplements and calcium-fortified foods. For example, a recent comparison in elderly men and women of calcium bioavailability from milk, calcium-fortified orange juice, and a calcium carbonate supplement found that all three calcium-rich products provided similar levels of bioavailable calcium (Martini and Wood, 2002). An important point is that we must find more effective ways not only to get more high-calcium products developed, but also to convince adult consumers to include them in their diets.

4.4   Dairy products and osteoporosis

With the apparent exception of intermittent parathyroid hormone (Neer *et al.*, 2001), treatments of osteoporosis do not result in increased BMD. Rather, most agents used to treat osteoporosis are aimed at slowing the rate of bone loss, thereby preserving what BMD is left. In this regard, high dietary calcium intakes have a weak anti-bone resorptive effect and slow the rate of bone loss in elderly persons, especially in the presence of adequate vitamin D intakes (Dawson-Hughes *et al.*, 1990, 1997; Dawson-Hughes, 1991). In perimenopausal women, the rapid decline in bone mass that occurs during the first 5 years after menopause cannot be prevented even with calcium supplementation supplying 1000 to 2000 mg Ca/d (Elders *et al.*, 1991). Although several studies have shown that high calcium intakes can increase BMD in young children, it is less certain whether the beneficial effects of calcium are merely to promote an earlier achievement of a person's genetically determined peak bone mass or actually enhances BMD. Regardless, it needs to be more generally recognized by the public that the cessation of high calcium intakes will probably result in a loss of accrued BMD (Lee *et al.*, 1996).

4.4.1   Racial differences in calcium intake
Variations in calcium intake across populations largely reflect differences in dairy food consumption. Populations living in areas where traditionally cows have been kept for producing milk have relatively high dietary intakes of calcium-rich milk and cheese. Milk consumption varies markedly between countries. High milk intakes are evident in North America and northern Europe and low intakes are generally found in Asia.

4.4.2   Milk as a dietary source of calcium
Since vitamin D-fortified milk provides both calcium and vitamin D in high amounts, it is a convenient food to achieve or maintain optimal nutritional bone health. The high calcium density of milk and milk-based food products can be an important source of dietary calcium during childhood and early adulthood, a time when the achievement of peak bone mass is determined.

One would expect that it might be relatively easy to demonstrate that high intake of dairy foods correlates with better bone health. However, this has not been the case, and a recent review of these studies indicates that the majority of studies that have investigated this question have found no effect of dairy food intake on bone health (Weinsier and Krumdieck, 2000). One point of concern regarding the potential of milk to enhance bone density is the relatively high protein content of milk, since a high protein intake increases urinary calcium loss (Kerstetter and Allen, 1990). However, in the elderly recent evidence suggests that increasing calcium intake by increasing the consumption of milk compared with other calcium-rich foods does not cause a greater increase in urinary calcium excretion (Martini and Wood, 2002).

4.5   Future trends: genetic markers of osteoporosis risk

Developments in the early part of the 21st century will probably centre around an increased understanding of the functional role of specific genes in the development of complex chronic diseases, including osteoporosis. In addition, nutritional sciences will further our understanding of the role of specific nutrients and functional food components in optimal bone health. It is likely that these findings will be translated into the development of novel milk-based products to promote bone health. Some of these products may be targeted to populations that are particularly at increased risk of developing osteoporosis.

Random mutations in DNA occur relatively frequently in the evolution of an organism. Some of these mutations are carried forward in the evolutionary development of a species. Comparing the genetic sequence of a person's DNA with those of others reveals that it contains many mutations in various genes. Some of these genetic variations are of practical physiological significance and some may prove useful as genetic markers of future disease risk. In these cases, the promise of genetic testing is that early identification of these disease-risk biomarkers will lead to early intervention to prevent or retard the development of the disease. Currently, these genetic markers are often measured as restriction fragment length polymorphisms (RFLPs) that are caused by random mutations in DNA that lead to variations in the ability of specific endonuclease enzymes to sever a given sequence of DNA. The aftermath of cutting a piece of DNA with these restriction enzymes then results in DNA fragments of various physical lengths that indicate the presence or absence of a given string of DNA code. Since mutations occur randomly throughout the genome, and do not necessarily occur within the protein-coding region of a gene, these RFLPs do not necessarily have a functional consequence to the gene in which they are detected. Nevertheless, they can potentially be useful as a marker of disease risk in certain circumstances, presum-ably because they are linked to other genetic mutations that are physiologically important. The advent of high-throughput genotyping methodologies will open up the prospect of screening a large number of these genetic polymorphisms initially for research purposes to identify disease relationships, and eventually to provide each person and their medical care providers with a disease–risk profile based on combinations of various gene markers.

Some examples of genetic markers that have been investigated to date in association studies of BMD or osteoporosis are the RFLPs in the vitamin D receptor (VDR) gene (using *Bsm*I, *Taq*I, *Apa*I and *Fok*I restriction enzymes), the estrogen receptor gene (using *Xba*I and *Pvu*II restriction enzymes) and nucleotide repeat polymorphisms in the Sp1 binding site in the collagen type I alpha I gene promoter (Wood and Fleet, 1998). Mutations in the Sp1 binding site of the collagen gene promoter could influence the regulation of collagen expression. Collagen is produced by bone osteoblast cells and provides the organic matrix into which inorganic calcium and phosphate are laid down to mineralize bone. In the case of the *Bsm*, *Taq* and *Apa* VDR RFLPs, there are no associated changes in the amino acid composition of the VDR protein, yet some studies suggest that there is a relation to BMD (Cooper and Umbach, 1996). However, a *Fok*I VDR RFLP is associated with the expression of a shortened form of the VDR lacking the three N-terminal amino acids (Gross *et al.*, 1996). The *Fok*I VDR polymorphism results from a C-to-T transition (ACG is mutated to ATG) and creates an initiation codon (ATG) that is three codons proximal to a downstream start site usually used for translation of the VDR protein. The presence of the *Fok*I site, designated by convention as lower case *f*, indicates a genetic sequence that creates an new initiation start site, allowing protein translation to initiate from the first ATG. The allele lacking the site (designated *F*), results in initiation of translation from the more distal second ATG site. Thus, protein translation products from these different alleles differ by three amino acids with the *f* variant elongated because of the creation of an earlier start site (Gross *et al.*, 1996). Moreover, the differing length of the two VDR protein variants has a physiological consequence. The *ff* *Fok*I polymorphism in the VDR gene has been associated with a 13% lower lumbar spine BMD and a greater rate of bone loss in the femoral neck (–4.7% vs –0.5%; *ff* vs *FF*) in postmenopausal Mexican–American women (Gross *et al.*, 1996). Likewise, in Caucasian premenopausal women (Harris *et al.*, 1997), the *ff* genotype is associated with a 4% lower total body BMD and 12% lower femoral neck BMD. However, not all studies have found an influence of the *Fok*I genotype (Eccleshall *et al.*, 1997). Nevertheless, these studies are harbingers of things to come. One day, when we have a far greater understanding of both the functional consequences and usefulness of various genetic markers of osteoporosis risk and the influence of various environmental factors on the expression of the phenotype, we may be in a position to read our genetic barcode to determine our nutritional requirements and disease–risk profile, thereby allowing us to design a more optimal program to maintain better health.

4.6   Future trends: redefining a nutritional prescription for optimal bone health

As our knowledge of nutrients and functional food components increases in the future, we will have an increased understanding of the nutritional ingredients needed for optimal bone health. To illustrate this potential trend, consider the case of vitamin K. Epidemiological studies have implicated vitamin K as a potentially important dietary factor that can affect BMD and the risk of bone fracture (Vermeer *et al.*, 1995). There is accumulating evidence that several vitamin K-dependent proteins (osteocalcin, matrix Gla protein, protein S, gas6) could play an important role in bone and cartilage metabolism (Ferland, 1998).

Currently, very few data exist on the adequacy of vitamin K status of different age groups in relation to bone health (Binkley and Suttie, 1995). Most of the evidence used to support a relationship between vitamin K and bone is based on epidemiologic associations between a biomarker of vitamin K status (percentage of undercarboxylated osteocalcin) and BMD or risk of hip fracture. Assessment of the percentage of undercarboxylated osteocalcin is considered a sensitive measure of vitamin K nutritional status (Sokoll and Sadowski, 1996). Szulc and colleagues (1993) reported that circulating undercarboxylated osteocalcin was higher in elderly women and was a significant risk factor of hip fractures. In a three-year follow-up of the same cohort, these investigators confirmed that undercarboxylated osteocalcin was still predictive of hip fracture after controlling for age and parathyroid hormone (PTH) (Szulc *et al.*, 1996). In a 22-month prospective cohort study of 7598 elderly women, undercarboxylated osteocalcin, a measure of vitamin K status, was predictive of hip fracture risk independent of femoral neck BMD, whereas total serum osteocalcin (a measure of bone turnover) was not (Vergnaud *et al.*, 1997). Knapen *et al.* (1989) recently reported a significant association between percentage of undercarboxylated osteocalcin and BMD among women within one to ten years of menopause, but not among other age groups. Significant associations between vitamin K intake or serum vitamin K levels and BMD have also been reported, although not consistently.

Kanai *et al.* (1997) reported that postmenopausal women with reduced BMD had significantly lower plasma vitamin K than postmenopausal women with normal bone density. In the Nurses' Health Study, women in the second quintile of vitamin K intake (dietary intakes of 107–142 µg/d) were at 40% lower risk of hip fracture compared with women in the lowest quintile of intake (Feskanich *et al.*, 1999). Vitamin K intakes among elderly men and women in the Framingham Osteoporosis Study (Framingham, Massachusetts) were not associated with BMD but did relate to fracture rate (Booth *et al.*, 2000). The above-mentioned epidemiological studies are highly suggestive of a role of vitamin K status on calcium and bone metabolism.

Currently, there is very little information concerning the effect of vitamin K supplementation on calcium and bone metabolism. Douglas *et al.* (1995) demonstrated in 20 postmenopausal women that 14 days of vitamin K supplementation increased carboxylation of osteocalcin, a risk factor for hip fracture. Orimo *et al.* (1992) reported increases in bone mass among 272 postmenopausal women following supplementation with menaquinones (vitamin K_2) for 24–48 weeks. No long-term (>12 months) clinical trials of the effects of vitamin K_1 (phylloquinone) supplementation on BMD or bone fracture have been reported. Ultimately, we will need to await the final outcomes of ongoing clinical trials of vitamin K supplementation and bone loss to strengthen our convictions concerning the importance of high vitamin K intakes in protecting bone mass and its possible role in preventing osteoporosis. However, the highly suggestive data of the relationship between vitamin K status and risk of osteoporotic fracture suggest that it may be an important new nutrient in the armamentarium to combat osteoporosis. Milk may be the ideal vehicle to fortify with vitamin K in the future.

4.7   Sources of further information and advice

National Osteoporosis Foundation
1232 22nd Street N.W.
Washington DC 20037-1292, USA
Internet address: www.nof.org

National Institutes of Health
Osteoporosis and Related Bone Diseases National Resource Center
1232 22nd Street, N.W.
Washington DC 20037-1292, USA
Internet address: www.osteo.org

Osteoporosis Society of Canada
33 Laird Drive
Toronto, Ontario M4G 3S9
Canada
Internet address: www.osteoporosis.ca

National Osteoporosis Society
Camerton
Bath BA2 0PJ, UK
Internet address: www.nos.org.uk

Osteoporosis Australia
Level 1, 52 Parramatta Road
Forest Lodge NSW 2037
GPO Box 121 Sydney NSW 2001
Australia
Internet address: www.osteoporosis.org.au

NIH Consensus Statements
Osteoporosis Prevention, Diagnosis, and Therapy
March 27–29, 2000
Vol. 17, No. 1
Internet address: www.consensus.nih.gov/cons/111/111_intro.htm

4.8   References

ALLEN L (1982), 'Calcium bioavailability and absorption: a review', *Am J Clin Nutr*, **35**, 783–808.

BARONDESS D A, NELSON D A and SCHLAEN S E (1997), 'Whole body bone, fat, and lean mass in black and white men', *J Bone Miner Res*, **12**(6), 967–971.

BAUER R L (1988), 'Ethnic differences in hip fracture: a reduced incidence in Mexican Americans', *Am J Epidemiol*, **127**(1), 145–149.

BAUER R L and DEYO R A (1987), 'Low risk of vertebral fracture in Mexican American women', *Arch Intern Med*, **147**(8), 1437–1439.

BINKLEY N C and SUTTIE J W (1995), 'Vitamin K nutrition and osteoporosis', *J Nutr*, **125**(7), 1812–1821.

BOOTH S, TUCKER K, CHEN H, HANNAN M, GAGNON D, CUPPLES L, WILSON P, ORDOVAS J, SCHAEFER E, DAWSON-HUGHES B and KIEL D (2000), 'Dietary vitamin K intakes are associated with hip fracture but not with bone mineral density in elderly men and women', *Am J Clin Nutr*, **71**, 1201–1208.

BULLAMORE J, WILKINSON R, GALLAGHER J, NORDIN B and MARSHALL D (1970), 'Effect of age on calcium absorption', *Lancet*, **2**(672), 535–537.

COOPER, C (1997), 'The crippling consequences of fractures and their impact on quality of life', *Am J Med*, 103(2A), 12S–17S; discussion 17S–19S.

COOPER C, CAMPION G and MELTON L J D (1992), 'Hip fractures in the elderly: a world-wide projection', *Osteoporos Int*, **2**(6), 285–289.

COOPER C, ATKINSON E J, JACOBSEN S J, O'FALLON W M and MELTON L J D (1993), 'Population-based study of survival after osteoporotic fractures', *Am J Epidemiol*, **137**(9), 1001–1005.

COOPER G S and UMBACH D M (1996), 'Are vitamin D receptor polymorphisms associated with bone mineral density? A meta-analysis [see comments]', *J Bone Miner Res*, **11**(12), 1841–1849.

CUMMINGS S R, CAULEY J A, PALERMO L, ROSS P D, WASNICH R D, BLACK D and FAULKNER K G (1994), 'Racial differences in hip axis lengths might explain racial differences in rates of hip fracture. Study of Osteoporotic Fractures Research Group', *Osteoporos Int*, **4**(4), 226–229.

DAWSON-HUGHES B (1991), 'Calcium supplementation and bone loss: a review of controlled clinical trials', *Am J Clin Nutr*, **54**(1 Suppl), 274S–280S.

DAWSON-HUGHES B, DALLAL G E, KRALL E A, SADOWSKI L, SAHYOUN N and TANNENBAUM S (1990), 'A controlled trial of the effect of calcium supplementation on bone density in postmenopausal women', *N Engl J Med*, **323**(13), 878–883.

DAWSON-HUGHES B, HARRIS S S, KRALL E A and DALLAL G E (1997), 'Effect of calcium and vitamin D supplementation on bone density in men and women 65 years of age or older', *N Engl J Med*, **337**(10), 670–676.

DOUGLAS A S, ROBINS S P, HUTCHISON J D, PORTER R W, STEWART A and REID D M (1995), 'Carboxylation of osteocalcin in post-menopausal osteoporotic women following vitamin K and D supplementation', *Bone*, **17**(1), 15–20.

ECCLESHALL T, GARNERO P, GROSS P, DELMAS P and FELDMAN D (1997), 'The start codon polymorphism of the vitamin D receptor gene is not associated with bone mineral density and bone turnover markers in premenopausal women', *J Bone Mineral Metab*, **12**(Suppl 1), S370.

ELDERS P J, NETELENBOS J C, LIPS P, VAN GINKEL F C, KHOE E, LEEUWENKAMP O R, HACKENG W H and VAN DER STELT P F (1991), 'Calcium supplementation reduces vertebral bone loss in perimenopausal women: a controlled trial in 248 women between 46 and 55 years of age', *J Clin Endocrinol Metab*, **73**(3), 533–540.

ELLIS K J, ABRAMS S A and WONG W W (1997), 'Body composition of a young, multiethnic female population', *Am J Clin Nutr*, **65**(3), 724–731.

FARMER M E, WHITE L R, BRODY J A and BAILEY K R (1984), 'Race and sex differences in hip fracture incidence', *Am J Public Health*, **74**(12), 1374–1380.

FERLAND G (1998), 'The vitamin K-dependent proteins: an update', *Nutr Rev*, **56**(8), 223–230.

FESKANICH D, WEBER P, WILLETT W C, ROCKETT H, BOOTH S L and COLDITZ G A (1999),

'Vitamin K intake and hip fractures in women: a prospective study', *Am J Clin Nutr,* **69**(1), 74–79.

GRIFFIN M R, RAY W A, FOUGHT R L and MELTON L J, 3rd (1992), 'Black–white differences in fracture rates', *Am J Epidemiol,* **136**(11), 1378–1385.

GROSS C, ECCLESHALL T R, MALLOY P J, V ILLA M L, MARCUS R and FELDMAN D (1996), 'The presence of a polymorphism at the translation initiation site of the vitamin D receptor gene is associated with low bone mineral density in postmenopausal Mexican–American women', *J Bone Miner Res,* **11**(12), 1850–1855.

HADDOCK L, ORTIZ V, VAZQUEZ M D, AGUILO F, BERNARD E, AYALA A and MEJIAS N (1996), 'The lumbar and femoral bone mineral densities in a normal female Puerto Rican population', *P R Health Sci J,* **15**(1), 5–11.

HARRIS S S, ECCLESHALL T R, GROSS C, DAWSON-HUGHES B and FELDMAN D (1997), 'The vitamin D receptor start codon polymorphism (*Fok*I) and bone mineral density in premenopausal American black and white women', *J Bone Miner Res,* **12**(7), 1043–1048.

HEANEY R P and WEAVER C M (1990), 'Calcium absorption from kale', *Am J Clin Nutr,* **51**(4), 656–657.

HO S C, BACON W E, HARRIS T, LOOKER A and MAGGI S (1993), 'Hip fracture rates in Hong Kong and the United States, 1988 through 1989', *Am J Public Health,* **83**(5), 694–697.

HO S C, LAU E M, WOO J, SHAM A, CHAN K M, LEE S and LEUNG P C (1999), 'The prevalence of osteoporosis in the Hong Kong Chinese female population', *Maturitas,* **32**(3), 171–178.

ITO M, LANG T F, JERGAS M, OHKI M, TAKADA M, NAKAMURA T, HAYASHI K and GENANT H K (1997), 'Spinal trabecular bone loss and fracture in American and Japanese women', *Calcif Tissue Int,* **61**(2), 123–128.

KANAI T, TAKAGI T, MASUHIRO K, NAKAMURA M, IWATA I and SAJI F (1997), 'Serum vitamin K level and bone mineral density in post-menopausal women', *Int J Gynaecol Obstet,* **56**(1), 25–30.

KERSTETTER J and ALLEN L (1990), 'Dietary protein increases urinary calcium', *J Nutr,* **120**, 134–136.

KNAPEN M H, HAMULYAK K and VERMEER C (1989), 'The effect of vitamin K supplementation on circulating osteocalcin (bone Gla protein) and urinary calcium excretion', *Ann Intern Med,* **111**(12), 1001–1005.

KNOX T A, KASSARJIAN Z, DAWSON-HUGHES B, GOLNER B B, DALLAL G E, ARORA S and RUSSELL R M (1991), 'Calcium absorption in elderly subjects on high- and low-fiber diets: effects of gastric acidity', *Am J Clin Nutr,* **53**, 1480–1486.

LAU E M and COOPER C (1996), 'The epidemiology of osteoporosis. The oriental perspective in a world context', *Clin Orthop,* **323**, 65–74.

LAUDERDAL D S, JACOBSEN S J, FURNER S E, LEVY P S, BRODY J A and GOLDBERG J (1997), 'Hip fracture incidence among elderly Asian–American populations', *Am J Epidemiol,* **146**(6), 502–509.

LEE W T, LEUNG S S, LEUNG D M and CHENG J C (1996), 'A follow-up study on the effects of calcium-supplement withdrawal and puberty on bone acquisition of children', *Am J Clin Nutr,* **64**(1), 71–77.

LOOKER A C, JOHNSTON JR C C, WAHNER H W, DUNN W L, CALVO M S, HARRIS T B, HEYSE S P and LINDSAY R L (1995), 'Prevalence of low femoral bone density in older U.S. women from NHANES III', *J Bone Miner Res,* **10**(5), 796–802.

LOOKER A C, ORWOLL E S, JOHNSTON JR C C, LINDSAY R L, WAHNER H W, DUNN W L, CALVO M S, HARRIS T B and HEYSE S P (1997), 'Prevalence of low femoral bone density in older U.S. adults from NHANES III', *J Bone Miner Res,* **12**(11), 1761–1768.

LUCKEY M M, WALLENSTEIN S, LAPINSKI R and MEIER D E (1996), 'A prospective study of bone loss in African–American and white women – a clinical research center study', *J Clin Endocrinol Metab,* **81**(8), 2948–2956.

MARTINI L and WOOD R (2002), 'Relative bioavailability of calcium-rich dietary sources in the elderly', *Am J Clin Nutr,* **76**, 1345–50.

NEER R M, ARNAUD C D, ZANCHETTA J R, PRINCE R, GAICH G A, REGINSTER J Y, HODSMAN A

B, ERIKSEN E F, ISH-SHALOM S, GENANT H K, WANG O and MITLAK B H (2001), 'Effect of parathyroid hormone (1-34) on fractures and bone mineral density in postmenopausal women with osteoporosis', *N Engl J Med*, **344**(19), 1434–1441.

ORIMO H, SHIRAKI M, FUJITA T, ONONMURA T and INOUE K (1992), 'Clinical evaluation of menatetrenone in the treatment of involutional osteoporosis: a double-blind multicenter comparative study with 1-hydroxy D3', *J Bone Miner Res*, **7**(Suppl 1), S122.

PATTANAUNGKUL S, RIGGS B L, YERGEY A L, VIEIRA N E, O'FALLON W M and KHOSLA S (2000), 'Relationship of intestinal calcium absorption to 1,25-dihydroxyvitamin D [1,25(OH)2D] levels in young versus elderly women: evidence for age-related intestinal resistance to 1,25(OH)2D action', *J Clin Endocrinol Metab*, **85**(11), 4023–4027.

PERRY H M 3RD, HOROWITZ M, MORLEY J E, FLEMING S, JENSEN J, CACCIONE P, MILLER D K, KAISER F E and SUNDARUM M (1996), 'Aging and bone metabolism in African American and Caucasian women', *J Clin Endocrinol Metab*, **81**(3), 1108–1117.

RAY N F, CHAN J K, THAMER M and MELTON L J 3RD (1997), 'Medical expenditures for the treatment of osteoporotic fractures in the United States in 1995: report from the National Osteoporosis Foundation', *J Bone Miner Res*, **12**(1), 24–35.

SHEIKH M, RAMIREZ A, EMMETT M, SANTA ANA C and FORDTRAN J (1988), 'Role of vitamin D-dependent and vitamin D-independent mechanisms in absorption of food calcium', *J Clin Invest*, **81**, 126–132.

SOKOLL L J and SADOWSKI J A (1996), 'Comparison of biochemical indexes for assessing vitamin K nutritional status in a healthy adult population', *Am J Clin Nutr*, **63**(4), 566–573.

SZULC P, CHAPUY M C, MEUNIER P J and DELMAS P D (1993), 'Serum undercarboxylated osteocalcin is a marker of the risk of hip fracture in elderly women', *J Clin Invest*, **91**(4), 1769–1774.

SZULC P, CHAPUY M C, MEUNIER P J and DELMAS P D (1996), 'Serum undercarboxylated osteocalcin is a marker of the risk of hip fracture: a three year follow-up study', *Bone*, **18**(5), 487–488.

VERGNAUD P, GARNERO P, MEUNIER P J, BREART G, KAMIHAGI K and DELMAS P D (1997), 'Undercarboxylated osteocalcin measured with a specific immunoassay predicts hip fracture in elderly women: the EPIDOS Study', *J Clin Endocrinol Metab*, **82**(3), 719–724.

VERMEER C, JIE K S and KNAPEN M H (1995), 'Role of vitamin K in bone metabolism', *Annu Rev Nutr*, **15**, 1–22.

WEAVER C M, PROULX W R and HEANEY R (1999), 'Choices for achieving adequate dietary calcium with a vegetarian diet', *Am J Clin Nutr*, **70**(3 Suppl), 543S–548S.

WEINSIER R L and KRUMDIECK C L (2000), 'Dairy foods and bone health: examination of the evidence', *Am J Clin Nutr*, **72**(3), 681–689.

WOOD R J (2000), 'Searching for the determinants of intestinal calcium absorption', *Am J Clin Nutr*, **72**(3), 675–676.

WOOD R J and FLEET J C (1998), 'The genetics of osteoporosis: vitamin D receptor polymorphisms', *Annu Rev Nutr*, **18**, 233–258.

WOOD R K, FLEET J C, CASHMAN K, BRUNS M E and DELUCA H F (1998), 'Intestinal calcium absorption in the aged rat: evidence of intestinal resistance to 1,25-dihydroxyvitamin D', *Endocrinology*, **139**(9), 3843–3848.

5

Probiotics and the management of food allergy

P.V. Kirjavainen, University of Turku, Finland

5.1 Introduction

A fundamental ability of the immune system is the recognition of self from harmful foreign substances. This task is particularly challenging in the gut where the gastrointestinal immune system must be able to find and distinguish the antigens of invading pathogens from the constant and enormous load of food antigens, environmental antigens and antigens originating from the resident microbiota. The immune system must be able to mount sufficient responses to eliminate the intruding pathogens while concomitantly restraining reactions to billions of harmless or even beneficial microorganisms and allowing the entry of nutrient molecules. The development of the immune system including the normal formation of oral tolerance is essentially dependent on microbial stimulus encountered at gut mucosa in early infancy (Bauer *et al.*, 1966; MacDonald and Carter, 1979; Pulverer *et al.*, 1990; Sudo *et al.*, 1997). In food allergies the formation of immunological tolerance to some food components is either not achieved or not maintained.

Prenatally the gastrointestinal tract is sterile but is *postpartum* rapidly colonised by thousands of bacteria originating from the mother, other external contacts (e.g. nurses) and the surrounding environment (Kirjavainen and Gibson, 1999). By the end of the first week of life, there are already millions of bacteria in a gram of faeces. Thereafter the number gradually increases and an adult colon is estimated to be inhabited by as many as 10^{14} total bacterial cells belonging to at least 500 different culturable species and weighing up to 2 kilograms (Kirjavainen and Gibson, 1999). As this burden of microbial antigens is separated from the body by only a single epithelial cell layer, it is not surprising that microbial stimulus has a major influence on gut-associated immunological homeostasis and gut barrier integrity. Indeed, it is because of the antigens of the microbiota that the fully matured gut-associated lymphoid tissue is our largest immunological organ containing a greater number of T cells than the rest of the body combined (Brandtzaeg *et al.*, 1998). The close host–microbe interaction provides the rationale for the view that the quality of microbial stimulus to which an atopic individual is exposed to, i.e. via the resident gut microbiota, sporadic infections, or other external bacteria such as probiotics, may affect the process of allergic sensitisation and expression of allergic symptoms (Strachan, 1989; Kalliomäki, 2001a, b; Kirjavainen *et al.*, 2001, 2002).

5.2   The mechanisms and symptoms of food allergy

By definition food allergies are immunologically mediated hypersensitivity responses to food antigens (Johansson *et al.*, 2001). Reactions that do not involve immunological mechanism are beyond the scope of this chapter. These reactions are referred to as non-allergic food hypersensitivities and include, for example, deficiencies in enzyme or transportation functions as in lactase deficiency and glucose/galactose malabsorption, respectively. Typically the immunological mechanism in food allergies is an immediate, IgE-mediated type I response, although in delayed symptoms immune complex-mediated type III and cell-mediated type IV responses may also be involved.

It is thought that IgE-mediated allergies can develop only in people with genetic predisposition to IgE type responsiveness. This genetic predisposition, known as atopy, may affect arguably 30–50% of the world population (Jarvis and Burney, 1998). The expression of atopy (i.e. the atopic phenotype) is characterised by high serum total IgE concentration and the presence of IgE antibodies specific to ordinarily harmless environmental antigens.

The IgE-mediated responses are characteristically divided in two phases:

- the **induction** phase during which the sensitisation to an allergen occurs;
- the **effector** phase during which the clinical manifestations of allergy are triggered and expressed (see Fig. 5.1).

It should be noted, however, that allergic sensitisation does not automatically lead to the development of allergic disease and expression of symptoms. Approximately 12–23% of children may be sensitised without symptoms. Furthermore, in early childhood the sensitisation is usually transient as up to 70% of sensitisations to food allergens appearing at the age of 1 have disappeared at the age of 6 years (Kulig *et al.*, 1999). These transient sensitisations to foods are commonly symptomless (van Asperen and Kemp, 1989).

5.2.1   Genetic predisposition
Atopy has shown to be associated with several chromosomal loci, including

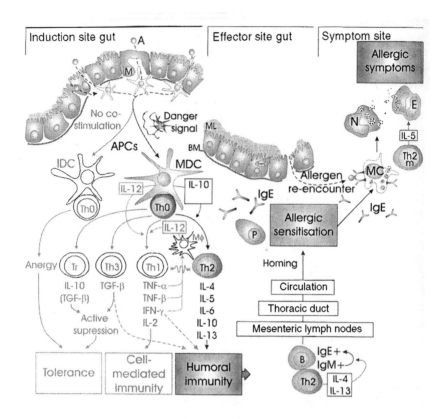

Fig. 5.1 Schematic depiction of the mechanism of food allergy. Luminal food allergens that gain access or are transported to the basolateral side of the epithelium (primarily) in the Peyer's patches are processed to peptide fragments by antigen-presenting cells (APCs) such as dendritic cells (DCs). A danger signal in conjunction with the antigen may drive the maturation of APCs, which, e.g. in DCs, results in increased expression of co-stimulatory molecules (CD80/86) and major histocompatibility complex class II (MHC II) molecules and production of cytokines such as IL-12. Presentation of processed allergen by mature APC results in differentiation of naïve T-helper cells to T-cell subsets with characteristic cytokine production patterns and consequently differential immunoregulatory effects. Allergic sensitisation may occur if the stimulatory conditions during the antigen presentation promote type 2 T-helper cell (Th2) activation: T-cell contact together with Th2 cytokines may induce heavy-chain switching in IgM-bearing B cells, which then (as all B and T cells primed in the Peyer's patches) migrate to mesenteric lymph nodes and via thoracic duct gain access to the systemic circulation. By homing mechanisms, they then re-enter the gut, where their maturation to IgE-producing plasma cells is completed. Allergic symptoms are provoked when on a subsequent encounter allergens cross-link a critical mass of IgE antibodies bound on effector cells, primarily mast cells, which consequently release pro-inflammatory mediators (see text for details). A = allergen, M = M cell, ML = mucous layer, BM = basement membrane, APCs = antigen-presenting cells, IDC = immature dendritic cell, MDC = mature dendritic cell, IL = interleukin, TGF = transforming growth factor, Th = T-helper, Tr = T-regulatory cell, MΦ = macrophage, TNF = tumour necrosis factor, B = B cell, P = plasma cell, MC = mast cell, E = eosinophil, N = neutrophil, m = memory cell.

chromosomes 5q31–33, 6p21, 12q and HLA-D region 11q13 (Barnes and Marsh, 1998; Barnes, 2000). Genes within these loci associated with atopy include *IL-4* (the primary cytokine promoting type 2 T-helper cell (Th2) function), *IFN-γ* (a Th1 function promoting cytokine), *IL-10* (powerful anti-inflammatory but also Th2-function maintaining cytokine), *CD14* (a receptor predominantly involved in the recognition of pathogen associated molecular patterns), *TGF-β* (powerful anti-inflammatory cytokine), *TNF* (proinflammatory cytokine family), *RANTES* (chemotactic molecule attracting, e.g. eosinophiles, basophiles and monocytes), the 4-series leukotrienes, the β-subunit of high-affinity IgE receptor (*FcεRIβ*), β2-adrenergic receptor (*ADRβ2*) and high-affinity IL-4 receptor (*IL-4R*).

5.2.2   Induction phase – allergic sensitisation

The food allergens are usually proteins (Ebo and Stevens, 2001). These can enter from the luminal side into the basolateral side of the epithelium through damaged epithelial cell layer or via active uptake by enterocytes or so-called M cells that are specialised to actively sample the gut contents by transporting macromolecules and particles from the lumen into organised lymphoid follicles known as Peyer's patches (MacDonald, 1998; see Fig. 5.1). In here these antigens are endocytosed by antigen-presenting cells, chiefly dendritic cells. Dendritic cells beneath the epithelium may also sample antigens directly by opening tight junctions between the epithelial cells and protruding its arms to the luminal side (Rescigno *et al.*, 2001). There is growing recognition that in addition to the contact with the antigen, activation of the antigen-presenting cells requires a danger signal indicative of host damage including molecules and molecular structures, released or produced by cells undergoing stress or abnormal cell death (Gallucci and Matzinger, 2001). In the case of allergens, their possible proteolytic activities may cause such signals.

The activated antigen-presenting cells present peptide fragments derived from the antigen to naïve CD4+ T-helper (Th) cells which consequently differentiate to effector cells (see Fig. 5.1). Antigens originating from pathogens tend to trigger the Th cell differentiation into Th1 cells capable of producing IFN-γ that activates macrophages and is thus a critical cytokine in mucosal host defence against invasive pathogens. The normal response to food antigens is the formation of oral tolerance, i.e. establishment and maintenance of unresponsiveness to antigens on the mucosal surfaces. Such unresponsiveness can be accomplished by three known mechanisms: (1) deletion of the effector cells by apoptosis, (2) rendering the effector cells anergic, i.e. functionally non-responsive, or (3) by active suppression due to generation of regulatory T cells. The regulatory T cells include Th3 cells producing high amounts of transforming growth factor-β (TGF-β), T regulatory type 1 (Tr1) cells producing high amounts of IL-10 and CD4+ CD25+ anergic T-cell population inhibiting the proliferation of responding T cells via cell–cell interactions (MacDonald, 1998; Groux, 2001; Maloy and Powrie, 2001).

The dose of the allergen is one determinant of the induced mechanism (Strober *et al.*, 1998). Tolerance to food antigens, which are usually encountered in relatively low numbers, is thought to be formed primarily via the regulatory T cells and the cytokines TGF-β and IL-10. TGF-β is a powerful immuno-suppressive cytokine that suppresses the acquisition of effector functions of naïve T cells. Thus under the presence of TGF-β, Th cells differentiate into pluripotent cells that are incapable of secreting effector cytokines immediately on activation but remain capable of differentiating into any effector T-cell subset once TGF-β is no longer present (Sad and Mosmann, 1994). TGF-β also prevents the maturation of dendritic cells, which as immature are capable of endocytosis but are not able to process an antigen. Thus in the presence of TGF-β, dendritic cells cannot stimulate T cells and may induce tolerance (Gorelik and Flavell, 2002). The immuno-suppressive properties of IL-10 are similar and it plays particularly important role in the differentiation of the regulatory T cells (Moore *et al.*, 2001). Tolerance to food antigens may also develop in the absence of danger signals stimulating the expression of co-stimulatory molecules on the antigen-presenting cells, as it has been suggested that antigen presentation by such cells results in the development of clonal anergy.

When allergic sensitisation occurs, the presentation of an allergen that should be harmless to the host does not result in the formation of oral tolerance but in the differentiation of the naïve Th cells into Th2 memory cells. The reasons for this immunological dysfunction are not fully understood but it probably reflects inherited aberrancies, e.g. in cytokine production and receptor expression associated with the atopic genotype. In addition, the quality of microbial exposure may play a key role here as discussed below in detail. The maturation of B cells to IgE-producing plasma cells is stimulated by contact with the Th2 cells in the presence of cytokines IL-4 and IL-13, produced by these cells.

5.2.3 Effector phase – expression of clinical symptoms

The symptomatic phase of allergies is triggered when on a re-encounter the allergen is recognised by the allergen-specific IgE antibodies (see Fig. 5.1). IgE binds to the high-affinity IgE receptor (FcϵRI) on mast cells, basophiles and dendritic cells and low-affinity IgE receptors (FcϵRII, CD23) on monocytes or macrophages and lymphocytes. A critical mass of allergen cross-linked IgE molecules bound on an effector cell results in the activation of that cell and initiation of an inflammatory cascade. Activation of the key participants in allergic inflammation, tissue mast cells, results (1) in immediate release of their granules containing, e.g. histamine, proteolytic enzymes and chemotactic factors attracting neutrophiles and eosinophiles to the site, (2) within 5–30 minutes in the production of lipid mediators such as leukotrien D_4 and prostaglandin D_2, and (3) within hours in the production of interleukins such as strongly proinflammatory TNF-α, IgE production-stimulating IL-4 and eosinophil proliferation-stimulating IL-5 (Broide, 2001). The activation of blood basophils results in similar response including release of histamine by the degranulation and production of cytokines TNF-α and IL-4 (Platts-Mills, 2001). These inflammatory activities result in the clinical symptoms of the allergic disease due to vasodilation, increased vascular permiability, inflammation, cell death and consequently tissue damage. Mast cell mediators can also contribute to late phase responses occurring four to eight hours after an immediate response and characterised by high eosinophilic infiltration.

The clinical symptoms of food allergy can be cutaneous, respiratory, gastrointestinal and cardiovascular. The cutaneous symptoms include pruritis, atopic dermatitis/eczema, urticaria, angioedema and non-specific erythrema. Gastrointestinal symptoms include nausea, vomiting, diarrhoea, abdominal cramps, bloating and gas and sometimes colic-like symptoms in children, usually in response to cows' milk. The respiratory symptoms such as sneezing, coughing, wheezing or asthma are less common and rare as the sole symptoms. The cardiovascular symptoms including hypotension, shock and cardiac dysrhythmias are associated with more severe cases and may cause death.

5.3   The prevalence of food allergy

The double-blind placebo-controlled food challenge is the golden standard for the diagnosis of food hypersensitivities. A positive skin prick test with the suspected allergen and/or the presence of immunoglobulin E antibodies specific to the allergen are indicative of an IgE-mediated mechanism and thus true food allergy.

Food allergies are a particular problem for infants and young children in whom the prevalence has been estimated to vary between 0.3 and 8% (Macdougall *et al.*, 2002). Most commonly children are allergic to cows' milk (2–3% of all children; Host *et al.*, 1995) but allergies to egg white, wheat and soy are also common. Multiple food allergies in infants may be associated with nutritional repercussions and growth retardation (Isolauri *et al.*, 1998). Fatal and severe food allergic reactions have also been documented in children and their incidence is thought to be rising. However, based on a recent retrospective analysis of fatalities in the United Kingdom, the risk that a child would die from a food allergic reaction was estimated to be as low as 1 in 800,000 per year (Macdougall *et al.*, 2002).

In adults, the prevalence of food allergies is relatively low. Although 3.8–25% of the general population perceive they are intolerant to foods, food allergy can be confirmed with food challenge and skin prick testing only in 0.8 to 2.4% of the subjects (Woods *et al.*, 2002). There is some indication that the number of cases of food allergy-induced anaphylactic shocks is increasing (Wuthrich and Ballmer-Weber, 2001). Adults tend to be allergic to fish, crustaceans, peanuts and tree nuts. However, increasing number of foods has been reported to be provoking allergies including tropical fruits, sesame seeds, psyllium, spices and condiments. These are commonly cross-allergies to an allergen derived from another source such as pollen or natural rubber latex (Ebo and Stevens, 2001). The triggering foods also differ between countries, probably in reflection of food consumption.

It remains arguable whether food allergies are increasing, although a dramatic increase in some other allergic diseases in the Western countries is evident. In the United States the prevalence of atopic eczema, the predominant atopic manifestation in the first years of life, increased from 3% in the 1960s to 10% in the 1990s (Horan *et al.*, 1992; United States Centers for Disease Control, 1998). Birth cohort surveys in the United Kingdom have shown that the prevalence of atopic eczema in children aged under 5 years has increased from 5% in people born in 1946 to 7% and 12%, respectively, for people born in 1958 and 1970 (Jarvis and Burney, 1998). Notably, in 39–63% of all infants and young children atopic eczema is triggered by one or more challenge-confirmed food allergies (Sampson, 1992; Isolauri and Turjanmaa, 1996; Burks et al., 1998).

The reasons for the increasing prevalence of allergies in the westernised countries are not known. In recent years most attention has received the hygiene hypothesis that was originally proposed by Strachan over a decade ago (Strachan, 1989). This hypothesis links the increasing prevalence with reduced microbial exposure early in life. He based this claim on observations that the prevalence of hay fever was inversely related to the number of children in household. Since then, a substantial amount of experimental data has provided mechanistic rationale to explain how the quality of early microbial stimulus and function could affect the expression of atopy and the development of allergic diseases.

5.4   Probiotics and food allergy: the clinical evidence

5.4.1   Reduction of the risk of developing allergic disease
There are two studies indicating that probiotics may have potential in reducing the risk of the development of allergies when administered in early infancy. In the first, the effects of neonatal administration of non-enteropathogenic *Escherichia coli* were followed by a questionnaire. The results showed that the administration was associated with reduced susceptibility for consequent development of allergic diseases. Subjects born prematurely and full-term were followed separately. Of the premature subjects, the proportion that was reported to have an allergic disease at 10 years of age was 60% smaller in those challenged to *E. coli* ($n = 77$) than in the controls ($n = 55$; 21% *vs* 53%). Of the subjects born full-term, the proportion of allergic subjects at 20 years of age was 29% smaller in the *E. coli*-challenged group ($n = 151$) than in the controls ($n = 179$; 36% *vs* 51%) (Lodinová-Žádníková and Cukrowská, 1999). In the second study, *Lactobacillus rhamnosus* strain GG (ATCC 53103) was given in double-blind placebo-controlled manner for two to four weeks to pregnant mothers from families where the mother herself, the father and/or the siblings of the unborn child had an atopic disease. After the birth, the probiotic/placebo was administered either to the mother or to the child for six months. Remarkably, in the children who were exposed to the probiotic during the perinatal period the occurrence of atopic eczema at two years old was 50% lower than in the placebo group (15/64 [23%] *vs* 31/68 [46%]; Kalliomäki *et al.*, 2001b).

5.4.2   Management of allergic manifestations
The first implication of the potential of using bacterial stimulus in the management of the symptoms of food allergies was reported in 1985 by Loskutova. In that study the administration of a mixture containing *Propionibacterium* sp. and *Lactobacillus*

acidophilus hastened the disappearance of food allergy manifestations. A year later Ciprandi and co-workers reported that patients ($n = 20$) receiving adjunctive treatment with *Bacillus subtilis* spores resulted in significant reduction of the frequency and severity of the symptoms of adults with urticaria/angioedema manifestations from food allergy (Ciprandi *et al.*, 1986). However, the safety of using such spores in immunocompromised patients has been brought into a question (Oggioni *et al.*, 1998). In a more recent partially blinded study by Trapp and co-workers (1993), volunteers given yoghurt had decreased concentrations of IgE in the serum and a lower frequency of allergies. Unfortunately, microbial characterisation of the yoghurt was not reported. Wheeler and co-workers (1997) studied the effect of yoghurt on cellular, humoral and phagocytic function in adults with atopic allergy. Consumption of yoghurt, fermented with *Lactobacillus bulgaricus* and *Streptococcus thermophilus*, induced no significant changes in any of the immune parameters investigated.

More prominent results were observed when infants with atopic eczema and cows' milk allergy were given an extensively hydrolysed whey formula supplemented with the *Lactobacillus* GG (Majamaa and Isolauri, 1997). In contrast to the placebo group ($n = 14$), in the infants who received the probiotic supplementation ($n = 13$) a significant improvement of clinical symptoms and alleviation of intestinal inflammation associated with food allergy was observed. These effects were later confirmed with another study population ($n = 27$) and the results were comparable whether *Lactobacillus* GG or *Bifidobacterium lactis* strain Bb-12 was used as supplementation (Isolauri *et al.*, 2000). With both strains the control of atopic eczema was achieved after two months of therapy, while in the placebo group it took six months. The effects on inflammatory markers were similar between the two strains: both treatments reduced the serum concentrations of soluble CD14 associated with chronic inflammation and urinary concentrations of eosinophil protein X (EPX) reflecting allergic inflammation. No effects were observed on the concentrations of cell surface molecules or chemotactic factors. Interestingly, the effects on the production of the anti-inflammatory cytokine TGF-β1 were opposite with the different probiotics, bifidobacteria-treatment resulting in decreased and lactobacilli-treatment increased serum TGF-β1 concentration. It has been later shown that feeding *Lactobacillus* GG to children expressing atopy results in induction of IL-10, another strongly immune suppressive cytokine (Pessi *et al.*, 2000). The anti-inflammatory action of *Lactobacillus* GG is also supported by a study where the administration of this strain to adults with cows' milk allergy was shown to down-regulate the overactive phagocytic functions associated with allergic inflammation (Pelto *et al.*, 1998).

Recently, the anti-allergenic effects of *Lactobacillus* GG were tested in a trial with teenagers and young adults who were allergic to birch pollen and apples (Helin *et al.*, 2002). The subjects received the probiotic for a total of 5.5 months prior, during and after the birch pollen season ($n = 18$). The treatment did not alleviate their symptoms or reduce their use of medication during the birch-pollen season or the subsequent months as compared to the placebo group ($n = 18$), nor did the treatment significantly affect the symptoms caused by apple in the oral challenge tests. This controversy to the positive effects of this strain in the management of cows' milk allergy and atopic eczema in infants demonstrates the limitations of a single probiotic strain and regimen in the management of different allergies during different periods of life.

5.5   Mechanisms of action: gut microbiota composition and food allergies

The mechanistic rationale for the probiotic therapy in the management of allergies is provided by the fundamental role of microbial stimulus and function in the normal development and regulation of immunological functions and gut barrier. The functions relevant in the context of allergies include stimulatory effects on immune functions that are antagonistic to Th2 activity, promotion of oral tolerance induction, anti-inflammatory action and effects reducing and modifying the host's exposure to allergens (see Fig. 5.2). The quality of microbial stimulus has been shown to influence functions in the both phases of allergy, i.e. induction and effector phase, and thus both phases are also rational targets of probiotic therapy.

5.5.1   Aberrancies in the composition of the gut microbiota associated with allergies

The predominant site for host–microbe interaction is in the gut and thus the compositional development of gut microbiota has been suggested to be the key determinant in whether or not the atopic genotype will be fully expressed and thus affect the development of allergic diseases. Therefore, determining the characteristics in the normal microbial compositions in the gut, and in the compositions associated with the expression of allergies, may help in understanding the aetiology of allergic diseases and point to the potential microbial targets of a probiotic (and prebiotic) therapy. Moreover, assessing the properties of the key microbes found may reveal direct immunological treatment targets and thus the immunomodulatory properties that a probiotic strain should have.

A study by Russian scientists in 1982 represents the first clear suggestion that allergies are associated with compositional aberrancy in the gut microbiota (Shaternikov et al., 1982). This study reported two years later in English was based on their assessment of 60 under 1-year-old infants with food allergy and atopic eczema. In this study it was claimed that the severity of the disease was in direct correlation with the stage of aberrancy in the faecal microbiota. This aberrancy was characterised as low prevalence of bifidobacteria and lactobacilli and high prevalence of Enterobactericeae, pathogenic species of staplylococci and streptococci as well as Candida species (Kuvaeva et al., 1984). Very similar observations on older subjects (10–45 years) were reported by Ionescu and co-workers (1986) another two years later: compared with healthy subjects ($n = 21$), those with atopic eczema ($n = 58$) had lower prevalence of lactobacilli, bifidobacteria and enterococci species but higher prevalence of Klebsiellae, Proteus, Staphylococcus aureus,

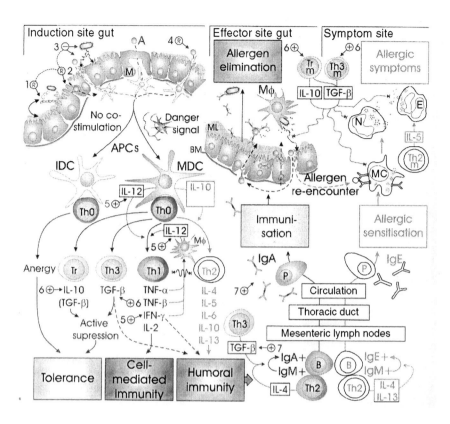

Fig. 5.2 Proposed functions of probiotic therapy in reducing the risk of sensitisation to food allergens and in the management of allergic symptoms. (1) Probiotics may interact with the host directly or by regulating the composition and function of endogenous gut microbiota and host–microbe cross-talk. (2) Reduction in antigen uptake by regulation of mucosal functions: permeability, production of digestive enzymes, antioxidative activity and angiogenesis, and (3) by exclusion of pathogens causing epithelial damage. (4) Antigen degradation and modification of antigen immunogenicity. (5) Induction of the production of Th1-favouring cytokines. (6) Induction of the production of anti-inflammatory cytokines, which in the induction arm promote the formation of tolerance by active suppression and in the effector arm maintain intestinal integrity by suppressing inflammatory responses (e.g. those mounted against pathogens) and down-regulate allergic inflammation and thus alleviate the symptoms. (7) Directing the humoral responses towards the production of IgA, which may result in reduced tendency for IgE switching and enhanced elimination of allergens and inflammatory microbes both at luminal and basolateral side of the epithelium. A = allergen, R = regulation by probiotics, –/+ = reduction/enhancement by probiotics, M = M cell, ML = mucous layer, BM = basement membrane, APCs = antigen-presenting cells, IDC = immature dendritic cell, MDC = mature dendritic cell, IL = interleukin, TGF = transforming growth factor, Th = T-helper, Tr = T-regulatory cell, MΦ = macrophage, TNF = tumour necrosis factor, B = B cell, P = plasma cell, MC = mast cell, E = eosinophil, N = neutrophil, m = memory cell.

Clostridium innocuum and *Candida* species. In 1990, Ionescu and co-workers published these results in detail (1990a) and very similar findings from a comparison of the faecal microbiota of 30 healthy subjects and 110 subjects with atopic eczema (1990b).

Unfortunately, it was a decade later that the topic received wider acknowledgement. A study by Bjorksten and co-workers (1999) provided confirmation for the previous findings by demonstrating that colonisation by lactobacilli was less common in both Estonian and Swedish 2-year-old children with food allergies ($n = 27$) than in the age-compatible healthy children ($n = 36$), while the opposite was true for coliforms and *S. aureus*. *Bacteroides* were shown to make up a larger proportion of the microbiota in healthy than in allergic infants. We compared the microbiota of healthy infants ($n = 10$) with that of infants with early onset atopic eczema and suspected cows' milk allergy ($n = 27$) during full breast-feeding. *Klebsiellae* species tended to be found more frequently and *Streptococcus* species less frequently in the faeces of allergic infants than in those of healthy infants (Kirjavainen *et al.*, 2001). Furthermore, the predominant anaerobic and facultatively anaerobic microbiota of allergic infants was characterised by lower prevalence of Gram-positive species. Lower prevalence of bifidobacteria was restricted to those allergic infants with gastrointestinal symptoms and higher prevalence of *Bacteroides* to infants whose cows' milk allergy was later confirmed by challenge. The connection between *Bacteroides* and cows' milk allergy is supported by our recently published data showing a positive correlation between faecal *Bacteroides* numbers and serum total IgE concentration within infants who were highly sensitive to cows' milk proteins as indicated by intolerance to extensively hydrolysed whey formula (Kirjavainen *et al.*, 2002). The faecal numbers of *E. coli* correlated similarly with total IgE in all the allergic infants, which is in accordance with the association suggested by the earlier studies between the high prevalence of coliforms and allergy.

Most interestingly, two studies have demonstrated differences in the faecal microbiota prior to the beginning of the expression of atopy and the development of allergic diseases, suggesting that these differences may not be secondary to the disease. In a study by our group some of the predominant microbial genera were enumerated with fluorescently labelled 16S RNA specific oligonucleotide probes (Kalliomäki *et al.*, 2001a). The neonatal faecal microbiota preceding the expression of atopy ($n = 12$) was characterised by the presence of lower numbers of bifidobacteria and higher numbers of bacteria belonging to the *Clostridium histolyticum* group than in the faeces of infants who did not express atopy ($n = 17$) (Kalliomäki *et al.*, 2001a). Bjorksten and co-workers (2001) demonstrated similar trends with plate culture methods with few additions. They found that bifidobacteria and enterococci were less frequently present and lactobacilli more frequently present in the neonatal faecal microbiota of infants who at 2 years had atopic eczema ($n = 18$) than in the microbiota of symptomless infants ($n = 26$). Later during the first year, a higher prevalence of *S. aureus*, higher numbers of clostridia and lower numbers of *Bacteroides* were other characteristics of the faecal microbiota of the infants symptomatic at 2 years. Nevertheless, further studies are required to confirm the causality between the microbial aberrancies in the gut microbiota and

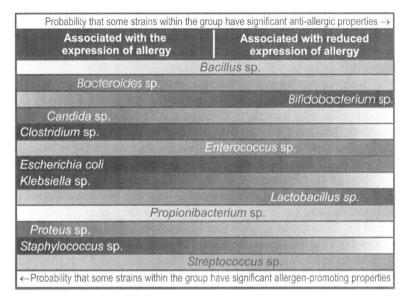

Fig. 5.3   Tentative mapping of gut bacterial groups for screening of potential probiotic strains for allergy treatment trials. The map presents a summary of the direct or indirect evidence of the associations between the bacterial groups and the expression of allergic disease as mentioned in the text. Map key: the darker the shading, the stronger the evidence. Remark: contradictory role within the bacterial group may be indicative of species-specific activity.

allergies (see Fig. 5.3 for a summary). This is because changes in the gut mucosa that influence the microbial colonisation process may precede the visible clinical symptoms and thus even these latter two studies do not exclude the possibility that such changes were present at the time of microbial evaluations.

5.6   Infant development and allergic sensitisation

Early infancy is the critical period in life in relation to the development of food allergies because of the presence of strong risk factors for allergic sensitisation, i.e. for the formation of Th2-polarised memory to food allergens. These risk factors include immature gut barrier allowing increased contacts between the immune system and potential allergens and Th2-skewed T-helper cell-mediated immune responses characteristic during the first months of life.

5.6.1   Th1/Th2 balance

The Th2 type responses prevail over the mutually antagonistic Th1 responses during early infancy probably as for protection from cell-mediated proinflammatory responses that could be detrimental for the developing foetus and neonate (Raghupathy, 2001). Normally the balance is shifted towards Th1 responses during the first year of life but the expression of atopic genotype appears to be associated with defects such as compromised IFN-γ production that disable the normal maturation of the Th1 function (Prescott *et al.*, 1999). Therefore, in these infants the immune system remains Th2 polarised and the high-risk window for allergic sensitisation is widened.

The bacteria encountered after birth are thought to be the major trigger for the shift away from the predominant Th2 responsiveness, as there appears to be a universal tendency for bacteria to stimulate the Th cell differentiation into Th1 effector cells. However, the responses triggered by the bacteria may be variable; e.g. an *in vitro* study suggests that Gram-positive bacteria tend to be more potent inducers of IL-12 than their Gram-negative counterparts, which were shown to induce more of IL-10 (Hessle *et al.*, 2000). The induction of IL-12 and IL-10 production *in vitro* has been demonstrated to vary also between different lactobacillus strains (von der Weid *et al.*, 2001; Christensen *et al.*, 2002). IL-12 is produced by, for example, dendritic cells and macrophages and is a key cytokine promoting the Th cell differentiation into Th1 cells. IL-10 is considered to be one of the most critical cytokines responsible for maintaining Th2 bias (Raghupathy, 2001); however, its strong anti-inflammatory properties are thought to prevent allergic sensitisation and responses (van den Biggerlaar *et al.*, 2000; Holt, 2000; Wills-Karp *et al.*, 2001; see Fig. 5.2).

Differences in the pathogen-associated molecular patterns (PAMPs) in bacteria may account for the variable innate immune responses mounted towards them. PAMPs are motifs that are not present on host cells but have been conserved invariant within a microorganism of given class due to their essential physiological function (Medzhitov, 2001) and as such are good targets for innate recognition. PAMPs include bacterial cell-wall components such as lipopolysaccharides (LPS) of Gram-negative bacteria and lipoteichoic acids of Gram-positive bacteria and unmethylated CpG DNA (Aderem and Ulevitch, 2000). A new family of essential pattern recognition receptors was discovered a few years ago: the Toll-like receptors (TLRs) expressed on antigen-presenting cells. Currently ten different Toll-like receptors have been found and they all are specialised for the recognition of different PAMPs, e.g. co-operation between TLR2 and TLR6 have been demonstrated to be essential for the cell activation in response to, for example, peptidoglycan from Gram-positive bacteria and lipoarabinomannan from mycobacterial cell wall, TLR4 (with the aid of another pattern recognition receptor, CD14) in response to, for example, LPS from Gram-negative bacteria and lipoteichoic acids from Gram-positive bacteria, and TLR9 in response to unmethylated CpG DNA (Ozinsky *et al.*, 2000; Tapping *et al.*, 2000; Takeuchi and Akira, 2001). Signalling downstream of these TLRs may result in activation of nuclear factor-κB (NF-κB) or mitogen-activated protein (MAP) kinases (c-Jun N-terminal kinase (JNK) and p38). These mediate the activation of genes in the nucleus encoding for proinflammatory cytokines, such as IL-12, chemokines and their receptors as well as other factors associated with dendritic cell maturation (Medzhitov, 2001).

The risk of allergic sensitisation may be strongly influenced by the ongoing recognition of bacteria by Toll-like receptors. Studies in mice with compromised

Toll-mediated signalling capacity indicate that antigen-specific Th1 responses to food allergens are dependent on simultaneously induced Toll-mediated activities, while similar dependency was not observed in Th2 responses (Schnare *et al.*, 2001). Furthermore, deficient signalling via Toll-like receptors may promote allergic sensitisation as re-exposing the mice to the allergen enhanced the production of IL-13 by T cells, a cytokine capable of inducing isotype class-switching of B cells to produce IgE (Schnare *et al.*, 2001). There is also increasing body of evidence that the downstream signalling may occur through different pathways when triggered by microbes. For example it is known that signalling via all TLRs can involve the adaptor protein MyD88, but signalling via TLR4 (and possibly via TLR3, which recognises viral double stranded DNA), but not TLR2 or TLR9, may also trigger the MyD88-independent pathway (Medzhitov, 2001). These different pathways have been shown to have different effects on dendritic cell activation which may be critical concerning the differentiation of Th cells, since signals via the MyD88-independent pathway do not trigger the production of IL-12 (Medzhitov, 2001). Since TLR4 is likely to be more important in the recognition of Gram-negative than Gram-positive bacteria, the probability that it activates the MyD88-independent pathway may also be higher than with Gram-positive species and this may contribute to the suggested differences in the IL-12 production between Gram-positive and Gram-negative bacteria. The importance of Toll-like receptors for Th maturation and the possibility that bacteria are recognised variably by the different TLRs provide the rationale for suggestions that qualitative differences in the gut microbiota could be reflected to the Th1/Th2 balance and thereby allergic sensitisation.

5.6.2   Formation of oral tolerance

The importance of microbial stimulus for the normal formation of oral tolerance is best demonstrated in a study by Sudo and co-workers (1997). They demonstrated that the Th2-mediated humoral responses but not Th1-mediated responses of germ-free mice were unsusceptible to the induction of oral tolerance but after the mice were colonised with a single *Bifidobacterium infantis* strain such tolerance was inducible. The explanation for this could be related to the necessity of the Toll-like signalling for the development of Th1 function as described above. The effect may also reflect microbial induction of cytokine production promoting active suppression. The possibility of such activity by lactic acid bacteria is supported by at least two studies. Firstly, by the aforementioned observation that *Lactobacillus* 66 treatment of allergic infants resulted in increased serum concentrations of TGF-β1 and, secondly, by an *in vitro* demonstration showing that different *Lactobacillus* strains provided different degree of stimulus for the formation of low proliferating Th cell population producing TGF-β and IL-10 (Von der Weid *et al.*, 2001).

5.6.3   Inflammation

Inflammatory responses play central role in allergic diseases. In the effector arm of allergy it is the inflammatory responses triggered by the allergen–IgE complex that cause the symptoms. In the inductor arm, inflammation can enhance the exposure to potential allergens by increasing the mucosal permeability and promote the maturation of dendritic cells and modulate the Th1/Th2/T regulatory cell balance (see Fig. 5.1; Crowe, 1999; Kirjavainen et al., 1999; Nagler-Anderson and Shi, 2001). The gut microbes and, as such, probiotics can affect the inflammatory responses in various ways and in the both arms of the allergic disease (see Fig. 5.2).

As described above there is a universal tendency to mount Th1-mediated proinflammatory responses to bacteria. Such reactions would be wasteful and detrimental if they were mounted against the innumerable commensal gut microbes and therefore the immune system has adapted to recognise and tolerate the harmless components of the gut microbiota while monitoring its composition via active uptake and mounting responses when potentially dangerous organisms are encountered (Neish et al., 2000; Nagler-Anderson, 2001; Hajjar et al., 2002). Pathogens may have a dual role concerning allergic inflammation. On the one hand, as they do not induce tolerance, they may have greater potential to stimulate the immunity away from the Th2-type responses than commensal bacteria towards which tolerance exists (Kirjavainen et al., 1999; Neish et al., 2000). Conversely, by triggering inflammatory responses or by toxin formation such bacteria may increase gut permeability and thus the exposure to potential allergens and the risk for allergic sensitisation (McDonel, 1979; Deitch et al., 1991; Duchmann et al., 1995; Obiso et al., 1995; Crowe, 1999; Madsen et al., 1999b; Lapa e Silva et al., 2000; Neish et al., 2000). Restraining the potentially inflammatory microbiota could therefore provide a potential target for probiotic therapy (Kirjavainen et al., 1999; Madsen et al., 1999a). Indeed, in our recent study it was shown that the allergic inflammation alleviating effects of bifidobacterial supplementation during weaning were accompanied by restraining action on Bacteroides and E. coli microbiota, the numbers of which correlated directly with the serum total IgE concentrations during breast-feeding (Kirjavainen et al., 2002).

Bifidobacteria and lactobacilli are antagonistic against many pathogens due to competitive exclusion, production of antimicrobial components (Gibson and Wang, 1994; Lievin et al., 2000) as well as via stimulation of the IgA production (Moreau et al., 1978; Kaila et al., 1992; Yasui et al., 1992; Majamaa et al., 1995). The latter effect can reduce the intestinal inflammation also by enhancing the exclusion of allergens and again reduce the risk of hypersensitivity reactions to such allergens in already sensitised subjects (Takahashi et al., 1998; Isolauri et al., 1993). As discussed above, specific probiotics may also reduce wide range of inflammatory responses by stimulating the production of IL-10 and TGF-β and therefore antigen non-specific bystander suppression (Pessi et al., 2000; Isolauri et al., 2000).

The degradation process is another defining parameter in relation to the inflammatory responsiveness mounted towards dietary antigens (Isolauri et al., 1993; Barone et al., 2000). Most antigens encountered are already processed when they contact the mucosal surface. Proteases produced by intestinal microbiota contribute to the processing of food antigens in the gut and modify their immunogenicity. Such an activity may also be used in probiotic therapy as demonstrated by finding that, in contrast to purified casein, which up-regulates IL-4 and INF-γ production in infants with cows' milk allergy, caseins degraded by probiotic bacterium-derived, *Lactobacillus* GG, enzymes were further shown to down-regulate IL-4 production with no effect on INF-γ release (Sütas *et al.*, 1996).

5.7   Selecting the right probiotic

Further clinical studies with larger study populations are required to confirm the beneficial effects of probiotics in the management of food allergy. However, taken together, the scant but prominent clinical demonstrations and the substantial number of experimental data providing the mechanistic rationale for bacterial therapy, the great potential of such an approach is undeniable. Therefore, in addition to the extended clinical studies with the current most studied probiotics, such as *Lactobacillus* GG and *Bifidobacterium* Bb-12, efforts should be focused on optimising the treatments, i.e. finding the best regimen and strains. This goal can be efficiently reached only by first assessing the specific properties that a probiotic strain should have so that it is able to carry out the desired effects. We have recently redefined probiotics as 'specific live or inactivated microbial cultures that have documented targets in reducing the risk of human disease or in their nutritional management', to address the importance of specific targets for specific strains (Isolauri *et al.*, 2002). Previously the screening for potential probiotic strains for clinical trials has not been focused enough on the disease-specific properties but more on selecting strains that have already been shown to have potential in the treatment of some other disease. These strains are commonly characterised by properties that are universally considered to be desirable for probiotic action, such as strong *in vitro* adhesion and viability, but on the other hand shown to not be always mandatory (Ouwehand and Salminen, 1998; Isolauri *et al.*, 2002). Such selection practice has been well justified for the preliminary studies and the reason is clear: little is known about specific properties of probiotics for specific diseases.

5.7.1   Microbiota-oriented approach

One approach to the identification of the desired properties has been the characterisation of important bacterial groups in the natural prevention of allergies by comparing the gut microbial composition in healthy and allergic individuals and in connection to the expressed phenotype of the disease. These data are somewhat variable, but there are also very clear common trends that do not appear to be associated with certain countries or time. Such trends include desirability of infant microbiota robust in bifidobacteria and enterococci and low in at least certain species of coliforms, clostridia and *S. aureus*. Data concerning lactobacilli and bacteroides are variable. In Fig. 5.3 these characteristics have been summarised in conjunction with the effects of allergy treatment trials with different probiotic strains. The variability in the results may reflect the differences in the study designs (particularly in the selection criteria for the study populations), but also that different species and strains within these genera have radically different effects in respect to the disease. The latter notion is supported by a recent study by Ouwehand and co-workers (2001) demonstrating that different bifidobacteria species tended to prevail in the microbiota of allergic infants ($n = 7$) compared with healthy infants ($n = 6$). Further and more detailed species- and even strain-specific characterisation of gut microbial composition in association with allergies is warranted.

5.7.2 Allergic immune phenotype-oriented approach

Another logical approach to studying the desired properties of probiotics is to characterise the immunological aberrancies associated with the development and manifestations of allergic disease and to investigate which properties in bacteria could stimulate counterbalancing responses. A good example of such an approach is a subsequent study by He and co-workers (2002) demonstrating that the bifidobacteria species prevailing in healthy infants were more potent inducers of IL-10 than the species more commonly found in the faeces of allergic infants, which conversely appeared to be more potent inducers of the proinflammatory cytokines tumour necrosis factor (TNF)-α, IL-6 and IL-12. As IL-10 may be one of the key cytokines promoting the formation of oral tolerance, and also key anti-inflammatory cytokine in restraining allergic responses, the characterisation of the components responsible for these differences could reveal a desirable property of a probiotic strain intended for the management of allergies. The recent advances in the understanding of bacterial recognition by the innate immune system, in particular the discovery of the Toll-like receptors and their specialised role in the recognition of different PAMPs, as well as improved understanding of the signal transduction pathways, may provide the focus points for such assessments.

The immunological aberrancies associated with allergies can also be looked at from a slightly different perspective: it is possible that the immunological aberrancies associated with the expression of atopic genotype set special requirements for microbial stimulus. In other words, exposure to the same bacteria does not necessarily result in similar immunological responses in individuals that express atopy and those that do not. An example of such aberrancy is low CD14 expression associated with atopy, which may compromise the bacterial recognition and therefore, for example, the ability of bacteria to promote the development of Th2 antagonising Th1 function. We have previously hypothesised that it may be possible to overcome this defect by quantitatively strong enough exposure to Gram-positive bacteria, the recognition of which appears to be less dependent on CD14 than that of Gram-negative bacteria. This suggestion is in agreement with our observation, which tentatively suggests that the early colonisation by Gram-positive bacteria may have been poor in infants who have developed an atopic disease (Kirjavainen et al., 2001).

5.7.3   Engineering the desirable probiotic strain

Once the key microbial functions and components in respective to the management of allergies have been identified, it may be possible to genetically implant these properties to bacterial strains that are otherwise most suitable for probiotic use, e.g. from the industrial point of view. It is also possible to genetically engineer recombinant bacteria with desired function such as production of anti-inflammatory cytokines IL-10 and TGF-β (Steidler *et al.*, 2000; Gao *et al.*, 2001). However, such approaches possess many unanswered questions of safety. These include the problems with preventing the spread of the genetically introduced property between other members of the microbiota and to microbiota of other individuals. A less controversial approach would be to screen for microorganisms with the desirable properties, e.g. for one strongly inducing IL-10 production, and then if necessary modify its technological properties without genetic engineering. This might be accomplished, for example, by exposing the microorganism to sublethal stress such as acidic conditions or heat, in order to improve the survival properties in the gastrointestinal tract and tolerance to stress, and thereby also to provide improved competitiveness against pathogens in the intestinal milieu (Klaenhammer and Kullen, 1999).

5.8   Conclusion and future trends

Probiotics have been successfully used in the risk reduction and management of allergic diseases but extended trials must be conducted before recommendations can be given for clinical practice. There are extensive experimental and clinical data on the effects by which probiotics may, by immunological or non-immunological means, reduce the risk of allergic sensitisation and development of allergic disease and, on the other hand, alleviate the symptoms in already allergic subjects. These mechanisms include immunomodulatory effects promoting Th2 antagonistic Th1 functions and the formation of oral tolerance, enhanced antigen elimination and down-regulated inflammatory responses, e.g. via stimulation of anti-inflammatory cytokines IL-10 and TGF-β, antagonism against pathogenic microbiota and effects seen as enhanced elimination, modified degradation, permeation and presentation of food antigens. Particular efforts should be made to characterise the properties in bacteria that are responsible for these mechanisms to facilitate the selection of optimal probiotics and optimise the treatment regimen.

5.9   Sources of further information and advice

In the beginning of this chapter, the basics of food allergies were briefly revised. A more comprehensive overview of the current understanding of food allergies, food allergens and their relationship with the gastrointestinal immune system has been recently published (Burks, 2002; Brandtzaeg, 2002; Samuel *et al.*, 2002). For in-depth understanding of food allergies, their mechanisms, diagnosis and treatments the reader is advised to consult a textbook such as *Food Allergy and Intolerance* by Brostoff and Challacombe (2002). Good sources for the up-to-the-minute information on food allergies are the specialised journals such as *Journal of Allergy and Clinical Immunology*, *Clinical and Experimental Allergy*, *Allergy* and *Current Opinion in Allergy and Clinical Immunology*.

The gastrointestinal immune system and the formation of oral tolerance are keystones in the aetiology of food allergies. There are number of good reviews covering these topics including those by Nagler-Anderson (2001), a very nicely written overview of the both subjects, Nagler-Anderson and Shi (2001), a comprehensive review on oral tolerance, Groux (2001) and Maloy and Powrie (2001), which review the current knowledge on regulatory T cells in detail. The review by Gorelik and Flavell (2002) presents comprehensively the relevant information on the importance of the key cytokine TGF-β and that by Wills-Karp and co-workers (2001) presents allergy related information on IL-10. An excellent reference source for information on different cytokines is the *Cytokines Online Pathfinder Encyclopaedia* in the Internet (Ibelgaufts, 2002).

The mechanisms of probiotic therapies are based on the normal host–microbe interactions. Further discussion concerning such interactions can be found from reviews including those of Hooper and Gordon (2001) and MacDonald and Pettersson (2000). Toll-like receptors and the signal transduction pathways they activate are the foundation for innate immune responses to bacteria and a number of excellent reviews can be found including those by Medzhitov (2001) and Takeuchi and Akira (2001).

5.10   References

ADEREM A and ULEVITCH R J (2000), 'Toll-like receptors in the induction of the innate immune response', *Nature*, **406**(6797), 782–787.

BARNES K C (2000), 'Atopy and asthma genes-where do we stand?', *Allergy*, **55**(9), 803–817.

BARNES K C and MARSH D G (1998), 'The genetics and complexity of allergy and asthma', *Immunol Today*, **19**(7), 325–332.

BARONE K S, REILLY M R, FLANAGAN M P and MICHAEL J G (2000), 'Abrogation of oral tolerance by feeding encapsulated antigen', *Cellular Immunol*, **199**, 65–72.

BAUER H, PARONETTO F, BURNS W A and EINHEBER A (1966), 'The enhancing effect of the microbial flora on macrophage function and the immune response. A study in germfree mice', *J Exp Med*, **123**, 1013–1024.

BJORKSTEN B, NAABER P, SEPP E and MIKELSAAR M (1999), 'The intestinal microflora in allergic Estonian and Swedish 2-year-old children', *Clin Exp Allergy*, **29**(3), 342–346.

BJORKSTEN B, SEPP E, JULGE K, VOOR T and MIKELSAAR M (2001), 'Allergy development and the intestinal microflora during the first year of life', *J Allergy Clin Immunol*, **108**(4), 516–520.

BRANDTZAEG P (2002), 'Current understanding of gastrointestinal immunoregulation and its relation to food allergy', *Ann NY Acad Sci*, **964**, 13–45.

BRANDTZAEG P, FARSTAD I N and HELGELAND L (1998), 'Phenotypes of T cells in the gut', *Chem Immunol*, **71**, 1–26.

BROIDE D H (2001), 'Molecular and cellular mechanisms of allergic disease', *J Allergy Clin Immunol*, **108** (2 Suppl), S65–71.

BROSTOFF J and CHALLACOMBE S J, EDS (2002), *Food Allergy and Intolerance*, W. B. Saunders, London.

BURKS W A, JAMES J M, HIEGEL A, WILSON G, WHEELER J G, JONES S M and ZUERLEIN N (1998), 'Atopic dermatitis and food hypersensitivity reactions', *J Pediatr,* **132**(1), 132–136.

BURKS W (2002), 'Current understanding of food allergy', *Ann NY Acad Sci,* **964**, 1–12.

CHRISTENSEN H R, FROKIAER H and PESTKA J J (2002), 'Lactobacilli differentially modulate expression of cytokines and maturation surface markers in murine dendritic cells', *J Immunol,* **168**(1), 171–178.

CIPRANDI G, SCORDAMAGLIA A, RUFFONI S, PIZZORNO G and CANONICA G W (1986), 'Effects of an adjunctive treatment with *Bacillus subtilis* for food allergy', *Chemioterapia,* **5**(6), 408–410.

CROWE S E (1999), 'Food allergies – immunophysiology aspects', in Ogra P L, Mestecky J, Lamm M E, Strober W, Bienenstock J and McGhee J R, *Mucosal Immunology,* San Diego, Academic Press, 1129–1139.

DEITCH E A, SPECIAN R D and BERG R D (1991), 'Endotoxin-induced bacterial translocation and mucosal permeability: role of xanthine oxidase, complement activation, and macrophage products', *Crit Care Med,* **19**(6), 785–791.

DUCHMANN R, KAISER I, HERMANN E, MAYET W, EWE K and MEYER ZUM BUSCHENFELDE K H (1995), 'Tolerance exists towards resident intestinal flora but is broken in active inflammatory bowel disease (IBD)', *Clin Exp Immunol,* **102**, 448–455.

EBO D G and STEVENS W J (2001), 'IgE-mediated food allergy–extensive review of the literature', *Acta Clin Belg,* **56**(4), 234–247.

GALLUCCI S and MATZINGER P (2001), 'Danger signals: SOS to the immune system', *Curr Opin Immunol,* **13**, 114–119.

GAO C F, KONG X T, GRESSNER A M and WEISKIRCHEN R (2001), 'The expression and antigenicity identification of recombinant rat TGF-beta1 in bacteria', *Cell Res,* **11**(2), 95–100.

GIBSON G R and WANG X (1994), 'Regulatory effects of bifidobacteria on the growth of the other colonic bacteria', *J Appl Bacteriol,* **77**, 412–420.

GORELIK L and FLAVELL R A (2002), 'Transforming growth factor-beta in T-cell biology', *Nature Rev Immunol,* **2**(1), 46–53.

GROUX H (2001), 'An overview of regulatory T cells', *Microbes Infect,* **3**(11), 883–889.

HAJJAR A M, ERNST R K, TSAI J H, WILSON C B and MILLER S I (2002), 'Human Toll-like receptor 4 recognizes host-specific LPS modifications', *Nat Immunol,* **3**(4), 354–359.

HE F, MORITA H, HASHIMOTO H, HOSODA M, KURISAKI J, OUWEHAND A C, ISOLAURI E, BENNO Y and SALMINEN S (2002), 'Intestinal *Bifidobacterium* species induce varying cytokine production', *J Allergy Clin Immunol,* **109**(6), 1035–1036.

HELIN T, HAAHTELA S and HAAHTELA T (2002), 'Effect of oral treatment with an intestinal bacterial strain, *Lactobacillus rhamnosus* (ATCC 53103), on birch-pollen allergy: a placebo-controlled double-blind study', *Allergy,* **57**(3), 243–246.

HESSLE C, ANDERSSON B and WOLD A E (2000), 'Gram-positive bacteria are potent inducers of monocytic interleukin-12 (IL-12) while Gram-negative bacteria preferentially stimulate IL-10 production', *Infect Immun,* **68**(6), 3581–3586.

HOLT P (2000), 'Parasites, atopy, and the hygiene hypothesis: resolution of a paradox?', *Lancet,* **356**(9243), 1699–1701.

HOOPER L V and GORDON J I (2001), 'Commensal host–bacterial relationships in the gut', *Science,* **292**(5519), 1115–1118.

HORAN R F, SCHNEIDER L C and SHEFFER A L (1992), 'Allergic disorders and mastocytosis', *JAMA,* **268**, 2858–2868.

HOST A, JACOBSEN H P, HALKEN S and HOLMENLUND D (1995), 'The natural history of cows' milk protein allergy/intolerance', *Eur J Clin Nutr,* **49**(1), S13–18.

IBELGAUFTS H (2002), *COPE Cytokines Online Pathfinder Encyclopaedia* [Online], Available: http://www.copewithcytokines.de/ [accessed 13 Aug. 2002].

IONESCU G, RADOVICIC D, SCHULER R, HILPERT R, NEGOESCU A, PREDA I, JURTHI E, ITZE W and ITZE L (1986), 'Changes in fecal microflora and malabsorption phenomena suggesting a contaminated small bowel syndrome in atopic eczema patients', *Microecology and Therapy*, **16**, 273.

IONESCU G, KIEHL R, ONA L and SCHULER R (1990a), 'Abnormal fecal microflora and malabsorption phenomena in atopic eczema patients', *J Adv Med*, **3**(2), 71–91.

IONESCU G, KIEHL R, WICHMANN-KUNZ F and LEIMBECK R (1990b), 'Immunobiological significance of fungal and bacterial infections in atopic eczema', *J Adv Med*, **3**(1), 47–58.

ISOLAURI E and TURJANMAA K (1996), 'Combined skin prick and patch testing enhances identification of food allergy in infants with atopic dermatitis', *J Allergy Clin Immunol*, **97**(1 Pt 1), 9–15.

ISOLAURI E, MAJAMAA H, ARVOLA T, RANTALA I, VIRTANEN E and ARVILOMMI H (1993), '*Lactobacillus casei* strain GG reverses increased intestinal permeability induced by cow milk in suckling rats', *Gastroenterology*, **105**(6), 1643–1650.

ISOLAURI E, SUTAS Y, SALO M K, ISOSOMPPI R and KAILA M (1998), 'Elimination diet in cows' milk allergy: risk for impaired growth in young children', *J Pediatr*, **132**(6), 1004–1009.

ISOLAURI E, ARVOLA T, SUTAS Y, MOILANEN E and SALMINEN S (2000), 'Probiotics in the management of atopic eczema', *Clin Exp Allergy*, **30**(11), 1604–1610.

ISOLAURI E, RAUTAVA S, KALLIOMÄKI M, KIRJAVAINEN P and SALMINEN S (2002), 'Role of probiotics in food hypersensitivity', *Curr Opin Allergy Clin Immunol*, **2**, 263–271.

JARVIS D and BURNEY P (1998), 'ABC of allergies. The epidemiology of allergic disease', *BMJ*, **316**(7131), 607–610.

JOHANSSON S G, HOURIHANE J O, BOUSQUET J, BRUIJNZEEL-KOOMEN C, DREBORG S, HAAHTELA T, KOWALSKI M L, MYGIND N, RING J, VAN CAUWENBERGE P, VAN HAGE-HAMSTEN M and WUTHRICH B (2001), 'A revised nomenclature for allergy. An EAACI position statement from the EAACI nomenclature task force', *Allergy*, **56**(9), 813–824.

KAILA M, ISOLAURI E, SOPPI E, VIRTANEN E, LAINE S and ARVILOMMI H (1992), 'Enhancement of the circulating antibody secreting cell response in human diarrhea by a human *Lactobacillus* strain', *Pediatr Res*, **32**, 141–144.

KALLIOMAKI M, KIRJAVAINEN P, EEROLA E, KERO P, SALMINEN S and ISOLAURI E (2001a), 'Distinct patterns of neonatal gut microflora in infants in whom atopy was and was not developing', *J Allergy Clin Immunol*, **107**(1), 129–134.

KALLIOMAKI M, SALMINEN S, ARVILOMMI H, KERO P, KOSKINEN P and ISOLAURI E (2001b), 'Probiotics in primary prevention of atopic disease: a randomised placebo-controlled trial', *Lancet*, **357**(9262), 1076–1079.

KIRJAVAINEN P V and GIBSON G R (1999), 'Healthy gut microflora and allergy: factors influencing development of the microbiota', *Ann Med*, **31**(4), 288–292.

KIRJAVAINEN P V, APOSTOLOU E, SALMINEN S J and ISOLAURI E (1999), 'New aspects of probiotics – a novel approach in the management of food allergy', *Allergy*, **54**(9), 909–915.

KIRJAVAINEN P V, APOSTOLOU E, ARVOLA T, SALMINEN S J, GIBSON G R and ISOLAURI E (2001), 'Characterizing the composition of intestinal microflora as a prospective treatment target in infant allergic disease', *FEMS Immunol Med Microbiol*, **32**(1), 1–7.

KIRJAVAINEN P V, ARVOLA T, SALMINEN S J and ISOLAURI E (2002), 'Aberrant composition of gut microbiota of allergic infants: a target of bifidobacterial therapy at weaning?', *Gut*, **51**(1), 51–55.

KLAENHAMMER T R and KULLEN M J (1999), 'Selection and design of probiotics', *Int J Food Microbiol*, **50**(1–2), 45–57.

KULIG M, BERGMANN R, KLETTKE U, WAHN V, TACKE U and WAHN U (1999), 'Natural course of sensitization to food and inhalant allergens during the first 6 years of life', *J Allergy Clin Immunol*, **103**(6), 1173–1179.

KUVAEVA I B, ORLOVA N G, VESELOVA O L, KUZNEZOVA G G and BOROVIK T E (1984), 'Microecology of the gastrointestinal tract and the immunological status under food allergy', *Nahrung*, **28**(6–7), 689–693.

LAPA E SILVA J R, POSSEBON DA SILVA M D, LEFORT J and VARGAFTIG B B (2000), 'Endotoxins, asthma, and allergic immune responses', *Toxicology*, **152**, 31–35.

LIEVIN V, PEIFFER I, HUDAULT S, ROCHAT F, BRASSART D, NEESER J R and SERVIN A L (2000), 'Bifidobacterium strains from resident infant human gastrointestinal microflora exert antimicrobial activity', *Gut*, **47**(5), 646–652.

LODINOVÁ-ŽÁDNÍKOVÁ R and CUKROWSKÁ B (1999), 'Influence of oral colonization of the intestine with a non-enteropathogenic *E. coli* strain after birth on the frequency of infectious and allergic diseases after 10 and 20 years', *Immunol Lett*, **69**, 64.

LOSKUTOVA I E (1985), 'Effectiveness of using Maliutka and Malysh adapted propionic–acidophilus mixtures in the combined treatment of congenital hypotrophy', *Vopr-Pitan*, **May–June**, 17–20.

MACDONALD T T (1998), 'T cell immunity to oral allergens', *Curr Opin Immunol*, **10**(6), 620–627.

MACDONALD T T and CARTER P B (1979), 'Requirement for a bacterial flora before mice generate cells capable of mediating the delayed hypersensitivity reaction to sheep red blood cells', *J Immunol*, **122**, 2624–2629.

MACDONALD T T and PETTERSSON S (2000), 'Bacterial regulation of intestinal immune responses', *Inflamm Bowel Dis*, **6**(2), 116–122.

MACDOUGALL C F, CANT A J and COLVER A F (2002), 'How dangerous is food allergy in childhood? The incidence of severe and fatal allergic reactions across the UK and Ireland', *Arch Dis Child*, **86**(4), 236–239.

MADSEN K L, DOYLE J S, JEWELL L D, TAVERNINI M M and FEDORAK R N (1999a), '*Lactobacillus* species prevents colitis in interleukin 10 gene-deficient mice', *Gastroenterology*, **116**(5), 1107–1114.

MADSEN K L, MALFAIR D, GRAY D, DOYLE J S, JEWELL L D and FEDORAK R N (1999b), 'Interleukin-10 gene-deficient mice develop a primary intestinal permeability defect in response to enteric microflora', *Inflamm Bowel Dis*, **5**(4), 262–270.

MAJAMAA H and ISOLAURI E (1997), 'Probiotics: a novel approach in the management of food allergy', *J Allergy Clin Immunol*, **99**, 179–185.

MAJAMAA H, ISOLAURI E, SAXELIN M and VESIKARI T (1995), 'Lactic acid bacteria in the treatment of acute rotavirus gastroenteritis', *J Pediatr Gastroenterol Nutr*, **20**, 333–338.

MALOY K J and POWRIE F (2001), 'Regulatory T cells in the control of immune pathology', *Nat Immunol*, **2**(9), 816–822.

MCDONEL J L (1979), 'The molecular mode of action of *Clostridium perfringens* enterotoxin', *Am J Clin Nutr*, **32**(1), 210–208.

MEDZHITOV R (2001), 'Toll-like receptors and innate immunity', *Nature Rev Immunol*, **1**(2), 135–145.

MOORE K W, DE WAAL MALEFYT R, COFFMAN R L and O'GARRA A (2001) 'Interleukin-10 and the interleukin-10 receptor', *Annu Rev Immunol*, **19**, 683–765.

MOREAU M C, DUCLUZEAU R, GUY-GRAND D and MULLER M C (1978), 'Increase in the population of duodenal IgA plasmocytes in axenic mice monoassociated with different living or dead bacterial strains of intestinal origin', *Infect Immun*, **21**, 532–539.

NAGLER-ANDERSON C and SHI H N (2001), 'Peripheral nonresponsiveness to orally administered soluble protein antigens', *Crit Rev Immunol*, **21**(1–3), 121–131.

NAGLER-ANDERSON C (2001), 'Man the barrier! Strategic defences in the intestinal mucosa', *Nature Rev Immunol*, **1**(1), 59–67.

NEISH A S, GEWIRTZ A T, ZENG H, YOUNG A N, HOBERT M E, KARMALI V, RAO A S and MADARA J L (2000), 'Prokaryotic regulation of epithelial responses by inhibition of I°B-± ubiquitination', *Science*, **289**, 1560–1563.

OBISO R J JR, LYERLY D M, VAN TASSELL R L and WILKINS T D (1995), 'Proteolytic activity of the *Bacteroides fragilis* enterotoxin causes fluid secretion and intestinal damage in vivo', *Infect Immun*, **63**(10), 3820–3826.

OGGIONI M R, POZZI G, VALENSIN P E, GALIENI P and BIGAZZI C (1998), 'Recurrent septicemia in an immunocompromised patient due to probiotic strains of *Bacillus subtilis*', *J Clin Microbiol*, **36**, 325–326.

OUWEHAND A C and SALMINEN S J (1998), 'The effects of cultured milk products with viable and non-viable bacteria', *Int Dairy J*, **8**, 749–758.

OUWEHAND A C, ISOLAURI E, HE F, HASHIMOTO H, BENNO Y and SALMINEN S (2001), 'Differences in *Bifidobacterium* flora composition in allergic and healthy infants', *J Allergy Clin Immunol*, **108**(1), 144–145.

OZINSKY A, UNDERHILL D M, FONTENOT J D, HAJJAR A M, SMITH K D, WILSON C B, SHROEDER L and ADEREM A (2000), 'The repertoire for pattern recognition of pathogens by the innate immune system is defined by cooperation between Toll-like receptors', *Proc Natl Acad Sci USA*, **97**(25), 13766–13771.

PELTO L, ISOLAURI E, LILIUS E M, NUUTILA J and SALMINEN S (1998), 'Probiotic bacteria down-regulate the milk-induced inflammatory response in milk-hypersensitive subjects but have an immunostimulatory effect in healthy subjects', *Clin Exp Allergy*, **28**(12), 1474–1479.

PESSI T, SÜTAS Y, HURME M and ISOLAURI E (2000), 'Interleukin-10 generation in atopic children following oral *Lactobacillus rhamnosus* GG', *Clin Exp Allergy*, **30**, 1804–1808.

PLATTS-MILLS T A (2001), 'The role of immunoglobulin E in allergy and asthma', *Am J Respir Crit Care Med*, **164**(8 Pt 2), S1–5.

PRESCOTT S L, MACAUBAS C, SMALLACOMBE T, HOLT B J, SLY P D and HOLT P G (1999), 'Development of allergen-specific T-cell memory in atopic and normal children', *Lancet*, **353**(9148), 196–200.

PULVERER G, KO H L, ROSZKOWSKI W, BEUTH J, YASSIN A and JELJASZEWICZ J (1990), 'Digestive tract microflora liberates low molecular weight peptides with immunotriggering activity', *Int J Med Microbiol*, **272**, 318–327.

RAGHUPATHY R (2001), 'Pregnancy: success and failure within the Th1/Th2/Th3 paradigm', *Semin Immunol*, **13**, 219–227.

RESCIGNO M, URBANO M, VALZASINA B, FRANCOLINI M, ROTTA G, BONASIO R, GRANUCCI F, KRAEHENBUHL J P and RICCIARDI-CASTAGNOLI P (2001), 'Dendritic cells express tight junction proteins and penetrate gut epithelial monolayers to sample bacteria', *Nat Immunol*, **2**(4), 361–367.

SAD S and MOSMANN T R (1994), 'Single IL-2-secreting precursor CD4 T cell can develop into either Th1 or Th2 cytokine secretion phenotype', *J Immunol*, **153**(8), 3514–3522.

SAMPSON H A (1992), 'The immunopathogenic role of food hypersensitivity in atopic dermatitis', *Acta Derm Venereol Suppl* (Stockh), **176**, 34–37.

SAMUEL B, LEHRER S B, AYUSO R and REESE G (2002), 'Current understanding of food allergens', *Ann NY Acad Sci*, **964**, 69–85.

SCHNARE M, BARTON G M, HOLT A C, TAKEDA K, AKIRA S and MEDZHITOV R (2001), 'Toll-like receptors control activation of adaptive immune responses', *Nat Immunol*, **2**(10), 947–950.

SHATERNIKOV V A, KUVAEVA I D, LADODO K S, ORLOVA N G and VESELOVA O L (1982), 'General and local humoral immunity and intestinal microflora in children with skin manifestations of food allergy', *Vopr-Pitan*, **Sep–Oct**, 51–56.

STEIDLER L, HANS W, SCHOTTE L, NEIRYNCK S, OBERMEIER F, FALK W, FIERS W and REMAUT E (2000), 'Treatment of murine colitis by *Lactococcus lactis* secreting interleukin-10', *Science*, **289**(5483), 1352–1355.

STRACHAN D P (1989), 'Hay fever, hygiene, and household size', *BMJ*, **299**(6710), 1259–1260.

STROBER W, KELSALL B and MARTH T (1998), 'Oral tolerance', *J Clin Immunol*, **18**(1), 1–30.

SUDO N, SAWAMURA S, TANAKA K, AIBA Y, KUBO C and KOGA Y (1997), 'The requirement of intestinal bacterial flora for the development of an IgE production system fully susceptible to oral tolerance induction', *J Immunol*, **159**, 1739–1745.

SÜTAS Y, SOPPI E, KORHONEN H, SYVAOJA E L, SAXELIN M, ROKKA T and ISOLAURI E (1996), 'Suppression of lymphocyte proliferation *in vitro* by bovine caseins hydrolysed with *Lactobacillus* GG-derived enzymes', *J Allergy Clin Immunol,* **98**, 216–224.

TAKAHASHI T, NAKAGAWA E, NARA T, YAJIMA T and KUWATA T (1998), 'Effects of orally ingested *Bifidobacterium longum* on the mucosal IgA response of mice to dietary antigens', *Biosci Biotechnol Biochem,* **62**, 10–15.

TAKEUCHI O and AKIRA S (2001), 'Toll-like receptors; their physiological role and signal transduction system', *Int Immunopharmacol,* **1**(4), 625–635.

TAPPING R I, AKASHI S, MIYAKE K, GODOWSKI P J and TOBIAS P S (2000), 'Toll-like receptor 4, but not toll-like receptor 2, is a signaling receptor for *Escherichia* and *Salmonella* lipopolysaccharides', *J Immunol,* **165**(10), 5780–5787.

TRAPP C L, CHANG C C, HALPERN G M, KEEN C L and GERSCHWIN M E (1993), 'The influence of chronic yoghurt consumption on population of young and elderly adults', *Int J Immunother,* **9**, 53–64.

UNITED STATES CENTERS FOR DISEASE CONTROL (1998), 'Forecasted state-specific estimates of self-reported asthma prevalence – 1998', *Morbidity and Mortality,* **47**, 1022–1025.

VAN ASPEREN P P and KEMP A S (1989), 'The natural history of IgE sensitisation and atopic disease in early childhood', *Acta Paediatr Scand,* **78**(2), 239–245.

VAN DEN BIGGELAAR A H, VAN REE R, RODRIGUES L C, LELL B, DEELDER A M, KREMSNER P G and YAZDANBAKHSH M (2000), 'Decreased atopy in children infected with *Schistosoma haematobium*: a role for parasite-induced interleukin-10', *Lancet,* **356**(9243), 1723–1727.

VON DER WEID T, BULLIARD C and SCHIFFRIN E (2001), 'Induction by a lactic acid bacterium of a population of CD4+ T cells with low proliferative capacity that produce transforming growth factor β and interleukin-10', *Clin Diagn Lab Immunol,* **8**, 695–701.

WHEELER J G, BOGLE M L, SHEMA S J, SHIRREL M A, STINE K C, PITTLER A J, BURKS A W and HELM R M (1997), 'Impact of dietary yoghurt on immune function', *Am J Med Sci,* **313**, 120–123.

WILLS-KARP M, SANTELIZ J and KARP C L (2001), 'The germless theory of allergic disease: revisiting the hygiene hypothesis', *Nature Rev Immunol,* **1**(1), 69–75.

WOODS R K, STONEY R M, RAVEN J, WALTERS E H, ABRAMSON M and THIEN F C (2002), 'Reported adverse food reactions overestimate true food allergy in the community', *Eur J Clin Nutr,* **56**(1), 31–6.

WUTHRICH B and BALLMER-WEBER B K (2001), 'Food-induced anaphylaxis', *Allergy,* **56**(Suppl 67), 102–104.

YASUI H, NAGAOKA N, MIKE A, HAYAKAWA K and OHWAKI M (1992), 'Detection of *Bifidobacterium* strains that induce large quantities of IgA', *Microb Ecol Health Dis,* **5**, 155–162.

6

Dairy products and the immune function in the elderly

H. Gill, Massey University, Palmerston North, New Zealand

6.1 Introduction

The population of elderly persons is growing rapidly in all countries. According to latest WHO estimates there are about 600 million people aged 60 years or older in the world and this figure is expected to rise to over 2 billion (20% of the world population) by the year 2050 (United Nations, 1999; WHO, 1999). Furthermore, the number of people over the age of 80 is expected to more than triple, and the number of centenarians is projected to increase by 15-fold within the next 50 years. This increase in life expectancy has been the result of improvements in diet, health-care and living conditions.

Ageing is associated with a decline or deterioration in many physiological functions including the immune system (Wick and Grubeck-Loebenstein, 1997). Decline in immune function with age is termed as immunosenescence and is found in all mammalian species. Immunosenescence is deemed to be responsible for enhanced morbidity and mortality in the aged. Epidemiological and clinical studies have revealed that elderly persons suffer more often from infectious diseases (respiratory tract and urinary tract infections, gastroenteritis, tuberculosis, endocarditis and septicaemia), cancers and autoimmune and atopic reactions than their younger counterparts (Kelley and Bendich, 1996). As a result, the care of elderly accounts for a significantly higher percentage of health-care costs relative to their population. For example, 11% of the elderly population use over 40% of health-care resources in the US (Chandra, 1989).

The effective means to correct this age-related immune dysfunction are highly sought and the years since 1980 have seen enormous interest in both fundamental and applied research that is aimed at improving geriatric health. It has been suggested that dietary supplementation could be used to improve and maintain immune health in the elderly. Being a non-invasive, cost-effective and practical means of health improvement, it is of special interest to both consumers and the health-care providers. The challenge, therefore, is to identify nutrients/food components or develop products that are able to improve immune health. This chapter will outline scientific evidence for immunological changes associated with ageing and then discuss evidence, both scientific and clinical, regarding the use of milk or milk products (including fermented products) for optimizing immune health in the elderly.

6.2   The immune system

To fully appreciate the significance of age-related alterations in immune function to human health it is important to understand how the immune system works. The immune system comprises a network of organs and cells that are strategically distributed throughout the body. The major function of the immune system is to defend the body against invasion by pathogenic organisms and against the development and metastasis of cancers. These protective effects are mediated by innate (non-specific) and adaptive (specific) immune systems. The major effectors of non-specific immunity include polymorphonuclear cells (PMN), mononuclear phagocytes, natural killer cells (NK cells) and the complement system. These responses represent the first line of host defence and operate non-selectively against foreign and abnormal antigens/materials. Inflammation is a form of non-specific host immune response where both humoral and cellular factors act in concert.

The specific immunity comprises antibody- and cell-mediated responses and is characterized by its specificity and memory; the responses are prompt and of greater intensity. Antibody-mediated immunity is accomplished by the production of specific antibodies by plasma cells (mature B cells). T cells are the central component of the cell-mediated immunity. T cells of the CD4$^+$ phenotype are helper/inducer cells whereas CD8 cells are suppressor/cytotoxic cells. These sub-populations of T cells act in concert to modulate and regulate the immune responses. Furthermore, CD4$^+$ cells can be divided into Th1 and Th2 cells based on their cytokine profile (Mosmann and Coffma, 1998; Ginaldi et al., 1999). Th1 cells produce IFN-γ, IL-2, IL-12 and L-18 and are vital for cell-mediated immunity. On the other hand, Th2 cells produce IL-4, IL-5, IL-6, IL-10 and IL-13 and promote antibody production including class switching from IgM to IgG1 and IgE; activation of Th2-type responses is associated with atopy. The development of an effective immune response requires a balance between Th1 and Th2 subsets and inappropriately skewed responses are associated with pathology (Crohn's disease, food allergies, diabetes, etc.).

Which of these mechanisms are activated depends on the nature of the disease-causing agent. Primarily, extracellular bacteria and viruses are eliminated/controlled by phagocytic cells, complement and antibodies. Intracellular bacteria, parasites and fungi are controlled by the activation of T cells and the secretion of cytokines such as IFN-γ that activate monocytes/macrophages. Cytotoxic T cells and NK cells eliminate cancerous cells and intracellular viruses. Most, if not all, immunocompetent cells and molecules produced by them have overlapping functions. Thus, an effective host response is dependent upon the successful interaction between various arms/components of the innate and specific immune systems. In the case of an over-activated immune system, such as in allergies and autoimmune diseases, modulation and regulation of the immune system are important in the resolution of the disease state.

6.3   Immunosenescence

Ageing is associated with profound and complex changes in immune function. Whether altered capability of the immune system with ageing simply reflects a general deterioration of the immune system or reshaping/realignment of the immune system is a subject of active debate (Weksler, 1995). A careful evaluation of the literature suggests that immunosenescence probably represents a combination of both, a decline and an active realignment of important functions. The evidence for the latter is more compelling as some immune responses are decreased, some increased, while others remain unchanged in the aged (Weksler, 1995). An increased prevalence of autoimmune, atopic and inflammatory reactions and the reduced ability of the elderly to mount effective immune responses to antigenic challenge further reflect this paradigm (Goodwin, 1995; Franceschi *et al.*, 1995). Studies on carefully selected healthy individuals have also showed that ageing does not affect immunocompetence in all individuals. Some centenarians were found to exhibit several immune functions comparable to young individuals (Franceschi *et al.*, 1995).

6.3.1   Cell-mediated immunity

T-cell function is more profoundly affected than humoral immunity by ageing (Franceschi *et al.*, 1995; Doria and Frasca, 1997; Lesourd, 1997). Aged individuals have more immature ($CD2^+$ $CD3^+$) and fewer mature ($CD3^+$) T cells (Lesourd and Mazari 1999). The absolute number of CD4 cells is slightly increased, while the number of CD8 cells is variously reported as normal, decreased or increased (Ginaldi *et al.*, 2001; Gill *et al.*, 2001b). T-cell proliferation responses, whether measured *in vitro* as a response to mitogens or antigen stimulation, mixed leucocyte reaction or *in vivo* as a delayed-type hypersensitivity (DTH) response to recall antigens (bacterial and fungal products and dinitrochlorobenzene (DNCB)) are significantly reduced (reviewed by Lesourd, 1997; Ginaldi *et al.*, 2001). T-cell helper activity for proliferation and maturation of B cells, and the generation of cytotoxic effectors also declines with age (reviewed in Miller, 1996). A decrease in the expression of CD28, an important co-stimulatory molecule, in the elderly has also been observed (Pawelec *et al.*, 1995). The ability of clonal cells to undergo clonal expansion is also reduced (Tyan, 1981).

Ageing is also accompanied by a progressive shift in T cells from naïve (CD25RA[+]) to memory cells (CD45RO[+]) (Lesourd and Meaume, 1994); this naïve to memory cell shift is observed in both CD4 and CD8 T cells (Ernst *et al.*, 1990a, b). The expression of CD3 is down-regulated on both naïve and memory cells (Ginaldi *et al.*, 2001).

Alterations in both functional and proliferative responses of naïve and memory T cells have been reported during senescence (Ginaldi *et al.*, 2000). A major consequence of these changes is that elderly are less able to mount an effective immune response to novel pathogens, and are less responsive to immunization strategies; an effective immune response to novel antigens or pathogens depends on the availability and optimal functional capacity of naïve T cells. Thymic involution has been deemed responsible for dramatic decline in T-cell function (Chandra *et al.*, 1997a, b).

The cytokine network also exhibits profound alterations with age. The production of some cytokines is increased while the production and utilization of others is decreased (Kubo and Cinader, 1990; Green-Johnson *et al.*, 1991; Ernst *et al.*, 1995; Cakman *et al.*, 1996). In general, the cytokine production shifts from Th1 to Th2 type. Compared with young individuals, the production of IL-2 and IFN-γ is reduced and the production of IL-4 and IL-10 is increased in the aged. These changes are thought to be due to a shift in T cells from naïve to memory phenotype; naïve T cells preferentially produce IL 2 and IFN-γ, while memory T cells produce IL-4 and IL-10 (Ernst *et al.*, 1990a, b).

The use of immune status as an indicator of general health (survival and disease risk) has been the subject of many studies. It has been suggested that a cluster of immune parameters including low CD4[+] T cell count, high CD8[+] T cell count, low B-cell numbers, low IL-2 production and low T-cell proliferation could have prognostic relevance to high subsequent mortality in the very old people; subjects that developed this high-risk profile showed poorer survival than those who did not (Ferguson *et al.*, 1995). Anergy in skin delayed-type hypersensitivity responses to recall antigens (Wayne *et al.*, 1990), and a decline in the numbers of both CD4[+] and CD16[+] (NK cells) at baseline has also been associated with increased mortality in subsequent years (Pawelec and Solana, 1997); individuals in the lowest quartile of CD4[+] and CD16[+] cell counts had about tenfold increased risk of dying in the first two years of follow-up compared with those in the highest quartile; low CD4[+] or CD16[+] cell counts carried a twofold risk.

6.3.2   Humoral immunity

Ageing affects both quality and quantity of antibody responses (LeMaoult *et al.*, 1997). The concentration of IgG in the serum is reduced while that of IgA is increased (Chandra, 1997a, b). The ability to mount antibody responses to new antigens and pathogens is significantly reduced (Moulias *et al.*, 1985). Booster responses remain unaffected, but it takes longer to reach peak levels. Antibody specificity shifts from foreign to auto, antibody isotype from IgG to IgM and antibody affinities from high to low (LeMaoult *et al.*, 1997). The number of B cells is not reduced but the frequency of B cells responsive to some, but not all, antigens is reduced (Miller, 1996). As a result, the elderly respond poorly to most vaccines. Memory responses induced by vaccination are also of short duration in old compared with young subjects (Poweres and Belshe, 1993). These changes in immunocompetence are attributed to impaired T-cell function in the elderly.

6.3.3   Innate immunity

The number of NK cells increases with ageing (Ales-Martinez *et al.*, 1988; Lesourd and Meaume, 1994). NK cells derived from the elderly (SENIEUR) exhibit increased expression of activation surface markers CD95 and HLA-DR, and reduced IL-2 induced proliferation and the expression of CD69, compared with cells from young donors. Studies on NK cell function have produced conflicting results (Lesourd, 1997; Chandra, 1989; Gill *et al.*, 2001b). A reduction in NK cell activity (Facchini *et al.*, 1987), and a reduced lytic activity on a per-cell basis in a subset of NK cells (recognized by GL 183 and EB 6) has been reported in the aged individuals (Franceschi *et al.*, 1995; Pawelec *et al.*, 1995). However, there is also evidence that the elderly subjects selected on the basis of strict health-performance criteria do not show reduction in NK cell numbers or activity (Franceshi *et al.*, 1995; Ligthart *et al.*, 1989; Lesourd and Meaume, 1994). The significance of these differences to health maintenance remains unclear. A recent study has shown that a persistently low NK cell activity is a predictor of impending morbidity (Franceschi *et al.*, 1995).

A body of evidence suggests that the function of macrophages and granulocytes is also impaired in the elderly (Chandra, 1989). The number of macrophages remains unchanged. However, more macrophages are required to elicit an optimal T-cell response to mitogens (Rich *et al.*, 1993). Reduced responsiveness of aged mice to pneumococcal vaccine is attributed to a defective macrophage function (Garg *et al.*, 1996). Organ-specific defects in macrophage function have also been reported in the aged (Higashimoto *et al.*, 1994). The ability of mononuclear cells from healthy elderly subjects and centenarians to produce proinflammatory cytokines such as IL-1, IL-6, IL-8 and TNF-α is enhanced (Rich *et al.*, 1993). However, the spontaneous release of IL-8 is decreased in the elderly compared with younger counterparts (Clark and Peterson, 1994). Monocytes from aged mice produce less hydrogen peroxide (Ding *et al.*, 1994) and higher amounts of prostaglandin E2 (Hayek *et al.*, 1997). Prostaglandin E_2 (PGE) is known to have a suppressive effect on T-cell proliferation and IL-2 production, and lymphocytes from aged subjects exhibit enhanced sensitivity to PGE (Goodwin, 1992).

The numbers of neutrophil and neutrophil precursors are not affected during ageing, but the proliferative responses of neutrophil precursor cells to G-CSF are lowered. Elderly patients with sepsis often fail to mount leucocytosis (Corberand *et al.*, 1981; McLaughlin *et al.*, 1986). Polymorphonuclear cells derived from the elderly show reduced migration ability. The capacity to phagocytose opsonized bacteria, yeast or zymosan particles is also reduced. Studies on the microbicidal activity of neutrophils from the elderly have produced conflicting data. However, a majority of the studies show an age-related decline in the microbicidal mechanisms of neutrophils against yeast and bacteria (Fulop *et al.*, 1986; Lipschitz *et al.*, 1988; Esperaza *et al.*, 1996). Whether these altered response arise in neutrophil progenitors or reflect the influence of cytokines on mature neutrophils in the circulation is not known (Butcher *et al.*, 2000).

The classical and complement activation pathways appear to be maintained/ preserved in the elderly. Ageing has little affect on antigen-processing pathways (Lesourd and Mazari, 1997). However, studies in old mice have shown that the ability of follicular dendritic cells to process and present antigen-containing immune complexes and the transport of antigens into lymph node germinal centres by migrating dendritic cells are impaired (Homes *et al.*, 1984). The significance of these deficits in the generation of immune responses is not known.

Neutrophils and macrophages play an important role in mediating protection against bacterial and fungal infections, which are largely responsible for increased morbidity and mortality in the elderly. Thus reduced neutrophil supply and/or function may account for increased susceptibility of the elderly to infection (Butcher *et al.*, 2000). NK cells play an important role in protection against neoplasms and the control of viral infections, as well as in the regulation of cell-mediated immunity via the secretion of interferons. Deficient NK cell function may account, at least in part, for enhanced susceptibility of the elderly to viral infections and cancers.

6.4   Nutrition and immune function in the elderly

Nutrition is a critical determinant of immunocompetence and malnutrition is the most common cause of immunodeficiency worldwide (Chandra, 1989, 1995). In the elderly, malnutrition and ageing exert a cumulative effect on immunocompetence and contribute significantly to increased morbidity and mortality (Lesourd, 1995; Heuser and Adler, 1997; Ginaldi *et al.*, 2001). Both protein energy malnutrition (PEM) and micronutrient (such as Zn, Se and vitamins (vitamin B6, folic acid, vitamin E)) deficiencies are common in the aged population. At least one-third of the elderly in developed countries exhibit nutritional deficiencies of some kind (Haller *et al.*, 1996); up to 18% of apparently healthy elderly (Lesourd *et al.*, 1998) and one-third of low-income elderly are reported to have vitamin B6 deficiency. In another study, Chandra (1992) reported vitamin E deficiency in 8.3% of healthy elderly subjects who lived in their own homes. Lack of variety, reduced energy intake, the presence of malabsorption in some individuals and drug–nutrient interaction are the major factors contributing to nutritional deficiencies in the aged population. The physically isolated, those living alone, the socially isolated, those with sensory or mental impairment or with chronic systemic diseases, the very poor and the very rich show the highest risk of malnutrition (Chandra, 1989).

Both PEM and micronutrient deficits are known to impair immune responsiveness (Chandra, 1985; Lesourd, 1995, 1997; Rasmussen *et al.*, 1996; Buzina-Suboticanec *et al.*, 1998). The magnitude of immunological impairment depends on the type of nutrient involved, its interaction with other nutrients, the severity of deficiency, the presence of concomitant infection and the age of the subject (Chandra, 1997a,b). Specific deficits produce specific effects. For example, low protein intake is associated with a sharp drop in immunity in aged animals and humans (Lipschitz *et al.*, 1988; Lesourd *et al.*, 1998; Chandra, 1990; Lesourd, 1990a, b, 1995; Lesourd and Meaume, 1994). PEM reduces both specific and non-specific immune responses. Lymphocyte counts (CD3), the proportions of CD4+ and CD8+ T cells and the ratio of CD4+ to CD8+ T cells are markedly reduced. Lymphocyte proliferation responses and DTH responses to recall and new antigens are impaired. The production of several cytokines such as IL-1, IL-2 and IFN-γ is also diminished. The efficacy of vaccines is significantly reduced. Antibody levels, sero-conversion rate as well as antibody affinity are reduced following vaccination in the elderly with PEM. A high level (about 75%) of unresponsiveness to vaccines in hospitalized elderly patients is related to undernutrition (Lesourd, 1997). PEM has little effect on antibody responses to T cell-independent antigens or antigens administered with adjuvants. However, the affinity of antibodies is reduced in patients with PEM. The concentration of secretory IgA to viral vaccines is also decreased. Phagocytic cell function and complement activity are also reduced. The magnitude of immune deficiency is correlated with the degree of nutritional deficit. Elderly subjects with reduced serum albumin levels (<30 g/l) show lymphopaenia and reduced blood CD4 counts (<400/mm^3) similar to that found in HIV infections (Lesourd, 1990a, b). Similarly, Zn deficiency reduces peripheral blood lymphocyte counts, T-lymphocyte proliferation responses, IL-2 production and T-cell cytotoxicity (Keen and Gershwin, 1990; Prasad *et al.*, 1997; Dardenne *et al.*, 1982).

Several recent studies have demonstrated that supplementation with single or multiple nutrients could be used to restore, at least partly, diminished immuno-competence in the institutionalized aged patients as well as self-sufficient, apparently healthy, elderly (reviewed by Chandra, 1989, 1995; Lesourd, 1997; High, 1999). Lesourd (1995) reported enhanced antibody responses to tetanus toxoid vaccine in undernourished elderly subjects following supplementation with 2580 kJ/day complete dietary formula (ready to eat). Similarly, supplementation with a complete dietary formula containing 1600–2100 kJ/day was found to correct protein energy-associated defects in cell-mediated immunity (lymphocyte proliferation to mitogens and DTH to recall antigens) in undernourished elderly patients (Lesourd *et al.*, 1990). Chandra (1992) reported the immunity-enhancing (immune function and clinical end-points) effect of multivitamin and mineral supplementation (at a level similar to the recommended daily allowance, RDA) in the elderly patients. After 12 months, the supplemented group showed significant enhancement in percentage or proportion of CD4+ cells, NK cell activity, proliferation responses to T cell mitogens, IL-2 production and IL-2 receptor expression. The incidence of infection-related illness was also substantially reduced in the supplemented group; the mean number of days of illness due to infection in supplemented group was 23 *vs* 48 in the placebo group. Antibiotic usage in supplemented group (18 days) was also lower than the placebo group (32 days).

In institutionalized elderly subjects, supplementation with defined oligovitamins

(vitamins A, C and E) increased numbers of circulating CD4[+] and CD8[+] T cells and proliferation responses to mitogens (Penn *et al.*, 1991). Zn supplementation has also been shown to increase the number of CD3[+], CD4[+], CD16 and CD56 lymphocytes, NK cell activity and DTH responses to recall antigens (Bogden *et al.*, 1990; Fortes *et al.*, 1998; High, 1999). Similarly supplementation with very high levels of vitamin E has been reported to enhance cellular and humoral immune responses in the aged (Meydani *et al.*, 1997). Together, these observations provide evidence that supplementation could be used to delay, prevent or restore age-related immune dysfunction and thus reduce morbidity and mortality due to immune-related diseases. However, the challenge is to identify nutrients/products that are able to provide these benefits.

6.5   Bovine milk and immunomodulation

Bovine milk contains a variety of proteins, carbohydrates, lipids, vitamins, minerals and growth factors. In addition to meeting the growth and developmental needs (nutritional requirements) of neonates, milk plays an important role in protecting the neonate until its own immune system becomes fully competent. Milk mediates its protective function by two distinct ways: first by providing passive protection through the transfer of molecules such as antibodies, and secondly by regulating the development of an effective and a balanced immune system.

The immunomodulatory effects of milk are attributed to a wide range of milk components, including whey proteins, caseins, lipids, hormones and growth factors (Gill, 2000; Cross and Gill, 2000). In addition, milk proteins contain a host of immunoactive peptides that can be released following enzymatic hydrolysis *in vitro*, commercial processing or normal digestion (Gill, 2000). Furthermore, fermented milk products contain lactic acid bacteria (LAB) that are able to influence a range of host immune responses (Gill, 1998; Erickson and Hubbard, 2000; Wold, 2001; Isolauri *et al.*, 2001; Gill and Cross, 2002).

Several *in vitro* and *in vivo* studies have shown that milk products (including fermented products) and milk components are able to influence aspects of both specific and non-specific host immune responses. These have been reviewed in detail by Gill (1998, 2000), Gill *et al.* (2000), Cross and Gill (2000), Erickson and Hubbard (2000), Korhonen *et al.* (2000); Wold (2001) and Isolauri *et al.* (2001). This review will focus only on observations derived from animal models or human studies. Immunomodulatory effects of fermented dairy products and probiotic LAB will be discussed in a separate section.

6.6   Milk proteins

6.6.1   Humoral immune response
The ability of milk proteins to enhance antibody responses to both T cell-dependent and independent antigens is well documented. Studies by Bounous and

colleagues (Bounous and Kongshawn, 1989; Bounous *et al.*, 1985, 1988) showed that dietary whey protein concentrate (WPC) is able to enhance antibody responses to systemic immunization with sheep red blood cells (a T cell-dependent antigen) and trinitrophenylated (TNP)-ficoll (a T cell-independent antigen); antibody responses to sheep red blood cells (SRBC) among WPC-fed mice were almost five times higher than mice fed equivalent diets of similar nutritional efficiency containing casein, soy, wheat, corn protein, egg albumin, beef/fish protein or Purina mouse chow. The immune-enhancing effect was manifested after two weeks of feeding and persisted for at least eight weeks (as long as the dietary treatment continued) (Bounous *et al.*, 1988).

Mixing whey protein with other protein sources (soy protein or casein) also resulted in enhanced antibody responses (Bounous *et al.*, 1988). WPC has also been found to enhance humoral responses to SRBC in CBA/N × DBA/2J F1 mice carrying the 'xid defect' (an accessory cell–cell interaction defect) (Bounous *et al.*, 1985). The enhancing effect of dietary whey on serum antibody responses to ovalbumin has also been reported (Wong and Watson, 1995; Gill *et al.*, 2002b).

Recent studies have provided evidence that milk proteins are also able to enhance mucosal antibody responses to orally administered antigens and vaccines. Mice fed diets containing whey protein concentrate, for 4–12 weeks, showed significantly higher mucosal antibody responses to ovalbumin, cholera toxin, polio and 'flu vaccine compared with mice fed normal chow (Gill *et al.*, 2002a, b, c, d; Low *et al.*, 2001). Potentiation of antibody responses occurred irrespective of when immunizations commenced (at the beginning, during the early stages or during the middle stages) during the dietary regimen (Low *et al.*, 2001). Inclusion of WPC into follow-on milk was also able to enhance both humoral and systemic antibody responses to orally administered antigens (Gill *et al.*, 2002b, c, d).

Whey contains a complex mixture of major and minor proteins. Diets containing individual major whey proteins have also been reported to increase systemic and mucosal antibody responses. Bounous and Kongshavan (1989) and Gill *et al.* (2002a) reported significantly higher serum and/or mucosal antibody responses in mice fed diets containing β-lactoglobulin, α-lactalbumin, BSA, lactoperoxidase or immunoglobulin than mice fed other protein sources. Mice fed lactoferrin (bLF) hydrolysates were also reported to exhibit higher concentrations of antigen-specific IgA antibody responses in bile and intestinal contents, following oral immunization with cholera toxin than mice fed control diets (Miyauchi *et al.*, 1997). The ability of caseinoglycopeptides (CGP), derived from κ-casein, to modulate antigen-specific serum antibody responses has also been demonstrated (Monnai *et al.*, 1998); CGP intake suppressed IgG antibody responses to orally administered antigens.

Bounous and Gold (1989) reported that immunoenhancing ability of WPC was contingent upon its undenatured state and is significantly greater for WPCs that contain higher concentrations of bovine Ig and serum albumin. However, recent studies in our laboratory have shown that immunomodulatory properties of WPC are resistant to commercial-type processing. Both native (commercially produced) and reconstituted (separated into major protein fractions and then recombined)

WPC were found to be effective at enhancing a range of humoral and cellular immune responses (Gill *et al.*, 2002a).

The effect of whey proteins on humoral immunity in human subjects has not been investigated. However, these results indicate that WPC or whey fractions enriched for various whey proteins could be effectively used as dietary adjuvants to enhance responses to vaccines in population groups, such as elderly, that show reduced ability to respond optimally to various vaccines.

6.6.2   Cellular immune responses

Cell-mediated immune responses play an important role in host protection against intracellular bacteria, viral infections and cancers. The ability of milk components to modulate cell-mediated immune responses has been the subject of several *in vitro* studies (Cross and Gill, 2000; Gill *et al.*, 2000). However, only a few studies have examined the effect of milk products on cell-mediated immune responses *in vivo*. Ingestion of bovine milk whey proteins, for five to eight weeks, was found to enhance footpad delayed-type hypersensitivity responses and *in vitro* concanavalin A-induced spleen cell proliferation in mice (Wong and Watson, 1995). Enhancement of lymphocyte proliferative responses to T-cell mitogens in mice fed WPC or increasing concentrations of whey protein hydrolysates, primed with *Mycobacterium bovis* BCG, one week prior to the assessment of immune function, has also been observed (Bounous *et al.*, 1983). In a later study, however, Bounous and Kongshavan (1985) found that whey protein had little effect on cell-mediated immunity as measured by graft *vs* host rejection, DTH and proliferation responses to T-cell mitogens. Whether it was due to different strains of mice, length of dietary supplementation/or methods employed is not known. Recent studies in our laboratory have shown that diets containing WPC are able to significantly enhance lymphocyte proliferative responses to T and B cell mitogens in mice (Gill *et al.*, 2002a,b,c). Fractionation of WPC further revealed that the lactoperoxidase-enriched fraction was able to significantly enhance both T and B cell mitogenesis, while the β-lactoglobulin- and immunoglobulin-enriched fractions were effective at enhancing mitogen-stimulated T-cell proliferation only (Gill *et al.*, 2002a). Casein-derived peptides have also been found to have enhancing effect on lymphocyte function. Spleen cells from mice fed κ-casein-derived CGP had significantly higher proliferative responses to stimulation with T-cell mitogens, phytohaemagglutinin (PHA); responses to B cell mitogens were not affected (Monnai *et al.*, 1989).

The ability of WPC to influence T-cell populations has also been reported. Mice fed undenatured whey protein concentrate for four weeks exhibited higher numbers of L3T4[+] cells (helper cells) and a higher ratio of L3T4[+]/Lyt-2[+] cells (helper: suppressor) compared with mice fed an isocaloric casein diet (Bounous and Gold, 1993; Bounous *et al.*, 1993). A significant increase in total white blood cells, CD4[+] and CD8[+] lymphocyte counts, and Con A-stimulated IFN-γ production by spleen cells in mice fed whey, compared with mice fed other proteins, have also been observed (Ford *et al.*, 2001). Ishida *et al.* (1992a) reported that oral administration of immune milk (from cows immunized with human intestinal bacteria) was also effective at enhancing cell-mediated immunity in mice; the mitogenic responses of mesenteric lymph node cells and the redirected cytotoxic activity of interepithelial lymphocytes to P815 tumour cells with anti-CD3mAb were significantly higher in mice fed immune milk compared with the control group.

Several minor whey components (such as lactoferrin, lactoperoxidase, peptides/growth factors) have also been found to enhance cell-mediated immunity. For example, Zimecki and Kruzel (2000) reported a dose-dependent enhancement of delayed-type hypersensitivity responses to a range of antigens (ovalbumin, BCG and SRBC) in mice, following oral or parenteral administration of bLF. In tumour-bearing mice, oral administration of bovine LF was reported to inhibit experimental metastasis and augment the numbers of CD4[+], CD8 [+]and NK cells (asialoGM1[+] cells) in the spleen and peripheral blood of tumour-bearing mice; the cytotoxic activities of these cells against Yac-1 and colon 26 carcinoma were also enhanced (Wang et al., 2000; Tsuda et al., 2002). Furthermore, the numbers of CD4[+] and CD8[+] cells, and the production of IL-18, interferon-γ and caspase-1 in the intestinal epithelium, were also increased in bLF-fed mice.

In cancer patients, intake of whey proteins has been reported to increase the number of blood leucocytes (Kennedy et al., 1995). Feeding diets containing milk proteins has also been shown to delay thymic involution. Rats fed whey protein-based diets for five weeks (from 6 to 11 weeks of age) had significantly greater thymic weight than rats fed a soy protein or normal diet (Register et al., 1996); thymic involution usually occurs between 8 and 15 weeks of age in rats. An increase in thymic weight and cellularity in mice fed whey protein was also noted by Parker and Goodrum (1990). Although the effect of dietary supplementation on immune function was not investigated in these studies, the results provide putative evidence that whey proteins may be able to slow the age-associated alterations in immune function.

Oral administration of immune milk has also been found to inhibit age-related decline in immune function (Ishida et al., 1992b). Eighteen-month-old mice, fed immune milk, exhibited significantly higher redirected cytotoxicity of intra-epithelial lymphocytes (IEL) and lymphocyte proliferation response to mitogenic or alloantigenic stimulation compared with the control group. Production of antigen-specific antibodies by spleen cells, following immunization, was also higher in mice fed immune milk. In fact, the responses at 18 months of age were not different from those observed at 2 months of age. In addition, a lower level of autoantibodies was observed at 8 and 18 months of age in the immune milk-fed group compared with the control group.

Previous studies have shown that a decline in immune capability of ageing mice is associated with a significant decrease in intracellular glutathione levels and that this immune deficit could be corrected with supplementation with glutathione (Furukawa et al., 1987). Bounous et al. (1988) reported that dietary whey proteins were able to significantly increase the level of tissue glutathione levels in aged mice. Therefore, it is likely that whey proteins may help preserve immuno-competence in the aged by maintaining high glutathione levels and thus enhance host resistance to infectious diseases and cancers. Enhanced resistance to infection and tumour development, and improved survival rate in aged mice fed whey protein-based diets, compared with controls fed the nutritionally equivalent Purina chow, has been reported previously (Gold *et al.*, 1989).

The mechanisms by which milk-derived components stimulate immunity are not known. It has been suggested that the immunoenhancing ability of whey proteins is related to its unique amino acid profile. Substitution of intact proteins with an equivalent free amino acid mixture, which duplicated the specific amino acid profile of the various proteins, was found to enhance humoral immunity, but elicited a lower response compared with the intact proteins (Bounous and Kongshavn, 1989; Bounous and Gold, 1989). Whey proteins have also been reported to enhance immune function by increasing levels of glutathione in immunocompetent cells (Bounous and Gold, 1989). It is also likely that some milk components exert their influence by directly interacting with immunocompetent cells; the presence of receptors for a number of milk-proteins/peptides on immuno-competent cells have been demonstrated (Gill and Rutherfurd, 1998).

6.6.3   Innate immune responses

Consumption of bovine milk-based diets has been found to potentiate phagocytic capacity of both neutrophils and monocytes/macrophages in animal models. Gill and colleagues observed significantly higher phagocytic activity of peripheral blood leucocytes and peritoneal macrophages in mice fed whey protein for four weeks, compared with those fed mouse chow (Gill *et al.*, 2002a,b,c). Oral administration of lactoferrin, lactoperoxidase-enriched whey fractions or milkfat conjugated linoleic acid (CLA) has also been reported to enhance phagocytic capacity of mouse blood leucocytes and peritoneal macrophages (Zhao, 1999; Gill *et al.*, 2002a). An association between enhanced phagocytic cell function and increased resistance to *Klebsiella* and *Salmonella* of mice fed WPC-based diets has also been suggested (Bounous *et al.*, 1981). Intake of immune milk has also been reported to enhance phagocytic cell function and protect mice against the lethal effects of radiation: the bactericidal activity of mesenteric lymph node (MLN) cells from the irradiated mice given immune milk was significantly higher than those from the mice given control milk prior to irradiation and post-irradiation (Ishida *et al.*, 1992a).

The ability of milk and milk products to enhance PMN and monocyte function in the aged individuals has also been demonstrated. Bosche and Plat (1995) reported enhanced phagocytic capacity of blood leucocytes in elderly subjects that consumed 0.5 litres of skim milk/day for four weeks. Subjects with higher pathological values (TLC, C-reactive protein, erythrcyte sedimentation rate (ESR)) showed the highest increase. Significantly enhanced microbicidal capacity of blood neutrophils in healthy elderly subjects following regular consumption of lactose-hydrolysed milk (2 glasses/day) for a period of six weeks has also been reported (Fig. 6.1; Arunachalam *et al.*, 2000). The improvements in bacterial-killing ability of granulocytes became obvious three weeks after continuous

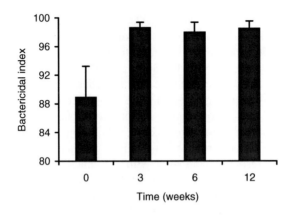

Fig. 6.1 Effect of dietary consumption of milk on polymorphonuclear (PMN) cell bactericidal activity in healthy, elderly subjects. Individuals consumed reconstituted low-fat milk powder for a six-week period (0–6 weeks). Bactericidal activity was assessed as the percentage of *Staphylococcus aureus* bacteria killed following culture with PMN. Responses at three, six and nine weeks were significantly higher ($P < 0.01$) than those observed prior to consumption of milk. Adapted from Arunachalam *et al.* (2000).

consumption of milk and the effect was maintained for 6 weeks after cessation of milk intake. A transient increase in immature neutrophils in blood and a decrease in the spontaneous release of IL-6 and TNF-α by blood cell cultures in human subjects given bLF for seven days has also been demonstrated (Zimecki *et al.*, 1999).

The ability of milk-derived components to exert anti-tumour effects is well documented (reviewed in Gill and Cross, 2000). Although the mechanisms by which these effects are mediated remain unclear, enhancement of NK cell activity has been suggested, at least in part, responsible for these anti-cancer effects. Studies by Nutter *et al.* (1983, 1990) demonstrated that anti-tumour immunity mediated by dietary milk proteins was associated with the enhancement of cytotoxic T cell and NK cell functions in mice. An increase in the number of NK cells in the blood, lymphoid tissue, lamina propria of the small intestine as well as enhanced NK cell activity in tumour-bearing mice fed bLF or bLF hydrolysates has also been demonstrated (Sekine *et al.*, 1997; Wang *et al.*, 2000). Similar observations on improved NK cell function and resistance to disease have been made by others. Shimizu *et al.* (1996) demonstrated that significant enhancement of T-cell mediated NK cell activity was associated with an increased resistance of bLF fed mice to cytomegalovirus infection. Supplementation with whey protein has also been shown to significantly enhance plasma glutathione levels and NK cell activity in patients with chronic hepatitis B (Watanabe *et al.*, 2000). Together, these observations suggest that milk components are not only able to enhance NK cell function but are also able to provide measurable improvements in disease resistance.

6.7   Antibodies and other protective agents in milk

The passive transfer of maternal immunity is pivotal for the survival of mammalian neonates. Species vary in the way this immunity is transferred to the offspring. In ruminants, the transfer of immunity from mother to offspring occurs postnatally via the colostrum. As a result, bovine colostrum and early milk contain a range of bioactive substances that can provide both specific and non-specific defences against infectious agents and foreign antigens to the neonate. The most important among these are immunoglobulins (Igs). IgG1 is the predominant antibody in bovine milk and colostrum.

Colostrum and early milk contain exceptionally high levels of antibodies (immunoglobulins), many of which exhibit specificities for common human pathogens (bacteria, viruses and parasites) (reviewed in Korhonen et al., 2000). It is therefore likely that these molecules could be effective in providing passive protection against a range of infectious agents in human subjects. Normally, the level of pathogen-specific antibodies in colostrum and milk is low, but could be enhanced significantly (by 100-fold) by immunizing cows against desired pathogens/antigens during pregnancy. The efficacy of milk/colostrum-derived Ig preparations against important human pathogens such as *Shigella flexneri*, *Helicobacter pylori* and *Vibrio cholerae* and rotavirus has been the subject of many clinical studies (Michalek et al., 1978; Ebina et al., 1985; Boesman-Finkelstein et al., 1989; Tacket et al., 1992; Casswall et al., 1998). While results have been variable, the evidence suggests that colostrum preparations, with high antibody titres against defined pathogens, can be effective in mediating protection against enteric infections in humans (Table 6.1). The use of hyper-immune colostrum containing high titres of specific antibody has been found to be more effective than normal Ig preparations, especially in the treatment of infant diarrhoea. A significantly reduction in the incidence of chronic diarrhoea among HIV-infected individuals given bovine colostral Ig concentrate has also been observed (Rump et al., 1992). These studies highlight opportunities for the development of Ig-containing milk or colostrum preparations with defined health benefits for specific population groups (children, hospitalized individuals, immuno-compromised, the elderly) that exhibit increased risk of infectious disease.

6.7.1   Immunoregulation

One of the unique features of the protection conferred by colostrum or milk is the virtual absence of host responses that could harm the recipient. This is due in part to the absence of phlogistic factors as well as the presence of immunoregulatory/anti-inflammatory factors in milk.

Owens and Nickerson (1989) reported the presence of a low molecular weight anti-inflammatory factor in the milk of hyperimmunized cows. Mice given anti-inflammatory factor (AIF) demonstrated significantly less mammary gland inflammation following infection with the mastitis-causing bacterium *Staphylococcus aureus*. In other studies, milk-derived AIF was found to suppress

Table 6.1   Efficacy of bovine milk-derived immunoglobulins (immune and non-immune) against bacterial, viral and parasitic infections: some examples

Target pathogen	Subjects	Effect	Reference
E. coli (enteropathogenic)	Preterms and infants	Reduced diarrhoea	Lodinova-Zadnikova et al. (1987)
E. coli (enteropathogenic) and rotavirus	Children	No effect on diarrhoea	Brunser et al. (1992)
E. coli (enterotoxigenic)	Healthy volunteers	Prevention of diarrhoea after experimental challenge	Freedman et al. (1998)
Helicobacter pylori	Adults	Attenuated symptoms of gastritis	Tarpila et al. (1994)
H. pylori	Adults	No effect	Opekun et al. (1999)
H. pylori	Children	Attenuated chronic inflammation of gastric antrum	Oona et al. (1997)
Shigella flexneri	Adults	Prevented shigellosis	Tacket et al. (1992)
Streptococcus mutans	Adults	Reduced number of S. mutans in dental plaque	Loimaranta et al. (1999)
Cryptosporidium	Adults	Reduced diarrhoea and oocyst secretion after experimental infection	Okhuysen et al. (1998)
Cryptosporidium	Adults (AIDS patients)	Reduced symptoms of intestinal inflammation and oocyst production	Stephan et al. (1990)

Adapted from Korhonen et al. (2000).

the host vs graft reaction, lymphocyte proliferative responses to ConA, and inflammatory response to carrageenin and *E. coli* (Ormrod and Miller, 1991, 1993). The AIF was effective following oral, subcutaneous, intramuscular, intraperitoneal or intravenous administration.

Anti-inflammatory properties of bLF are also well documented. Zimecki *et al.* (1998) demonstrated that oral treatment of rats with bLF (five doses, 10 mg each, on alternate days) was able to inhibit caragenin-induced inflammation by 50%, compared with the control treatment. The inhibition was associated with a substantial reduction in the ability of spleen cells to produce IL-6 and TNF-α. A significant reduction in serum TNF-α in mice given bLF, 24 h prior to intravenous challenge with LPS (a potent inflammatory agent) has also been noted (Machnicki *et al.*, 1993). In human studies, oral administration of bLf was found to reduce spontaneous release of pro-inflammatory cytokines IL-6 and TNF-α by peripheral blood mononuclear cells. IL-1β, TNF-α and IL-6 mediate not only inflammatory and immune reactions, but also the development of cachexia. The balance between these pro-inflammatory cytokines and anti-inflammatory cytokines (IL-10, TGF-β) is pivotal for the fine-tuning and resolution of inflammation.

The presence of transforming growth factor (TGF)-β (a potent anti-inflammatory agent) in bovine milk in biologically significant amounts has also been demonstrated; cheese whey and milk contain 3.7 ± 0.7 ng and 4.3 ± 0.8 ng TGF-β/ml. Kulkarni and Karlsson (1993) demonstrated that neonatal mice genetically deficient for TGF develop chronic inflammation of the lower intestine, and remain viable only as long as they received maternal milk containing immunosuppressive TGF. Fell *et al.* (2000) reported that the administration of a casein-based polymeric diet, containing high levels of TGF-β, to children with Crohn's disease was able to ameliorate chronic inflammation of the intestine and reduce the levels of proinflammatory cytokines IL-1β, IL-8 and IFN-γ. These beneficial effects were attributed to the ability of TGF-β to suppress inflammatory cytokine production. It is important to note that the bioactivity of growth factors such as TGF-β is not affected by pasteurization. Low pH (pH 2.0) also has little effect on TGF-β activity.

Allergic diseases result from an imbalance in Th1- and Th2-type immune responses and are characterized by increased levels of reaginic antibody (IgE) and inflammation due to the release of vasoactive amines from mast cells. The ability of bovine milk-derived factors to suppress allergic responses has also been reported. Watson *et al.* (1992) reported that a whey fraction, isolated from bovine colostrum, could suppress systemic IgE antibody responses in mice. Otani and Yamada (1995) demonstrated that bovine lactoferrin and κ-casein could inhibit the release of histamine from mast cells *in vitro*, suggesting that these proteins might also be effective in limiting allergic inflammation *in vivo*.

The ability of a whey-derived growth factor to regulate the development of oral tolerance has also been demonstrated (Penttila *et al.*, 2001). Oral administration of a growth factor-enriched whey protein extract to suckling rat pups, between four and nine days postnatally, significantly inhibited immune activation of spleen cells to an orally administered food antigen, ovalbumin. Treatment with whey extract at the time of oral sensitization to ovalbumin resulted in an increased secretion of TGF-β into the culture supernatants of spleen cells stimulated with ovalbumin. TGF-β is known to play an important role in tolerance induction. Inhibition of anti-CD3 antibody-induced proliferation of human lymphocytes by a milk growth factor (containing TGF-β1 and TGF-β2) has also been reported (Stoeck *et al.*, 1989).

Furthermore, studies on sheep colostrum have led to the identification of a proline-rich polypeptide (PRP, also known as colostrinin) with immunoregulatory properties (Janusz *et al.*, 1974, 1981). PRP has been found to influence the maturation and differentiation of thymocytes into T-helper/T-suppressor cells (Zimecki *et al.*, 1984a, b), to induce release of cytokines by immunocompetent cells (Inglot *et al.*, 1996), to stimulate growth and differentiation of resting B cells (Julius *et al.*, 1988) and to regulate autoimmune host responses (Zimecki *et al.*, 1982; Kundu *et al.*, 2000). In clinical studies, oral administration of ovine PRP has been shown to exert psycho-stimulatory effects and significantly improve the outcomes of Alzheimer's disease patients with mild to moderate dimentia (Leszek *et al.*, 1999). Specific receptors for PRP on murine thymocytes have also been described (Janusz *et al.*, 1986). The presence of PRP in human milk has also been reported. Little is known about the nature and functions of PRP derived from bovine colostrum or early milk.

Although the immunomodulatory effects of most milk-derived products in humans remain to be investigated, the evidence discussed above suggests that milk products could be used for optimizing immune function in the elderly (e.g. arthritis, inflammatory bowel disease (IBD), diabetes, Alzhiemer's disease); that is to down-regulate as well as enhance immunity in population groups with less inadequate immune function (e.g. the elderly, critically ill, immunocompromised).

6.8   Fermented dairy products and probiotic LAB

The human gastrointestinal tract (GIT) harbours an extremely complex and diverse microbial ecosystem. The GIT of an adult human contains approximately 1.6–2 kg of bacteria (10^{14} cfu), representing over 400 different species. In other words, there are ten times more microbial cells than the number of eukaryotic cells in the body. The microbial density is relatively sparse in the small intestine (10^2–10^3/g) but becomes increasingly abundant in the lower gut (10^{10}–10^{12}/g). While a majority of these bacteria are non-pathogenic/beneficial, some of these organisms possess the potential to cause disease. In healthy individuals, there is a definite balance between good and bad bacteria. However, perturbations in the homeostasis of the normal microflora due to factors such as antibiotic therapy, stress and infection result in proliferation of undesirable or pathogenic microbes and enhanced predisposition to a number of clinical disorders such as cancers, inflammatory diseases, allergies, ulcerative colitis and infectious illnesses.

Metchnikoff (1907) was perhaps the first to recognize the health-enhancing attributes of fermented products containing microbes. He hypothesized that the harmful effects of the undesirable bacteria could be overcome through ingesting LAB (probiotics) and that the longevity of Bulgarians was due to the consumption of large quantities of lactobacilli-fermented milk. In the ensuing 90 years, there has been a plethora of published studies to support Mechnikoff's hypothesis. It is now well recognized that a healthy (balanced) microflora is pivotal to optimum health and that supplementation with probiotics could be used to shift the balance of the gut microflora away from potentially harmful/pathogenic bacteria towards a beneficial or health-promoting microorganisms, like lactobacilli, bifidobacteria. Probiotics are defined as live bacterial feed/food supplements that promote host health by balancing its intestinal microflora. Consumption of probiotics is suggested to confer a range of health benefits including modulation of the immune system. Most probiotics are LAB, such as lactobacilli and bifidobacteria, and are commonly consumed in the form of dairy products (yoghurt or other fermented products).

6.9   Immunomodulatory effects of fermented milk products and LAB

The effects of LAB on the immune system have been the focus of active research since the 1980s. The results of these studies have been varied due mainly to differences in species and strains of LAB, dose, delivery medium and experimental protocols. However, these studies have provided unequivocal evidence that certain strains of LAB are able to stimulate as well as regulate several aspects of immune function. Several excellent reviews on immunomodulatory effects of probiotics in animals and humans have been published in recent years (reviewed by Gill, 1998; Erickson and Hubbard, 2000; Meydani and Ha, 2000; Isolauri et al., 2001; Wold, 2001) and therefore the following section will focus mainly on evidence from human studies.

6.9.1   Immune enhancement

The effect of LAB consumption on natural immunity has been the subject of many human studies (Table 6.2). The results of these studies have shown that the consumption of specific strains of LAB is able to enhance:

- phagocytic activity of blood leucocytes (PMN and mononuclear cells);
- oxidative defence mechanisms, including production of reactive oxygen species and secretion of cytokines by phagocytic cells;
- NK cell activity.

Schiffrin et al. (1995) reported that consumption of fermented milk containing Lb. johnsonii La1 or B. lactis Bb12 for three weeks was able to enhance phagocytic activity of peripheral blood leucocytes (PMN and monocytes) in human volunteers. The PMN showed the greater increases in phagocytic cell function than mononuclear cells. The increase in phagocytic capacity persisted for six weeks after cessation of fermented milk consumption. Mikes et al. (1995) reported a significant increase in the ability of neutrophils from human subjects given Entercoccus faecium for six weeks to produce oxygen radical following stimulation with zymosan and phorbol myristate acetate (PMA). Intake of milk containing Lactobacillus GG has also been reported to significantly increase the expression of CR1, CR3, FcγRI and FcαR in neutrophils (Pelto et al., 1998). Augmentation of phagocytic cell function and NK cell activity, and increases in the percentage of NK cells in the peripheral blood of human volunteers following regular consumption of yoghurt or milk containing probiotics has also been reported by others (De Simone et al., 1989; Gill et al., 2001a,c,d).

As noted previously, phagocytic cells play a central role in protection against microbial infections, while NK cells are effective in combating viruses and cancers. Therefore, it can be concluded from the previously described studies that consumption of some LAB strains could stimulate aspects of natural immunity and that this may impact positively on human health.

Table 6.2   Examples of immunoenhancing effects of fermented products or probiotic LAB in humans

Yoghurt/ probiotic strain	Study subjects	Effect	Reference
Yoghurt	25–40 years old	↑ IFN-γ production	Halpern *et al.* (1991)
		↑ 2–5 A synthetase in blood mononuclear cells	Solis-Pereyra *et al.* (1997)
	20–40 years old	No effect on IFN-γ production and blood leukocyte counts ↓ Allergic symptoms	Trapp *et al.* (1993)
	20–47 years old	↑ IFN-γ levels in serum	De Simone *et al.* (1989)
Bifidobacterium lactis (DR10™)	60–83 years old	↑ Phagocytosis by PMN ↑ IFN-α	Arunachallam *et al.* (2000)
		↑ Phagocytosis by PMN and monocytes ↑ NK cell activity ↑ CD3⁺, CD4⁺, CD25⁺ and CD56⁺ (NK cells) cells	Gill *et al.* (2001a, b)
		↑ Phagocytosis by PMN and monocytes ↑ NK cell activity	Chiang *et al.* (2000)
L acidophilus La1	23–62 years old	↑ Phagocytic capacity of blood leucocytes (PMN and mononuclear cells)	Schiffrin *et al.* (1995)
		No effect on serum or mucosal IgA levels	Marteau *et al.* (1997)
	21–57 years old	↑ Phagocytic activity of PMN	Donnet-Hughes *et al.* (1999)
L. casei Shirota	40–65 years	No effect on Ig levels, NK cell activity and lymphocyte cytokine production	Spanhaak *et al.* (1998)
L. rhamnosus GG	22–50 years old, tolerant or non-tolerant to milk	Up-regulation of CR1 and CR3 expression on PMN in healthy subjects and down-regulating effect in allergic subjects	Pelto *et al.* (1998)
	2–5-month-old children	Superior antibody response to oral rotavirus vaccination ↑ IgM antibody secreting cells	Isolauri *et al.* (1995)
	Children with acute gastoentritis	↑ Rotavirus-specific IgA antibody-secreting cells	Kaila *et al.* (1992)
L. rhamnosus GG or *Lactococcus lactis*	Healthy volunteers	No difference in antibody responses to oral *Salmonella typhi* vaccine ↑ Expression of CR3, and to a lesser extent CR1 on PMN in *L. lactis* group	He *et al.* (2000)

Table 6.2 cont'd

Yoghurt/ probiotic strain	Study subjects	Effect	Reference
B. lactis Bb12	15–31 month-old children; all vaccinated with poliovirus by 12 months of age	↑ Anti-polio IgA levels in faeces	Fukushima *et al.* (1998)
	23–62 years of age	↑ Phagocytic capacity of blood leucocytes (PMN and mononuclear cells)	Schiffrin *et al.* (1995)
Fermented milk containing *L. acidophilus* La1 and *B. bifidum* Bb12	19–45 years old	Slight increase in anti-*S typhi* antibody response ↑ Serum total IgA level	Link-Amster *et al.* (1994)
Milk with *L. rhamnosus* HN001	Adults/elderly	↑ Phagocytosis by PMN ↑ NK cell activity	Gill *et al.* (2001a, b, c) Sheih *et al.* (2001)

Feeding specific strains of LAB to human subjects has also been shown to augment several aspects of specific immune responses. For example, Kaila *et al.* (1992) and Majamaa *et al.* (1995) reported higher levels of specific mucosal and serum antibody responses in children with acute rotavirus diarrhoea given *Lactobacillus* GG than the control group. Feeding of fermented milk containing *Lb. acidophilus* La1 and *Bifidobacteria* Bb12 for three weeks significantly enhanced the effectiveness of oral *Salmonella typhi* vaccine in human subjects (Link-Amster *et al.*, 1994). The serum IgA response to *S. typhi* Ty21a was significantly higher in the probiotic-fed group than in the control group. Enhanced immunogenicity of an oral rotavirus vaccine in humans fed *Lactobacillus* GG has been reported (Isolauri *et al.*, 1995). Supplementation with probiotics has also been reported to increase the percentages of T helper (Sawamura *et al.*, 1994; Gill *et al.*, 2001) and B cells (De Simone *et al.*, 1991) in humans.

The ability of LAB and fermented milk products to modulate cytokine production has also been demonstrated. De Simone *et al.* (1989) reported significantly higher levels of IFN-γ in serum of subjects fed lyophilized lactobacilli and yoghurt compared with the control group. Enhanced concentration of 2–5 A synthetase (an IFN-γ inducible protein) in subjects consuming yoghurt than in subjects consuming milk was observed by Solis-Pereyra and Lemonnier (1991). Similar effects of yoghurt consumption on IFN-γ production in humans have been reported by Halpern *et al.* (1991) and by Aottouri and Lemonnier (1997). Modulatory effect of LAB consumption on IFN-α production in human subjects has also been described (Kishi *et al.*, 1996; Arunachalam *et al.*, 2000). Contrary to these studies, however,

Trapp *et al.* (1993) found no effect of yoghurt consumption for a year on the levels of IFN-γ in plasma. The reason for this discrepancy is not known. Long-term consumption of yoghurt has also been shown to increase production of IL1β, IL-6, IL-10, IFN-γ and TNF-α (Halpern *et al.*, 1991; Solis-Pereyra and Lemonnier 1993; Solis-Pereyra *et al.*, 1997; Miettinen *et al.*, 1996; Aottouri and Lemonnier 1997).

Animal and human studies have also provided evidence that probiotic intake could be effective in restoring the age-related decline in immune function. Healthy elderly subjects fed milk containing *Lb. rhamnosus* HN001 or *Bifidobacterium lactis* HN019) for three to six weeks showed significantly higher leucocyte (neutrophils and monocytes) phagocytic capacity and NK cell activity than subjects fed milk without probiotics (Fig. 6.2; Gill *et al.*, 2001a,c,d). It was further observed that individual with poor pre-intervention immunity had consistently greater relative increases in immune function than did those with adequate pre-intervention immunity (Fig. 6.3; Gill *et al.*, 2001d). Furthermore, increases in immune function were also correlated with age, with subjects older than 70 years experiencing significantly greater improvements in immune function than those under 70 years (Fig. 6.4; Gill *et al.*, 2001c). In aged mice, intake of probiotics was reported to restore interferon (IFN-γ and IFN-α) producing capacity (Muscettola *et al.*, 1994). Together, these observations suggest that dietary consumption of certain strains of probiotics may be useful in combating the deleterious effects of immunosenescence on cellular immunity.

6.9.2   Increased disease resistance

A relationship between LAB-induced immunostimulation and enhanced resistance to disease has also been demonstrated in several human studies. Kaila *et al.* (1992) reported that *Lactobacillus* GG promotes recovery from rotavirus diarrhoea in children with acute rotavirus gastroenteritis through augmentation of host immune responses. Children receiving *Lactobacillus* GG had reduced duration of diarrhoea and a higher anti-rotavirus IgA response. A similar association between the protective effects of *Lactobacillus* GG and enhancement of specific immune responses was reported by Majamaa *et al.* (1995). Several other studies have reported enhanced resistance to gastrointestinal infections/diarrhoeal diseases in humans following consumption of some dietary LAB. However, in many cases the definitive evidence – that an increased resistance to disease was the direct consequence of an enhanced immune response to dietary consumption of LAB – is often lacking. Several other mechanisms by which LAB may contribute to host protection against pathogenic microorganisms have been suggested. These include competition for adhesion sites, interbacterial competition and production of antimicrobial substances. However, little is known about the relative contribution of these mechanisms in host defence. It is likely that stimulation of the host immune responses acts in concert with other LAB-mediated mechanisms to effect resistance.

There is also evidence that yoghurt or probiotic consumption may reduce the

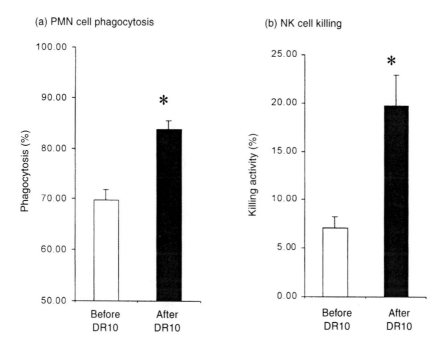

Fig. 6.2 Dietary consumption of *Bifidobacterium lactis* HN019 (DR10™) enhances natural cellular immune responses: three weeks after switching from a dietary supplement of lactose-hydrolysed low-fat milk (LH-LFM) to LH-LFM containing *B. lactis* HN019, enhanced PMN cell phagocytic activity and NK cell activity was observed among 23 healthy, middle-aged and elderly subjects. Immune measurements were made at each time point. Asterisks refer to significant increases ($P < 0.05$) following the ingestion of DR10™
Adapted from Chiang *et al.* (2000).

risk of colon and bladder cancer (reviewed by Parodi, 1999). Whether these anti-tumour effects are simply due to alterations in the microflora, which plays a role in the aetiology of the cancer, or are the result of immune stimulation is not known. However, LAB intake associated increases in anti-cancer immune responses, such as NK cell numbers and cytotoxicity, in cancer patients and experimental animals, suggests that stimulation of the immune system may be responsible, at least in part, for anti-cancer effects of probiotics. Kato *et al.* (1994) demonstrated that oral administration of *Lb. casei* (Shirota) was able to enhance immune cell function in mice following the resection of primary tumours, and subsequently reduce the growth of secondary tumours compared with the control mice. Clinical studies in patients with bladder cancer have also indicated that *Lb. casei* Shirota can act as an effective biotherapeutic agent against secondary tumours (Aso *et al.*, 1995), and can enhance anti-tumour (NK cell activity) cellular immune responses (Sawamura *et al.*, 1994).

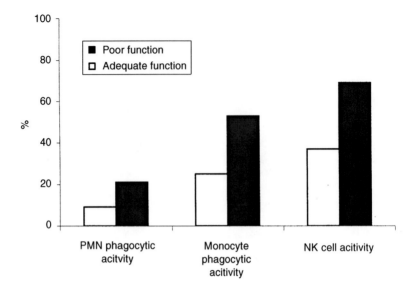

Fig. 6.3 Percentage increase in phagocytic activity and NK cell activity in elderly subjects with poor or adequate pre-intervention immune function. Subjects consumed lactose-hydrolysed low-fat milk containing 5×10^{10} *Bifidobacterium lactis* HN019 organisms/day for three weeks. Pre-intervention immunity was defined on the basis of *in vitro* immune responses (phagocytic activity and NK cell function); responses were stratified into poor or adequate subgroups (lowest tertile and middle and highest tertiles combined, respectively). Significant differences ($P \leq 0.02$) were observed for PMN and monocyte phagocytic activity.

6.9.3 Immunoregulation

A balance between Th1- and Th2-type responses is essential for immune system homeostasis. A range of immune-mediated disease such as allergies, autoimmune and immunoinflammatory disorders are the consequence of an imbalance (overactivation/down-regulation) in Th1- versus Th2-type immune responses. Studies have shown that some strains of LAB are able to down-regulate over-activated immune responses (Table 6.3). For example, some *Lactobacillus* strains can suppress IgE antibody responses in mice (Shida *et al.*, 1998; Matsuzaki and Chin, 2000), and can reverse pro-allergy immune responses following oral delivery (Matsuzaki *et al.*, 1998; Murosaki *et al.*, 1998). In human clinical studies, long-term consumption of yoghurt or fermented milk products has been shown to reduce levels of serum IgE and alleviate some of the symptoms of nasal allergy among elderly subjects or atopic eczema among infants (Halpern *et al.*, 1991; Trapp *et al.*, 1993; Kalliomaki *et al.*, 2001). Reduction of intestinal inflammation in infants with food allergies, following consumption of a whey formula supplemented with *Lb. rhamnosus* GG, has also been demonstrated (Majamaa and Isolauri, 1997). In subjects with milk allergy, probiotic supplementation has been

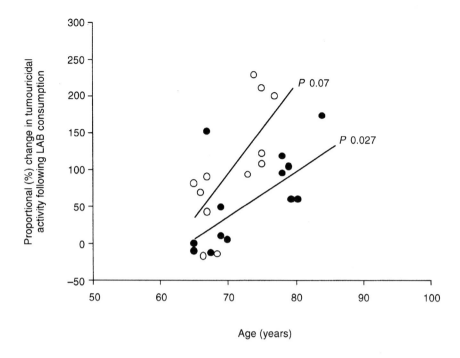

Fig. 6.4 Relationship between subjects' age and relative change in peripheral blood mononuclear cell tumoricidal activity following consumption of milk containing *Lactobacillus rhamnosus* HN001 (O) or *Bifidobacterium lactis* HN019 (●). Adapted from Gill *et al.* (2001c).

shown to suppress the activation of neutrophils and monocytes; milk challenge enhanced the expression of CR1, CR3, FcγR1 and FcαR in neutrophils, and CR1, CR3 and FcαR in monocytes in hypersensitive subjects, whereas the intake of milk containing *Lactobacillus GG* prevented the increases in the expression of these receptors. The possible mechanisms by which probiotics mediate their anti-allergy effects include, induction of high-level expression of pro-Th1 cytokines such as IL-12 (Murosaki *et al.*, 1998), stabilization of gut mucosal barrier against macromolecular sensitization (Majamaa and Isolauri, 1997) and down-regulation of proinflammatory receptors on leucocytes (Pelto *et al.*, 1998). In the case of milk allergy, the release of immunoregulatory peptides from milk substrate by lactobacilli enzymes has been suggested to account for suppression of allergic responses; consumption of casein hydrolysed with *Lb. rhamnosus* was reported to reduce the production of IL-4 by antigen-stimulated PBL from milk sensitive subjects compared with subjects given milk (Rokka *et al.*, 1997).

Down-regulatory effect of some probiotics in subjects with IBD (Crohn's disease, ulcerative colitis, pouchitis) has also been reported. IBD, chronic or recurrent inflammation of the intestine, is a disorder of unknown aetiology. It is believed that an abnormal host response to lumenal bacteria (members of the gut

Table 6.3 Effect of probiotic consumption on immune-mediated/immunoinflammatory disorders in humans: some examples

Condition	Subjects	Treatment/ study subjects	Effect	Reference
Asthma and/or rhinitis	13–15 years old with history of asthma and/or rhinitis	Yoghurt with *L. acido-philus*	No effect on asthma ↑ IFN-γ and IL-2 production by PBL ↑ Oxidative burst	Wheeler *et al.* (1997)
Allergy	20–40 and 55–70 years	Yoghurt	↓ Serum IgE levels ↓ Symptoms of allergy	Trapp *et al.* (1993)
Allergic symptoms	22–55 years old, hypersensitive to milk	Milk fermented with *L. rhamnosus* GG	↓ Expression of CR1, CR3 and FcγR on PMN	Pelto *et al.* (1998)
Milk allergy		Hydrolysed milk formula with *L. rhamnosus* GG	↓ TNF-α in faeces ↓ Clinical symptoms	Majamaa and Isolauri (1997)
Crohn's disease/ juvenile chronic arthritis (JCA)	5–17 years old (Crohn's disease) and 1–14 years old with JCA	*L. rhamnosus* GG	↑ IgA antibody-secreting cells to food antigens in subjects with Crohn's disease Normalization of faecal urease activity in subjects with JCA	Malin *et al.* (1996)
Pouchitis	Patients	VSL#3	↓ Relapses rate (15% in treated group *vs* 100% in control group)	Gionchetti *et al.* (2000)

VSL#3 (CSL, Milan, Italy) contains 300 billion viable lactobacilli (*L. casei, L. plantarum, L. acidophilus, L. delbruecki*).

microflora) and/or a defective gut mucosal barrier may be responsible for the initiation and perpetuation of IBD. Several recent clinical studies have shown that administration of certain probiotic strains could be effective in the management of IBD (Table 6.3); patients given probiotics showed a marked reduction in relapses of ulcerative colitis and a significant reduction in the severity of clinical signs (reviewed in Marteau and Boutron-Ruault, 2002).

Furthermore, probiotic treatment has been shown to exhibit anti-diabetic (non-insulin-dependent (NIDDM) and insulin-dependent diabetes mellitus (IDDM) effects in animal models. The oral administration of *L. casei* to 4-week-old KK-A^y mice (a spontaneous diabetic model of NIDDB) or raising the mice on an *L casei*-containing diet significantly decreased the plasma glucose levels at eight to ten weeks compared with the control group (Matsuzaki *et al.*, 1997a). Analysis of host immune responses revealed that anti-diabetic effects of *L. casei* administration were associated with an increase in the proportions of CD4+ cells and a decrease in the production of IFN-γ and IL-2. In non-obese diabetic (NOD) mice (an IDDM model), oral feeding of *L. casei* was found to inhibit the occurrence of diabetes and regulate host immune responses; the proportion of CD45R+ B cells in the spleen was increased, and the proportion of CD8 cells and IFN-γ production were decreased in mice fed *L. casei* compared with the control mice (Matsuzaki *et al.*, 1997b).

The ability of lactobacilli to reduce subjective symptoms of arthritis among rheumatoid patients (Nenonen *et al.*,1998) and the severity of collagen-induced arthritis in mice (cited by Erickson and Hubbard, 2000) has also been reported.

6.10   Future trends

Ageing is associated with profound alterations in immune function. There is compelling evidence to suggest that the increased susceptibility of the elderly to infectious diseases, malignancies and autoimmune disorders is, at least in part, the result of this immune dysfunction. Nutritional deficiencies, common in the elderly, are known to further impair immune responsiveness. Results of dietary intervention studies highlight the importance of nutrition in optimizing immune function in the aged individuals. It has been demonstrated that supplementation with single or multiple nutrients could restore or delay the age-related decline in immune function as well as increase resistance to infectious diseases. This offers unique opportunities to develop novel food products or supplements that are able to correct nutritional deficits as well as provide specific immune benefits. Dietary approaches are especially attractive as these are more practical and cost-effective, and could be implemented easily on a larger scale. In this context, the unique nutritional and immunomodulatory properties of milk and milk products need careful consideration. In addition to being a rich source of nutrients (such as proteins, growth factors, vitamins and minerals), bovine milk contains a host of components with potent immunomodulatory properties (immunostimulatory as well as immunoregulatory). Several animal studies have shown that milk-derived factors are able to provide passive protection against a range of bacterial, viral and parasitic infections, enhance immune function and down-regulate over-activated host immune responses (inflammatory/autoimmune).

Consumption of fermented milk products containing specific strains of lactic acid bacteria has also been reported to enhance aspects of natural and acquired immune responses and regulate dysfunctional immune responses (allergies, chronic inflammatory diseases such as IBD) in human subjects. However, relatively little is known about the effect of immunomodulatory milk or milk-derived products in the elderly. Well-designed and controlled studies, involving healthy elderly, institutionalized elderly and the elderly who are afflicted by various immune-mediated and inflammatory diseases are required to evaluate the potential of milk-based products for optimizing immune health in the elderly. These studies need to be conducted in conjunction with studies designed to understand mechanisms by which milk-derived products may influence the immune system. Without a mechanistic approach, the significance and relevance of this will be compromised. Identification of phenotypic markers of immune function is of limited value unless these are of functional value. Thus the assays used to assess immune status must be functionally relevant to clinical outcomes. Study design should also reflect the real world. Also, there is a need for establishing dose–response relationships, for both the antigenic challenge and the nutrient/product under investigation. The future will see the development of a range of milk and milk-based products that are able to correct nutritional deficits as well as restore, prevent or slow the age-related decline in immune function and thus increase resistance to diseases and responsiveness to vaccines. The potential of milk and milk-based products to contribute to the improvement and maintenance of geriatric health is under-explored and untapped.

6.11   References

ALES-MARTINEZ J E, ALVAREJ-MON M, MERINO F, BONILLA F, MARTINEZ C, DURANTEZ C and DE LA HERA A (1988). Decreased TcR-CD3-T cell numbers in healthy aged humans. Evidence that T cell defects are marked by a reciprocal increases of TcR-CD3-CD2$^+$ natural killer cells. *Eur J Immunol*, **18**, 1827–1830.

AOTTOURI N and LEMONNIER D (1997), Production of interferon induced by *Streptococcus thermophilus*: role of CD4$^+$ and CD8$^+$ lymphocytes. *Nutr Biochem*, **8**, 25–31.

ARUNACHALAM K, GILL H S and CHANDRA R K (2000), Enhancement of natural immune function by dietary consumption of *Bifidobacterium lactis* (HN019). *Eur J Clin Nutr*, **54**, 1–5.

ASO Y, AKAZA H, KOTAKE T, TSUKAMOTO T, IMAI K and NAITO S (1995), Preventive effect of a *Lactobacillus casei* preparation on the recurrence of superficial bladder cancer in a double-blind trial. The BLP Study Group. *Eur Urol*, **27**, 104–109.

BOESMAN-FINKELSTEIN M, WALTON N E and FINKELSTEIN R E (1989), Bovine lactogenic immunity against cholera toxin-related enterotoxins and *Vibrio cholerae* outer membranes. *Infect Immunity*, **57**, 1227–1235.

BOGDEN J D, OLESKE J M, LAVENHAR M A, MUNVES E M, KEMP F W, BRUENING K S, HOLDING K J, DENNY T N, GUARINO M A and HOLLAND B K (1990), Effects of one year of supplementation with zinc and other micronutrients on cellular immunity in the elderly. *J Am Coll Nutr*, **9**, 214–225.

BOUNOUS G and GOLD P (1989), Whey protein composition, a method for producing it and use of composition. New Zealand Patent 231865.

BOUNOUS G and GOLD P (1993), Treatment of HIV-sero-positive individuals – using undenatured whey protein concentrate to increase T-helper cell concentration and blood glutathione content. Patent Number 9320831 and 93–351356/44.

BOUNOUS G, STEVENSON M M and KONGSHAVN P A (1981), Influence of dietary lactalbumin hydrolysate on the immune system of mice and resistance to salmonellosis. *J Infect Dis*, **144**, 281–290.

BOUNOUS G and KONGSHAVN P A L (1985), Differential effect of dietary protein type on the B-cell and T-cell immune responses in mice. *J Nutr*, **115**, 1403–1408.

BOUNOUS G and KONGSHAVN P A L (1989), Influence of protein type in nutritionally adequate diets on the development of immunity. In: *Absorption and Utilisation of Amino Acids*, Vol. II, M Freedman, Ed; Boca Raton, Florida, CRC Press, 219–233.

BOUNOUS G, LETOURNEAU L and KONGSHAVN P A L (1983), Influence of dietary protein type on the immune system of mice. *J Nutr*, **113**, 1415–1421.

BOUNOUS G, SHENOUDA N, KONGSHAVN P A and OSMOND D G (1985), Mechanism of altered B-cell response induced by changes in dietary protein type in mice. *J Nutr*, **115**, 1409–1417.

BOUNOU G, KONGSHAVN P A AND GOLD P (1988), The immunoenhancing property of dietary whey protein concentrate. *Clin Invest Med*, **11**, 271–278.

BOUNOUS G, BARUCHEL S, FALUTZ J AND GOLD P (1993), Whey proteins as a food supplement in HIV-seropositive individuals. *Clin Invest Med*, **16**, 204–209.

BROSCHE T and PLATT D (1995), Nutritional factors and age-associated changes in cellular immunity and phagocytosis: a review. *Ageing Immunol Infect Dis*, **6**, 29–40.

BRUNSER O, ESPINOZA J, ARAYA M, SPENCER E, HILPERT H, LINK-AMSTER H and BRUSSOW H (1992) Field trial of an infant formula containing anti-rotavirus and anti-*Escherichia coli* milk antibodies from hyperimmunized cows. *J Pediatr Gastroent Nutr*, **15**, 63–72.

BUTCHER S, CHAHEL H and LORD J M (2000), Ageing and the neutrophils: no appetite for killing? *Immunology*, **1000**, 411–416.

BUZINA-SUBOTICANEC K, BUZINA R, STAVLJENIC A, FARLEY T M, HALLER J, BERGMAN-MARKOVIC B and GORAJSCAN M (1998), Ageing, nutritional status and immune response. *Int J Vit Nutr Res*, **68**, 133–141.

CAKMAN I, ROHWER J, SCHUTZ R M, KIRCHNER H and RINK L (1996), Dysregulation between Th1 and Th2 T cell subpopulations in the elderly. *Mech Age Dev*, **87**, 197–209.

CASSWALL T H, SARKER S A, ALBERT M J, FUCHS G J, BERGSTROM M, BJORCK L and HAMMARSTROM L (1998), Treatment of *Helicobacter pylori* infection in infants in rural Bangladesh with oral immunoglobulins from hyperimmune bovine colostrum. *Aliment Pharmacol Ther*, **12**, 563–568.

CHANDRA R K (1985),Trace element regulation of immunity and infection. *J Am Coll Nutr*, **4**, 5–16.

CHANDRA R K (1989), Nutrition, regulation of immunity and risk of infection in old age. *Immunology*, **67**, 141–147.

CHANDRA R K (1990), Nutrition is an important determinant of immunity in old age. In: *Nutrition and Ageing*, D M Prinsley and H H Standsteas, Eds; New York, Alan R Liss, 321–334.

CHANDRA R K (1992), Effect of vitamin and trace-element supplementation on immune responses infection in elderly subjects. *Lancet*, **340**, 1124–1127.

CHANDRA R K (1995), Nutrition and immunity in the elderly: Clinical significance. *Nutr Rev*, **53**, S80–83.

CHANDRA R K (1997a), Graying of the immune system: can nutrient supplements improve immunity in the elderly? *JAMA*, **277**, 1398–1399.

CHANDRA R K (1997b), Nutrition and the immune system: an introduction. *Am J Clin Nutr*, **66**, 460S–463S.

CHIANG B L, SHEIH Y H, WANG L H, LIAO C K and GILL H S (2000), Enhancing immunity by dietary consumption of a probiotic lactic acid bacterium (*Bifidobacterium lactis* HN019): optimization and definition of cellular immune responses. *Eur J Clin Nutr*, **54**, 849–855.

CLARK J A AND PETERSON T C (1994), Cytokine production and ageing: overproduction of IL-8 in elderly males in response to lipopolysaccharide. *Mech. Ageing Dev*, **77**, 127–139.

CORBERAND J, NGYEN F, LAHARRAGUE P, FONTANILLES A M, GLEYZES B, GYRARD E and SENEGAS C (1981), Polymorphonuclear function and ageing in humans. *J Am Geriatr Soc*, **29**, 391–397.

CROSS M L and GILL H S (2001a), Immunomodulatory properties of milk. *Br J Nutr*, **84** (Supplement No. 1), S81–S89.

DARDENNE M, PLEAU J M, NABARRA B, LEFRANCIER P., DERRIEN M, CHOAY M and BATCH J F (1982), Contribution of Zn and other metals to the biologica activity of the serum thymic factor, *Proc Natl Acad Sci USA*, **79**, 5370–5373.

DE SIMONE C, BIANCHI SALVADORI B, JIRILLO E, BALDINELLI L, BITONTI F and VESELY R

(1989), Modulation of immune activities in humans and animals by dietary lactic acid bacteria. In: *Yogurt Nutritional and Health Properties*, R C Chandan, Ed; London, John Libbey Eutotext, 201–213.

DE SIMONE C, ROSATI E, MORETTI S, BIANCHI SALVADORI B, VESELY R and JIRILLO E (1991), Probiotics and stimulation of the immune response. *Europ J Clin Nutr*, **45** (Suppl.), 32–34.

DING A, HWANG S and SCHWAB R (1994), Effect of ageing on murine macrophages: diminished response to IFN-gamma for enhanced oxidative metabolism. *J. Immunol*, **153**, 2146–2152.

DONNET-HUGHES A, ROCHAT F, SERRANT P, AESCHLIMANN J M and SCHIFFRIN E J (1999), Modulation of nonspecific mechanisms of defense by lactic acid bacteria: effective dose. *J Dairy Sci*, **82**, 863–869.

DORIA G and FRASCA D (1997), Genes, immunity, and senescence: looking for a link. *Imunol Rev*, **160**, 159–170.

EBINA T, SATO A, UMEZU K, ISHIDA N, OHYAMA S, OIZUMI A, AIKAWA K, KATAGIRI S, KATSUSHIMA N, IMAI A *et al.* (1985), Prevention of rotavirus infection by oral administration of cow colostrum containing antihuman rotavirus antibody. *Med Microbiol Immunol (Berl)*, **174**, 177–185.

ERICKSON K L and HUBBARD N E (2000), Probiotic immunomodulation in health and disease. *J. Nutr*, **130**, 403S–409S.

ERNST D N, WEIGLE W O, NOONAN D J, MCQUITTY D N and HOBBS M V (1990a), Differences in the expression profile of CD45RB, Pgp-1 and 3G11 membrane antigens and in the patterns of lymphokine secretion by splenic CD4+ T cells from young and aged mice. *J Immunol*, **145**, 1295–1302.

ERNST D N, WEIGLE W O, NOONAN D J, MCQUITTY D N and HOBBS M V (1990b), Differences in the subset composition of CD4+ T cell populations from young and old mice. *Ageing Immunol Infect Dis*, **2**, 105–109.

ERNST D N, WEIGLE O and HOBBS M V (1995), Aging and lymphokine gene expression by T cell subsets. *Nutr Rev*, **53**(II), S18–S24.

ESPERAZA B, SANCHEZ M, RUIZ M, BARRANQUERO M, SABINO E and MERINO F (1996), Neutrophil function in elderly persons assessed by flow cytometry. *Immunol Invest*, **25**, 185–190.

FACCHINI A, MARIANI E, MARIANI A R, PAPA S, VITALE M and MANZOLI F A (1987), Increased number of circulating Leu 11+ (CD16) large granular lymphocytes and decreased NK activity during ageing. *Clin Exp Immunol*, **68**, 340–347.

FELL, J M, PAINTIN M, ARNAUD-BATTANDIER F, BEATTIE R M, HOLLIS A, KITCHING P, DONNET-HUGHES A, MACDONALD T T and WALKER-SMITH J A (2000), Mucosal healing and a fall in mucosal pro-inflammatory cytokine mRNA induced by a specific oral polymeric diet in paediatric Crohn's disease. *Aliment Pharmacol Ther*, **14**, 281–289.

FERGUSON F G, WIKBY A, MAXON P, OLSSON J and JOHNSSON B (1995), Immune parameters in a longitudinal study of a very old population of Swedish people: a comparison between survivors and nonsurvivors. *J Gerontol*, **50**, B378–B382.

FORD J T, WONG C W and COLDITZ I G (2001), Effects of dietary protein type on immune responses and levels of infection with *Eimeria vermiformis* in mice. *Immunol Cell Biol*, **79**, 23–27.

FORTES C, FORASTIERE F, AGABITI N, FANO V, PACIFICI R, VIRGILI F, PIRAS G, GUIDI L, BARTOLONI C, TRICERRI A, ZUCCARO P, EBRAHIM S and PERUCCI C A (1998), The effect of zinc and vitamin A supplementation on immune response in an older population. *J Am Ger Soc*, **46**, 19–26.

FRANCESCHI C, MONTI D, SANSONI P and COSSARIZZA A (1995), The immunology of individuals: the lesson of centenarians. *Immunol Today*, **16**, 12–16.

FREEDMAN D J, TACKET C O, DELEHANTY A, MANEVAL D R, NATARO J and CRABB J H (1998), Milk immunoblobulin with specific activity against purified colonization factor antigens can protect against oral challenge with enterotoxigenic *Escherichia coli. J Infect Dis*, **177**, 662–667.

FUKUSHIMA Y, KAWATA Y, HARA H, TERADA A and MITSUOKA T (1998), Effect of a probiotic formula on intestinal immunoglobulin A production in healthy children. *Int J Food Microbiol*, **42**, 39–44.

FULOP T, KOMAROMI I, FORRIS G, WORUM I and LEOVY A (1986), Age-dependent variations of intra-lysosomal release from human PMN leukocytes under various stimuli. *Immunobiology*, **171**, 302–307.

FURUKAWA T, MEYDANI S N and BLUMBERG J B (1987), Reversal of age-associated decline in immune responsiveness by dietary glutathione supplementation in mice. *Mech Ageing Develop*, **38**, 107–117.

GARG M, LUO W, KAPLAN A M and BONDADA S (1996), Cellular basis of decreased immune response to pneumococcal vaccine in aged mice. *Infect Immunol*, **64**, 4456–4462.

GILL H S (1998), Stimulation of the immune system by lactic cultures. *Int Dairy J*, **8**, 535–544.

GILL H S (2000), Dairy products and immune health. In: *Functional Foods 2000*, F Angus and C Miller, Eds, Leatherhead Publishing, Surrey, 268–284.

GILL H S and CROSS L (2000), Anti-cancer properties of milk. *Brit J Nutr*, **84**, S161–S166.

GILL H S and CROSS L (2002), Probiotics and immune function. In: *Nutrition and Immune Function*, P Calder P, S Fields and H S Gill, Eds, London, CABI Publishing, 251–272.

GILL H S and RUTHERFURD K J (1998), Immunomodulatory properties of bovine milk. *Bull Int Dairy Fed*, **336**, 31–35.

GILL H S, DOULL F, RUTHERFURD K J and CROSS M L (2000), Immunoregulatory peptides in milk. *Br J Nutr*, **84** (supplement 1), 111–117.

GILL H S, CROSS M L, RUTHERFURD K J and GOPAL, P K (2001a), Dietary probiotic supplementation to enhance cellular immunity in the elderly. *Brit J Biomed Sci*, **58**, 94–96.

GILL H S, DARRAGH A J and CROSS M L (2001b), Optimising immunity and gut function in the elderly. *J Nutr, Health Ageing*, **5**, 80–91.

GILL H S, RUTHERFURD K J and CROSS M L (2001c), Dietary probiotic supplementation enhances natural killer cell activity in the elderly: an investigation of age-related immunological changes. *J Clin Immunol*, **21**, 264–271.

GILL H S, RUTHERFURD K J, CROSS M L and GOPAL P K (2001d), Enhancement of immunity in the elderly by dietary supplementation with the probiotic *Bifidobacterium lactis* HN019. *Am J Clin Nutr*, **74**, 833–839.

GILL H S, RUTHERFURD K J, FRAY L and BROOMFIELD A (2002a), Bovine whey proteins: immunomodulatory effects in mice. Unpublished.

GILL H S and colleagues (2002b), Dose response immunoenhancing effects of whey protein concentrate in follow-on milk. Unpublished.

GILL H S and colleagues (2002c), Effect of an immunostimulating whey protein concentrate delivered to mice in milk, based on a growing-up milk formula format. Unpublished.

GILL H S and colleagues (2002d), Immunostimulatory effects of bovine whey proteins in infant milk formula. Unpublished.

GINALDI L, DE MARTINIS M, MODESTI M, LORETO M F and QUAGLINO D (1999), The immune system in the elderly. II. Specific cellular immunity. *Immunol Res*, **20**, 109–115.

GINALDI L, DE MARTINIS M, MODESTI M, LORETO M F, CORSI M P and QUAGLINO D (2000), Immunophenotypical changes of T lymphocytes in the elderly. *Gerontology*, **46**, 242–248.

GINALDI L, LORETO M F, CORSI M P, MODESTI M and DE MARTINUS M (2001), Immunosenescence and infectious diseases. *Microbes Infection*, **3**, 851–857.

GIONCHETTI P, RIZZELLO F, VENTURI A, BIRGIDI P, MATTEUZZI D, BAZZOCCHI G, POGGIOLI G, MIGLIOLI M and CAMPIERI M (2000), Oral bacteriotherapy as maintenance treatment in patients with chronic pouchitis: a double-blind, placebo-controlled trial. *Gastroenterology*, **119**, 305–309.

GOLD P, BOUNOUS G and KONGSHAVN P A L (1989), Lactalbumin as food supplement (effect of dietary whey protein (lactalbumin) on the immune response to sheep red blood cells, host resistance to bacterial infections, development of tumors and the process of ageing). European Patent 0339656.

GOODWIN J S (1992), Changes in lymphocyte sensitivity to prostaglandin E2, histamines, hydrocortisone, and X irradiation with age: studies in a healthy elderly population. *Clin. Exp Immunol Immunopathol*, **25**, 243–251.

GOODWIN J S (1995), Decreased immunity and increased morbidity in the elderly. *Nutr Rev*, **53**(II), S41–45.

GREEN-JOHNSON J M, HAQ J A and SZEWCZUK M R (1991), Effects of aging on the production of cytoplasmic interleuikn-4 and 5, and interferon-gamma, by mucosal and systemic lymphocytes after activation with phytohemagglutinin. *Aging Immunol Infect Dis*, **3**, 43–47.

HALLER J, WEGGEMANS ER M, LAMMI-KEEFI C J and FERRY M (1996), Changes in the vitamin status of elderly Europeans: plasma vitamins A, E, B-6, B-12, folic acid and carteniods. *Eur J Clin Nutr*, **50** (Suppl. 2), S32–S45.

HALPERN G M, VRUWINK K G, VAN DE WATER J, KEEN C L and GERSHWIN M E (1991), Influence of long-term yoghurt consumption in young adults. *Int J Immunotherapy*, **7**, 205–210.

HAYEK G M, MURA C, WU D, BEHRKA A A, HAN S N, PAULSON E, HWANG D and MEYDANI S N (1997), Enhanced expression of cyclooxygenase with age in murine macrophages. *J Immunol*, **159**, 2445–2451.

HE F, TUOMOLA L A, ARVILOMMI H and SALMONEN S (2000), Modulation of humoral immune responses through intake of probiotics. *FEMS Immunol Med Microbiol*, **29**, 47–52.

HEUSER M D and ADLER W H (1997), Immunological aspects of aging and malnutrition: consequences and intervention with nutritional immunomodulators. *Clin Geriatr Med*, **13**, 697–715.

HIGASHIMOTO Y, OHATA M, UETANI K *et al.* (1994), Influence of age on pulmonary alveolar macrophage clonal growth. *Japan J Geriatr*, **31**, 854–859.

HIGH K P (1999), Micronutrient supplementation and immune function in the elderly. *Clin Infect Dis*, **28**, 717–722.

HOMES K L, SCHNIZLEIN C T, PERKINS E H and TEW J C (1984), The effect of age on nitrogen retention in lymphoid follicles and in collagenous tissues of mice. *Mech Ageing Dev*, **25**, 243–255.

INGLOT A. JANUSZ M and LISOWSKI J (1996), Colostrinine: a proline rich polypeptide from ovine colostrum is a modest cytokine inducer in human leukocytes. *Arch Immune Ther Exp*, **44**, 215–224.

ISHIDA A, YOSHIKAI Y, MUROSAKI S, HIDAKA Y and NOMOTO K (1992a), Administration of milk from cows immunized with intestinal bacteria protects mice from radiation-induced lethality. *Biotherapy*, **5**, 215–225.

ISHIDA A, YOSHIKAI Y, MUROSAKI S, KUBO C, HIDAKA Y and NOMOTO K (1992b), Consumption of milk from cows immunized with intestinal bacteria influences age-related changes in immune competence in mice. *J. Nutr*, **122**, 1875–1883.

ISOLAURI E, JOENSUS J, SUOMALAINEN H, LUOMALA M and VESIKARI, T (1995), Improved immunogenicity of oral D XRRV reabsorbant rotavirus vaccine by *Lactobacillus casei* GG. *Vaccine*, **13**, 310–312.

ISOLAURI E, SUTAS Y, KANKAANPAA P, ARVILOMMI H and SALMINEN S (2001), Probiotic effect on immunity. *Am J Clin Nutr*, **73**, 444S–450S.

JANUSZ M, LISOWSKI J and FRANEK F (1974), Isolation and characterization of a proline-rich polypeptide from ovine colostrum. *FEBS Lett*, **49**, 276–281.

JANUSZ M, ZIMECKI M and LISOWSKI J (1981), Chemical and physical characterization of a proline-rich peptide from sheep colostrum. *Biochem J*, **199**, 9–15.

JANUSZ M, STAROSCIK K, ZIMECKI M, WIECZOREK Z and LISOWSKI J (1986), A proline-rich polypeptide PRP with immunoregulatory properties isolated from ovine colostrum murine thymocytes have on their surface receptor specific for PRP. *Arch Immune Ther Exp*, **34**, 427–436.

JULIUS M H, JANSUZ M and LISOWSKI J (1988), A colostral protein that induces the growth and differntiation of resting B lymphocytes. *J Immunol*, **140**, 1366–1371.

KAILA M, ISOLAURI E, SOPPI E, VIRTANEN E, LAINE S and ARVILOMMI H (1992), Enhancement of the circulating antibody secreting cell response in human diarrhea by a human *Lactobacillus strain*. *Pediatr Res*, **32**, 141–144.

KALLIOMAKI M, SALMINEN S, ARVILOMMI H, KERO P, KOSKINEN P and ISOLAURI E (2001), Probiotics in primary prevention of atopic disease: a randomised placebo-controlled trial. *Lancet*, **357**, 1076–1079.

KATO I, ENDO K YOKOKURA T (1994), Effects of oral administration of *Lactobacillus casei* on antitumor responses induced by tumor resection in mice. *Int J Immunopharmacol*, **16**, 29–36.

KEEN C L and GERSHWIN M E (1990), Zinc deficiency and immune function. *Ann Rev Nutr*, **10**, 415–431

KELLEY D S and BENDICH A (1996), Essential nutrients and immunologic functions. *Am J Clin Nutr*, **63**, 994S–996S.

KENNEDY R S, KONOK P, BOUNOUS G, BARUCHEL S and LEE T D G (1995), The use of whey protein concentrate in the treatment of patients with metastatic carcinoma: a phase I–II clinical study. *Anticancer Res*, **15**, 2643–2650.

KISHI A, UNO K, MATSUBARA Y, OKUDA C and KISHIDA T (1996), Effect of the oral administration of *Lactobacillus brevis* subsp. *coagulans* on interferon-α producing capacity in humans. *J Am Coll Nutr*, **15**, 408–412.

KORHONEN H, MARNILA P and GILL H S (2000), Bovine milk antibodies for health. *Brit J Nutr*, **84**, SS135–S146.

KUBO M and CINADER B (1990), Polymorphism of age-related changes in interleukin (IL) production: differential changes of T-helper subpopulations, synthesizing IL-2, IL-3 and IL-4. *Eur J Immunol*, **24**, 133–136.

KULKARNI A B and KARLSSON S (1993), Transforming growth factor β1-knockout mice. A mutation in one cytokine gene causes a dramatic inflammatory disease. *Am J Pathol*, **143**, 3–9.

KUNDU B, PURI A, SINGH G, SAHAI R, TRIPATHI L M and SRIVASTAVA V M L (2000), Immunomodulatory activities of hexapeptides related to proline rich peptide from colostrum. *Bioorg Med Chem Lett*, **10**, 1181–1183.

LEMAOULT J, SZABO P and WEKSLER M E (1997), Effect of age on humoral immunity, selection of the B-cell repertoire and B-cell development. *Immunol Rev*, **160**, 115–126.

LESOURD B (1990a), Protein malnutrition: major cause of malnutrition in the elderly. *Age Nutr*, **1**, 132–138.

LESOURD B M (1990b), Le vieillissement immunologique: influence de la denutrition (Immunological ageing: influence of undernutrition). *Ann Biol Clin (Paris)*, **48**, 309–318.

LESOURD B M (1995), Protein undernutrition as the major cause of decreased immune function in the elderly: Clinical and functional implications. *Nutr Rev*, **53**(II), S86–S92.

LESOURD B M (1997), Nutrition and immunity in the elderly: modification of immune responses with nutritional treatments. *Am J Clin Nutr*, **66**, 478S–484S.

LESOURD B and MAZARI L (1999), Nutrition and immunity in the elderly. *Proc Nutr Soc*, **58**, 685–695.

LESOURD, B M and MEAUME S (1994), Cell mediated immunity changes in ageing: relative importance of cell subpopulation switches and of nutritional factors. *Immunol Lett*, **40**, 235–242.

LESOURD B M, FAURE-BECONE, M and THIOLLET (1990), Action immunostimulante d'une supplementation orale complete chez des sujets ages dentris (Immunostimulatory action of a complete oral supplementation in undernourished elderly patients). *Age Nutr*, **1**, 41–51.

LESOURD B, MAZARI L and FERRY M (1998), The role of nutrition in immunity in the aged. *Nutr Rev*, **56**, S113–S125.

LESZEK J, INGLOT A D, JANSUZ M, LISOWSKI J, KRUKOWSKA K and GEORGEAIDES J A (1999), Clostrinin®: A proline-rich polypeptide (PRP) complex isolated from ovine colostrum for treatment of Alzheimer's disease: a double-blind, placebo-controlled study. *Arch Immune Ther Exp*, **47**, 377–385.

LIGTHART G J, SCHUIT H R and HIJMANS W (1989), Natural killer cell function is not diminished in the healthy aged and is proportional to the number of NK cells in the peripheral blood. *Immunology*, **68**, 396–402.

LINK-AMSTER H, ROCHAT F, SAUDAN K Y, MIGNOT O and AESCHLIMANN J M (1994), Modulation of a specific humoral immune response and changes in intestinal flora mediated through fermented milk intake. *FEMS Immunol Med Microbiol*, **10**, 55–64.

LIPSCHITZ D A, UDUPA K D and BOXER L A (1988), The role of calcium in age-related decline of neutrophils function. *Blood*, **71**, 659–665.

LODINOVA-ZADNIKOVA R, KORYCH B and BARTAKOVA Z (1987), Treatment of gastrointestinal infections in infants by oral administration of colostral antibodies. *Die Nahrung*, **31**, 465–467.

LOIMARANTA V, LAINE M, SODERLING E, VASARA E, ROKKA S, MARNILA P, KORHONEN H, TOSSAVAINEN O and TENOVUO J (1999), Effects of bovine immune- and non-immune whey preparations on the composition and pH response of human dental plaque. *Europ J Oral Sci*, **107**, 244–250.

LOW P, RUTHERFURD K J, CROSS M L and GILL H S (2001), Enhancement of mucosal antibody responses by dietary whey protein concentrate. *Food Agric Immunol*, **13**, 255–264.

MACHNICKI M, ZIMECKI M and ZAGULSKI T (1993), Lactoferrin regulates the release of tumor necrosis factor alpha and interleukin 6 *in vitro*. *Int J Exp Pathol*, **74**, 433–439.

MAJAMAA H and ISOLAURI E (1997), Probiotics: a novel approach in the management of food allergy. *J Allergy Clin Immunol*, **99**, 179–185.

MAJAMAA H, ISOLAURI E, SAXELIN M and VESIKARI T (1995), Lactic acid bacteria in the treatment of acute rotavirus gastroenteritis. *J Pediatric Gastroent Nutr*, **20**, 333–338.

MALIN M, SOUMALAINEN H, SAXELIN M and ISOLAURI E (1996), Promotion of IgA in patients with Crohn's disease by oral bacterial therapy with *Lactobacillus* GG. *Ann Nutr Metab*, **40**, 137–145.

MARTEAU P and BOUTRON-RAULT M C (2002), Nutritional advantages of probiotics and prebiotics. *Br J Nutr*, **87**, S153–S157.

MARTEAU P, VAERMAN J P, DEHENNIN J P, BORD S, BRASSART D and POCHART P (1997), Effect of intrajejunal perfusion and chronic ingestion of *Lactobacillus johnsonni* strain La1 on serum concentrations and jejunal secretions of immunoglobulins and serum proteins in healthy humans. *Gastroenterol Clin Biol*, **21**, 293–308.

MATSUZAKI T and CHIN J (2000), Modulating immune responses with probiotic bacteria. *Immunol Cell Biol*, **78**, 67–73.

MATSUZAKI T, YAMAZAKI R, HASHIMOTO S and YOKOKURA T (1997a), Antidiabetic effect of *Lactobacillus casei* in a non-insulin-dependent diabetes mellitus (NIDDM) model using KK-A^y mice. *Endocrine J*, **44**, 357–365.

MATSUZAKI T, NAGARA Y, KADO S, UCHIDA K, KATO I, HASHIMOTO S and YOKOKURA T (1997b), Prevention of onset in an insulin-dependent diabetes mellitus model, NOD mice, by oral feeding *Lactobacillus casei*. *APMIS*, **105**, 643–649.

MATSUZAKI T, YAMAZAKI R, HASHIMOTO S and YOKOKURA T (1998), The effect of oral feeding of *Lactobacillus casei* strain Shirota on immunoglobulin E production in mice. *J Dairy Sci*, **81**, 48–53.

MCLAUGHLIN B, O'MALLEY K and COTTER T G (1986), Age-related differences in granulocyte chemotaxis and degranulation. *Clin Sci*, **70**, 59–64.

METCHNIKOFF E (1907), *The Prolongation of Life*. London, Heinemann.

MEYDANI S N and HA W-K (2000), Immunologic effects of yoghurt. *Am J Nutr*, **71**, 861–872.

MEYDANI S N, MEYDANI M, BLUMBERG J B, LEKA L S, SIBER G, LOSZEWSKI R, THOMPSON C, PEDROSA M C, DIAMOND R D and STOLLAR B D (1997), Vitamin E supplementation and *in vivo* immune response in healthy elderly subjects. *JAMA*, **277**, 1380–1386.

MICHALE S M, MCGHEE J R, ARNOLD R R and MESTECKY J (1978), Effective immunity to dental caries: selective induction of secretory immunity by oral administration of *Streptococcus mutans* in rodents. *Adv Exp Med Biol*, **107**, 261–269.

MIETTINEN M, VUOPIO-VARKILA J and VARKILA K (1996), Production of human tumour necrosis factor alpha, interleukin-6, and interleukin-10 is induced by lactic acid bacteria. *Infect Immun*, **64**, 5403–5405.

MIKES Z, FERENCIK M, JAHNOVA E, EBRINGER L and CIZNAR I (1995), Hypocholesterolemic and immunostimulatory effects of orally applied *Enterococcus faecium* M-74 in man. *Folia Microbiol*, **40**, 639–646.

MILLER R A (1996), The ageing immune system. *Science*, **273**, 70–74.

MIYAUCHI H, KAINO A, SHINODA I, FUKUWATARI Y and HAYASAWA H (1997), Immunomodulatory effect of bovine lactoferrin pepsin hydrolysate on murine splenocytes and Peyer's patch cells. *J Dairy Sci*, **80** (10):2330–2339.

MONNAI M, HORIMOTO Y and OTANI H (1998), Immunomodulatory effect of dietary bovine kappa-caseinoglycopeptide on serum antibody levels and proliferative responses of lymphocytes in mice. *Milchwissenschaft*, **53** (3):129–132.

MOSMANN T R and COFFMA R L (1989), Heterogeneity of cytokine secretion patterns and function of helper T cells. *Adv Immunol*, **46**, 111–115.

MOULIAS R, DEVILLECHABROLLE A, LESOURD B, PROUST J, MARESCOT M R, DOUMERC S, FAVRE-BERRONE M, CONGY F and WANG A (1985), Respective roles of immune and nutritional factors in priming of the immune response of the elderly. Mech. *Ageing Dev*, **31**, 123–137.

MUROSAKI S, YAMOMOTO Y, ITO K, INOKUCHI T, KUSAKA H, IKEDAH and YOSHIKAI Y (1998), Heat-killed *Lactobacillus plantarum* L-137 suppresses naturally fed antigen-specific IgE production by stimulation of IL-12 production in mice. *J Allergy Clin Immunol*, **102**, 57–64.

MUSCETTOLA M, MASSAI L, TANGANELLI C and GRASSO G (1994), Effects of lactobacilli on interferon production in young and aged mice. *Ann NY Acad Sci*, **717**, 226–232.

NENONEN M T, HELVE T A, RAUMA A L and HANNINEN O O (1998), Cooked, lactobacilli-rich, vegan food and rheumatoid arthritis. *Br J Rheumatol*, **37**, 274–281.

NUTTER R L, GRIDLEY D S, KITTERING J D, ANDRES M L, APRECIO R M and SLATER J M (1983), Modification of a transplantable colon tumor and immune responses in mice fed different sources of protein, fat and carbohydrate. *Cancer Lett*, **18**, 49–62.

NUTTER R L, KITTERING J D, APRECIO R M, WEEKS D A and GRIDLEY D S (1990), Effects of dietary fat and protein on DMH-induced tumor development and immune responses. *Nutr Cancer*, **13**, 141–152.

OKHUYSEN P C, CHAPPELL C L, CRABB J, VALDER L M, DOUGLASS E T and DUPONT H L (1998), Prophylactic effect of bovine anti-*Cryptosporidium parvum*. *Clin Infect Dis*, **26**. 1324–1329.

OONA M, RAGO T, MAAROOS H, MICKELSAAR M, LOIVUKENE K, SALMINEN S and KORHONEN H (1997), *Helicobacter pylori* in children with abdominal complaints: has immune bovine colostrum some influence on gastritis? *Alpe Adria Microbiol J*, **6**, 49–57.

OPEKUN A R, EL-ZAIMAITY H M, OSATO M S, GILGER M A, MALATY H M, TERRY M, HEADON D R and GRAHAM D Y (1999), Novel therapies for *Helicobacter pylori* infection. *Alimentary Pharmacol and Therapeut*, **13**, 35–42.

ORMROD D J and MILLER T E (1991), The anti-inflammatory activity of a low molecular weight component derived from the milk of hyperimmunised cows. *Agents Actions*, **32**, 160–166.

ORMROD D J and MILLER T E (1993), Milk from hyperimmunised dairy cows as a source of a novel biologic response modifier. *Agents Actions*, **38**, 146–149.

OTANI, H and YAMADA Y (1995), Effects of bovine kappa-casein and lactoferrin on several experimental models of allergic diseases. *Milchwissenschaft*, **50**, 549–53.

OWENS W E and NICKERSON S C (1989), Evaluation of an anti-inflammatory factor derived from hyperimmunized cows. *Proc Soc Exp Biol Med*, **190**, 79–86.

PARKER N T and GOODRUM K J (1990), A comparison of casein lactalbumin and soy protein effect on immune response to T-dependent antigen. *Nutr Res*, **10**, 781–792.

PARODI P W (1999), The role of intestinal bacteria in the causation and prevention of cancer: modulation by diet and probiotics. *Aust J Dairy Tech*, **54**, 103–121.

PAWELEC G and SOLANA R (1997), Immunosenescence. *Immunol Today*, **18**: 514–516.

PAWELEC G, ADIBZADEH M, POHLA H and SCHAUDT K (1995), Immunosenescence: ageing of the immune system. *Immunol Today*, **16**, 420–423.

PELTO L, ISOLAURI E, LILIUS E M, NUUTILA J and SALMINEN S (1998), Probiotic bacteria down-regulate the milk-induced inflammatory response in milk-hypersensitive subjects but have an immunostimulatory effect in healthy subjects. *Clin Exp Allergy Immunol*, **28**, 1474–1479.

PENN N D, PURKINS L, KELLEHER J, HEATLEY R V, MASCIE-TAYLOR B H and BELFIELD P W (1991), The effect of dietary supplementation with vitamins A, C and E on cell-mediated immune function in elderly long-stay patients: a randomized controlled trial. *Age Ageing*, **20**, 169–174.

PENTTILA I A, ZHANG M F, BATES E, REGISTER G, READ L C and ZOLA H (2001), Immune modulation in suckling rat pups by a growth factor extract derived from milk whey. *J Dairy Res*, **68**, 587–599.

POWERS D C and BELSHE R B (1993), Effect of ageing on cytotoxic T lymphocyte memory as well as serum and local antibody responses elicited by inactivated influenza virus vaccine. *J Infect Dis*, **163**, 584–592.

PRASAD A S, KAPLAN J, BECK F W, PENNY H S, SHAMSA F H, SALWEN W A, MARKS S C and MATHOG R H (1997), Trace elements in head and neck cancer patients: zinc status and immunologic functions. *Otolaryngol Head Neck Surg*, **116**, 624–629.

RASMUSSEN L B, KIENS B, PEDERSEN B K and RICHTER E A (1996), Effect of diet and plasma fatty acid composition on immune status in elderly men. *Am J Clin Nutr*, **59**, 572–577.

REGISTER G O, MCINTOSH G J, LEE V W L and SMITHERS G W (1996), Whey proteins as nutritional and functional food ingredients. *Food Australia*, **48**, 123–127.

RICH E, MINCEK M, ARMITAGE K *et al.* (1993), Accessory function and properties of monocytes from healthy elderly humans for T lymphocyte response to mitogens and antigens. *Gerontology*, **39**, 93–108.

ROKKA T, SYVAOJA E L, TUOMINE J and KORHONEN H (1997), Release of bioactive peptides by enzymatic proteolysis of *Lactobacillus* GG fermented UHT milk. *Milchwissenschaft*, **52**, 675–678.

RUMP J A, ARNDT R, ARNOLD A, BENDICK C, DICHTELMULLER H, FRANKE M, HELM E B, JAGER H, KAMPMANN B, KOLB P *et al.* (1992), Treatment of diarrhoea in human immunodeficiency virus-infected patients with immunoglobulins from bovine colostrum. *Clin Investig*, **70**, 588–594.

SAWAMURA A., YAMAGUCHI Y, TOGE T, NAGATA N, IKEDA H, NAKANISHI K and ASAKURA A (1994), Enhancement of immuno-activities by oral administration of *Lactobacillus casei* in colorectal cancer patients. *Biotherapy*, **8**, 1567–1572.

SCHIFFRIN E J, ROCHAR F, LINK-AMSTER H, AESCHLIMANN J M and DONNET-HUGHES A (1995), Immunomodulation of human blood cells following the ingestion of lactic acid bacteria. *J Dairy Sci*, **78**, 491–497.

SEKINE K, WATANABE E, NAKAMURA J, TAKASUKA N, KIM D J, ASAMOTO M, KRUTOVSKIKH V, BAB-TORIYAMA H, OTA T, MOORE M, MASUDA M, SUGIMOTO H, NISHINO H, KAKIZOE T and TSUDA H (1997), Inhibition of azoxymethane-initiated colon tumor by bovine lactoferrin administration in F344 rats. *Jap J Cancer Res*, **88**, 523–526.

SHEIH Y H, CHIANG B L, WANG L H, LIAO C K and GILL H S (2001), Systemic immunity-enhancing effects in healthy subjects following dietary consumption of the lactic acid bacterium *Lactobacillus rhamnosus* HN001. *J Am Coll Nutr*, **20**, 149–156.

SHIDA K, MAKINO K, MORISHITA A, TAKAMIZAWA K, HACHIMURA S, AMETANI A, SATO T, KUMAGAI Y, HABU S and KAMINOGAWA S (1998), *Lactobacillus casei* inhibits antigen-induced IgE secretion through regulation of cytokine production in murine splenocyte cultures. *Int Arch Allergy Immunol*, **115**, 278–287.

SHIMIZU K, MATSUZAWA H, OKADA K, TAZUME S, DOSAKO S, KAWASAKI Y, HASHIMOTO K and KOGA Y (1996), Lactoferrin-mediated protection of the host from murine cytomegalovirus infection by a T-cell-dependent augmentation of natural killer cell activity. *Arch Virol*, **141**, 1875–1889.

SOLIS-PEREYRA B and LEMONNIER D (1991), Induction of 2–5A synthetase activity and interferon in humans by bacteria used in dairy products. *Europ Cytokine Network*, **2**, 137–140.

SOLIS-PEREYRA B and LEMONNIER D (1993), Induction of human cytokines by bacteria used in dairy foods. *Nutr Res*, **13**, 1127–1140.

SOLIS-PEREYRA B, AATTOURI N and LEMONNIER D (1997), Role of food in the stimulation of cytokine production. *Am J Clin Nutr*, **66**, 521S–525S.

SPANHAAK S, HAVENAAR R and SCHAAFMA G (1998), The effect of consumption of milk fermented by *Lactobacillus casei* strain Shirota on the intestinal microflora and immune parameters in humans. *Eur J Clin Nutr*, **52**, 899–907.

STEPHAN W, DICHTELMULLER H and LISSNER R (1990), Antibodies from colostrum in oral immunotherapy. *J Clin Chem Clin Biochem*, **28**, 19–23.

STOECK M, RUEGG C, MIESCHER S, CARREL S, COX D, VON FLIEDNER V and ALKAN S (1989), Comparison of the immunosuppressive properties of milk growth factor and transforming growth factor $\beta 1$ and $\beta 2$. *J Immunol*, **143**, 3258–3265.

TACKET C O, BINION S B, BOSTWICK E, LOSONSKY G, ROY M J and EDELMAN R (1992), Efficacy of bovine milk immunoglobulin concentrate in preventing illness after *Shigella flexneri* challenge. *Am J Trop Med Hyg*, **47**, 276–283.

TARPILA S, KORHONEN H and SALMINEN S (1994), Immune colostrums in the treatment of *Helicobacter pylori* gastritis, In: *Abstract book of 24th International Dairy Congress, 18–22 Sept. 1995, Melbourne, Australia*, 293.

TRAPP C L, CHANG C C, HALPERN G M, KEEN C L and GERSHWIN M E (1993), The influence of chronic yogurt consumption on populations of young and elderly adults. *Int J Immunother*, **9**, 53–64.

TSUDA H, SEKINE K, FUJITA K and LIGO M (2002), Cancer prevention by lactoferrin and underlying mechanisms – a review of experimental and clinical studies. *Biochem Cell Biol*, **80**, 131–136.

TYAN M L (1981), Marrow stem cells during development and ageing. In: *Handbook of Immunology and Ageing* M M B Kay and T Makinodan, Eds; New York, CRC Press, 87–102.

UNITED NATIONS (1999), *Population Ageing 1999*. Population Division, United Nations.

WANG W P, IIGO M, SATO J, SEKINE K, ADACHI I and TSUDA H (2000), Activation of intestinal mucosal immunity in tumor-bearing mice by lactoferrin. *Jpn J Cancer Res*, **91**, 1022–1027.

WATANABE A, OKADA K, SHIMIZU Y, WAKABAYASHI H, HIGUCHI K, NIIYA K, KUWABARA Y, YASUYAMA, T, ITO H, TSUKISHIRO T, KONDOH Y, EMI N and KOHRI H (2000), Nutritional therapy of chronic hepatitis by whey protein (non-heated). *J Med*, **31**, 283–302.

WATSON D L, FRANCIS G L and BALLARD F J (1992), Factors in ruminant colostrum that influence cell growth and murine IgE antibody responses. *J Dairy Res*, **59**, 369–380.

WAYNE S J, RHYNE R L, GARRY P J and GOODWIN J S (1990), Cell-mediated immunity as a predictor of morbidity and mortality in the aged. *J Gerontol Ser A Biol Sci Med Sci*, **45**, M45–M48.

WEKSLER M E (1995), Immune senescence: deficiency or dysregulation? *Nutr Rev*, **53**(II): S3–S7.

WHEELER J G, SHEMA S, BOGLE M L, SHIRRELL A, BURKS A W, PITTLER A and HELM R M (1997), Immune and clinical impact of *Lactobacillus acidophilus* on asthma. *Ann Allergy, Asthma Immunol*, **79**, 229–233.

WICK G and GRUBECK-LOEBENSTEIN B (1997), Primary and secondary alterations of immune reactivity in the elderly: impact of dietary factors and disease. *Immunol Rev*, **160**: 171–184.

WOLD A E (2001), Immune effects of probiotics. *Scand J Nutr*, **45**, 76–85.

WONG C W and WATSON D L (1995), Immunomodulatory effects of dietary whey proteins in mice. *J Dairy Res*, **62** (2):359–368.

WHO (1999), *Health and Development in the 20th Century*. World Health Report, World Health Organisation, Geneva, Switzerland.

ZHAO H (1999), Effect of synthetic and bovine milk conjugated linoleic acid (CLA) on immune function. MSc thesis, Massey University, Palmerston North, New Zealand.

ZIMECKI M and KRUZEL M L (2000), Systemic or local co-administration of lactoferrin with sensitizing dose of antigen enhances delayed type hypersensitivity in mice. *Immunol Lett*, **74**, 183–188.

ZIMECKI M, MIEDZYBRODZKI R and SZYMANIEC S (1998), Oral treatment of rats with bovine lactoferrin inhibits carageenan-induced inflammation: correlation with decreased cytokine production. *Arch Immunol Therapiae Exp*, **46**, 361–365.

ZIMECKI M., JANUSZ M, STAROSCIK K, LISOWSKI J and WIECZOREK Z (1982), Effect of a proline-rich polypeptide on donor cells in graft-versus-host reaction. *Immunology*, **47**, 141–146.

ZIMECKI M, LISOWSKI J, HARABA T, WIECZOREK Z, JANUSZ M and STAROSCIK K (1984 a), The effect of a propine-rich polypeptide (PRP) on the humoral immune response I. Distinct effect of PRP on the T-cell properties of mouse glass-nonadherent (NAT) and glass-adherent (GAT) thymocytes in thymectomised mice. *Arch Immune Ther Exp*, **32**, 191–196.

ZIMECKI M, LISOWSKI J, HARABA T, WIECZOREK Z, JANUSZ M and STAROSCIK K (1984 b), The effect of a propine-rich polypeptide (PRP) on the humoral immune response II. PRP induces differentiation of helper cells from glass-nonadherent thymocytes (NAT) and suppressor cells from glass-adherent thymocytes (GAT). *Arch Immune Ther Exp*, **32**, 197–201.

ZIMECKI M, SPIEGEL K, WLASZCZYK A, KUBLER A and KRUZEL M L (1999), Lactoferrin increases the output of neutrophil precursors and attenuates the spontaneous production of TNF-alpha and IL-6 by peripheral blood cells. *Arch Immunol Ther Exp (Warsz)*, **47**, 113–118.

7

The therapeutic use of probiotics in gastrointestinal inflammation

F. Shanahan, University College Cork, Ireland

7.1 Introduction

If one considers the collective metabolic capacity of the intestinal microbiome, the potential benefits of manipulating the microflora by naturally occurring food grade bacteria is limited only by one's imagination. Although knowledge of the metabolic activity of the intestinal flora is still relatively superficial, it is this aspect of the flora that probably offers the greatest potential for health benefit and therapeutic exploitation.

Modification of the gastrointestinal bacterial flora has become an attractive therapeutic strategy for several inflammatory, infectious and neoplastic intestinal disorders. Interest in this area has been generated by growing evidence for the role of the commensal flora in the pathogenesis of Crohn's disease, ulcerative colitis and colorectal cancer. In addition, the revelation of *Helicobacter pylori* and its pathogenic role in peptic ulcer disease has served as a sobering lesson on the role of intestinal bacteria in chronic diseases.

Appropriate interpretation and handling of the intestinal microbial environment by the host requires that the bacterial residents within the lumen be constrained, without excessive immune reactivity, while retaining the capacity for effective immune responses to episodic challenge with pathogens. This entails exquisitely precise mucosal immune regulation and discriminatory accuracy in relation to commensal and pathogenic bacterial signals from the lumen.[1–8] Disruption of these processes underlies chronic inflammatory bowel disease (IBD; Crohn's disease and ulcerative colitis).

In this chapter the evidence for abnormal host–microbe interactions in IBD will be reviewed and the therapeutic potential of manipulating the bacterial flora with functional dairy products, such as probiotics, will be summarised. First, to appreciate the rationale for using functional dairy probiotic products, the composition of the human gut flora and its physiological and pathophysiological importance will be examined.

7.2   Bacteria in the gut

Some extraordinary statistics have been cited to portray the human gut flora. Approximately 1–2 kg of bacteria reside within the adult gut, with bacterial cells outnumbering the total number of mammalian cells by a factor of ten. There are apparently more bacterial cells within the human body than the total number of people who have ever lived on the planet! Furthermore, if one considers the metabolic activity of a single *Escherichia coli*, it can be appreciated that the collective metabolic activity of the gut flora eclipses that of the liver and is tantamount to a virtual or hidden organ. Indeed, the combined number of genes within the intestinal microbiome is up to a hundred times the size of the human genome!

Notwithstanding their quantitative importance, comparatively little is known of the metabolic activity of the bacterial residents within the gut. Most of the flora cannot be cultured by conventional methods and requires molecular strategies for detailed study. In humans, the composition of the flora appears to be stable but individual. Molecular bacterial fingerprinting techniques suggest that the individuality of the flora may be subject to host genetic influences,[9] whereas environmental influences such as sanitary or dietary variables appear to exert their influence on early colonisation and probably also determine the induction of bacterial enzymes and metabolic activity of the established flora.[3,6,8]

The quantitative and qualitative complexity of the resident commensal flora is greatest at either end of the alimentary tract. In the oral cavity, there are about 200 different bacterial species, whereas the stomach contains less than 103 colony forming units/ml (cfu/ml). Intestinal bacterial density increases distally with the greatest gradient being across the ileocaecal valve where there are approximately 10^8 bacteria per gram of ileal content *vs* up to 10^{12} per gram of colonic content.[3,5] The composition of the flora at the mucosal surface differs from that within the lumen and faeces, with the ratios of anaerobes to aerobes being lower at mucosal surfaces.

7.3   Studying gut flora

Numerically, obligate anaerobes are the predominant components of the gastrointestinal flora and this poses methodological difficulties for conventional microbiological approaches to the study of the flora. These include difficulties with sampling, transport and strorage, reviewed elsewhere.[10] Most of the indigenous bacteria cannot be cultured by traditional methods. This has prompted a shift from culture-dependent methods towards molecular-dependent study and a trend favouring genotyping over traditional phenotyping.

While molecular approaches are unlikely to replace culture-based technology, they have made a major contribution to the knowledge of the unculturable bacteria.[11,12] Thus, a profile of the composition of the flora can be obtained by extraction of bacterial nucleic acid from faecal or mucosal biopsy samples, amplification of 16SrDNA using universal primers spanning conserved and variable regions and separation of the hypervariable DNA fragments by chemical gradient or temperature gradient gel electrophoresis (DGGE and TGGE). Full denaturation of the DNA fragments is impeded by incorporating a GC-rich 5′ end to one of the primers (GC clamp). Variations in migration distance through the denaturing gradients reflect the diversity of 16S 'species' in the sample. The more dominant the organism within the sample, the more intense the specific amplified product, making the technique semiquantitative. Identification of individual bacterial strains can also be achieved without a conventional culture step by cutting amplified bands from the gel, further amplification, cloning and sequencing. In addition, the technique can be refined by incorporating species-specific primers into the amplification reaction. Potential confounding factors with this technology include variations in efficiency of extraction of nucleic acid from different populations of bacterial cells, mutation artefacts and uncertainty in the fidelity of the amplification reactions.[13]

Molecular profiling of the flora has been facilitated by the increasing availability of sequence data, from which species-specific molecular probes can be designed. Such probes enable identification of bacteria using various techniques, such as fluorescent *in situ* hybridisation analysis (FISH), alone or in combination with flow cytometry (FISH-FLOW) and DNA microarray and DNA chip analyses.

In gastrointestinal inflammatory disorders such as Crohn's disease and ulcerative colitis, immune reactivity against the flora can be used to identify the microbes involved in the pathogenesis of disease. Thus, anti-neutrophil cytoplasmic antibodies (pANCA) associated with ulcerative colitis have been used to identify colonic bacteria expressing a pANCA-related epitope.[14,15] Candidate microbes have also been pursued by looking for unique bacterial nucleic acid sequences associated with particular locations or lesions using subtractive cloning and genomic representational difference analysis. This technology has been reported to have identified a novel bacterial gene linked with lesions of Crohn's disease compared with adjacent non-lesional mucosa. The sequence was found to represent a bacterial transcription factor from an apparent commensal organism, *Pseudomonas fluorescens*.[16–18]

7.4   Gut flora and intestinal function

Comparative studies of germ-free and conventionally colonised animals illustrate well the influence of the commensal flora on intestinal structure and function. Life without a gut flora is associated with reductions in each of the following: mucosal cell turnover, digestive enzyme activity, cytokine production, lymphoid tissue, lamina propria cellularity, vascularity, muscle wall thickness and motility. In contrast, enterochromaffin cell area and caloric intake required to sustain body mass are increased. This implies that the flora produce regulatory signals to condition development and function of epithelial and subepithelial structures within the gut. The identity of these signals is currently being pursued with modern technology such as laser capture microdissection and gene array analysis.[3,8]

In the absence of bacterial signalling from the lumen, the mucosal-associated lymphoid tissue (MALT) is rudimentary and associated with defective cell-mediated immunity.[19] This can be accounted for, in part, by lack of microbial induction of IL-12, but also defective education of the mucosal immune system due to lack of microbial stimulation early in development. In addition, the microbial flora is critical for the fine tuning of T cell repertoires and T_H1/T_H2 cytokine profiles.[20] Since dietary and other environmental variables influence the composition of the colonising flora, they also influence the mucosal immune system and may also account for apparent species variability in mucosal immune responses. Thus, the normal default mucosal immune response to innocuous luminal antigens is thought to be biased towards a T_H2/T_H3 profile of cytokine production in mice,[20] but in humans mucosal T-cell responses are TH1-biased.[21]

Several beneficial effects of the commensal flora within the gut may be deduced from comparisons of germ-free and re-colonised animals.[22] Chief of these is probably defence against infection by competition for nutrients and epithelial binding sites and production of antimicrobial factors, such as lactic acid and bacteriocins effective against pathogens. Defence is also bolstered by the priming effect of the flora on the mucosal immune response which is maintained in state of 'controlled physiological inflammation' on ready alert. In addition to mucosal defence, the flora produce short chain fatty acids – a major energy source for colonic epithelia – from dietary fermentable carbohydrates. Other metabolic activities that favourably affect the host include breakdown dietary carcinogens and synthesis of biotin, folate and K vitamins. The bacterial flora also influence the bioavailability of drugs. For example, clinicians treating patients with ulcerative colitis for over half a century have exploited colonic bacterial azoreductases to metabolise the prodrug sulphasalazine, releasing the active aminosalicylate moiety where it can act locally.

For some individuals, under certain circumstances, components of the flora may become a risk factor for development of disease. For example, bacterial metabolism of bile acids, dietary fat and preservatives such as nitrates has been implicated in the pathogenesis of colorectal and other cancers.[23] Another risk is bacterial translocation into the systemic circulation, which may occur in the setting of overgrowth syndromes, immunodeficiency, or disruption of mucosal barrier function.[3,24] Finally, as discussed in detail below, the flora represent an essential ingredient of the pathogenesis of Crohn's disease and ulcerative colitis.

7.5   Gut immune function

To contain the indigenous bacterial flora and to eliminate infections with patho-gens, the immune response must be capable of sampling and accurately interpreting the local microenvironment.[1] Immunological sampling of the intestinal lumen occurs at three major sites:

- M cells overlying lymphoid follicles transport particulate and some microbial antigens to antigen-presenting cells (dendritic cells, B cells and macrophages) within the follicle;[25]
- surface enterocytes transport soluble antigens and may produce defensins and chemokines which recruit inflammatory cells to the site of any breach in the mucosal barrier by invasive pathogens;[26,27]
- dendritic cells which serve an immunosensory function by extending dendrites into the lumen between the enterocytes without disrupting tight junctions.[28] Dendritic cells are antigen-presenting cells that are critical for the initiation of immune responses. They are capable of sensing diverse pathogens and exhibit pathogen-specific responses.[29,30]

The signals and host receptors underlying commensal/probiotic interactions with the mucosal immune system are incompletely understood. Bacterial signals that have been identified include formylated peptides such as f-met–leu–phe,[31] lipopolysaccharide (LPS) and cell wall constituents such as peptidoglycans and bacterial nucleotides. The host discriminates pathogens from commensals by pattern recognition receptors, known as Toll-like receptors (TLRs), which are expressed by epithelial, dendritic and other cell types.[32] Epithelial expression of TLRs has been reported to be differentially altered in ulcerative colitis and Crohn's disease,[33,34] and it might be anticipated that one or more of the susceptibility genes for IBD will be found to be associated with polymorphisms in the molecular recognition of luminal bacteria by TLRs.

The host immune system uses multiple TLRs to detect several features of a microbe simultaneously: TLR2 recognises lipoproteins and peptidoglycans and triggers the host response to Gram-positive bacteria and yeast; TLR4 mediates responses to LPS primarily from Gram-negative bacteria; TLR1 and TLR6 participate in activation of macrophages by Gram-positive bacteria; whereas TLR5 and TLR9 recognise flagellin and bacterial (CpG) DNA, respectively.[32,35–37]

Transduction of bacterial signals into host immune responses is not fully understood, but nuclear factor κB (NF-κB) has been established as a key regulator of epithelial responses to pathogens, such as invasive salmonella.[38] In contrast, the counter-regulatory factor to NF-κB seems to be exploited by some non-pathogenic components of the flora which may attenuate proinflammatory responses by delaying the degradation of the inhibitory IkB.[40] Whether probiotic bifidobacteria and lactobacilli use similar mechanisms to inhibit mucosal inflammation is not known.

Colonisation of the gut with the bacterial flora activates various immune-associated genes, the functional significance of which for the bacteria is unclear, but it is advantageous for host defence. However, unrestrained mucosal T-cell activation poses a risk of tissue destruction and IBD.[41] Termination of immune responses occurs by induction of apoptosis in activated immune cells, but in Crohn's disease there is resistance of mucosal T cells to apoptosis and this leads to accumulation and perpetuation of inflammation.[42–44] Control of activated mucosal T cells also appears to be defective at the level of regulatory T cells in Crohn's disease.[45–47] Regulatory T cells within the intestinal mucosa appear to be similar to regulatory T cells within most peripheral tissues that are responsible for preventing autoimmunity to tissue-specific self-antigens. Their regulatory activities are mediated by IL-10 and TGF-β and involve signalling through a negative regulator of T-cell activation – the cytotoxic T lymphocyte-associated antigen (CTLA-4).[48] The impact of the flora and probiotics on regulatory T cells is currently being explored.

7.6   Microbial subversion of intestinal immunosensory function

As with any sampling system, sampling the intestinal microbial environment is vulnerable to episodic uptake of bacterial and viral pathogens. This occurs at the level of the M cells[25] and at non-M cell enterocytes. For example, the human immunodeficiency virus (HIV) exploits galactosylceramide and the chemokine receptor (CCR5) on the surface of small intestinal epithelial cells, as co-receptors, to cross the mucosal barrier and initiate a cascade of infection of mucosal T cells.[49,50] Bacterial pathogens also exploit mucosal immunosensory functions using various intriguing mechanisms. Enteropathogenic *Escherichia coli* deploy a bifunctional protein, intimin, the ligand for which is inserted into the host enterocyte to facilitate adherence. Intimin also binds to β1-integrin on T cells and co-stimulates primed mucosal T cells, which leads to mucosal thickening, crypt cell hyperplasia and shedding. The net effect of this is increased surface area and epithelial renewal, thereby facilitating fresh colonisation and transmission.[51] Pathogenic species of *Yersinia* can also modify the host immune response by blocking activation of the transcription factor, NF-κB, and disrupts the signal transduction process within epithelial cells for tranducing microbial signals and alerting the immune system to invasive pathogens.[38,52] Finally, pathogens such as *Cryptosporidium parvum* may prolong their survival within infected epithelia by inhibiting apoptosis of the infected cell.[53,54]

7.7   Bacterial translocation

Bacterial translocation is an important consideration for those recommending use of probiotic-containing dairy products. It may defined as the passage of viable bacteria from the gastrointestinal tract to extraintestinal sites, including the mesenteric lymph node (MLN), liver, spleen and other endorgans.[3,24] Low levels of spontaneous translocation of commensal bacteria are thought to occur continuously in healthy individuals. These migrating bacteria are eliminated *en route* and may actually be beneficial to the host by stimulating protective immunity. Excessive bacterial translocation arises in three major situations: intestinal bacterial overgrowth syndromes, increased mucosal permeability with defective barrier function, and deficiencies of the host immune system. Animal studies have indicated that resident bacteria translocate by an intraepithelial route and thence via lymph to the mesenteric lymph node. In the setting of physical disruption to the mucosal epithelium, direct passage to the portal circulation may also occur. Bacterial translocation has been confirmed in human surgical patients and appears to be more common in the elderly, particularly in the presence of distal bowel obstruction, and in the setting of disorders requiring urgent abdominal surgery.[55]

Immunological containment of bacterial translocation involves both the innate and acquired immune systems, but the rapidity of innate responses appears to be more important. While translocation seems to be increased in T-cell deficient mice and in IgA-deficient mice, systemic sepsis does not usually occur in these conditions,[24] nor does it occur in humans with IgA deficiency.[56] However, systemic sepsis from translocating bacteria does occur when there is defective granulocyte-mediated microbicidal activity.[56,57]

The rate and efficiency with which commensal bacteria translocate from the gut are variable. Gram-negative, facultative anaerobes such as the Enterobacteriaceae, *E. coli*, *Klebsiella pneumoniae* and *Proteus mirabilis* translocate at a greater rate than other indigenous bacteria, whereas obligate anaerobes and Gram-positive bacteria appear to be less efficient translocators. Thus, the bacterial species most likely to translocate are generally the same as that which commonly cause septicaemia in elderly hospitalised patients.[55]

7.8   Intestinal bacteria and IBD

Observational and experimental evidence in animals and humans has implicated the indigenous flora as an essential component of the pathogenesis of Crohn's disease and ulcerative colitis.[58,59] The most compelling evidence is derived from different spontaneously occurring or genetically engineered (knock-out or transgenic) animal models of IBD.[60–62] Irrespective of the underlying genetic defect/susceptibility, colonisation with commensal flora is required for expression of the inflammatory disease. In some animal models, the bowel disease has been adoptively transferred to immune-deficient recipients when reconstituted with T cells from a diseased donor that were sensitised to the enteric bacterial flora.[63] Furthermore, modification of the flora by administration of probiotics has delayed the disease onset and attenuated the inflammatory process.[64,65]

In humans, several circumstantial lines of evidence implicate the indigenous flora in the pathogenesis of IBD:

• First, the inflammatory disease is greatest in areas of highest bacterial numbers.

- Second, diverting the faecal stream reduces the inflammatory activity and relapse is inevitable with restoration of the faecal stream.
- Third, probiotic and some antibiotic regimens are therapeutically effective.
- Fourth, immune reactivity against the bacterial flora is common in patients with both colitis and Crohn's disease and appears to reflect a loss of immunological tolerance to indigenous flora.[66-69]
- Fifth, patients with defective phagocytic microbicidal function, such as glycogen storage disease type 1b, chronic granulomatous disease and Hermansky–Pudlak syndrome, develop Crohn's-like lesions that respond to antibiotics or treatment of the underlying immune defect.[70]
- Sixth, experimental instillation of faecal material into normal-appearing loops of bowel in patients with Crohn's disease, suggests that bacterial products within the faecal stream are directly linked to the mechanism of disease.[71,72]
- Finally, it appears that patients with IBD do not handle resident bacteria normally, and have increased numbers of bacteria within the intestinal mucosa compared with non-inflamed and inflammatory disease controls.[73-77]

Whether the mucosal flora in IBD are also qualitatively abnormal is less clear.[76] Reports on the composition of the flora have been conflicting and various species including including sulphate-reducing bacteria, non-*H. pylori* helicobacters and other species will require more detailed scrutiny.[78]

An important insight into the interaction between genes, bacteria and immunity has been provided by the discovery of the *NOD2* susceptibility gene for Crohn's disease, on chromosome 16.[79,80] The gene product, *NOD2* protein, serves as an intracytoplasmic receptor for bacterial products such as lipopolysaccharide (LPS). *NOD2* is expressed in monocytes and appears to be linked to the immunoinflammatory cascade by activating the transcription factor, NF-κB. The precise mechanism by which *NOD2* mutations predispose to Crohn's disease and whether there is over- or under-activation of NF-κB has not been resolved but it is noteworthy that NF-κB may also have an anti-inflammatory role *in vivo* in the resolution of inflammation.[81]

In patients with ulcerative colitis, the construction of an ileal pouch following colectomy represents a human 'model' showing the contribution of genes, bacteria and immune mechanisms to the pathogenesis. The contribution of bacteria to the pathogenesis of pouchitis is shown by the efficacy of both antibiotic and probiotic therapy.[82] Endogenous disease modifiers may also interact with the flora to influence disease activity in susceptible individuals. For example, stress appears to alter intestinal barrier function, favouring entry of luminal bacteria which may activate previously sensitised T cells and trigger relapse of inflammatory disease.[83]

The importance of the environmental contribution to the pathogenesis of IBD is illustrated by incomplete concordance rates in monozygotic twins for Crohn's disease (<50%) and ulcerative colitis (<10%).[84] In addition, the striking increases in incidence and prevalence of IBD worldwide have been too rapid to be accounted for on the basis of changes in population susceptibility genes. While an environmental influence could suggest a transmissible agent as a cause of IBD, and this is difficult to exclude in the light of the once unexpected role of *H. pylori* as a cause of peptic ulceration, there may be a more subtle explanation. Thus, the environmental influence may relate to changes in immune perception of the microbial environment.

Evidence against a transmissible infection as a cause of IBD includes the absence of horizontal or vertical transmission of either Crohn's disease or ulcerative colitis, the response of both conditions to suppression of the host immune response, and the fact that environmental conditions, such as poor sanitation, endemic parasitism and overcrowding, which should favour a transmissible agent, actually appear to protect against Crohn's disease.[85,86] In addition, evidence for specific infections such as *Mycobacterium paratuberculosis* and measles virus in IBD has been conflicting and difficult to reconcile with the efficacy of anti-TNF-α (tumour necrosis factor) therapies.[87,88]

An explanation for the increasing frequency of IBD as countries become industrialised might be that the susceptibility genes for IBD in developed countries are the same as those that protect against infection in an unsanitary environment. The survival advantage created by enhanced mucosal immune reactivity in a developing country may become a risk factor for immune-mediated disease as lifestyle and environmental conditions change.[89] It is particularly noteworthy that changes in prevalence of IBD have occurred simultaneously with similar trends in other chronic inflammatory disorders such as allergies, asthma, multiple sclerosis and insulin-dependent diabetes mellitus. Immune-mediated tissue injury is a feature of all these conditions and it seems plausible that environmental factors may act at the level of immune regulation rather than as transmissible infections of multiple different target organs.

The immune system can be regarded as the sense of danger within the microenvironment.[1] As with the other senses, deprivation or inappropriate input will adversely affect immune education, learning and adaptation. At birth, mucosal immune development is incomplete and continues throughout childhood. Immunological education, fine tuning of T-cell repertoires and cytokine balance are determined by environmental contact with childhood infections and commensal flora within the gut.[20] Lifestyle and environmental changes associated with industrialisation include improved sanitation and hygiene, decline in endemic parasitism, increased antibiotic usage, vaccinations, delayed exposure to childhood infections, smaller family size, life on concrete with reduced exposure to soil microbes in urban areas and reduced consumption of fermented food products. Collectively, these may influence immune education and perception and represent a risk factor for IBD, in genetically susceptible individuals.

7.9   Modifying the gut flora: probiotics in practice

Commensal bacteria within the gut vary in their capacity to drive the inflammatory process.[58] Because lactobacilli, bifidobacteria and other bacterial species have no apparent proinflammatory activity, they have been explored for probiotic potential.

Probiotics are essentially commensal organisms, which when consumed in adequate amounts (usually as food-grade live microorganisms), confer a health effect on the host. The emphasis on live microbes contrasts with prebiotics, which are non-digestable food ingredients, often of an oligo- or polysaccharide nature, that beneficially affect the health of the host by selectively stimulating the growth or activity of certain indigenous bacterial species. Mixtures of probiotics and prebiotics are referred to as synbiotics.[90] While probiotics may be ingested in various formats, consumption as a natural dairy product such as yoghurt has been the most traditional mode of delivery of probiotics to the gut.

Evidence supporting a role for probiotic and prebiotic strategies in IBD and non-IBD conditions such as atopy, infection and cancer have been reviewed elsewhere.[58,90,91] The efficacy of probiotic feeding in animal models of IBD has been shown by several investigators.[64,65] In humans, a non-pathogenic strain of *E. coli* appears to have efficacy equivalent to that of mesalazine in ulcerative colitis.[92,93] The most compelling evidence for probiotics in IBD has been reported with a probiotic cocktail of eight bacterial strains in the maintenance of remission of pouchitis.[82] However, the optimum combination of probiotic organisms has not been defined and, indeed, the possibility that some combinations might be antagonistic rather than synergistic has not been determined. Another strategy that is being explored is rigorous definition of microbial, immunological and functional characteristics of individual probiotic strains so that mechanisms can be clarified and then appropriate combinations designed if necessary.[94,95] Furthermore, probiotic performance should be defined in terms of the health benefit that is required, e.g. anti-inflammatory or anti-infective or both. Furthermore, in complex disorders such as IBD, there may be subset-specific indications requiring strain-specific prescriptions.[58]

7.10   Future trends

Potential problems with probiotic usage that require resolution have already been alluded to, but there are additional confounding factors that need to be addressed. First, there is a lack of fidelity of *in vitro* assays for predicting probiotic performance *in vivo*. Second, optimal probiotic dose and faecal recovery yield need to be settled; indeed, the target microbial niche within the gut may vary in different clinical settings. Third, currently, there are poor standards of verification and regulation of probiotic product stability. Fourth, there is a remarkable lack of rigorous probiotic strain–strain comparisons in different clinical settings. Finally, even though the mechanism of action of probiotics is unclear and probably variable, specific functional activity can be acquired by engineering non-pathogenic microorganisms to deliver anti-inflammatory cytokines or other biologically active molecules such as vaccines to the intestinal mucosa. Proof of this principle has been demonstrated with the food-grade *Lactococcus lactis*, engineered to secrete IL-10 within the gut. This was shown to be effective in two animal models of IBD.[96] However, before this approach can be applied to humans, individual and public health safety concerns must first be resolved.[97]

7.11   Sources of further information and advice

The bibliography cited within the text and referenced below contains several reviews that outline early work on functional foods, in general, and probiotics in particular. The following two easily accessible sources are particularly noteworthy:

1. Report of a Joint FAO/WHO Expert Consultation (2001) Health and nutritional properties of probiotics in food including powder milk and live lactic acid bacteria. http://www.fao.org/es/ESN/Probio/report.
2. Working group report. Guidelines for the evaluation of probiotics in food. Joint Food and Agriculture Organization of the United Nations and the World Health Organisation. http://www.fao.org/es/ESN/Probio/probio.htm.

7.12   Acknowledgement

The author is supported in part by the Health Research Board (HRB) of Ireland, the Higher Education Authority (HEA) of Ireland, the European Union (PROGID QLK-2000-00563).

7.13   References

1. SHANAHAN F (2000), Mechanisms of immunologic sensation of intestinal contents. *Am J Physiol (Gastrointest Liver Physiol)*, **278**, G191–G196.
2. MACDONALD T T and PATTERSSON S (2000), Bacterial regulation of intestinal immune responses. *Inflammatory Bowel Diseases*, **6**, 116–122.
3. BERG R D (1996), The indigenous gastrointestinal microflora. *Trends Microbiol*, **4**, 430–435.
4. FRENCH N and PETTERSSON S (2000), Microbe-host interactions in the alimentary tract: the gateway to understanding inflammatory bowel disease. *Gut*, **47**, 162–163.
5. BOCCI V (1992), The neglected organ: bacterial flora has a crucial immunostimulatory role. *Perspect Biol Med*, **35**, 251–260,.
6. BENGMARK S (1998), Ecological control of the gastrointestinal tract. The role of probiotic flora. *Gut*, **42**, 2–7.
7. GORDON J I, HOOPER L V, MCNEVIN S M, WONG M and BRY L (1997), Epithelial cell growth and differentiation III. Promoting diversity in the intestine: conversations between the microflora, epithelium, and diffuse GALT. *Am J Physiol*, **273** (Gastrointest Liver Physiol), G565–G570.
8. HOOPER L V and GORDON J I (2001), Commensal host–bacterial relationships in the gut. *Science*, **292**, 1115–1118.
9. VAN DE MERWE J P, STEGEMAN J H and HAZENBERG M P (1983). The resident faecal flora is determined by genetic characteristics of the host. Implications for Crohn's disease? *Antonie van Leeuwenhoek*, **49**, 119–124.
10. BORRIELLO S P, HUDSON M and HILL M (1978), Investigation of the gastrointestinal bacteria flora. In: R D Russell, Ed, *Clinics in Gastroenterology*, Vol. 7, WB Saunders, Philadelphia, 329–349.
11. VAUGHAN E E, SCHUT F, HEILIG H G H J, ZOETENDAL E G, DE VOS A M and AKKERMANS A D L (2000), A molecular view of the intestinal ecosystem. *Curr Issues Intest Microbiol*, **1**, 1–12.

12. AKKERMANS A D L, ZOETENDAL E G, FAVIER C F, HEILIG H G H J, AKKERMANS-VAN VLIET W M and DE VOS W M (2000), Temperature and denaturing gradient gel electrophoresis analysis of 16S rRNA from human faecal samples. *Bioscience Microflora*, **19**, 93–98.
13. QIU S, WU L, HUANG L *et al.* (2001), Evaluation of PCR-generated chimeras, mutations, and heteroduplexes with 16S rRNA gene based cloning. *Appl Environ Microbiol*, **67**, 880–887.
14. DALWADI H, WEI B and BRAUN J (2000), Defining new pathogens and non-culturable infectious agents. *Curr Opin Gastroenterol*, **16**, 56–59.
15. COHAVY O, BRUCKNER D, GORDON L K, MISRA R, WEI B, EGGENA M E, TARGAN S R and BRAUN J (2000), Colonic bacteria express an ulcerative colitis pANCA-related protein epitope. *Infect Immun*, **68**, 1542–1548.
16. WEI B, DALWADI H, GORDON L K, LANDERS C, BRUCKNER D, TARGAN S R and BRAUN J (2001), Molecular cloning of a *Bacteroides caccae* TonB-linked outer membrane protein identified by an inflammatory bowel disease marker antibody. *Infect Immunol*, **69**, 6044–6054.
17. DALWADI H, WEI B, KRONENBERG M, SUTTON C L and BRAUN J (2001), The Crohn's disease-associated bacterial protein I2 is a novel enteric T cell superantigen. *Immunity*, **15**, 149–158.
18. SUTTON C L, KIM J, YAMANE A, DALWADI H, WEI B, LANDERS C, TARGAN S R and BRAUN J (2000), Identification of a novel bacterial sequence associated with Crohn's disease. *Gastroenterology*, **119**, 23–31.
19. MACDONALD T T and CARTER P B (1979), Requirement for a bacterial flora before mice generate cells capable of mediating the delayed hypersensitivity reaction to sheep red blood cells. *J Immunol*, **122**, 2426–2429.
20. ROOK G A W and STANFORD J L (1998), Give us this day our daily germs. *Immunol Today*, **19**, 113–116.
21. MACDONALD T T (2001), Monteleone G. Interleukin-12 and Th1 immune responses in human Peyer's patches. *Trends Immunol*, **22**, 244–247.
22. MIDTVEDT T (1999), Microbial functional activities. In: L A Hanson and R H Yolken (Eds) *Intestinal Microflora*, Vol 42, Nestle Nutrition Workshop Series, Lippincott-Raven, Philadelphia, 79–96.
23. DUGAS B, MERCENIER A, LENOIR-WIJNKOOP I, ARNAUD C, DUGAS N and POSTAIRE E (1999), Immunity and probiotics. *Immunol Today*, **20**, 387–390.
24. BERG R D (1999), Bacterial translocation from the gastrointestinal tract. *Adv Exp Med Biol*, **473**, 11–30.
25. NEUTRA M R (1988), Role of M cells in transepithelial transport of antigens and pathogens to the mucosal immune system. *Am J Physiol Gastrointestinal Liver Physiol*, **274**, G785–G791.
26. KAGNOFF M F and ECKMANN L (1997), Epithelial cells as sensors for microbial infection. *J Clin Invest*, **100**, 6–10.
27. OUELLETTE A J (1999), Paneth cell antimicrobial peptides and the biology of the mucosal barrier. *Am J Physiol (Gastrointest Liver Physiol)*, **277**, G257–G261.
28. RESCIGNO M, URBANO M, VALZASINA B, FRANCOLINI M, ROTTA G, BONASIO R, GRANUCCI F, KRAEHENBUHL J-P and RICCIARDI-CASTAGNOLI P (2001), Dendritic cells express tight junction proteins and penetrate gut epithelial monolayers to sample bacteria. *Nat Immunology*, **2**, 361–367.
29. HUANG Q, LIU F, MAJEWSKI P *et al.* (2001), The plasticity of dendritic cell responses to pathogens nd their components. *Science*, **294**, 870–875.
30. IWASAKI A KELSALL B L (1999), Freshly isolated Peyer's patch but not spleen, dendritic cells produce interleukin 10 and induce the differentiation of T helper type 2 cells. *J Exp Med*, **190**, 229–239.
31. ANTON P, O'CONNELL J, O'CONNELL D *et al.* (1998), Mucosal binding sites for the bacterial chemotactic peptide, formyl-methionyl-leucyl-phenylalanine (FMLP). *Gut*, **42**, 374–379.

32. AKIRA S, TAKEDA K and KAISHO T (2001), Toll-like receptors: critical proteins linking innate and acquired immunity. *Nat Immunol*, **2**, 675–680.
33. CARIO E, ROSENBERG I M, BRANDWEIN S L *et al.* (2000), Lipopolysaccharide activates distinct signalling pathways in intestinal epithelial cell lines expressing Toll-like receptors. *J Immunol*, **164**, 966–972.
34. CARIO E and PODOLSKY D K (2000), Differential alteration in intestinal epithelial cell expression of toll-like receptor 3 (TLR3) and TLR4 in inflammatory bowel disease. *Infect Immunol*, **68**, 7010–7017.
35. KRIEG A M and WAGNER H (2000), Causing a commotion in the blood: immunotherapy progresses from bacteria to bacterial DNA. *Immunol Today*, **21**, 521–526.
36. HEMMI H, TAKEUCHI O, KAWAI T *et al.* (2000), A Toll-like receptor recognizes bacterial DNA. *Nature*, **408**, 740–745.
37. BAUER S, KIRSCHNING C K, HACKER H *et al.* (2001), Human TLR9 confers responsiveness to bacterial DNA via species-specific CpG motif recognition. *Proc Nat Acad Sci USA*, **98**, 9237–9242.
38. ELEWAUT D, DIDONATO J A, KIM J M *et al.* (1999), NF-kB is a central regulator of the intestinal epithelial cell innate immune response induced by infection with enteroinvasive bacteria. *J Immunol*, **163**, 1457–1466.
39. GERWITZ A T, RAO A S, SIMON P O *et al.* (2000), *Salmonella typhimurium* induces epithelial IL-8 expression via Ca^{++}-mediated activation of the NF-kB pathway. *J Clin Invest*, **105**, 79–92.
40. NEISH A S, GEWIRTZ A T, ZENG H, YOUNG A N, HOBERT M E, KARMALI V, RAO A S and MADARA J L (2000), Prokaryotic regulation of epithelial responses by inhibition of IκB-α ubiquitination. *Science*, **289**, 1560–1563.
41. SHANAHAN F (2002), Crohn's disease. *Lancet*, **359**, 62–69.
42. BOIRIVANT M, MARINI M, DI FELICE G *et al.* (1999), Lamina propria T cells in Crohn's disease and other gastrointestinal inflammation show defective CD2 pathway-induced apoptosis. *Gastroenterology*, **116**, 557–565.
43. INA K, ITOH J, FUKUSHIMA K *et al.* (1999), Resistance of Crohn's disease T cells to multiple apoptotic signals is associated with a Bcl-2/Bax mucosal imbalance. *J Immunol*, **163**, 1081–1090.
44. ATREYA R, MUDTER J, FINOTTO S *et al.* (2000), Blockade of interleukin 6 *trans* signalling suppresses T-cell resistance against apoptosis in chronic intestinal inflammation: evidence in Crohn's disease and experimental colitis *in vivo*. *Nat Med*, **6**, 583–588.
45. MALOY K J and POWRIE F (2001), Regulatory T cells in the control of immune pathology. *Nat Immunol*, **2**, 816–822.
46. POWRIE F (1995), T cells in inflammatory bowel disease: protective and pathogenic roles. *Immunity*, **3**, 171–174.
47. KRONENBERG M and CHEROUTRE H (2000), Do mucosal T cells prevent intestinal inflammation? *Gastroenterology*, **118**, 974–977.
48. READ S, MALMSTRÖM V and POWRIE F (2000), Cytotoxic T lymphocyte-associated antigen 4 plays an essential role in the function of CD25+CD4+ regulatory cells that control intestinal inflammation. *J Exp Med*, **192**, 295–302.
49. BOMSEL M (1997), Transcytosis of infectious human immunodeficiency virus across a tight human epithelial cell line barrier. *Nat Med*, **3**, 42–47.
50. MENG G, WEI X, WU X *et al.* (2002), Primary intestinal epithelial cells selectively transfer R5 HIV-1 to CCR5+ cells. *Nat Med*, **8**, 150–156.
51. HIGGINS L M, FRANKEL G, CONNERTON I *et al.* (1999), Role of bacterial intimin in colonic hyperplasia and inflammation. *Science*, **285**, 588–591.
52. MEIJER L K, SCHESSER K, WOLF-WATZ H *et al.* (2000), The bacterial protein YopJ abrogates multiple signal transduction pathways that converge on the transcription factor CREB. *Cell Microbiol*, **2**, 231–238.
53. MCCOLE D F, ECKMANN L, LAURENT F and KAGNOFF M F (2000), Intestinal epithelial cell apoptosis following *Cryptosporidium parvum* infection. *Infect Immunol*, **68**, 1710–1713.

182   Functional dairy products

54. FAN T, LU H, HU H et al. (1998), Inhibition of apoptosis in chlamydia-infected cells: blockade of mitochondrial cytochrome c release and caspase activation. *J Exp Med*, **187**, 487–496.
55. O'BOYLE C J, MACFIE J, MITCHELL C J, JOHNSTONE D, SAGAR P M and SEDMAN P C (1988), Microbiology of bacterial translocation in humans. *Gut*, **42**, 29–35.
56. SHANAHAN F and TARGAN S (1999), Immunologic diseases: gastrointestinal manifestations of immunodeficiency, hypersensitivity, and graft versus host disease. In: T Yamada, D Alpers, C Owyang, D W Powell amd L Laine (Eds) *Textbook of Gastroenterology*, 3rd edn. JB Lippincott Company, Philadelphia, 2547–2563.
57. SHILOH M U, MACMICKING J D, NICHOLSON S et al. (1999), Phenotype of mice and macrophages deficient in both phagocyte oxidase and inducible nitric oxide synthase. *Immunity*, **10**, 29–38.
58. SHANAHAN F (2000), Probiotics and inflammatory bowel disease: is there a scientific rationale? *Inflamm Bowel Dis*, **6**, 107–115.
59. SHANAHAN F (2001), Inflammatory bowel disease: immunodiagnostics, immunotherapeutics, and ecotherapeutics. *Gastroenterology*, **120**, 622–635.
60. BLUMBERG R S, SAUBERMANN L J and STROBER W (1999), Animal models of mucosal inflammation and their relation to human inflammatory bowel disease. *Curr Opin Immunol*, **11**, 648–656.
61. WIRTZ S and NEURATH M F (2000), Animal models of intestinal inflammation: new insights into the molecular pathogenesis and immunotherapy of inflammatory bowel disease. *Int J Colorectal Dis*, **15**, 144–160.
62. KOSIEWICZ M M, NAST C C, KRISHNAN A et al. (2001), Th1-type responses mediate spontaneous ileitis in a novel murine model of Crohn's disease. *J Clin Invest*, **107**, 695–702.
63. CONG Y, BRANDWEIN S L, MCCABE R P et al. (1998), CD4+ T cells reactive to enteric bacterial antigens in spontaneously colitic C3H/HeJBir mice: increased T helper cell type 1 response and ability to transfer disease. *J Exp Med*, **187**, 855–864.
64. MADSEN K L, DOYLE J S, JEWELL L D, TAVERNINI M M and FEDORAK R N (1999), *Lactobacillus* species prevents colitis in interleukin 10 gene-deficient mice. *Gastroenterology*, **116**, 1107–1114.
65. O'MAHONY L, FEENEY M, O'HALLORAN S et al. (2001), Probiotic impact on microbial flora inflammation and tumor development in IL-10 knock out mice. *Aliment Pharmacol Ther*, **15**, 1219–1225.
66. DUCHMANN R, KAISER I, HERMANN E, MAYET W, EWE K and MEYER ZUM BÜSCHENFELDE K-H (1995), Tolerance exists towards resident intestinal flora but is broken in active inflammatory bowel disease (IBD). *Clin Exp Immunol*, **102**, 448–455.
67. DUCHMANN R, SCHMITT E, KNOLLE P et al. (1996), Tolerance towards resident intestinal flora in mice is abrogated in experimental colitis and restored by treatment with interleukin-10 or antibodies to interleukin-12. *Eur J Immunol*, **26**, 934–938.
68. DUCHMANN R, MAY E, HEIKE M et al. (1999), T cell specificity and cross reactivity toward enterobacteria, *Bacteroides*, *Bifidobacterium*, and antigens from resident intestinal flora in humans. *Gut*, **44**, 812–818.
69. MACPHERSON A, KHOO U Y, FORGACS I et al. (1996), Mucosal antibodies in inflammatory bowel disease are directed against intestinal bacteria. *Gut*, **38**, 365–375.
70. SHANAHAN F and BERNSTEIN C (1993), Odd forms of inflammatory bowel disease – what can they tell us? *Gastroenterology*, **104**, 327–329.
71. HARPER P H, LEE E G G, KETTLEWELL M G W et al. (1985), Role of the faecal stream in the maintenance of Crohn's colitis. *Gut*, **26**, 279–284.
72. D'HAENS G R, GEBOES K, PEETERS M et al. (1998), Early lesions of recurrent Crohn's disease caused by infusion of intestinal contents in excluded ileum. *Gastroenterology*, **114**, 262–267.
73. SCHULTSZ C, VAN DEN BERG F, TEN KATE F W et al. (1999), The intestinal mucus layer

from patients with inflammatory bowel disease harbors high numbers of bacteria compared with controls. *Gastroenterology*, **117**, 1089–1097.

74. DARFEUILLE-MICHAUD A, NEUT C, BARNICH N *et al.* (1998), Presence of adherent *Escherichia coli* strains in ileal mucosa of patients with Crohn's disease. *Gastroenterology*, **115**, 1405–1413.

75. SWIDINSKI A, LADHOFF A, PERNTHALER A *et al.* (2002), Mucosal flora in inflammatory bowel disease. *Gastroenterology*, **122**, 44–54.

76. BRAUN J (2002), Unsettling facts of life: bacterial commensalism, epithelial adherence, and inflammatory bowel disease. *Gastroenterology*, **122**, 228–230.

77. GLASSER A-L, BOUDEAU J, BARNICH N *et al.* (2001), Adherent invasive Escherichia coli strains from patients with Crohn's disease survive and replicate within macrophages without inducing host cell death. *Infect Immunol*, **69**, 5529–5537.

78. FOX J G (2002), The non-*H. pylori* helicobacters: their expanding role in gastrointestinal and systemic diseases. *Gut*, **50**, 273–283.

79. HUGOT J P, CHAMAILLARD M, ZOUALI H *et al.* (2001), Association of NOD2 leucine-rich repeat variants with susceptibility to Crohn's disease. *Nature*, **411**, 599–603.

80. OGURA Y, BONEN D K, INOHARA N *et al.* (2001), A frameshift mutation in *NOD2* associated with susceptibility to Crohn's disease. *Nature*, **411**, 603–606.

81. LAWRENCE T, GILROY D W, COLVILLE-NASH P R and WILLOUGHBY D A (2001), Possible new role for NF-kB in the resolution of inflammation. *Nat Med*, **7**, 1291–1297.

82. GIONCHIETTI P, RIZZELLO F, VENTURI A *et al.* (2000), Oral bacteriotherapy as maintenance treatment in patients with chronic pouchitis: a double blind, placebo-controlled trial. *Gastroenterology*, **119**, 305–309.

83. QIU B S, VALLANCE B A, BLENNERHASSETT P A and COLLINS S M (1999), The role of CD4[+] lymphocytes in the susceptibility of mice to stress-induced reactivation of experimental colitis. *Nat Med*, **5**, 1178–1182.

84. TYSK C, LINDBERG E, JARNEROT G and FLODERUS-MYRHED B (1988), Ulcerative colitis and Crohn's disease in an unselected population of monozygotic and dizygotic twins: a study of heritability and the influence of smoking. *Gut*, **29**, 990–996.

85. GENT A E, HELLIER M D, GRACE R H *et al.* (1994), Inflammatory bowel disease and domestic hygiene in infancy. *Lancet*, **343**, 766–767.

86. ELLIOTT D E, URBAN J F JR, ARGO C K and WEINSTOCK J V (2000), Does the failure to acquire helminthic parasites predispose to Crohn's disease? *FASEB J*, **14**, 1848–1855.

87. KEANE J, GERSHON F, WISE R P *et al.* (2001), Tuberculosis associated with infliximab, a tumor necrosis factor α-neutralising agent. *N Engl J Med*, **345**, 1098–1104.

88. ROBERTSON D J and SANDLER R S (2001), Measles virus and Crohn's disease: a critical appraisal of the current literature. *Inflamm Bowel Dis*, **7**, 51–57.

89. TAYLOR K D, ROTTER J I and YANG H, Genetics of inflammatory bowel disease. In: *Inflammatory Bowel Disease: From Bench to Bedside*, 2nd edn, S R Targan, F Shanahan and L C Karp, (Eds), Kluwer Academic Publishers, Dordrecht (in press).

90. Report of a Joint FAO/WHO Expert Consultation (2001), Health and nutritional properties of probiotics in food including powder milk and live lactic acid bacteria. http://www.fao.org/es/ESN/Probio/report. Pdf file.

91. DUNNE C and SHANAHAN F (2002), Role of probiotics in the treatment of intestinal infections and inflammation. *Curr Opin Gastroenterol*, **18**, 40–45.

92. KRUIS W, SCHÜTZ E, FRIC P, FIXA B, JUDMAIERS F and STOLTE M (1997), Double-blind comparison of an oral *Escherichia coli* preparation and mesalazine in maintaining remission of ulcerative colitis. *Aliment Pharmacol Ther*, **11**, 853–858.

93. REMBACKEN B J, SNELLING A M, HAWKEY P M, CHALMERS D M and AXON A T R (1999), Non-pathogenic *Escherichia coli* versus mesalazine for the treatment of ulcerative colitis: a randomised trial. *Lancet*, **354**, 635–639.

94. COLLINS J K, MURPHY L, MORRISSEY D *et al.* (2002), A randomised controlled trial of a probiotic *Lactobacillus* strain in healthy adults: assessment of its delivery, transit, and

influence on microbial flora and enteric immunity. *Microbial EcolHealth and Dis*, **14**, 81–89.

95. DUNNE C, MURPHY L, FLYNN S *et al.* (1999), Probiotics: from myth to reality. Demonstration of functionality in animal models of disease and in human clinical trials. *Antonie von Leeuwenhook*, **76**, 279–292.

96. STEIDLER L, HANS W, SCHOTTE L *et al.* (2000), Treatment of murine colitis by *Lactococcus lactis* secreting interleukin-10. *Science*, **289**, 1352–1355.

97. SHANAHAN F (2000), Therapeutic manipulation of gut flora. *Science*, **289**, 1311–1312.

Part II

Functional dairy ingredients

8

Caseinophosphopeptides (CPPs) as functional ingredients

R.J. FitzGerald, University of Limerick, Ireland, and
H. Meisel, Institut für Chemie und Technologie der Milch, Germany

8.1 Introduction

Caseinophosphopeptides (CPPs), as the name suggests, are casein-derived peptides
that have phosphorus bound via monoester linkages to seryl residues. The term
appears to have been first introduced by Mellander in 1950 to describe a group of
phosphorylated casein peptides that enhanced bone calcification independently of
vitamin D in rachitic children. Given their highly negatively charged structures
arising from phosphorylation, CPPs have the ability to bind a range of
macroelements such as calcium, magnesium and iron, and trace minerals such as
zinc, barium, chromium, nickel, cobalt and selenium. It has long been thought that
the enhanced dietary bioavailability of calcium from milk and dairy products may
be attributed to the presence of CPPs. This is due to the observation that *in vitro*
CPPs possess the ability to bind and solubilise calcium at high pH, which
corresponds to the alkaline conditions pertaining in the small intestine where
maximal passive vitamin D-independent calcium adsorption takes place.

The functional food ingredient role of CPPs is mainly linked to their potential
ability to enhance the bioavailability of dietary calcium. A number of food
ingredient companies currently market CPP containing/enriched milk protein
hydrolysate products. A general challenge to the functional food ingredients sector
is to demonstrate the efficacy and safety of food products or components therein
that are claimed to have beneficial health effects. The current challenge in relation
to CPPs is to unequivocally demonstrate their efficacy in enhancing dietary
mineral bioavailability in humans. This chapter will outline the structure of CPPs,
their production, their potential contribution to calcium bioavailability, their safety
and the general ingredient roles of CPPs.

8.2   Structural characteristics and production of CPPs

The structures of CPPs have been previously reviewed (West, 1986; Reynolds, 1994; FitzGerald, 1998). The individual caseins, i.e. α_{s1}-, α_{s2}-, β- and κ-casein, are phosphorylated to different extents. Bovine α_{s1}-casein can have up to 13 phosphoseryl residues whereas κ-casein has only one phosphoseryl group. The specificity of casein kinases in the mammary gland appear to be dictated by the amino acid sequence around the sites of phosphorylation, which is manifested in the so-called triplet anionic region (Mercier, 1981). The amino sequence around the main regions of phosphorylation in bovine caseins is summarised in Fig. 8.1. A common feature in the regions of high phosphorylation is the occurrence of three phosphoseryl followed by two glutamic acid residues, i.e. SerP–SerP–SerP–Glu–Glu (Fig. 8.1). This specific sequence occurs at residues (66–70) in α_{s1}-casein, (8–12) and (56–60) in α_{s2}-casein and (17–21) in β-casein. These highly polar regions represent the binding sites for minerals. However, not all seryl residues in the caseins are phosphorylated. Furthermore, the extent of phosphorylation is also related to sequence polymorphisms as evidenced from analysis of the primary sequences of different bovine milk protein variants. The N-terminal tryptic fragment, f(1–25), of bovine β-casein variant A possesses four phosphoseryl groups whereas the corresponding D variant contains only three. This is due to the substitution of a serine residue in the A variant with lysine in the D variant (Fig. 8.1).

α_{s1}-casein
gln-met-glu-ala-glu-ser(P)-ile-ser(P)-ser(P)-ser(P)-glu-glu-ile-val-   f(59–79)
pro-asn-ser(P)-val-glu-gln-lys
val-pro-asn-ser(P)-ala-glu-glu-arg   f(112–119)

α_{s2}-casein
glu-his-val-ser-ser(P)-ser(P)-glu-glu-ser-ile-ile-ser(P)-gln-glu   f(5–18)
asn-pro-ser(P)-lys-glu-asn   f(29–34)
gly-ser(P)-ser(P)ser(P)glu-glu-ser(P)-ala-glu-val   f(55–64)
gln-leu-ser(P)-thr-ser(P)-glu-glu-asn-ser-lys-lys-thr-val-asp-met-   f(127–147)
glu-ser(P)-thr-glu-val-phe

β-casein
ile-val-glu-ser(P)-ser(P)-ser(P)-glu-glu-ser-ile-lys   Variant A f(12–23)
ile-val-glu-ser(P)-lys-ser(P)-glu-glu-ser-ile-lys   Variant D f(12–23)

κ-casein
glu-ala-ser(P)-pro-glu-val-ile   f(147–153)

Fig. 8.1   Sequence of different phosphorylated casein regions found in bovine milk (ser(P) = serine phosphate) (taken from FitzGerald, 1998).

8.2.1   Production of CPPs

The basic protocol for the production of CPPs involves enzymatic digestion of whole casein or fractions enriched in specific individual caseins. A range of proteinase activities from mammalian, bacterial and plant sources have been employed for casein hydrolysis (Manson and Annan, 1971; Brule *et al.*, 1982, 1989; Berrocal *et al.*, 1989; Juillerat *et al.*, 1989; Reynolds, 1992; Adamson and Reynolds, 1995; McDonagh and FitzGerald, 1998). To date, crude and relatively purified pancreatic proteinase preparations containing trypsin and chymotrypsin, and combinations of pancreatic activities (pancreatin) have been mainly described. However, proteinases from a range of bacterial, fungal and plant sources containing, for example, thermolysin, subtilisin, pronase, and papain activities have also been successfully used in the hydrolysis of casein for CPP production.

Generally, hydrolysis is carried out at neutral to alkaline pH (pH 7.0–8.0). Adjustment of hydrolysates to pH 4.6 results in the appearance of insoluble material that can be removed subsequently by centrifugation. The CPPs present in the resulting clarified hydrolysate supernatants can be aggregated on inclusion of a range of mineral salts around neutral pH. Calcium has preferentially been used for CPP aggregate formation; however, barium, iron, zinc, magnesium, copper, manganese and cobalt have all been mentioned in the literature as potential aggregating agents (Reeves and Latour, 1958; Manson and Annan, 1971; Brule *et al.*, 1982, 1989; Berrocal *et al.*, 1989; Juillerat *et al.*, 1989; Reynolds, 1992; Adamson and Reynolds, 1995; McDonagh and FitzGerald, 1998).

The aggregated CPPs can be precipitated with hydrophilic solvents such as ethanol, methanol, propanol, butanol and acetone (Reeves and Latour, 1958; Adamson and Reynolds, 1995; McDonagh and FitzGerald, 1998) and have been collected by centrifugation. Ultrafiltration through defined molecular mass cut-off membranes has also been used at pilot-scale to separate aggregated CPPs from non-phosphorylated peptides (Brule *et al.*, 1982, 1994; Reynolds, 1992). Subsequently, diafiltration has been employed to remove excess mineral salt from ultrafiltered CPPs (Reynolds, 1993). Alternatively, CPPs have been isolated/enriched from calcium/ethanol precipitated aggregates using a range of ion-exchange procedures (Berrocal *et al.*, 1989; Juillerat *et al.*, 1989; Kunst, 1990; Koide *et al.*, 1991; Lihme *et al.*, 1994). Fractions containing isolated/enriched CPPs are then subjected to spray- or freeze-drying. The yield of CPPs produced by the above protocols is reported to range from 6 to 20% (Gerber and Jost, 1986; Juillerat *et al.*, 1989; Kunst, 1990; McDonagh and FitzGerald, 1998; Ellegård *et al.*, 1999).

Phosphopeptides are susceptible to dephosphorylation during exposure to alkaline and high heat treatments. These conditions result in β-elimination of phosphoserine, resulting in the generation of dehydroalanine (Lorient, 1979). Lysinoalanine formation between dehydroalanine and peptide/protein bound lysine in turn may lead to reduced lysine bioavailability (Manson and Carolan, 1980; Hasegawa *et al.*, 1981; De Koning and Van Rooijen, 1982). Therefore, the processing conditions during CPP production, especially for example, during spray-drying, may have a significant impact on CPP quality. Heating sodium caseinate and the pH 4.6 soluble portion of a tryptic hydrolysate of sodium caseinate at 170 °C, a typical inlet temperature during spray-drying, resulted in significant losses in peptide bound phosphorus in both samples (Meisel *et al.*, 1991a).

8.3   CPPs and mineral (calcium) bioavailability

The functional food ingredient potential of CPPs is mainly related to the ability of CPPs to bind and solubilise minerals such as calcium. CPPs prevent the precipitation of calcium in the presence of phosphate at alkaline pH *in vitro* (Berrocal *et al.*, 1989; Sato *et al.*, 1991; Gaucheron *et al.*, 1995; Holt *et al.*, 1996, 1998; Aoki *et al.*, 1998; Holt, 2001). Holt (2001) has demonstrated that calcium and phosphate form soluble nanoclusters in the presence of CPPs. Furthermore, CPPs inhibit the formation of hydroxyapatite and brushite crystals (Reynolds, 1993; Holt *et al.*, 1996). The highly anionic phosphorylated regions in CPPs are responsible for mineral binding. However, the amino acid sequence around this anionic hydrophilic domain also seems to play a significant role in mineral binding (Reynolds, 1994; McDonagh and FitzGerald, 1998). Dephosphorylated peptides/proteins do not bind minerals such as calcium, zinc and iron as efficiently as their equivalent phosphorylated derivatives (Harzar and Kauer, 1982; Sato *et al.*, 1983; Gerber and Jost, 1986; Berrocal *et al.*, 1989; Bouhallab *et al.*, 1991; Meisel and Olieman, 1998; Yeung *et al.*, 2001).

The calcium-binding properties of CPPs have been studied using calcium selective electrodes (Berrocal *et al.*, 1989; Meisel *et al.*, 1991b), micro-ultrafiltration methods in combination with atomic absorption (Perich *et al.*, 1992), competitive assays with Chelex resin (Meisel *et al.*, 1989) and capillary zone electrophoresis (Meisel and Olieman, 1998). The calcium binding constants for heterogeneous CPP preparations is reported to be in the order of 10^2–10^3 M^{-1} (Berrocal *et al.*, 1989; Sato *et al.*, 1991; Schlimme and Meisel, 1995). Fractioned CPPs are reported to have significantly greater binding affinities for calcium, e.g. 0.32 mM^{-1} for α_{s1}-casein (59–79)5P at pH 8.0 (Meisel and Olieman, 1998). Lee *et al.* (1980) reported that one mole of CPP could bind approximately 40 moles of calcium. The calcium solubilising ability of CPP-enriched fractions derived from casein hydrolysates digested with bacterial, fungal, plant and animal origin ranged from 7.4 to 24.0 mg calcium per mg CPP (McDonagh and FitzGerald, 1998). The calcium solubilising ability of CPPs may be affected by genetic polymorphism. Han *et al.* (2000) reported that the calcium solubilising ability of CPPs derived from Korean cattle containing the H variant of β-casein had approximately 23% increased calcium solubilising ability than CPPs derived from casein that did not contain the H variant of β-casein.

The (1–25) phosphopeptide fragment of β-casein was reported to bind 4 mol iron per mol peptide. Enzymatic dephosphorylation of the β-casein (1–25) rendered the bound iron completely dialysable (Bouhallab *et al.*, 1991). The presence of CPP-bound iron was also shown to significantly inhibit enzymatic dephosphorylation (Bouhallab *et al.*, 1991; Yeung *et al.*, 2001).

All the calcium phosphate in milk exists in the form of nanometre-sized clusters, or nanoclusters. Calcium phosphate nanoclusters are reported to consist of a core (radius = 2.4 nm) containing ~350 hydrated calcium phosphate ion pairs. This core is surrounded by a shell 1.6 nm thick comprising about 50 phosphopeptide chains (Holt *et al.*, 1996, 1998). Soluble nanoclusters containing otherwise insoluble minerals and essential trace elements such as calcium, magnesium, zinc, iron, manganese and iodine can be formed in the presence of CPPs. By uniformly raising the pH in, for example, a calcium phosphate solution via the liberation of ammonia using jack bean urease in the presence of urea, it was reported to be possible to develop stable solutions containing up to 1 M calcium in the presence of β-casein (1–25)4P and β-casein (1–42)5P (Holt, 2001).

8.4   Human studies with CPPs

Milk and dairy products are abundant sources of bioavailable calcium with about 70% of dietary calcium needs being supplied by milk and dairy products in developed Western society (IDF, 1997). The excellent bioavailability of dietary calcium during the consumption of dairy products has been attributed to the presence of CPPs in casein-containing diets (Gammelgård-Larsen, 1991; Kitts and Yuan, 1992; Schaafsma, 1997). CPPs *in vitro* can maintain calcium soluble in the presence of phosphate at alkaline pH (Reeves and Latour, 1958; Sato *et al.*, 1991; Ono *et al.*, 1994; Holt *et al.*, 1996, 1998; Holt, 2001; Aoki *et al.*, 1998). These are the conditions that pertain in the small intestine where maximal passive vitamin D-independent calcium absorption takes place (Bronner, 1987). This ability of CPP to maintain calcium solubility in the conditions as found in the distal ileum is therefore proposed to be the basis for enhanced absorption of dietary calcium (Lee *et al.*, 1980, 1983; Mykännen and Wasserman, 1980; West, 1986; Meisel and Frister, 1988; Li *et al.*, 1989; Kitts and Yuan, 1992; Kitts *et al.*, 1992).

If the above hypothesis is correct, it is clear that in order to act as mineral carriers or enhancers of mineral solubility, CPPs must display the ability to resist, or at least to partially resist, enzymatic digestion during intestinal passage. Therefore, several animal and two human studies have reported to date on the production of CPPs *in vivo*. Analysis of the intestinal fluid of minipigs fed a bovine casein-containing diet, for example, demonstrated the presence of CPPs (Meisel and Frister, 1988). CPPs were detected in the stomach and duodenal contents of adult humans after ingestion of milk or yoghurt (Chabance *et al.*, 1998). It has recently been demonstrated, for the first time, that phosphopeptides can survive the prolonged intestinal passage to the distal small intestine (ileum) in humans following ingestion of diets containing milk or peptide preparations enriched in CPPs (Meisel *et al.*, 2001, 2002). Interestingly, these studies show better survival during intestinal passage by CPPs derived from consumption of milk as opposed to consumption of CPPs *per se*.

Numerous animal and a very limited number of human studies have been performed on the role of CPPs in mineral, mainly calcium, bioavailability (for reviews see FitzGerald, 1998; Scholz-Ahrens and Schrezenmeir, 2000 and references therein). To date most studies have been performed on animals. Dietary studies on CPPs with rats, chickens and piglets have yielded conflicting results which have, in part, been attributed to differences in methodology used to quantify extrinsic and intrinsic calcium absorption in addition to differences in the quality of the CPP preparations under study (Kitts and Yuan, 1992; FitzGerald, 1998). The rat study of Bennett *et al.* (2000) reported that the true fractional absorption of calcium was in fact decreased on inclusion of high CPP levels (200–500 g CPP/kg meal). These results suggested that high levels of dietary CPPs may chelate all available dietary calcium, thereby making it unavailable for subsequent absorption.

Co-ingestion of CPPs with calcium by postmenopausal women having low basal absorptive performance was reported to significantly enhance calcium absorption (Heaney *et al.*, 1994). The human studies of Hansen *et al.* (1997a) reported a 30% (approximate) increase in calcium and zinc absorption by adults from a rice-based infant gruel supplemented with CPPs. This effect was not observed in the presence of phytate-containing meals (Hansen *et al.*, 1996, 1997b). Preliminary results from a more recent human study indicate that consumption of CPPs in beverage format does not significantly increase the fractional absorption of dietary calcium (Teucher *et al.*, 2002). Further human studies with CPPs are required to fully elucidate the potential role of these peptides in modulating mineral bioavailability. One of the difficulties in studying CPP effects on mineral bioavailability in animals or humans arises from the presence of complex inter-actions with other meal constituents (Rossander *et al.*, 1992; Hansen *et al.*, 1996). Furthermore, long-term stringent control during human feeding experiments is difficult to achieve.

A recent study by Erba *et al.* (2001) using everted small intestine of rats clearly demonstrated that the presence of CPPs can limit the inhibitory effect of phosphate on calcium availability and increase calcium transport across the distal ileum of rat. However, the extent of calcium absorption was dependent on the Ca:Pi ratio, i.e. in the presence of CPPs, calcium transport was decreased by 40% at Ca:Pi = 1:1 whereas 60% inhibition of calcium transport was observed at Ca:Pi = 1:2. These studies clearly demonstrate that not only is the interaction of other meal constituents important, but also that the concentration-dependent effects of other constituents on mineral bioavailability plays a role. Furthermore, the format in which CPPs are consumed by humans appears to dictate the ability of CPPs to reach the distal ileum (Meisel *et al.*, 2001, 2002).

8.5   Effect of CPPs on mineral uptake in specific cell systems

In vitro studies with the human intestinal cell line, HT-29, demonstrated in the presence of extracellular calcium that both a commercial preparation containing a mixture of CPPs and an individual CPP (β-casein (1–25) 4P) induced a transient rise in intracellular calcium, which did not influence ATP-induced release of calcium from intracellular stores. This effect was not observed in the absence of extracellular calcium. Pre-treatment of the cells with thapsinargin, which completely depletes intracellular calcium stores, did not abolish the cell response to CPPs. Furthermore, repetitive stimulation of the cells with CPPs always resulted in a transient rise in intracellular calcium, indicating that the cells did not become desensitised to the addition of CPPs (Ferraretto et al., 2001). The authors concluded, since the data indicated that the process influenced by CPPs is non-saturating, that it does not involve a membrane receptor or an endogenous ion channel system. They also suggested that CPPs may act on transmembrane calcium flux by either inserting themselves into the plasma membrane to form calcium selective channels or as calcium carrier peptides, which could be rapidly internalised via endocytosis or other processes, resulting in the delivery of calcium to the cytosol.

The presence of CPPs in fertilisation medium was shown to increase the rate of boar spermatozoa penetration of *in vitro* matured pig oocytes. Spermatozoa also retained the ability to penetrate oocytes for longer in the presence of CPPs. Calcium uptake by spermatozoa was increased on the inclusion of CPPs in the fertilisation medium (Mori *et al.*, 1996).

CPPs influenced zinc absorption from aqueous phytate-containing solutions by human colon carcinoma (Caco–2) cells. In a control system, phytate reduced binding and uptake of zinc by 79% in Caco-2 cells. Inclusion of low levels of CPPs increased zinc binding and uptake by 94%. Higher levels of CPPs, however, inhibited zinc binding and uptake (Hansen *et al.*, 1996). These results again emphasise the importance of CPP concentration on specific test parameters.

The presence of CPPs was reported to increase calcification in cultured rat bone embryonic explants. This may, in part, be linked to the observed increase in solubility of calcium in the culture medium (Gerber and Jost, 1986).

8.6   Cytomodulatory effects

Based on various cytochemical studies, there is increasing evidence of the possible involvement of milk protein-derived peptides as specific signals that can trigger viability of cancer cells. Purified peptides, corresponding to bioactive sequences of casein, revealed modulation of cell viability, i.e. proliferation and apoptosis, in different human cell culture models (Meisel and Günther, 1998; Hartmann *et al.*, 2000). Accordingly, apoptosis of human leukaemia cells (HL-60) was induced by the CPP β-CN(f1–25)4P. Since the dephosphorylated derivative β-CN(f1–25)0P showed comparable modulating activity, the effects are obviously not mediated by phosphoserine residues. Cheese contains phosphopeptides as natural constituents, and peptides from a lyophilised extract of Gouda cheese inhibited proliferation of leukaemia cells, even at concentrations of 1 pmol/l (Meisel and Günther, 1998). The antiproliferative effect of Gouda extract was shown to be a result of peptide-induced apoptosis. It is interesting to note that cancer cell lines were more reactive to peptide-induced apoptotic stimulation than non-malignant cells.

Effects on both cell viability and immune cell function (*see* Section 8.8.4 on immunomodulation) may be a mechanism through which cytomodulatory CPPs exert protective effects in cancer development. The most probable target site for cytomodulatory action seems to be the gastrointestinal tract. The antiproliferative effects of milk-protein derived peptides as well as other minor milk components in colon cancer cell lines suggests that they could have a role in the prevention of colon cancer by blocking hyperproliferation of the epithelium and by promoting apoptosis (MacDonald *et al.*, 1994; Ganjam *et al.*, 1997; Meisel *et al.*, 1998).

Besides a possible positive action there may be adverse effects of CPPs that would impose restrictions to their application in functional foods for human nutrition. However, cytochemical experiments carried out on CPP preparations using a variety of human cell culture model systems did not point to a significant cytotoxic potential of CPPs towards human cells, provided that the orally ingested doses correspond to reasonable concentrations for use in functional foods. Thus, CPPs can be supposed not to provoke a cytotoxic response in cells of the respective types *in vivo* (Hartmann *et al.*, 2000; Hartmann and Meisel, 2002).

8.7   Safety assessment of CPPs

In order that CPPs can receive acceptance as functional food ingredients or as nutraceuticals it is essential to demonstrate their overall safety following human consumption. Information regarding the safety of consuming CPPs or fractions enriched in CPPs is required by clinicians, legislators and the food industry in order to prepare a case for the widespread utilisation of CPPs by consumers. Given that CPPs have been consumed by humans for centuries as part of diets rich in dairy products, it would be expected that consumption of CPPs should pose little or no risk to the general population.

8.7.1   Antigenic and allergenic potential of CPPs

Studies with synthetic peptides corresponding to phosphorylated regions of α_{s1}-casein demonstrated that α_{s1}-casein (63–70)4P was more strongly recognised by animal-derived polyclonal antisera than the corresponding dephosphorylated peptide (Perich *et al.*, 1992). Studies with polyclonal anti-α_{s1}-casein antibodies also indicated that the antigenicity of α_{s1}-casein was lost on dephosphorylation (Otani *et al.*, 1987). No significant modification in the antigenicity of animal antisera raised against purified bovine α_{s2}- and β-casein was observed following enzymatic dephosphorylation of these proteins or by deletion (through studies with deletion variants) of an associated major phosphorylation site (Bernard *et al.*, 2000). It was concluded from these findings that most of the polyclonal antibodies raised against pure α_{s2}- and β-casein were not directed against the major phosphorylation sites on these molecules and that immunoreactivity was not affected by phosphorylation of the serine residues. Furthermore, it was concluded that the immunoreactivity of these caseins appears to be unaffected by any potential conformational change induced by dephosphorylation.

Studies with antisera derived from human volunteers who display allergenic reactions to casein showed that dephosphorylation or deletion of phosphorylated regions (as demonstrated using deletion variants) led to decreased recognition of α_{s2}- and β-casein by the IgE in these sera. It was concluded that dephosphorylation probably lowers the affinity of IgE for the caseins rather than lowering the specific IgE titre by comparison of the results of direct *vs* competitive inhibition enzyme-linked immunosorbent assay (ELISA). From supporting studies with isolated phosphorylated and dephosphorylated β-casein (1–25), it was concluded that it was very likely that at least part of the anti-casein IgE is directed against a major site of phosphorylation. It was also suggested that this finding may, in part, explain the observed cross-reactivity between the individual caseins (Bernard *et al.*, 2000). Therefore, the allergenic potential of CPPs, following their consumption by the segment of the human population that displays allergenic reactions to caseins, should be considered in the exploitation of CPPs as functional food ingredients/ nutraceuticals.

8.8   Potential ingredient applications of CPPs

The potential ingredient applications of CPPs have been previously reviewed (FitzGerald, 1998) and are summarised in Table 8.1.

Table 8.1   Potential applications of caseinophosphopeptides (CPPs) (reproduced from FitzGerald, 1998. Copyright Elsevier Science 1998; reproduced with permission)

Application	Rationale	Product
Prevention of osteoporosis	Increase absorption of Ca by including Ca–CPP as functional food ingredient for target populations, allowing maximum deposition of bone in early life	Ca–CPP complexes
Recalcification of bones after fracture	Increase serosal [Ca], aiding bone mineralisation	Ca–CPP complexes
Calcic addition during treatment of rickets	Increase serosal [Ca]	Ca–CPP complexes
Hypertension	Increased serosal [Ca] may aid reduction of hypertension	Ca–CPP complexes
Prevention of dental caries	Inclusion of Ca–CPP as ingredients in mouthwashes, toothpastes	Ca–CPP complexes.
Mg deficiency in pregnancy and old age	Aid Mg supplementation	Mg-CPP complexes, ingredients
Anaemia	Aid Fe supplementation, increase serosal [Fe]	Fe–CPP complexes.
Humanisation of bovine milk	Increase phosphorus levels	CPPs
Oligoelement supplementation	Aid supplementation of Zn, Cu, Cr, Ni, Co, Mn and Se	Oligoelement–CPP complexes

8.8.1 Mineral supplementation

The application for which CPPs are probably best linked is in the area of mineral supplementation. It is thought that CPPs may aid dietary mineral absorption by enhancing solubility. A whole range of minerals including calcium, magnesium and iron are required to maintain health (Passmore et al., 1974; Fairweather-Tait, 1988; Buttriss, 1990; Osborne et al., 1996). Inadequate uptake of these minerals may lead to conditions such as osteoporosis, dental caries, hypertension and anaemia. Various foods have been cited in the patent literature as potential carriers for CPPs. These include confectionary products, breakfast cereals, juices, milk, dairy products and sports drinks (Reynolds, 1993; Han et al., 1996). However, as already stated herein, there is a lack of extensive evidence linking consumption of CPPs and enhanced mineral bioavailability in humans.

On the other hand, the recent patent by Holt (2001) signifies the potential application of CPPs in the development of nanoclusters containing various minerals and trace elements. Solubilisation of minerals and essential trace micronutrients in nanocluster structures by the inclusion of CPPs allows an efficient means of delivering high doses of essential nutrients in a small volume. These complexes have been shown to be stable following sterilisation and freeze-drying. The potential exploitation of CPP-containing nanoclusters has been cited for food, beverage, dental-care and pharmaceutical applications.

8.8.2 Anticaries effects

The development of dental caries is associated with demineralisation of teeth as a result of the production of organic acids by odontopathogenic dental plaque-forming bacteria during the fermentation of dietary sugars. The anticariogenic activity of CPPs is related to their potential ability to localise calcium phosphate at the tooth surface and as a consequence their ability to help prevent demineralisation and aid in remineralisation of tooth enamel. The anticariogenic effects of CPPs have been shown in animal and human experiments (Reynolds, 1997, 1998, 1999; Rose, 2000a). The anticariogenic effect of CPPs may also be related to the ability of CPPs to compete with dental plaque-forming bacteria such as Streptococcus mutans for calcium (Rose, 2000b). Claims have been made in the patent lierature for the incorporation of CPPs into dental hygiene products (Reynolds, 1993; Han et al., 1996; Holt, 2001). The inclusion of CPP-amorphous calcium phosphate nanoclusters into sugar-free chewing gum has been shown to increase enamel remineralisation in a human in situ model (Shen et al., 2001).

8.8.3 Humanisation of bovine milk

The concept of supplementing bovine milk with CPPs in order to 'humanise' bovine milk has been patented. This arises from the fact that bovine milk has an organic phosphorus : total phosphorus ratio of 0.34, whereas human milk has a ratio of 0.89 (Brule et al., 1989).

8.8.4   Immunomodulation

Both peptides and proteins from milk may contribute to an overall immunostimulatory response. Recent studies have focused on immunoenhancing properties of caseinophosphopeptides. Hata *et al.* (1998, 1999) reported on the immunostimulatory action of caseinophosphopeptides β-CN(f1–25)4P, β-CN(f1–28)4P, α_{S1}-CN(f59–79)5P and α_{S2}-CN(f1–32)4P which enhanced immunglobulin G production in mouse spleen cell cultures. Moreover, the level of serum and intestinal antigen-specific IgA was higher in mice fed the caseinophosphopeptide preparation than those fed the control diet (Otani *et al.* 2000). The oral ingestion of immunomodulatory caseinophosphopeptides is claimed to enhance mucosal immunity (Otani *et al.*, 2000). Hartmann *et al.* (2000) found that the treatment of human lymphocytes (peripheral blood lymphocyte, PBL) with different CPP preparations resulted in a significant increase in IgG production *in vitro*, even exceeding the response mediated by known mitogens.

The molecular mechanism by which CPPs exert their immunomodulatory effects is not yet defined. It is known that lymphocytes and macrophages express receptors for many biologically active mediators. The immunostimulatory activity of CPPs was attributable to phosphoseryl residues (Hata *et al.* 1998, 1999).

8.9   Summary and future trends

The functional food ingredient/nutraceuticals potential of CPPs revolves around their ability to bind and solubilise minerals such as calcium. Preparative-scale production procedures for the generation of CPPs are available. Peptide preparations enriched in CPPs are available from multinational food ingredient suppliers. Furthermore, dental-care/hygiene products containing CPPs, targeted to have an anticariogenic effect, are currently available in the marketplace. A number of patents exist on the production and application of CPPs as functional food ingredients/nutraceuticals and as dental care/hygiene agents.

Despite numerous animal studies and a very limited number of human studies, unequivocal proof that consumption of CPPs enhances dietary calcium bioavailability is as yet unavailable. The difficulty in obtaining such information arises from complex interactive effects between dietary minerals such as calcium and other dietary constituents such as phytate which strongly influence bioavailability. In this regard, it is interesting to note that the format in which CPPs are consumed, e.g. either in milk or as isolated CPPs, has been shown to strongly influence the survivability of CPPs during gastrointestinal transit (Meisel *et al.*, 2001, 2002). The effect of consuming minerals in nanocluster format on subsequent dietary mineral bioavailability is worthy of further study.

In vitro cell culture studies with isolated CPPs and peptide fractions enriched in CPPs demonstrate no adverse cytotoxicological effects associated with CPPs (Hartmann and Meisel, 2002; Hartmann *et al.*, 2000). Interestingly, the fact that CPPs have been shown *in vitro* to have an immunomodulatory effect may represent a new biological function of ingested phosphorylated peptides (Hata *et al.*, 1999; Hartmann and Meisel, 2002). The observation that IgE displays a higher binding affinity for phosphorylated caseins than the corresponding dephosphorylated derivatives is relevant to consumers who display allergenic reactions to milk proteins (Bernard *et al.*, 2000) and has significant consequences for the appropriate labelling of consumer foods and other products containing CPPs.

Finally, the continued study of CPP structure and function will undoubtedly lead to a greater understanding of the functional and metabolic role, and potential applications, of CPPs. This will contribute to the development and expansion of CPPs as functional food ingredients in the prevention of diseases and as nutraceuticals in the treatment of particular disease states.

8.10   References

ADAMSON N J and REYNOLDS E C (1995), 'Characterisation of tryptic casein phosphopeptides prepared under industrially relevant conditions', *Biotechnol Bioeng*, **45**, 196–204.

AOKI T, NAKANO T, IWASHITA T and others (1998), 'Preparation and characterisation of micellar calcium phosphate-casein phosphopeptide complex', *J Nutr Sci Vitaminol*, **44**, 447–456.

BENNET T, DESMOND A, HARRINGTON M, MCDONAGH D, FITZGERALD R J, FLYNN and CASHMAN K D (2000), 'The effect of high intakes of casein phosphopeptides on calcium absorption in the rat', *Br J Nutr*, **83**, 673–680.

BERNARD H, MEISEL H, CREMMINON C and WAL J-M (2000), 'Post-translational phosphoryla-tion affects the IgE binding capacity of caseins', *FEBs Lett*, **467**, 239–244.

BERROCAL R, CHANTON S, JUILLERAT M A, PAVILLARD B, SCHERZ J-C and JOST R (1989), 'Tryptic phosphopeptides from whole casein: II. Physicochemical properties related to the solubilisation of calcium', *J Dy Res*, **56**, 335–341.

BOUHALLAB S, LÉONIL J and MAUBOIS J L (1991), 'Complexation du fer par le phosphopeptide (91–25) de la caséine β; action de l'alcalase et de la phosphatase acide', *Lait*, **71**, 435–443.

BRONNER F (1987), 'Intestinal calcium absorption: mechanisms and applications', *J Nutr*, **117**, 1347–1352.

BRULE G, ROGER L, FAUQUANT J and PIOT M (1982), 'Phosphopeptides from casein-based material', US Patent No. 4,358,465.

BRULE G, ROGER L, FAUQUANT J and PIOT M (1989), 'Casein phosphopeptide composition', US Patent No. 4,816,398.

BRULE G, ROGER L, FAUQUANT J and PIOT M (1994), 'Nutrient composition containing non-phosphorylated peptides from casein-based material', US Patent No. 5,334,408.

BUTTRISS J (1990), 'The role of calcium in a balanced diet', *J Soc Dy Technol*, **43**(1), 1–3.

CHABANCE B, MARTEAU P, RAMBAUD J C, MIGLIORE-SAMOUR D, BOYNARD M, PERROTIN P, JOLLÈS P and FIAT A M (1998), 'Casein peptide release and passage to the blood in humans during digestion of milk or yoghurt', *Biochimie*, **80**, 155–165.

DE KONING P J and VAN ROOIJEN P J (1982), 'Aspects of formation of lysoalanine in milk and milk products', *J Dy Res*, **49**, 725–736.

ELLEGÅRD K H, GAMMELGÅRD-LARSEN C, SØRENSEN E S and FEDESOV S (1999), 'Process-scale chromatographic isolation, characterisation and identification of tryptic bioactive casein phosphopeptides', *Int Dy J*, **9**(9), 639–652.

ERBA D, CIAPPELLANO S and TESTOLIN G (2001), 'Effect of caseinophosphopeptides on inhibition of calcium intestinal absorption due to phosphate', *Nutr Res*, **21**, 649–656.

FAIRWEATHER-TAIT S J (1988), 'Zinc in human nutrition', *Nutr Res,* **1**, 23–27.

FERRARETTO A, SIGNORILE A, RAVAGHI C, FIORILLI A and TETTAMANTI G (2001), 'Casein-phosphopeptides influence calcium uptake by cultured human intestinal HT-29 tumor cells', *J Nutr,* **131**, 1655–1661.

FITZGERALD R J (1998), 'Potential uses of caseinophosphopeptides', *Int Dy J,* **8**, 451–457.

GAMMELGÅRD-LARSEN C (1991), 'Casein phosphopeptide, CPP – a promising new way to enhance calcium utilisation and absorption', *Proceedings of Food Ingredients (Asia) Conference,* Singapore, April, 1991, Maarssen Expoconsult, 53–58.

GANJAM L S, THORNTON W H, MARSHALL R T and MACDONALD R S (1997), 'Antiproliferative effects of yoghurt fractions obtained by membrane dialysis on cultured mammalian intestinal cells', *J Dy Sci,* **80**, 2325–2329.

GAUCHERON F, LE GREAT Y, SINBANDHIT S, GUENOT P and BRULE G (1995), 'Binding of calcium by β-casein in the presence of inorganic phosphate', *Sci des Aliments* **15**, 481–489.

GERBER H W and JOST R (1986), 'Casein phosphopeptides: their effect on calcification of *in vitro* cultured embryonic rat bone', *Cal Tissue Int,* **38**, 350–357.

HAN S K, SHIN Y C and PARK J H (1996), 'Casein phosphopeptide, casein containing same and a process for the preparation thereof', World Patent, WO 96/29340.

HAN S H, SHIN Y C and BYUN H D (2000), 'Biochemical, molecular and physiological charaterisation of a new beta-casein variant detected in Korean cattle', *Anim Genet,* **31**(1), 49–51.

HANSEN M, SANDSTRÖM B and LONNERDAL B (1996), 'The effect of casein phosphopeptides on zinc and calcium absorption from high phytate infant diets assessed in rat pups and Caco-2 cells', *Ped Res,* **40**(4), 547–552.

HANSEN M, SANDSTRÖM B, JENSEN M and SØRENSEN S S (1997a), 'Casein phosphopeptides improve zinc and calcium absorption from rice-based but not from whole-grain infant cereals', *J Ped Gastroent Nutr,* **24**(1), 56–62.

HANSEN M, SANDSTRÖM B, JENSEN M and SØRENSEN S S (1997b), 'Effect of casein phosphopeptide on zinc and calcium absorption from bread meals', *J Tr Elem Med Biol,* **11**(3), 143–149.

HARTMANN R and MEISEL H (2002), 'Cytochemical assessment of phosphopeptides derived from casein as potential ingredients for functional food', *Nahrung/Food* (submitted).

HARTMANN R, GUNTHER S, MARTIN D, MEISEL H, PENTZIEN A-K, SCHLIMME E and SCHOLZ N (2000), 'Cytochemical model systems for the detection and characterisation of potentially bioactive milk components', *Kiel Milch Forsch,* **52**, 61–85.

HARZAR G and KAUER H (1982), 'Binding of zinc to casein', *Am J Clin Nutr,* **35**, 981–987.

HASEGAWA K, OKAMOTO N, OZAWA H, KITAJIMA S and TAKADO Y (1981), 'Limits and sites of lysoalanine formation in lysozyme, α-lactalbumin and α_{s1}- and β-casein by alkali treatment', *Agric Biol Chem,* **45**, 1645–1651.

HATA I, HIGASHIYAMA S and OTANI H (1998), 'Identification of a phosphopeptide in bovine α_{s1}-casein digests as a factor influencing proliferation and immunoglobulin production in lymphocyte cultures', *J Dy Res,* **65**, 569–578.

HATA I, UEDA J and OTANI H (1999), 'Immunostimulatory action of a commercially available casein phosphopeptide preparation, CPP-III, in cell cultures', *Milchwissen,* **54**, 3–7.

HEANEY R P, SAITO Y and ORIMO H (1994), 'Effect of caseinophosphopeptides on absorbability of co-ingested calcium in normal postmenopausal women', *J Bone Min Metab,* **12**(1), 77–81.

HOLT C (2001), 'Calcium phosphate nanoclusters and their application', World Patent, WO0144106.

HOLT C, WAHLGREN N M and DRAKENBURG T (1996), 'Ability of a β-casein phosphopeptide to modulate the precipitation of calcium phosphate by forming amorphous dicalcium phosphate nanoclusters', *Biochem J,* **314**, 1035–1039.

HOLT C, TIMMINS P A, ERRINGTON, N *et al.* (1998), 'A core-shell model of calcium phosphate nanoclusters derived from sedimentation equilibrium and small angle X-ray and neutron scattering measurements', *Eur J Biochem*, **252**, 73–78.

IDF (1997), 'Dietary calcium in health', *Bull No. 322*, International Dairy Federation, Brussels.

JUILLERAT M A, BAECHLER R, BERROCAL R, CHANTON S, SCHERZ J-C and JOST R (1989), 'Tryptic phosphopeptides from whole casein I; Preparation and analysis by FPLC', *J Dy Res*, **56**, 603–611.

KITTS D D and YUAN Y V (1992), 'Caseinophosphopeptides and calcium bioavailability', *Trends Fd Sci Technol*, **3**, 31–35.

KITTS D D, YUAN Y V, NAGASAWA T and MORIYAMA Y (1992), 'Effect of casein, casein phosphopeptides and calcium intake on ileal ^{45}Ca disappearence and temporal systolic blood pressure in spontaneously hypertensive rats', *Br J Nutr*, **68**, 765–781.

KOIDE K, ITOYAMA K, FUKUSHIMA T, MIYAZAWA F and KUWATA T (1991), 'Method for separation and concentration of phosphopeptides', Eur Patent Applic, EP 0443718A2.

KUNST A (1990), 'Process to isolate phosphopeptides', Eur Patent Applic, EP 0476199A1.

LEE Y S, NOGUCHI T and NAITO H (1980), 'Phosphopeptides and soluble calcium in the small intestine of rats given a casein diet', *Br J Nutr*, **43**, 457–467.

LEE Y S, NOGUCHI T and NAITO H (1983), 'Intestinal absorption of calcium in rats given diets containing casein or amino acid mixture; the role of casein phospho-peptides', *Br J Nutr*, **49**, 67–76.

LI Y, TOME D and DESJEUX J F (1989), 'Indirect effect of casein phosphopeptides on calcium absorption in rat ileum *in vitro*', *Reprod Nutr Devel*, **29**, 227–233.

LIHME A O F, AAGESEN M I, GAMALGARD-LARSEN C and ELLEGARD K H (1994), 'Method for isolating biomolecules by ion exchange', World Patent, WO 94/06822.

LORIENT D (1979), 'Covalent bonds formed in proteins during milk sterilisation: Studies on caseins and casein peptides', *J Dy Res*, **46**, 393–396.

MACDONALD R S, THORNTON W H and MARSHALL R T (1994), 'A cell culture model to identify biologically active peptides generated by bacteial hydrolysis of casein', *J Dy Sci*, **77**, 1167–1175.

MANSON W and ANNAN W D (1971), 'The structure of a phosphopeptide derived from β-casein', *Arch Biochem Biophys*, **145**, 16–26.

MANSON W and CAROLAN T (1980), 'Formation of lysoalanine from individual bovine caseins' *J Dy Res*, **47**, 193–198.

MCDONAGH D and FITZGERALD R J (1998), 'Production of caseinophosphopeptides (CPPs) from sodium caseinate using a range of commercial proteases', *Int Dy J*, **8**, 39–45.

MEISEL, H and FRISTER H (1988), 'Chemical characterisation of a caseino-phosphopeptide isolated in the *in vivo* digests of a casein diet', *Biol Chem Hoppe-Seyler*, **369**, 1275–1279.

MEISEL H and GÜNTHER S (1998), 'Food proteins as precursors of peptides modulating human cell activity', *Nahrung/Food*, **42**, 175–176.

MEISEL H and OLIEMAN C (1998), 'Estimation of calcium binding constants of casein phosphopeptides by capillary zone electrophoresis', *Acta Chim Acta*, **372**, 291–297.

MEISEL H, FRISTER H and SCHLIMME E (1989), 'Biologically active peptides in milk proteins, *Z. Ernährungswiss*, **28**, 267–278.

MEISEL H, ANDERSSON H B, BUHL K, ERBERSDOBLER H F and SCHLIMME E (1991a), 'Heat-induced changes in casein-derived phosphopeptides', *Zeit Ernahrung*, **30**, 227–232.

MEISEL H, BEHRENS S and SCHLIMME E (1991b), 'Calcium-Bindungsstudie an Phosphopeptidfraktionen aus der in vitro-Proteolyse von Casein', *Kiel Milchwirtschaftl Forschungsber*, **43**, 199–211.

MEISEL H, GÜNTHER S, MARTIN D and SCHLIMME E (1998), 'Apoptosis induced by modified ribonucleosides in human cell culture systems', *FEBS Lett*, **433**, 265–268.

MEISEL H, BERNARD H, FAIRWEATHER-TAIT S, FITZGERALD R J, HARTMANN R, LANE C N, MCDONAGH D, TEUCHER B and WAL J-M (2001), 'Nutraceutical and functional food ingredients for food and pharmaceutical applications', *Br J Nutr*, **85**, 1.

MEISEL H, BERNARD H, FAIRWEATHER-TAIT S, FITZGERALD R J, HARTMANN R, LANE C N,

MCDONAGH D, TEUCHER B and WAL J-M (2002), 'Detection of caseinophosphopeptides (CPPs) in the distal ileum of human', *Br J Nutr*, (in press).

MELLANDER O (1950), 'The physiological importance of the casein phosphopeptide calcium salts. 11. Peroral calcium dosage of infants', *Acta Soc Med Uppsala*, **55**, 247–255.

MERCIER J-C (1981), 'Phosphorylation of caseins, present evidence for an amino acid triplet code postranslationally recognised by specific kinases' *Biochimie*, **63**, 1–17.

MORI T, HIRAYAMA M, SUSUKI K, SHIMIZU H and NAGAI T (1996), 'Effect of casein phospho peptides and Ca^{+2} on penetration of boar spermatozoa into pig oocytes matured *in vitro*', *Biol Reprod*, **55**(2), 364–369.

MYKKÄNEN H M and WASSERMAN R H (1980), 'Enhanced absorption of calcium by casein phosphopeptides in rachitis and normal chicks, *J Nutr*, **119**, 2141–2148.

ONO T, OHOTAWA T and TAKAGI Y (1994), 'Complexes of casein phosphopeptide and calcium phosphate prepared from casein micelles by tryptic digestion', *Biosci, Biotech Biochem*, **58**, 1376–1380.

OSBORNE C G, MCTYRE R B, DUDEK J, ROCHE K E, SCHEUPLEIN R, SILVERSTEIN B, WEINBERG M S and SALKELD A A (1996), 'Evidence for the relationship of calcium to blood pressure', *Nutr Rev*, **54**, 365–381.

OTANI H, HORI H and HOSONO A (1987), 'Antigenic reactivity of dephosphorylated α_{s1}-casein, phosphopeptide from β-casein and *O*-phospho-L-serine toward the antibody to native α_{s1}-casein', *Agric Biol Chem*, **51**, 2049–2054.

OTANI H, KIHARA Y and PARK M (2000), 'The immunoenhancing property of a dietary casein phosphopeptide preparation in mice', *Food Agric Immunol*, **12**, 165–173.

PASSMORE R, NICOL B M and RAO M N (1974), 'Handbook on human nutritional requirement', *FAO Nutr Study No. 28*, WHO Monograph Series No. 61.

PERICH J W, KELLY D P and REYNOLDS E C (1992), 'Efficient solution-phase synthesis of phosphopeptides related to casein and stratherin', *Int J Pep Prot Res*, **40**, 81–88.

REEVES R E and LATOUR N G (1958), 'Calcium phosphate sequestering phosphopeptide from casein', *Science*, **128**, 472.

REYNOLDS E C (1992), 'Production of phosphopeptides from casein', World Patent Applic, WO 92/18526.

REYNOLDS E C (1993), 'Phosphopeptides for the treatment of dental calculus', World Patent, WO 93/03707.

REYNOLDS E C (1994), 'Anticariogenic phosphopeptides', In: *Proc 24th Int Dy Congr*, Melbourne, Australia, International Dairy Federation, Brussels, Record No. 7698–7796/10379.

REYNOLDS E C (1997), 'Remineralisation of enamel subsurface lesions by casein phosphopeptide-stabilised calcium phosphate solutions' *J Dent Res*, **76**(9), 1587–1595.

REYNOLDS E C (1998), 'Anticariogenic complexes of amorphous calcium phosphates stabilised by caseinophosphopeptides: a review', *SCD Sp Care Dent*, **18**, 6–16.

REYNOLDS E C (1999), 'Anticariogenic casein phosphopeptides. *Prot Pep Letts*, **6**, 295–303.

ROSE R K (2000a), 'Effect of anticariogenic casein phosphopeptide on calcium diffusion in streptococcal model dental plaques', *Arch Oral Biol*, **45**(7), 569–575.

ROSE R K (2000b), 'Binding characteristics of *Streptococcus mutans* for calcium and casein phosphopeptide', *Caries Res*, **34**(5), 427–431.

ROSSANDER L, SANDBERG A S and SANDSRÖM B (1992), 'The influence of dietary fibre on mineral adsorption and utilisation'. In: T F Schweizer and C A Edwards, Eds, *Dietary Fibre – A Component of Food*, London, Springer-Verlag, 197–216.

SATO R, NOGUCHI T and NAITO H (1983), 'The necessity for the phosphate portion of casein molecules to enhance Ca absorption from the small intestine', *Agric Biol Chem*, **47**, 2415–2417.

SATO R, SHINDO M, GUNSHIN H, NOGUCHI T and NAITO H (1991), 'Characterisation of phosphopeptide derived from bovine β-casein: an inhibitor to intra-intestinal precipitation of calcium phosphate', *Biochim Biophys Acta*, **1077**, 413–415.

SCHAAFSMA G (1997), 'Bioavailability of calcium'. In: *Dietary Calcium and Health*, Bulletin of the International Dairy Federation No. 322, 20–24.

SCHLIMME E and MEISEL H (1995), 'Bioactive peptides derived from milk proteins. Structural, physiological and analytical aspects', *Die Nahrung, 39*, 1–20.

SCHOLZ-AHRENS K and SCHREZENMEIR J (2000), 'Effects of bioactive substances in milk on mineral and trace element metabolism with special reference to casein phosphopeptides', *Br J Nutr,* **84**(S1), 147–153.

SHEN P, CAI F, NOWICKI A, VINCENT J and REYNOLDS E C (2001), 'Remineralisation of enamel subsurface lesions by sugar-free chewing gum containing casein phosphopeptide-amorphous calcium phosphate', *J Dent Res, 80*(12), 2066–2070.

TEUCHER B, MAJSAK-NEWMAN G, DAINTY J R, MCDONAGH D and FAIRWEATHER-TAIT S J (2002), 'The effect of caseinophosphopeptides on calcium absorption in adults' (*submitted for publication*).

WEST D W (1986), 'Structure and function of the phosphorylated residues of casein', *J Dy Res,* **53**, 333–352.

YEUNG A C, GLAHN R P and MILLER D D (2001), 'Dephosphorylation of sodium caseinate, ennzymatically hydrolysed casein and casein phosphopeptides by intestinal alkaline phosphatase: implications for iron availability', *J Nutr Biochem,* **12**, 292–299.

9

Oligosaccharides

G. Boehm and B. Stahl, Numico Research Germany, Germany

9.1 Introduction

Since 1980, according to IUB-IUPAC terminology (IUB-IUPAC; Joint Commission on Biochemical Nomenclature (JCBN), 1980), oligosaccharides have been defined as carbohydrates with a degree of polymerisation up to 10. However, as there is no physiological reason for this definition, oligosaccharides have also been defined as ranging from a degree of polymerisation of 2 up to 20 and more (British Nutrition Foundation, 1990; Kunz *et al.*, 2000). Recently the IUB-IUPAC Joint Commission on Biochemical Nomenclature stated that the borderline between oligo- and polysaccharides cannot be drawn strictly. However, the term oligosaccharide is commonly used to refer to defined structures as opposed to a polymer of unspecified length (IUB-IUPAC, 1997).

Free oligosaccharides are natural constituents of all placental mammals' milk and can also be found in bacteria, fungi, plants, etc. They derive from hydrolysis of dietary polymers during digestion. Technologically, they can be extracted from natural sources, and can be synthesised from monomers and/or small oligosaccharides, or derived from hydrolysis of natural polymers. Human milk contains considerable amounts of oligosaccharides (Thurl *et al.*, 1996; Kunz *et al.*, 2000), indicating that, at least during postnatal development, they play an important physiological role. It is widely accepted that human milk oligosaccharides contribute to the establishment of a particular intestinal flora (dominated by bifidobacteria), the postnatal stimulation of the immune system, to the defence effect of human milk against viral and bacterial infections, and the enhancement of the bioavailability of minerals. Although human milk is the specific diet during the postnatal period, there is evidence that effects of oligosaccharides observed during the postnatal period can also be found in later life. Human milk oligosaccharides can be seen as a natural example of a complex oligosaccharide mixture with several beneficial effects for humans.

9.2   Structural aspects of free oligosaccharides

9.2.1   Human milk oligosaccharides (HMOS)

Human milk contains 0.7–1.2 g oligosaccharides/l, making the oligosaccharide fraction a major component of human milk (Thurl *et al.*, 1996; Kunz *et al.*, 2000). The monomers of human milk oligosaccharides are D-glucose (Glc), D-galactose (Gal), *N*-acetyglucosamine (GlcNAc), L-fucose (Fuc) and sialic acid (*N*-acetyl neuraminic acid, Neu5Ac). The basic structure of the human milk oligosaccharide is given in Fig. 9.1. With few exceptions (Table 9.1, Nos. 79–87), all core molecules carry lactose at their reducing end (Table 9.1 No. 1–78). Further variations occur due to activity of several fucosyltransferases and sialyltransferases, resulting in an attachment of Fuc and/or NeuAc at a different position in the core molecule in different linkages (Fig. 9.1).

As a result, human milk contains a great variety of oligosaccharide structures. Table 9.1 summarises the known complex structures of human milk oligo-saccharides. In particular, the activity of the fucosyltransferases shows genetically determined differences, resulting in differences in the pattern of oligosaccharides among individuals. In Caucasians, who are classified as secretors, the $\alpha1$–2 fucosyltransferase is dominant, resulting in more complex oligosaccharides, all possessing Fuc$\alpha1$–2Galβ 1–3GlcNAc residues. Another fucosyltransferase is Lewis gene dependent and attaches Fuc residues in $\alpha1$–3/4 linkages to GlcNAc residues in the core molecule. A further fucosyltransferase attaches fucose residues exclusively in $\alpha1$–3 linkages. Due to the activity of these fucosytransferases coded by secretor and/or Lewis genes, four different groups of oligosaccharide patterns were detected in individual human milk samples (Thurl *et al.*, 1997). During lactation, the total concentration of oligosaccharides decreases. Compared to early

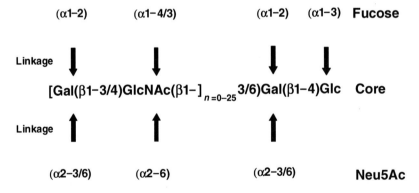

Fig. 9.1 Basic structure of human milk oligosaccharides. The arrows mark the possible location of the glycosidic linkages of the respective fucose and sialic acid residues. Please note that not all possible linkages might occur in a distinct oligosaccharide.

Table 9.1 Structures of neutral and acidic human milk oligosaccharides

No.	Abbreviation	Oligosaccharide	Lx/y-z	Reference
Neutral human milk oligosaccharides				
1	2'-FL	Fuc (α1-2) Gal (β1-4) Glc	L0/1-0	Kuhn et al. (1956c)
2	3-FL	Gal (β1-4) Glc ⎜ Fuc (α1-3)	L0/1-0	Montreuil (1956)
3	DF-L	Fuc (α1-2) Gal (β1-4) Glc ⎜ Fuc (α1-3)	L0/2-0	Kuhn and Gauhe (1958a)
4	LNT	Gal (β1-3) GlcNAc (β1-3) Gal (β1-4) Glc	L1/0-0	Kuhn and Baer (1956a)
5	LNneoT	Gal (β1-4) GlcNAc (β1-3) Gal (β1-4) Glc	L1/0-0	Kuhn and Gauhe (1962)
6	LNFP I	Fuc (α1-2) Gal (β1-3) GlcNAc (β1-3) Gal (β1-4) Glc	L1/1-0	Kuhn et al. (1956c)
7	LNFP II	Gal (β1-3) GlcNAc (β1-3) Gal (β1-4) Glc ⎜ Fuc (α1-4)	L1/1-0	Kuhn et al. (1958b)
8	LNFP III	Gal (β1-4) GlcNAc (β1-3) Gal (β1-4) Glc ⎜ Fuc (α1-3)	L1/1-0	Kobata and Ginsburg (1969b)
9	LNFP V	Gal (β1-3) GlcNAc (β1-3) Gal (β1-4) Glc ⎜ Fuc (α1-3)	L1/1-0	Ginsburg et al. (1976)
10	LNDH I	Fuc (α1-2) Gal (β1-3) GlcNAc (β1-3) Gal (β1-4) Glc ⎜ Fuc (α1-4)	L1/2-0	Kuhn and Gauhe (1958a)
11	LNDH II	Gal (β1-3) GlcNAc (β1-3) Gal (β1-4) Glc ⎜ ⎜ Fuc (α1-4) Fuc (α1-3)	L2/0-0	Kuhn and Gauhe (1960)

Table 9.1 cont'd

No.	Abbreviation	Oligosaccharide	Lx/y-z	Reference
12	LNH	Gal (β1-4) GlcNAc (β1-6)\ Gal (β1-4) Glc Gal (β1-3) GlcNAc (β1-3)/	L2/0-0	Kobata and Ginsburg (1972a)
13	LNneoH	Gal (β1-3) GlcNAc (β1-3)/ Gal (β1-4) Glc Gal (β1-4) GlcNAc (β1-6)\	L2/0-0	Kobata and Ginsburg (1972b)
14	paraLNH	Gal (β1-4) GlcNAc (β1-3)/ Gal (β1-3) GlcNAc (β1-3) Gal (β1-4) GlcNAc (β1-3) Gal (β1-4) Glc	L2/1-0	Yamashita et al. (1977)
15	paraLNneoH	Gal (β1-4) GlcNAc (β1-3) Gal (β1-4) GlcNAc (β1-3) Gal (β1-4) Glc	L2/1-0	Yamashita et al. (1977)

16	F-LNH I	Gal (β1-4) GlcNAc (β1-6)\
		                                          Gal (β1-4) Glc
		Fuc (α1-2) Gal (β1-3) GlcNAc (β1-3)/
		                       |
		                     Fuc (α1-3)
			L1/2-0	Yamashita et al. (1977)

17	F-LNH II	Gal (β1-4) GlcNAc (β1-6)\ Gal (β1-4) Glc Gal (β1-3) GlcNAc (β1-3)/ Gal (β1-4) GlcNAc (β1-6)\ Fuc (α1-	L2/0-0	Dua et al. (1985)
18	F-LNneoH	Gal (β1-4) GlcNAc (β1-3)/ ⌐Gal (β1-4) GlcNAc (β1-4) GlcNAc (β1-3) Gal (β1-4) Glc	L2/1-0	Kobata and Ginsburg (1972b)
19	F-paraLNH	Fuc (α1-3) Gal (β1-3) GlcNAc (β1-3) Gal (β1-4) GlcNAc (β1-3) Gal (β1-4) Glc	L2/1-0	Sabharwal et al. (1988a)

20	F-paraLNH	Gal (β1-3) GlcNAc (β1-3) Gal (β1-4) GlcNAc (β1-3) Gal (β1-4) Glc
		       |
		   Fuc (α1-4)
			L2/1-0	Bruntz et al. (1988)

21 DF-LNH

```
            Fuc (α1-3)
               |
Gal (β1-4) GlcNAc (β1-6)\
                         Gal (β1-4) Glc
Gal (β1-3) GlcNAc (β1-3)/
   |
Fuc (α1-4)
Fuc (α1-3)
```

L2/2-0   Dua et al. (1985)

21 DF-LNHa

```
Gal (β1-4) GlcNAc (β1-6)\
                              Gal (β1-4) Glc
Fuc (α1-2) Gal (β1-3) GlcNAc (β1-3)/
               |
            Fuc (α1-3)
```

L2/2-0   Yamashita et al. (1977)

23 DF-LNneoH

```
Gal (β1-4) GlcNAc (β1-6)\
                         Gal (β1-4) Glc
Gal (β1-4) GlcNAc (β1-3)/
   |
Fuc (α1-3)
Fuc (α1-3)
```

L2/2-0   Haeuw-Fievre et al. (1993)

24 DF-LNneoH

```
Gal (β1-4) GlcNAc (β1-6)\
                              Gal (β1-4) Glc
Gal (β1-4) GlcNAc (β1-3)/
               |
            Fuc (α1-3)
```

L2/2-0   Haeuw-Fievre et al. (1993)

25 DF-para-LNH

```
Gal (β1-3) GlcNAc (β1-3) Gal (β1-4) GlcNAc (β1-3) Gal (β1-4) Glc
                            |
                         Fuc (α1-4)
```

L2/2-0   Yamashita et al. (1977)

26 DF-para-LNneoH

```
Gal (β1-4) GlcNAc (β1-3) Gal (β1-4) GlcNAc (β1-3) Gal (β1-4) Glc
   |                        |
Fuc (α1-4)               Fuc (α1-3)
```

L2/2-0   Yamashita et al. (1977)

Table 9.1 cont'd

No.	Abbreviation	Oligosaccharide	Lx/y-z	Reference

27	TF-LNH	Fuc (α1-3)
		    |
		Gal (β1-4) GlcNAc (β1-6)\\
		                                 Gal (β1-4) Glc
		Fuc (α1-2) Gal (β1-3) GlcNAc (β1-3)/
			L2/3-0	Sabharwal *et al.* (1988)

28	TF-paraLNH	Fuc (α1-4) | Fuc (α1-2)Gal (β1-3) GlcNAc (β1-3) Gal (β1-4) GlcNAc (β1-3) Gal (β1-4) Glc	L2/3-0	Strecker *et al.* (1988)
29	TF-paraLNH	Fuc (α1-3) Fuc (α1-3) | | Gal (β1-3) GlcNAc (β1-3) Gal (β1-4) GlcNAc (β1-3) Gal (β1-4) Glc Fuc (α1-4)	L2/	Bruntz *et al.* (1988)
30	TF-paraLNneoH	Fuc (α1-3) Fuc (α1-3) | | Gal (β1-4) GlcNAc (β1-3) Gal (β1-4) GlcNAc (β1-3) Gal (β1-4) Glc	L2/	Bruntz *et al.* (1988)
31	LNO	Gal (β1-4) GlcNAc (β1-6)\\ Gal (β1-4) Glc Gal (β1-3) GlcNAc (β1-3)/	L3/0-0	Yamashita *et al.* (1976b)
32	LNneoO	Gal (β1-4) GlcNAc (β1-3)/ Gal (β1-4) Glc Gal (β1-4) GlcNAc (β1-6)\\	L3/0-0	Yamashita *et al.* (1976b)
33	Iso-LNO	Gal (β1-4) GlcNAc (β1-3)/ Gal (β1-4) Glc Gal (β1-3) GlcNAc (β1-6)\\	L3/0-0	Yamashita *et al.* (1976b)
34	Para-LNO	Gal (β1-3) GlcNAc (β1-3)/ Gal (β1-3) GlcNAc (β1-4) Gal (β1-4) Glc -GlcNAc (β1-3) Gal (β1-4) Glc	L3/0-0	Haeuw-Fievre *et al.* (1993)

35	F-LNO	<pre> Fuc (α1-3)
Gal (β1-4) GlcNAc (β1-3) Gal (β1-4) GlcNAc (β1-6)\
Gal (β1-4) Glc
Gal (β1-3) GlcNAc (β1-3)/</pre>	L3/1-0	Yamashita *et al.* (1976b)
36	F-LNneoO	<pre>Gal (β1-3) GlcNAc (β1-4) GlcNAc (β1-6)\
Gal (β1-4) Glc
Gal (β1-4) GlcNAc (β1-3)/
Fuc (α1-3)</pre>	L3/1-0	Yamashita *et al.* (1976b)
37	DF-LNO I	<pre>Gal (β1-4) GlcNAc (β1-3) Gal (β1-4) GlcNAc (β1-6)\
Gal (β1-4) Glc
Gal (β1-3) GlcNAc (β1-3)/
Fuc (α1-3)</pre>	L3/2-0	Tachibana *et al.* (1978)
38	DF-LNO II	<pre>Gal (β1-4) GlcNAc (β1-3) Gal (β1-4) GlcNAc (β1-6)\
Gal (β1-4) Glc
Gal (β1-3) GlcNAc (β1-3)/
Fuc (α1-4)
Fuc (α1-3)
Fuc (α1-4)</pre>	L3/2-0	Tachibana *et al.* (1978)
39	DF-LNneoO I	<pre>Gal (β1-3) GlcNAc (β1-3) Gal (β1-4) GlcNAc (β1-6)\
Gal (β1-4) Glc
Gal (β1-4) GlcNAc (β1-3)/
Fuc (α1-3)</pre>	L3/2-0	Tachibana *et al.* (1978)
40	DF-LNneoO II	<pre>Gal (β1-3) GlcNAc (β1-3) Gal (β1-4) GlcNAc (β1-6)\
Gal (β1-4) Glc
Gal (β1-4) GlcNAc (β1-3)/
                        Fuc (α1-3)</pre> | L3/2-0 | Tachibana *et al.* (1978) |

Table 9.1 cont'd

No.	Abbreviation	Oligosaccharide	Lx/y-z	Reference
41	TF-LNO	(see structure below)	L3/3-0	Tachibana *et al.* (1978)
42	TF-LNneoO	(see structure below)	L3/3-0	Tachibana *et al.* (1978)
43	TF-isoLNO	(see structure below)	L3/3-0	Strecker *et al.* (1992)
44	TetraF-isoLNO	(see structure below)	L3/4-0	Haeuw-Fievre *et al.* (1993)
45	TetraF-paraLNO	(see structure below)	L3/4-0	Haeuw-Fievre *et al.* (1993)

41 TF-LNO
```
                              Fuc (α1-3)
                                  |
Gal (β1-4) GlcNAc (β1-3) Gal (β1-4) GlcNAc (β1-6)\
                                                  Gal (β1-4) Glc
                 Gal (β1-3) GlcNAc (β1-3)/
                               |
                          Fuc (α1-4)
```

42 TF-LNneoO
```
Gal (β1-3) GlcNAc (β1-3) Gal (β1-4) GlcNAc (β1-6)\
                                                  Gal (β1-4) Glc
                           Gal (β1-4) GlcNAc (β1-3)/
                                |          |
                           Fuc (α1-3)  Fuc (α1-3)
```

43 TF-isoLNO
```
Fuc (α1-2) Gal (β1-3) GlcNAc (β1-3) Gal (β1-4) GlcNAc (β1-6)\
                                                             Gal (β1-4) Glc
           Fuc (α1-2) Gal (β1-3) GlcNAc (β1-3)/
                                     |
                                Fuc (α1-3)
```

44 TetraF-isoLNO
```
Fuc (α1-2) Gal (β1-3) GlcNAc (β1-3) Gal (β1-4) GlcNAc (β1-6)\
                                                             Gal (β1-4) Glc
           Fuc (α1-2) Gal (β1-3) GlcNAc (β1-3)/
                                     |
                                Fuc (α1-3)
```

45 TetraF-paraLNO
```
                          Fuc (α1-4)
                              |
Fuc (α1-2) Gal (β1-3) GlcNAc (β1-3) Gal (β1-4) GlcNAc (β1-3) Gal (β1-4) GlcNAc (β1-3) Gal (β1-4) Glc
     |                                               |
 Fuc (α1-4)                                     Fuc (α1-3)
     |
 Fuc (α1-3)
```

No.	Abbreviation	Structure	Code	Reference
46	PentaF-isoLNO	Fuc (α1-4) Fuc (α1-3) | Fuc (α1-2) Gal (β1-3) GlcNAc (β1-4) Gal (β1-4) GlcNAc (β1-6)\ Fuc (α1-2) Gal (β1-3) GlcNAc (β1-3)/ Gal (β1-4) Glc | Fuc (α1-4) Gal (β1-4) GlcNAc (β1-6)\	L3/5-0	Haeuw-Fievre et al. (1993)
47	LND	Gal (β1-4) GlcNAc (β1-6)\ Gal (β1-4) Glc Gal (β1-3) GlcNAc (β1-3)/	L4/0-0	Bruntz et al. (1988)

Acidic human milk oligosaccharides

No.	Abbreviation	Structure	Code	Reference
48	3´-SL	Neu5Ac (α2-3) Gal (β1-4) Glc	L0/0-1	Kuhn and Brossmer (1959)
49	6´-SL	Neu5Ac (α2-6) Gal (β1-4) Glc	L0/0-1	Kuhn (1959)
50	F-SL	Neu5Ac (α2-3) Gal (β1-4) Glc | Fuc (α1-3)	L0/1-1	Grönberg et al. (1989)
51	LSTa	Neu5Ac (α2-3) Gal (β1-3) GlcNAc (β1-3) Gal (β1-4) Glc	L1/0-1	Kuhn and Gauhe (1962)
52	LSTb	Gal (β1-3) GlcNAc (β1-3) Gal (β1-4) Glc | Neu5Ac (α2-6)	L1/0-1	Kuhn and Gauhe (1962)
53	LSTc	Neu5Ac (α2-6) Gal (β1-4) GlcNAc (β1-3) Gal (β1-4) Glc	L1/0-1	Kuhn and Gauhe (1962)
54	F-LSTa	Neu5Ac (α2-3) Gal (β1-3) GlcNAc (β1-3) Gal (β1-4) Glc | Fuc (α1-4)	L1/1-1	Wieruszeski et al. (1985)
55	F-LSTb	Fuc (α1-2) Gal (β1-3) GlcNAc (β1-3) Gal (β1-4) Glc | Neu5Ac (α2-6)	L1/1-1	Wieruszeski et al. (1985)
56	F-LSTc	Neu5Ac (α2-6) Gal (β1-4) GlcNAc (β1-3) Gal (β1-4) Glc | Fuc (α1-3)	L1/1-1	Smith et al. (1987)

Table 9.1 cont'd

No.	Abbreviation	Oligosaccharide	Lx/y-z	Reference
57	S-LNH	Neu5Ac (α2-6) Gal (β1-4) GlcNAc (β1-6)\ Gal (β1-3) GlcNAc (β1-3)/ Gal (β1-4) Glc	L2/0-1	Kobata and Ginsburg (1972a)
58	S-LNneoH I	Neu5Ac (α2-6) Gal (β1-4) GlcNAc (β1-6)\ Gal (β1-4) GlcNAc (β1-3)/ Gal (β1-4) Glc	L2/0-1	Kobata and Ginsburg (1972a)
59	S-LNneoH II	Neu5Ac (α2-6) Gal (β1-4) GlcNAc (β1-3)/ Neu5Ac (α2-6) Gal (β1-4) GlcNAc (β1-6)\ Gal (β1-4) Glc	L2/0-1	Grönberg *et al.* (1989)
60	FS-LNH	Fuc (α1-2) Gal (β1-3) GlcNAc (β1-3)/ Fuc (α1-3) ┘ Gal (β1-4) GlcNAc (β1-6)\ Gal (β1-4) Glc	L2/1-1	Yamashita *et al.* (1977)
61	FS-LNH I	Gal (β1-3) GlcNAc (β1-3)/ Neu5Ac (α2-6) Fuc (α1-3) ┘ Gal (β1-4) GlcNAc (β1-6)\ Gal (β1-4) Glc	L2/1-1	Grönberg *et al.* (1992)
62	FS-LNH II	Neu5Ac (α2-3) Gal (β1-3) GlcNAc (β1-3)/ Neu5Ac (α2-6) Gal (β1-4) GlcNAc (β1-6)\ Gal (β1-4) Glc	L2/1-1	Grönberg *et al.* (1992)
63	FS-LNH III	Gal (β1-3) GlcNAc (β1-3)/ Fuc (α1-4) ┘ Gal (β1-4) Glc	L2/1-1	Grönberg *et al.* (1992)

No.	Name	Structure	Code	Reference

64	FS-LNneoH I	Fuc (α1-3)
		    │
		Gal (β1-4) GlcNAc (β1-6)\\
		                      Gal (β1-4) Glc
		Neu5Ac (α2-6) Gal (β1-4) GlcNAc (β1-3)/
			L2/1-1	Grönberg et al. (1989)

65	FS-LNneoH II	Neu5Ac (α2-6) Gal (β1-4) GlcNAc (β1-6)\\ Fuc (α1- Gal (β1-4) Glc Neu5Ac (α2-6) Gal (β1-4) GlcNAc (β1-3)/	L2/1-1	Kobata and Ginsburg (1972)

66	DFS-LNH	Gal (β1-4) Glc
		Fuc (α1-2) Gal (β1-3) GlcNAc (β1-3)/
		             │
		        Fuc (α1-4)
		        Fuc (α1-3)
			L2/2-1	Grönberg et al. (1992)

67	DFS-LNneoH	Fuc (α1-2) Gal (β1-4) GlcNAc (β1-6)\\ Gal (β1-4) Glc Gal (β1-4) GlcNAc (β1-4) Neu5Ac (α2-6) Gal (β1-4) GlcNAc (β1-3)/	L2/2-1	Grönberg et al. (1992)

68	FS-LNO	Gal (β1-4) GlcNAc (β1-6)\\
		                           Gal (β1-4) Glc
		Neu5Ac (α2-3) Gal (β1-3) GlcNAc (β1-3)/
		               │
		          Fuc (α1-4)
		          Fuc (α1-3)
			L3/1-1	Kitagawa et al. (1993)

69	TFS-isoLNO	Fuc (α1-2) Gal (β1-3) GlcNAc (β1-4) GlcNAc (β1-6)\\
		                                         Gal (β1-4) Glc
		Neu5Ac (α2-3) Gal (β1-3) GlcNAc (β1-3)/
		               │
		          Fuc (α1-4)
			L3/3-1	Kitagawa et al. (1993)

Table 9.1 cont'd

No.	Abbreviation	Oligosaccharide	Lx/y-z	Reference
70	DS-LNT	Neu5Ac (α2-6) | Neu5Ac (α2-3) Gal (β1-3) GlcNAc (β1-3) Gal (β1-4) Glc | Neu5Ac (α2-6)	L1/0-2	Grimmonprez and Montreuil (1968)
71	FDS-LNT	Neu5Ac (α2-3) Gal (β1-3) GlcNAc (β1-3) Gal (β1-4) Glc | Fuc (α1-4)	L1/1-2	Kitagawa *et al.* (1991)
72	DS-LNH I	Neu5Ac (α2-6) Gal (β1-4) GlcNAc (β1-6)\\ Gal (β1-4) Glc Neu5Ac (α2-3) Gal (β1-3) GlcNAc (β1-3)/ Gal (β1-4) GlcNAc (β1-6)\\	L2/0-2	Kitagawa *et al.* (1991)

73	DS-LNH II	Neu5Ac (α2-3) Gal (β1-3) GlcNAc (β1-3)/
		                                  Gal (β1-4) Glc
		                |
		Neu5Ac (α2-6)
			L2/0-2	Kitagawa *et al.* (1991)

74	DS-LNneoH	Neu5Ac (α2-6) Gal (β1-4) GlcNAc (β1-6)\\ Gal (β1-4) Glc	L2/0-2	Grönberg *et al.* (1992)
75	FDS-LNH I	Neu5Ac (α2-6) Gal (β1-4) GlcNAc (β1-3)\\ Fuc (α1-3) Gal (β1-4) GlcNAc (β1-6)\\ Gal (β1-4) Glc Neu5Ac (α2-3) Gal (β1-3) GlcNAc (β1-3)/ Neu5Ac (α2-6) Fuc (α1-3)	L2/1-2	Yamashita *et al.* (1976a)
76	FDS-LNH II	Gal (β1-4) GlcNAc (β1-6)\\ Gal (β1-4) Glc Neu5Ac (α2-3) Gal (β1-3) GlcNAc (β1-3)/ | Neu5Ac (α2-6)	L2/1-1	Kitagawa *et al.* (1991)

No.	Name	Structure	Code	Reference

77	FDS-LNneoH	Neu5Ac (α2-3/6) Gal (β1-4) GlcNAc (β1-6)\
		                       Gal (β1-4) Glc
		 Neu5Ac (α2-3/6) Gal (β1-4) GlcNAc (β1-3)/
		               |
		             Fuc (α1-3)
			L2/1-2	Yamashita *et al.* (1976a)

78	TS-LNH	Neu5Ac (α2-6) Gal (β1-4) GlcNAc (β1-6)\
		                      Gal (β1-4) Glc
		 Neu5Ac (α2-3) Gal (β1-3) GlcNAc (β1-3)/
		            |
		          Neu5Ac (α2-6)
			L2/0-3	Fievre *et al.* (1991)

79	3'-GL	Gal (β1-3) Gal (β1-4) Glc	–	Donald and Feeney (1988)
80	4'-GL	Gal (β1-4) Gal (β1-4) Glc	–	Snow Brand Co. (1995)
81	6'-GL	Gal (β1-6) Gal (β1-4) Glc	–	Yamashita and Kobata (1974)

82	–	Neu5Ac (α2-3) Gal (β1-3) GlcNAc
		             |
		         Fuc (α1-4)
			–	Kitagawa *et al.* (1990)

83	–	Neu5Ac (α2-3) Gal (β1-3) GlcNAc (β1-3) Gal
		             |
		         Fuc (α1-3)
			–	Kitagawa *et al.* (1990)

84	–	Gal (β1-4) GlcNAc (β1-6)\ Gal (β1-4) Glc Neu5Ac (α2-3) Gal (β1-3)/	–	Grönberg *et al.* (1992)

Table 9.1 cont'd

No.	Abbreviation	Oligosaccharide	Lv/y-z	Reference

85	–	Fuc (α1-2)Gal (β1-3) GlcNAc (β1-3) Gal (β1-4) GlcNAc (β1-3) Gal (β1-4) Glc
		                6S 6S
		                        Fuc (α1-3)
			–	Guérardel *et al.* (1999)

86	–	Gal (β1-3) GlcNAc (β1-3) Gal (β1-4) GlcNAc (β1-3) Gal (β1-4) Glc
		                           6S
		    Fuc (α1-4) Fuc (α1-3)
			–	Guérardel *et al.* (1999)

87	–	Fuc (α1-2)Gal (β1-3) GlcNAc (β1-3) Gal (β1-4) GlcNAc (β1-3) Gal (β1-4) Glc Fuc (α1-4) Fuc (α1-3)	–	Guérardel *et al.* (1999)

stages of lactation, this concentration is between five and ten times lower (Kunz *et al.*, 2000, 2002; Coppa *et al.*, 1999). Early in lactation, α1-2 linked fucosylated oligosaccharides dominate over α1-3/4 Fuc linkage, whereas in later lactation they converge to approximately equal concentrations (Chaturvedi *et al.*, 2001).

9.2.2   Animal milk oligosaccharides

Compared to human milk, the concentration of oligosaccharides of milk of the most important domestic mammalian animals is much lower (Urashima *et al.*, 1997b). Also, on a structural perspective, there are significant differences of the oligosaccharide pattern between human milk and milk of domestic animals. Fucosylated oligosaccharides, which represent the majority in the oligosaccharides in human milk, could not be detected in the milk of cow, sheep, goat, and horse (Finke, 2000). On the other hand, 3'- and 6'-galactosyl-lactose and 3'- and 6'-sialyl-lactose could be detected in all animal milks as well as in human milk. The basic scheme of animal milk oligosaccharides is given in Fig. 9.2. Also in animal milk, the concentration of oligosaccharides decreases during lactation (Kuhn and Brossmer, 1956; Finke, 2000). However, in late lactation cows' milk, an increase in sialylated oligosaccharides was found (Martin, 2001). Additionally, there are different oligosaccharide patterns among the different species. The known oligosaccharides of the most important domestic mammalian animals are summarised in Table 9.2. An excellent review on oligosaccharides of milk and colostrum in a variety of non-human mammals is given by Urashima *et al.* (2001).

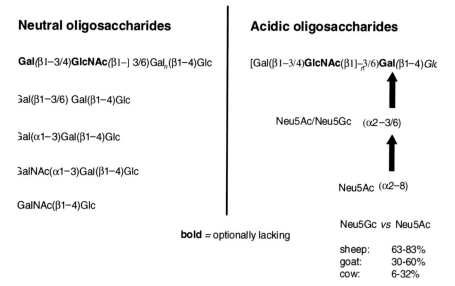

Fig. 9.2   Basic structure of milk oligosaccharides of some domestic animal (cow, sheep, goat, and horse) oligosaccharides (after Finke, 2000). Optionally lacking monosaccharides are depicted in bold letters.

Table 9.2 Structures of neutral and acidic milk oligosaccharides of some domestic animals

No.	Abbreviation	Oligosaccharides	Source	Reference
Neutral animal milk oligosaccharides				
1	–	GalNAc (β1-4) Glc	Cow colostrum	Saito *et al.* (1984)
2	–	Gal (β1-4) GlcNAc	Cow colostrum	Saito *et al.* (1984)
3	2'-FL	Fuc (α1-2) Gal (β1-4) Glc	Goat colostrum	Urashima *et al.* (1994a)
			Cow colostrums	Urashima *et al.* (1991b)
4	–	Gal (α1-3) Gal (β1-4) Glc	Goat colostrum	Urashima *et al.* (1994a)
			Sheep colostrum	Urashima *et al.* (1989a)
			Cow colostrum	Saito *et al.* (1987)
5	3'-GL	Gal (β1-3) Gal (β1-4) Glc	Goat colostrum	Urashima *et al.* (1994a)
			Sheep colostrum	Urashima *et al.* (1989a)
			Horse colostrum	Urashima *et al.* (1989b)
			Cow colostrum	Saito *et al.* (1987)
6	6´-GL	Gal (β1-6) Gal (β1-4) Glc	Goat colostrum	Urashima *et al.* (1994a)
			Sheep colostrum	Urashima *et al.* (1989a)
			Horse colostrum	Urashima *et al.* (1989b)
7	–	Gal (β1-4) GlcNAc | Fuc (α1-3)	Cow colostrum	Saito *et al.* (1987)
8	–	GalNAc (α1-3) Gal (β1-4) Glc	Cow colostrum	Urashima *et al.* (1991b)
9	–	GlcNAc (β1-6) Gal (β1-4) Glc	Goat milk	Chaturvedi and Sharma (1988)
10	–	Gal (β1-4) GlcNAc (β1-6) Gal (β1-4) Glc	Goat milk	Chaturvedi and Sharma (1990) Urashima *et al.* (1991a)
			Horse colostrum	
11	LNneoT	Gal (β1-4) GlcNAc (β1-3) Gal (β1-4) Glc	Horse colostrum	Urashima *et al.* (1991b)
12	–	Gal (β1-4) GlcNAc (β1-6) Gal (β1-4) Glc | Fuc (α1-3)	Goat milk	Chaturvedi and Sharma (1990)

13	–	Gal (β1-3) GlcNAc (β1-6) Gal (β1-4) Glc
		       |
		    Fuc (α1-3)
			Goat milk	Chaturvedi and Sharma (1990)

14	–	Gal (β1-4) GlcNAc (β1-6)\ Gal (β1-4) Glc Gal (β1-3)/	Cow colostrum Horse colostrum	Urashima *et al.* (1991b) Urashima *et al.* (1989b)
15	–	Gal (β1-4)\ GlcNAc (β1-3) Gal (β1-4) Glc Gal (β1-3)/	Goat milk	Chaturvedi and Sharma (1990)
16	LNneoH	Gal (β1-4) GlcNAc (β1-6)\ Gal (β1-4) Glc Gal (β1-4) GlcNAc (β1-3)/	Horse colostrum	Urashima *et al.* (1991a)
Acidic animal milk oligosaccharides				
17	–	Neu5Ac (α2-3) Gal	Cow colostrum	Urashima *et al.* (1997c)
18	3'-SL	Neu5Ac (α2-3) Gal (β1-4) Glc	Cow colostrum Goat colostrum Sheep colostrum	Schneir and Rafelson (1966) Urashima *et al.* (1997b) Nakamura *et al.* (1998)
19	6'-SL	Neu5Ac (α2-6) Gal (β1-4) Glc	Cow colostrum Goat colostrum	Schneir and Rafelson (1966) Urashima *et al.* (1997b)

Table 9.2 cont'd

No.	Abbreviation	Oligosaccharides	Source	Reference
			Cow colostrum	Kuhn and Gauhe (1965)
20 –		Neu5Gc (α2-3) Gal (β1-4) Glc	Goat colostrum	Urashima *et al.* (1997b)
			Sheep colostrum	Nakamura *et al.* (1998)
			Goat colostrum	Veh *et al.* (1981)
21 –		Neu5Gc (α2-6) Gal (β1-4) Glc	Cow colostrum	Urashima *et al.* (1997b)
			Sheep colostrum	Nakamura *et al.* (1998)
22 –		O-Acetyl-Neu5Ac (α2-3) Gal (β1-4) Glc	Cow colostrum	Urashima *et al.* (1997c)
23 6'-SLn		Neu5Ac (α2-6) Gal (β1-4) GlcNAc	Cow colostrum	Kuhn and Gauhe (1965)
			Goat colostrum	Urashima *et al.* (1997b)
24 –		Neu5Gc (α2-6) Gal (β1-4) GlcNAc	Cow colostrum	Kuhn and Gauhe (1965)
25 6'-SLn-1-P	Neu5Ac (α2-6) Gal (β1-4) GlcNAc-1-PO$_4$		Cow colostrum	Kuhn and Gauhe (1965)
26 6'-SLn-6-P	Neu5Ac (α2-6) Gal (β1-4) GlcNAc-6-PO$_4$		Cow colostrum	Kuhn and Gauhe (1965)
27 3'-SGL	Neu5Ac (α2-3) Gal (β1-3) Gal (β1-4) Glc		Cow colostrum	Parkinnen and Finne (1987)
28 –		Gal (β1-3) Gal (β1-4) Glc | Neu5Ac (α2-6)	Goat milk	Viverge *et al.* (1997)
29 –		Gal (β1-6) Gal (β1-4) Glc | Neu5Ac (α2-3)	Goat milk	Viverge *et al.* (1997)
30 DS-L	Neu5Ac (α2-8) Neu5Ac (α2-3) Gal (β1-4) Glc		Cow colostrum	Kuhn and Gauhe 1(965)

9.2.3   Oligosaccharides of non-milk origin

Because the composition and structure of human milk oligosaccharides cannot be reproduced and until now large-scale preparations of animal milk oligosaccharides have not been commercially available, oligosaccharides of more simple structures than human milk oligosaccharides have been used as components in several dietary products to mimic the beneficial effects of human milk oligosaccharides (Gibson and Roberfroid, 1995; Roberfroid and Delzenne, 1998; Loo *et al.*, 1999; Jenkins *et al.*, 1999; Gibson, 1999). The most important prebiotic (see below, 9.4) oligosaccharides, larger than disaccharides, of non-milk origin, their structure and the method of preparation are given in Table 9.3. The most intensive studies exist for fructans and, to a lesser extent, for galactooligosaccharides (Loo *et al.*, 1999). Fructans are linear or branched fructose-polymers, which are either β2-1-linked inulins or β2-6-linked levans. Because the inulin-type fructans can be easily extracted from plant sources (e.g. asparagus, garlic, leek, onion, artichoke, chicory roots, etc.) they have been widely used as an ingredient in dietary products. The basic structure of fructans is given in Table 9.3.

Galactooligosaccharides are synthesised from lactose via enzymatic transgalactosylation using ß-galactosidases mainly of bacterial origin (e.g. *Bacillus circulans*) (Tanaka and Matsumoto, 1998). The commercial use of β-galactosidase in food processing is extensively reviewed elsewhere (Kinsella and Taylor, 1995). Galactooligosaccharides consist of a chain of galactose molecules usually with a glucose molecule at the reducing terminus, varying in chain length (degree of polymerisation range 3–8) and linkages (for basic structure, see Table 9.3).

Table 9.3 Prebiotic oligosaccharides of non-milk origin

Prebiotic oligosaccharides of non-milk origin with a significant effect on bifidobacteria in human faeces, their chemical structure and the principal method of their preparation (after Roberfroid and Slavin, 2000, and Crittenden and Playne, 1996)

Trivial name	Structure	Preparation	Estim. yearly production in t
Galacto-OS	$[\text{Gal}(\beta 1-]_n 3/4/6)\text{Gal}(\beta 1-4)\text{Glc}$	Enzymatic synthesis from lactose	15,000
Fructo-OS/ Inulin	$[\text{Fru}(\beta 2-]_n 1)\text{Glc}$	Extraction from natural sources Enzymatic hydrolysis of natural polymers Enzymatic synthesis from sucrose	12,000
Palatinose/ Isomaltulose-OS	$\text{Glc}(\alpha 1-6)\text{Fru}$ $[\text{Glc}(\alpha 1-]6)\text{Glc}(\alpha 1-6)\text{Fru}$	Enzymatic synthesis from sucrose	5,000
Soybean OS	$[\text{Gal}(\alpha 1-]_n 6)\text{Glc}(\alpha 1-2)\text{Fru}$	Extraction from natural sources	2,000
Lactosucrose	$\text{Gal}(\beta 1-4)\text{Glc}(\alpha 1-2)\text{Fru}$	Enzymatic synthesis from lactose	1,600
Xylo-OS	$[\text{Xyl}(\beta 1-]_n 4)\text{Xyl}$	Enzymatic hydrolysis of e.g. corncob xylan	300

9.3 Physiological functions of dietary oligosaccharides

The extremely wide variation of oligosaccharide structures in human milk indicates that oligosaccharides are involved in many functional effects related to the gastrointestinal tracts as well as to systemic processes. Although many attractive and promising concepts exist, information about the relationship between oligosaccharide structure and physiological function is still rather limited.

In 1998, a group of European experts analysed physiological effects of non-digestible oligosaccharides (Loo *et al.*, 1999) based on published *in vitro* studies as well as on clinical trials in adults. They came to the conclusion that there is considerable evidence that oligosaccharides affect intestinal flora and bowel habit. They also decided that there is promising evidence that oligosaccharides might affect mineral absorption and lipid metabolism and that end-products of bacterial metabolism might play a role in colon cancer prevention. The interaction of oligosaccharides and the immune system was not part of the project. However, in the mean time new data have been published indicating that prebiotic oligosaccharides directly or via modulating the intestinal flora can influence the immune system.

Additionally, there are some reports on a possible influence of dietary oligosaccharides on brain development. This was discussed for galactose-containing molecules, with galactose as source for galactocerebroside, the predominant glycolipid in myelin and, thus especially important for the development of the infant brain (Kunz *et al.*, 1999).

Based on studies with newborn rats, it was discovered that oral administration of sialic acid increases the cerebral and cerebellar sialic acid content (Carlson and House, 1986). However, contradicting results were obtained from studies with mice and rats receiving *N*-acetylneuraminic acid orally and intravenously (Nöhle and Schauer, 1981). Therefore, further research is needed to prove a possible influence of dietary mono- and oligosaccharides, respectively, on brain development.

9.4   Effect on intestinal flora: prebiotic role

It has been known for more than one hundred years that *bifidobacteria* dominate the faecal flora of breast-fed infants (Tissier, 1899). Around 50 years ago, human milk oligosaccharides were identified as one of the important bifidogenic factors found in human milk (György, *et al.*,1954; Kunz, *et al.*, 2000). There is a wide consensus that *bifidobacteria*, along with *lactobacilli*, can be classified as health-promoting bacteria, which is also the basis for their use as probiotics. Oligosaccharides seem to be an important factor in the defence of the newborn infant against infection (Kunz *et al.*, 2000).

The basic assumption of the prebiotic concept is that these dietary ingredients are non-digestible, reach the colon and can be utilised by the health-promoting colonic bacteria (Gibson and Roberfroid, 1995). The human intestine lacks enzymes able to hydrolyse β-glycosidic linkages with the exception of lactose, which makes oligosaccharides with β-linkages non-digestible (Rivero-Urgell and Santamaría-Orleans, 2001; Engfer *et al.*, 2000; Gnoth *et al.*, 2000; Roberfroid and Delzenne, 1998). Thus, human milk oligosaccharides and most animal milk oligosaccharides, as well as the non-milk oligosaccharides, can be considered to be indigestible.

Another prerequisite for a prebiotic effect is that the dietary ingredient that reaches the colon can be utilised by the intestinal flora to stimulate those bacteria that are naturally part of the ecosystem in the colon. Most data concerning the utilisation of oligosaccharides by faecal bacteria come from *in vitro* studies and are related to non-milk oligosaccharides. Some common non-milk oligosaccharides that show significant bifidogenic effect in human faeces are summarised in Table 9.3. Additionally, clinical trials in human infants and adults have demonstrated the stimulation of *bifidobacteria* and *lactobacilli* by several non-milk oligosaccharides, in particular galactooligosaccharides and fructans (Tables 9.4 and 9.5). More recently, it has been demonstrated in infants that a mixture of galactooligosaccharides and fructans is very effective in stimulating a *bifidobacteriae* and changing the stool characteristics to become closer to those found in breast-fed infants (Moro *et al.*, 2002; Boehm *et al.*, 2002), indicating that these oligosaccharides of non-milk origin can, at least partially, mimic the effects of human milk oligosaccharides.

Table 9.4   Physiological effects of galactooligosaccharides in human studies

Detected effect	Study protocol	Reference
Gastrointestinal symptoms and faecal frequency (>). Effect on intestinal microecosystem	15 g GOS/d in yoghurt, 3 weeks, 12 healthy adults	Teuri *et al.* (1998)
Faecal concentrations of bifidobacteria (>), alters the fermentative activity of colonic flora in humans	10 g TOS/d, 21 days, 8 healthy adults	Bouhnik *et al.* (1997)
Faecal pH (<), pattern of organic acid in faeces closer to that of breast-fed infants	1 or 2% glactosyllactose in infant formula, 27 newborn infants	Yahiro *et al.* (1982)
Faecal concentrations of bifidobacteria (>), lactobacilli slightly (>)	2.5 g, 5 g or 10 g GOS/d, 7 days, 12 healthy adults	Ito *et al.* (1990)
Faecal concentrations of bifidobacteria (>), lactobacilli (>), bacteroides (<)	10 g TOS/d, 1 week, 11 healthy male adults	Dombo *et al.* (1997)
Seems to relieve constipation in most elderly people without causing gastro-intestinal symptoms	9 g GOS/d in yoghurt, 2 weeks, 14 elderly female subjects	Teuri *et al.* (1998)
Faecal concentrations of bifidobacteria (>), *Bacteroidaceae* and *Candida* spp. (<), faecal pH (<), concentrations of faecal ammonia, *p*-cresol, and indole, propionic acid, isobutyric acid, isovaleric acid and valeric acid (<)	15 g transgalactosylated disaccharide/d, 6 days, 12 healthy adults	Ito *et al.* (1993)
Good tolerance	15 g GOS/d, 3 weeks, 12 healthy men	van Dokkum *et al.* (1995)
Faecal concentrations of bifidobacteria (>), pathogens (−). Faecal nitro-reductase activity, and concentrations of indole and isovaleric acid (<)	2.5 g GOS/d, 3 weeks, 12 healthy adults in whom the number of indigenous bifidobacteria were comparatively low	Ito *et al.* (1993)
Faecal concentrations of bifidobacteria (>). 10 g TOS/d: lactobacilli (>), pathogens (<)	3 g or 10 g TOS/d, 2 weeks (week 1: 3 g TOS, week 2: 10 g TOS), 5 healthy adults	Tanaka *et al.* (1983)
Improved constipation condition, *p*-cresol (<)	10 g/d GOS, 18 days, 50 young volunteers	Ito *et al.* (1994)
Defecation frequency and stool softness (>). Gastrointestinal symptoms (>) (10 g GOS/d)	5 g, or 10 g GOS/d, 1 week, 128 healthy adults with constipation tendency	Deguchi *et al.* (1997)
Faecal concentrations of bifidobacteria (>), bacteroides (<), faecal pH (<)	4' galactosyllactose, 10 healthy human volunteers	Ohtsuka *et al.* (1989)
True calcium absorption (>) (not accompanied by increased urinary Ca excretion)	20 g/d TOS, two 9-day treatment periods, 19-day washout period, 12 post-menopausal women	van der Heuvel *et al.* (2000)

> increase.
< decrease.
− no effect.

Table 9.5 Physiological effects of fructans in human studies

Detected effect	Study protocol	Reference
10 g, or 20 g FOS/d: Faecal concentrations of bifidobacteria (>)	0, 2.5, 5, 10, 20 g FOS/d, 7 days, 40 healthy adults	Bouhnik *et al.* (1999)
Fractional calcium absorption (>)	10 g oligofructose/d, two g-day treatment periods, 19-day washout period, 12 healthy adolescents	van den Heuvel *et al.* (1999)
Serum total cholesterol, LDL-cholesterol and LDL/HDL-ratio (<), serum HDL, triglycerides and blood glucose (–)	2.5% FOS in 375 ml/d yoghurt fermented by *Lactobacillus acidophilus*, two 3 week treatment periods, 1 week washout period, 30 healthy adults	Schaafsma *et al.* (1998)
Faecal concentrations of bifidobacteria (>). Oligofructose: Faecal concentrations of bacteroides, clostridia, and fusobacteria (<). Inulin: gram-positive cocci (<). Stool frequency (>). Faecal wet and dry matter, nitrogen and energy excretion (>)	15 g oligofructose/d, 15 g inulin, 15 days each, 8 healthy adults	Gibson *et al.* (1995)
Stool frequency and faecal weight (>), faecal water pH tended to (<), breath hydrogen (>)	14.3 g FOS, three 7-day treatment periods separated by 7-day washout periods, 16 patients with ileal pouch-anal anastomosis	Alles *et al.* (1997)
Faecal concentrations of bifidobacteria (>), enterococci and enterobacteria (<). Faecal SCFAs and lactate (–), faecal pH, beta-glucosidase and beta-glucuronidase activities (–). Functional constipation (<) with only mild discomfort	20–40 g inulin/d, 19 days, 10 elderly constipated persons	Kleessen *et al.* (1997)
Improved constipation	10 g/d FOS, 18 days, 50 young volunteers	Ito *et al.* (1994)
Faecal concentrations of bifidobacteria (>), lactobacilli tended to (>), enterobacteriaceae (<). Faecal pH (<), improved consistency of stools	8 g FOS/d, 2 weeks, 23 elderly patients	Mitsuoka *et al.* (1987)
Faecal concentrations of bifidobacteria (>), beta-fructosidase activity (>). Faecal pH, activities of nitroreductase, azoreductase, and beta-glucuronidase, concentrations of bile acids and neutral sterols (–)	12.5 g FOS/d, three 12-day periods, 20 healthy adults	Bouhnik *et al.* (1996)
Apparent absorption and balance of calcium (>), Mg, Fe and Zn (–)	40 g/d inulin, three 28-day periods, 9 healthy young men	Coudray *et al.* (1997)

Table 9.5 cont'd

Detected effect	Study protocol	Reference
Faecal concentrations of bifidobacteria (>), faecal pH (<), little intestinal discomfort	8 g/d Fn-type chicory oligofructose, 5 weeks, 8 healthy adults	Menne *et al.* (2000)
Number of stools (>), stool pH (–), faecal concentrations of bifido-bacteria (–), frequency of nappy rash and colics (–)	1; 2; 3 g/d FOS, 2 weeks, 34 infants 7 to 20 days of age	Guesry *et al.* (2000)
Severity of diarrhoeal disease (<), stool consistency (–), discomfort or flatus (–)	0.55 g oligofructose/15 g of cereal, 6 months, 63 infants (aged 4 months to 24 months)	Saavedra *et al.* (1999)
GOS and inulin did not have a negative effect on iron and calcium absorption	15 g FOS or inulin/d, 3 weeks, 12 healthy adults	van den Heuvel *et al.* (1998)

> increase.
< decrease.
– no effect.

Owing to the difficulties of separation and purification of milk oligosaccharides, only a limited number of studies using human milk fractions as test substances have been performed (Finke, 2000). However, there are several studies of breast-fed infants regarding the effect of breast feeding on the intestinal flora. With respect to the possible dominance of *bifidobacteria* in the faecal flora, most, but not all studies, could demonstrate the development of an intestinal flora dominated by *bifidobacteria* and *lactobacilli*, most probably due to the oligosaccharide content of human milk. Studies performed in Scandinavia in particular could not demonstrate the bifidogenic effect of breast feeding, indicating that factors other than the diet can influence the intestinal flora (Adlerberth, 1997). Although the data in breast-fed infants are not consistent, a dominance of *bifidobacteria* and *lactobacilli* and a reduction of *clostridia* are accepted as indicators of a balanced intestinal flora.

Although many of the studies demonstrated an effect of prebiotics on *bifidobacteria*, simultaneous effects of these prebiotics on other groups of bacteria are rarely detected (Loo *et al.*, 1999; Moro *et al.*; 2002, Boehm *et al.*, 2002) indicating that the establishment of a balanced intestinal flora in the colon is only one factor of a system of dietary factors that reduce the incidence of infections in breast-fed infants in comparison to bottle-fed infants.

9.5   Effect on intestinal infections and mineral absorption

The anti-infective effect of human milk has been partly attributed to the high amount of free oligosaccharides as well as glyco-conjugates because these structures

Table 9.6 Carbohydrate ligands of pathogens and pathogenic substances

Pathogen	Specificity	Reference
Bacteria		
Escherichia coli (Typ 1- Fimbriae)	Man	Duguid and Gillies (1957)
E. coli (S- Fimbriae)	Neu5Ac(α2-3)	Parkinnen *et al.* (1983)
E. coli (K 99)	Neu5Gc	Ouadia *et al.* (1992)
E. Coli (P-Fimbriae)	Gal(α1-4)Gal	Strömberg *et al.* (1990)
Campylobacter pylori	Monosialylated gangliosides	Emödy *et al.* (1988)
C. pylori	NeuAc(α2-3) Gal	Evans *et al.* (1988)
Streptococcus pneumoniae	Gal(β1-4)GlcNAc/Gal(β1-3)GlcNAc	Andersson *et al.* (1986)
S. pneumoniae	GalNAc(β1-3)Gal/GalNAc(β1-4)Gal	Cundell and Tuomanen (1994)
S. suis	Neu5Ac(α2-3)Gal(β1-4) GlcNAc(β1-3)Gal	Liukkonen *et al.* (1992)
S. suis	Gal(α1-4)Gal	Haataja *et al.* (1993)
S. sanguis	NeuAc(α2-3)Gal(β1-4)Glc	Murray *et al.* (1982)
Helicobacter pylori	Neu5Ac(α2-3)	Hirmo *et al.* (1996)
H. pylori	Neu5Ac(α2-3) Gal	Miller-Prodraza *et al.* (1996)
Vibrio cholerae	Fuc	Holmgren *et al.* (1983)
Pseudomonas aeruginosa	Gal(β1-4)GlcNAc/Gal(β1-3)GlcNAc	Ramphal *et al.* (1991)
Chlamydia trachomatis	$(Man)_{8-9}$	Cho-chou Kuo *et al.* (1996)
Mycoplasma pneumoniae	Neu5Ac(α2-3)Gal with polylactosamines	Loveless and Feizi (1989)
Viruses		
Influenza A	Neu5Ac(α2-6)Gal(β1-4)Glc	Weis *et al.* (1988)
Influenza B	Neu5Ac(α2-3)Gal(β1-4)Glc	Weis *et al.* (1988)
Influenza C	Neu5,9Ac₂	Rogers *et al.* (1986)
Polyomavirus	Neu5Ac(α2-3)Gal(β1-3)GalNAc	Cahan and Paulson (1980)
Rotavirus	Neu5Ac	Svensson (1992)
Toxins		
Vibrio cholerae toxin	GM 1 Ganglioside/Neu5Ac(α2-3)- -Gal(β1-4)Glc	Idota *et al.* (1995)
Hast *E. coli* toxin	Fuc	Cravioto *et al.* (1991)
Pertussis toxin	Neu5Ac	Brennan *et al.* (1988)
Yeasts		
Candida albicans	Fuc(α1-2)Gal β	Brassart *et al.* (1991)
Protozoa		
Trypanosoma cruzi	Neu5Ac	Schenkman and Eichinger (1993)

might prevent intestinal attachment of infectious agents by acting as receptor analogues (Beachey, 1981; Coppa, 1990; Zopf and Roth, 1996; Newburg, 1999). Table 9.6 summarises the receptor analogues so far identified in human milk and the respective infective agents. Similar observations have been made with cows' milk oligosaccharides, which reduced the adhesion of enterotoxic *Escherichia coli* strains of the calf (Mouricout *et al.*, 1990).

There are also first *in vitro* results available that fructans interfere with the adhesion of *Salmonella enterica* sv. Typhimurium and non-pathogenic *Escherichia coli* in the pig small intestine (Naughton *et al.*, 2001)

Pectin, and in particular, pectin hydrolysate are also able to reduce the adhesion of bacteria to intestinal epithelial cells *in vitro* and *in vivo* (Guggenbichler *et al.*, 1997). This effect of pectin hydrolysate can explain the therapeutic effect of carrot soup, which has been used for 100 years by European paediatricians as treatment against diarrhoea (Moro, 1908).

In summary, there is evidence that the prevention or reduction of adhesion of infectious agents to epithelial surfaces is an important mechanism of dietary oligosaccharides in increasing the defence capacity against intestinal infections (Dai *et al.*, 2000).

9.5.1   Effect on mineral absorption

The animal studies as well as clinical trials in humans have been focused on the absorption of calcium and magnesium, and in few experiments the effect of dietary oligosaccharides on the metabolism of iron and iodine has been studied. In human infants, the calcium absorption from human milk is significantly higher than from any formula (Carnielli *et al.*, 1996). There are many factors in human milk that might influence calcium absorption. One of the important factors is the composition of lipids which reduces the formation of calcium soaps in the colon (Carnielli *et al.*, 1996). Thus, the difference in calcium absorption between human milk and formula cannot easily be attributed to the content of oligosaccharides.

There are several ways in which dietary oligosaccharides might influence mineral absorption. Most likely, the increase of bacterial synthesis of short chain fatty acids in the colon plays a key role, but also increasing the passive and active calcium transport in the small intestine or increased transport of calcium from the small intestine to the colon may be involved in the stimulation of mineral absorption by oligosaccharides (Scholz-Ahrens *et al.*, 2001). In rat experiments with dietary galactooligosaccharides (Chonan and Watanuki, 1996) measuring complete calcium absorption, including the colonic component, indicated that the dietary intake of galactooligosaccharides increased calcium absorption. In a study of postmenopausal women, dietary galactooligosaccharides increased calcium absorption significantly (van den Heuvel *et al.*, 2000). Fructooligosaccharides are also able to stimulate calcium absorption (Ohta *et al.*, 1998; Roberfroid and Delzenne,1998; Frank, 1998). More recently, a preterm infant formula, supplemented with a mixture of galactooligosaccharides (90%) and inulin (10%), resulted in higher renal excretion of calcium compared to that found in infants fed a non-supplemented formula, indicating an increased calcium absorption (Lidestri *et al.*, 2002).

There are also some reports that magnesium, iodine and iron metabolism can be improved by dietary oligosaccharides (Roberfroid and Delzenne, 1998; Frank, 1998). However, the data from the literature are not consistent, indicating that the experimental design as well as the type of oligosaccharide and the mineral intake influence the results.

9.6   Effect on the immune system and other physiological effects

Human milk protects the infant against infection, and oligosaccharides play an important role among other factors (Xanthou, 1998). The development of a balanced intestinal microflora seems to be one of the main mechanisms for the host defence activity of oligosaccharides. This discussion has been greatly stimulated by the observation of a Finnish group (Kalliomäki *et al.*, 2001) that administration of a probiotic strain *Lactobacillus GG* to a group of infants with a high risk of developing atopy reduced the frequency of atopic symptoms. Additionally, differences in the intestinal microbiota composition between allergic and healthy infants have been demonstrated (Ouwehand *et al.*, 2001; Böttcher *et al.*; 2000). Observations of the effect of prebiotic oligosaccharides on the immune system are rare. If the internal microbiota is the acting factor in the postnatal immune modulation, then prebiotic oligosaccharides could have an effect on the immune system. Although the idea is attractive, no clinical studies demonstrating such a prebiotic effect on the postnatal developement in infants have been published. Very recently, it was reported that elderly persons receiving a prebiotic mixture (fructans of different molecular size) experienced significantly higher influenza A antibody titres as a response to vaccination than patients receiving a placebo (Bunout *et al.*, 2002). Only a few reports on the direct effect of human milk oligosaccharides on immune functions have been reported so far. Human milk oligosaccharide structures like LNFPIII and LNneoT (see Table 9.1) show an effect on IL10 production (Velupillai and Harn, 1994). It is further discussed that human milk is involved in the generation of anti-inflammatory mediators that suppress Th-1-type and inflammatory responses (Terrazas *et al.*, 2001). The interaction of human milk oligosaccharides with selectins in cell–cell-interactions of e.g. leucocytes and lymphocytes was described recently by Rudloff *et al.* (2002).

In summary, the possibility that prebiotic oligosaccharides could influence the immune response of the host would make prebiotics dietary intervention in reduction of atopic diseases an attractive choice to probiotics, However, the current data do not allow any conclusions to be drawn on the issue.

Other promising effects are the influence of prebiotic oligosaccharides on the lipid metabolism (for review see Roberfroid and Delzenne, 1998) and brain development (Kunz *et al.*, 1999; Carlson *et al.*, 1986). The clinical relevance of the observed effects is still under discussion and much more research is needed to prove these effects (Meyer *et al.*, 2001).

9.7   Analytical methods

Depending on the type, size and structure as well as the (biological) source of oligosaccharides, a variety of separation techniques and methods have to be applied for the characterisation of the molecules of interest. Numerous publications and reviews are available and only a selection of literature can be discussed in this chapter. The choice of a suitable analytical method depends strongly on the size, composition and physicochemical properties of the carbohydrate of interest. For general information on oligosaccharides and glycoconjugates the reader is referred to Varki *et al.* (1999) and Ernst *et al.* (2000). Food-derived carbohydrates are excellently described by Nakakuki (1994) as well as by Cho *et al.* (1999). General information on dietary fibers and strategies on the detection of dietary fibers and prebiotic oligosaccharides has been provided by Southgate (1995) and Spiller (2001). For information on the complex group of heteropolysaccharides, especially those of lactic acid bacteriae from dairy products, the reader is referred to De Vuyst and Degeest (1999). For information on milk composition, the reader is referred to Jensen (1995). Milk oligosaccharides are described in detail by Newburg and Neubauer (1995) and Kunz *et al.* (2000). For general analytical techniques and methods in the field of glycobiology, the reader is referred to Ziad el Rassi (1995) and Lennarz and Hart (1994).

Briefly, the major techniques for the analyses of glycans (pure carbohydrates) and glycoconjugates (such as glycoproteins and glycolipids) are chromatography, mass spectrometry (MS), spectroscopy, electrophoretic methods and separation techniques like crystallisation and filtration. Chromatography plays a major role in separation and analysis of oligosaccharides. Apart from ion exchange chromato-graphy, affinity chromatography (by lectins or antibodies) and gel permeation chromatography (GPC), high-performance liquid chromatography (HPLC) is the most frequently used technique. As stationary phases, normal phase (e.g. amino phases), reversed phases (e.g. C-18) and graphite carbon columns are used. Refractive index, UV and fluorescence radioactivity are used as detection principles, with evaporative light scattering detection as a technique with great potential. High pH anion exchange chromatography with pulsed amperometric detection (HPAEC-PAD) is a powerful method widely used for carbohydrate analysis (Lee, 1990, 1996). Many of the liquid chromatography methods mentioned can be scaled up for separation of larger amounts of samples. Special continuous preparative chromatographical techniques, such as simulated moving bed chromatography (Finke *et al.*, 2000) and annular chromatography (Finke *et al.*, 2002) have to be mentioned here. Other chromatographic methods such as thin layer chromatography and paper chromatography are less frequently used and not discussed here.

Filtration techniques (Sarney *et al.*, 2000), centrifugation and precipitation can be used for sample preparation, e.g. removal of salts, lipids, proteins and polysaccharides, whereas crystallisation may be used for purification and preparation of distinct oligosaccharide structures (first described by Polonowski and Lespagnol, 1933). For methods such as capillary electrophoresis (Monnig and

Kennedy *et al.*, 1994), fluorophor-assisted carbohydrate electrophoresis (FACE; Jackson,. 1994) and thin layer chromatography (TLC; Sherma, 1994) nuclear magnetic resonance (NMR; Haw, 1992; Vliegenthart *et al.*, 1983) the reader is referred to the mentioned reviews as well as to the monographs listed at the beginning of this chapter.

For structural analyses of distinct oligosacharides, NMR spectroscopy is a very powerful method but if only minute amounts of samples are available, mass spectrometry (MS) plays an important role for sequencing of molecules. Methylation analyses and gas chromatography mass spectrometry (GC–MS) of monosaccharide compounds are widely used for determination of structural features of oligosaccharides. Fast atom bombardment MS (FAB–MS) enable the structural analyses of mainly derivatised oligosaccharides (Egge and Peter-Katalinic, 1987; Dell, 1987; Egge *et al.*, 1991) and more recent techniques like matrix-assisted laser desorption/ionisation mass spectrometry (MALDI-MS; Stahl *et al.*, 1991; Mock *et al.*, 1991) and (nano) electrospray ionisation mass spectrometry (ESI-MS; Bahr *et al.*, 1997) enable the characterisation of native (underivatised) carbohydrates. MALDI-MS of glycans and glycoconjugates is excellently reviewed by Harvey (Harvey, 1999). Nano-ESI-MS is especially well suited for analyses of oligosaccharides, glycoproteins and other glycoconjugates (Karas *et al.*, 2000).

Alongside the structures characterised so far (see Table 9.1) human milk oligosaccharides contain a large number of new high molecular weight neutral oligosaccharides. These molecules have been detected especially by combination of liquid chromatography coupled off-line with MALDI–MS (Stahl *et al.*, 1994). More recently, also high molecular weight acidic oligosaccharides have been detected by a similar strategy (Finke *et al.*, 1999). All high mass structures found so far show a composition according to the following formula:

$$L\ X/Y–Z$$

with:

L	=	one lactose unit per HMOS molecule (Gal-Glc) at the reducing end
X	=	no. of *N*-acetylglucosamine–galactose units (Gal–GlcNAc)
Y	=	no. of fucose units (Fuc)
Z	=	no. of sialic acid units (Neu5Ac)
		(with X, Y, Z = 0,1,2,3,4...)

The largest group of molecules (around 10 kDa) detected by MALDI–MS consists of more than 50 monosaccharides, therefore the term 'human milk polysaccharides' seems to be more suited than 'human milk oligosaccharides' for these compounds (Fig. 9.3). In the formula above five fucose (Y) residues resemble the molecular mass increment of two Gal–GlcNAc units (X); the difference being only 0.042 Da. The difference between the mass increment of two fucose units (Y) *vs* one sialic acid is 1.028 Da. Use of common MALDI–MS instruments does not enable this ambiguity to be resolved for high-mass molecules. One solution is the use of post-source delay MALDI–MS PSD–MALDI (Spengler *et al.*, 1991); by analysing the metastable fragment of a molecule of interest. Metastable fragmentation happens during the flight of the desorbed ions in the field-free drift region of the

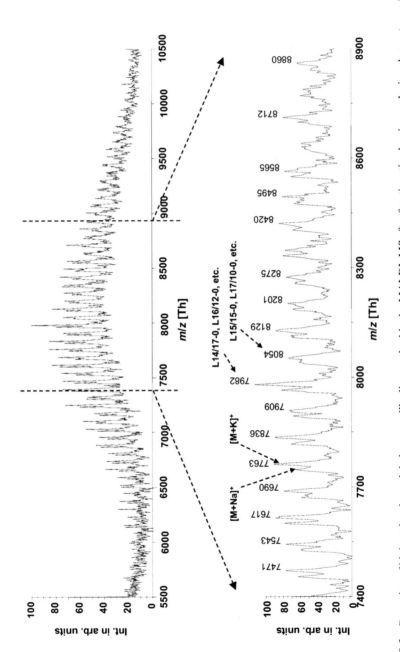

Fig. 9.3 Detection of high molecular weight human milk oligosaccharides by MALDI–MS after fractionation by size exclusion chromatography. Sum of 100 single spectra in positive ion mode; ions detected as cationided molecules (i.e. sodium- and potassium-adduct ions, respectively. Laser wavelength 337 nm, mass accuracy: 0.05%, UV-absorbing matrix: 5-chloro-2-mercaptobenzothiazole 10 g/l in water:tetrahydrofurane:ethanol = 1:1:1(v/v/v). The bottom panel shows the section of the upper mass spectrum between the dashed lines. For the nomenclature of the oligosaccharides, please refer to the text.

MALDI–TOF (time of flight) instrument. This fragmentation can be differentiated against prompt fragmentation during desorption or hydrolysis or a naturally occurring mixture of molecules by scanning the reflectron potential revealing a 'sequence' of the oligosaccharide of interest.

As an alternative to PSD–MALDI, negative ion nano-ESI-ion trap MS of native underivatised oligosaccharides is an even more powerful technique, since in the MS^n-mode of fragmentation the real sequence from the reducing end of the molecules can be generated with additional information on position of the glycosidic linkage of the released fragment (Pfenninger *et al.*, 2002a, b). The use of MALDI–Fourier Transform Ion Cyclotron Resonance Mass Spectrometry (FTICR–MS) offers a direct detection of the distinct unambiguous composition of human milk oligosaccharides by high resolution measurements (Pfenninger *et al.*, 2002c). All these techniques also can be applied to animal milk oligosaccharides. The oligosaccharide pattern of domestic animal milks is much simpler and less concentrated than that of human milk oligosaccharides (Table 9.2). The composition of these oligosaccharides does not follow the above-mentioned formula LX/Y–Z. Thus, an even larger variety of new structures follows from that, because for these oligosaccharides isomeric molecules with identical masses also exist.

Because composition and structure especially of human milk have not commercially been available until now, oligosaccharides of more simple structures, e.g. like the prebiotic oligosaccharides depicted in Table 9.3 have been used as components to mimic the beneficial effects of human milk oligosaccharides. For the selection of prebiotic oligosaccharides for human consumption it is recommended that analytical techniques are used that are capable of resolving individual components of usually complex mixtures beside the characterisation and quantitation via their monosaccharide components. Especially for the widely used prebiotic oligosaccharides, the classical dietary fibre analyses have been insufficient since these molecules show good solubility compared to typical fibres, the so-called non-starch polysaccharides. For the analyses of prebiotics such as fructans (oligofructoses, inulin) and galactooligosaccharides, AOAC methods (Association of Official Analytical Chemists) have been recently published. Fructans are extracted and hydrolysed by specific enzymes and the resulting monosaccharides are analysed by HPAEC-PAD (Hoebregs *et al.*, 1997), whereas galacto-oligosaccharides are analysed by a different HPAEC-PAD method after extraction of the oligosaccharides and hydrolysis by β-galactosidase (De Slegte *et al.*, 2002).

As described above HPAEC is also suited for the detection of native underivatised oligo-/polysaccharides. As an example a mixture of galactooligosaccharides (GOS) and long chain inulin (FOS) analysed by HPAEC is shown in Fig. 9.4. This mixture aims to resemble the molecular weight profile of human milk oligosaccharides as the weight of the size of prebiotic oligosaccharides determines, among others, the fermentation rate and thereby the location of fermentation in the gut. The GOS are in a tenfold excess over the FOS and the latter can be seen only after increasing the intensity scale of the chromatogram. Nevertheless, each individual oligosaccharide can be detected in one single liquid chromatography run.

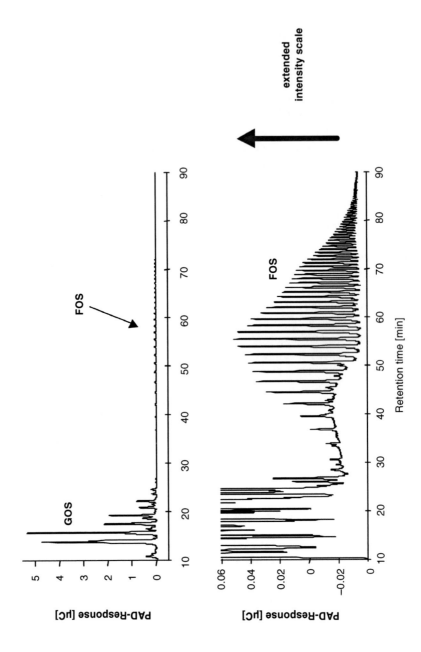

Fig. 9.4   HPAE chromatogram of a prebiotic mixture of galactooligosaccharides and long chain fructooligosaccharides.

9.8   Future trends

Although the understanding of the molecular mechanisms of dietary oligosaccharides is weak, there are some health benefits that can be directly attributed to these ingredients. These are stimulating attempts to investigate the structural basis of these functions. In fact, there is some possibility of producing particular 'designed' oligosaccharide molecules. Therefore, the development of test systems to investigate the biological effects of oligosaccharides *in vitro* as well as *in vivo* has a high priority. Because the possible health benefits of dietary oligosaccharides are species dependent, animal studies have a limited importance in elucidating the beneficial effects of oligosaccharides to humans.

The intestinal microbiota play a key role as mediator of the physiological effects of dietary oligosaccharides. During the last decade, enormous progress has been made in improving the analytical techniques to measure faecal flora (Millar *et al.*, 1996). Unfortunately, the intestinal microbiota in humans can only be studied using faecal samples, which does not adequately reflect the intestinal microbial ecosystem. Therefore, non-invasive methods will be developed to investigate the metabolic activity of the intestinal microbiota using stable isotope techniques.

In summary, dietary oligosaccharides of different origins offer a variety of functional effects which might have significant benefits to the general health of humans. Because oligosaccharides are safe, quite stable and can easily be processed in the dairy and food industries, there is a great potential for them to contribute as functional dietary ingredients in various functional products.

9.9   Acknowledgements

The authors thank Ms Beate Müller-Werner for her excellent HPAEC analyses and Mr Marco Mank Dipl. Ing (FH) for his excellent MALDI–MS analyses.

9.10   References

ADLERBERTH I (1997), 'The establishment of a normal intestinal microflora in the newborn infant' in L A Hanson and R B Yolken (Eds): *Probiotics, Other Nutritional Factors, and Intestinal Flora*, 42nd Nestlé Nutrition Workshop, New York, Raven Press, 63–78.

ALLES M S, KATAN M B, SALEMANS J M J I, LAERE VAN K M J, GERICHHAUSEN M J W, ROZENDAAL M J and NAGENGAST F M (1997), 'Bacterial fermentation of fructooligosaccharides and resistant starch in patients with an ileal pouch-anal anastomosis', *Am J Clin Nutr*, **66**, 1286–1292

ANDERSSON B, PORRAS O, HANSON L A, LAGERGARD T and SVANBORG-EDEN C (1986), 'Inhibition of attachment of *Streptococcus pneumoniae* and *Haemophilus influenzae* by human milk and receptor oligosaccharides', *J Infect Dis*, **153**, 232–237.

BABA M, SNOECK R, PAUWELS R and DECLERQ E (1988), 'Sulfated polysaccharides are potent and selective inhibitors of various enveloped viruses, including herpes simplex virus, cytomegalovirus, vesicular stomatitis virus, and human immunodeficiency virus', *Antimicrob Agents Chemother*, **32**, 1742–1745.

BAHR U, PFENNINGER A, KARAS M and STAHL B (1997), 'High sensitive analysis of neutral underivatised oligosaccharides by nanoelectrospray mass spectrometry', *Anal Chem*, **69**, 4530–4535.

BEACHEY E H (1981), 'Bacterial adherence: adhesin–receptor interactions mediating the attachment of bacteria to mucosal surface', *J Infect Dis*, **143**, 325–345

BOEHM G, LIDESTRI M, CASETTA P, JELINEK J, NEGRETTI F, STAHL B and MARINI A (2002), Supplementation of a bovine milk formula with an oligosaccharide mixture increases counts of faecal bifidobacteria in preterm infants', *Arch Dis Child*, **86**, F0–F4.

BÖTTCHER M F, NORDIN E K, SANDIN A, MIDTVEDT T and BJÖRKSTÉN B (2000), 'Microflora-associated characteristics in faeces from allergic and non-allergic infants', *Clin Exp Allergy*, **30**, 1590–1596.

BOUHNIK Y, FLOURIE B, D'AGAY-ABENSOUR L, POCHART P, GRAMET G, DURAND M and RAMBAUD J C (1997), 'Administration of transgalacto-oligosaccharides increases faecal bifidobacteria and modifies colonic fermentation metabolism in healthy humans', *J Nutr*, **127**, 444–448.

BOUHNIK Y, FLOURIE B, RIOTTOT M, BISETTI N, GAILING M F, GUIBERT A, BORNET F and RAMBAUD J C (1996), 'Effects of fructo-oligosaccharides ingestion on faecal bifidobacteria and selected metabolic indexes of colon carcinogenesis in healthy humans', *Nutr Cancer*, **26**, 21–29

BOUHNIK Y, VAHEDI K, ACHOUR L, ATTAR A, SALFATI J, POCHART P, MARTEAU P, FLOURIE B, BORNET F and RAMBAUD J C (1999), 'Short-chain fructo-oligosaccharide administration dose-dependently increases fecal bifidobacteria in healthy humans', *J Nutr*, **129**, 113–116.

BRASSART D, WOLTZ A, GOLLIARD M and NEESER J R (1991), '*In vitro* inhibition of adhesion of *Candida albicans* clinical isolates to human buccal epithelial cells by Fucα1→2Galβ-bearing complex carbohydrates', *Infect Immun*, **59**, 1605–1613.

BRENNAN M J, DAVID J L, KENIMER J G and MANCLARK C R (1988), 'Lectin-like binding of pertussis toxin to a 165-kilodalton Chinese hamster ovary cell glycoprotein', *J Biol Chem*, **263**, 4895–4899.

BRITISH NUTRITION FOUNDATION (1990), *Complex Carbohydrates in Foods: Report of the British Nutrition´s Task Force*, London, Chapman and Hall.

BRUNTZ R, DABROWSKI U, DABROWSKI J, EBERSHOLD A, PETER-KATALINIC J and EGGE K (1988),' Fucose containing oligosaccharides from human milk from a donor of blood group 0 lea nonsecretor', *Biol Chem Hoppe-Seyler*, **369**, 257–273.

BUNOUT D, HIRSCH S, DE LA MAZA M P and MUNOZ C (2002), 'Effects of prebiotics on the immune response to vaccination in the elderly', *J Clin Nutr*, **75**, 419.

CAHAN L D and PAULSON J C (1980), 'Polyoma virus adsorbs to specific sialyloligo-saccharide receptors on erythrocytes', *Virology*, **103**, 505–509.

CARLSON S E and HOUSE S G (1986),' Oral and intraperitoneal administration of N-acetylneuraminic acid: effect on rat cerebral and cerebellar *N*-acetylneuraminic', *J Nutr*. **116**, 881–886.

CARNIELLI V P, LUIJENDIJK I H T, GOUDOEVER J B, SULKERS E J, BOERLAGE A A, DEGENHART H J and SAUER P J J (1996), 'Structural position and amount of palmitic acid in infant formulas: effects on fat, fatty acid, and mineral balance', *J Pediatr Gastroenterol Nutr*, **23**, 553–560.

CHATURVEDI P and SHARMA C B (1988), 'Goat milk oligosaccharides: purification and characterization by HPLC and high-field ^{1}H-NMR spectroscopy', *Biochim Biophys Acta*, **967**, 115–121.

CHATURVEDI P and SHARMA C B (1990), 'Purification, by high-performance liquid chroma-tography, and characterisation, by high-field [1]H-n.m.r. spectroscopy, of two fucose-containing pentasaccharides of goat's milk', *Carbohydr Res*, **203**, 91–101.

CHATURVEDI P, WARREN C D, ALTAYE M, MORROW A L, RUIZ-PALACIOS G, PICKERING L K and NEWBURG D S (2001), 'Fucosylated human milk oligosaccharides vary between individu-als and over the course of lactation', *Glycobiology*, **11**, 365–372.

CHO S S, PROSKY L and DREHER M (Eds) (1999), *Complex Carbohydrates in Foods*, Marcel Dekker.

CHONAN O and WATANUKI M (1996), 'The effect of 6'-galactooligosaccharides on bone mineralisation of rats adapted to different levels of dietary calcium', *Int J Vitam Nutr Res*, **66**, 244–249.

COPPA G V, GABRIELLI O, GIORGI P, CATASSI C, MONTANARI M P, VARALDO P E and NICHOLS B L (1990), 'Preliminary study of breastfeeding and bacterial adhesion to uroepithelial cells', *Lancet*, **335**, 569–571.

COPPA G V, PIERANI P, ZAMPINI L, CARLONI I, CARLUCCI A and GABRIELLI O (1999), 'Oligosaccharides in human milk during different phases of lactation', *Acta Paediatr*, **430**, 89–94.

COUDRAY C, BELLANGER J, CASTIGLIA-DELAVAUD C, RÉMÉSY C, VERMOREL M and RAYSSIGUIER Y (1997), 'Effect of soluble or partly soluble dietary supplementation on absorption and balance of calcium, magnesium, iron and zinc in healthy young men', *Eur J Clin Nutr*, **51**, 375–80.

CRAVIOTO A, TELLO A, VILLAFAN H, RUIZ J, DE VEDOVO S and NEESER J R (1991), 'Inhibition of localised adhesion of enteropathogenic *Escherichia coli* to HEP-2 cells by immunoglobulin and oligosaccharide fractions of human colostrum and breast milk', *J Infect Dis*, **163**, 1247–1255.

CRITTENDEN R G and PLAYNE M J (1996), 'Production, properties and applications of food-grade oligosaccharides', *Trends in Food Science and Technology*, **7**, 353–361.

CUNDELL D R and TUOMANEN E I (1994), 'Receptor specificity of adherence of *Streptococcus pneumoniae* to human type-II pneumocytes and vascular endothelial cells *in vitro*', *Microbiol Pathogenesis*, **17**, 361–374.

DAI D, NANTHKUMAR N N, NEWBURG D S and WALKER W A (2000), 'Role of oligosac-charides and glycoconjugates in intestinal host defense', *J Pediatr Gastroenterol Nutr*, **30**, 23–33.

DEGUCHI Y, MATSUMOTO K, ITO A and WATANUKI M (1997), 'Effects of beta 1–4 galactooligosaccharides administration on defecation of healthy volunteers with constipation tendency', *Jpn J Nutr*, **55**, 13–22

DE SLEGTE J, BAKX E, BASTIAANS H J A P, BENET S, BRUNT K, GRENIER-LUSTALOT M F, HEUTINK M A T, HISCHENHUBER C, KARPPINEN S, MENGERINK G W, MÜLLER-WERNER B, NIEMINEN A, NIJBOER F N, PIRO R, PRANDIN L, ROHRER J S, SAARINEN K, SANDERS P, SANGIORGI E, SCHOLSH A, SCHOTERMAN H C, SIPONEN S, STAHL B, STÖBER P, THOMAS D H and VISCOGLIOSI H (2002), 'Determination of trans-galactooligosaccharides in selected food products by ion exchange chromatography: collaborative study', *J AOAC Intern*, **85**(2), 417–423.

DE VUYST L and DEGEEST B (1999), 'Heteropolysaccharides from lactic acid bacteria', *FEMS Microbial Rev*, **23**, 153–177.

DELL A (1987), 'Advances in carbohydrate chemistry and biochemistry', *FAB Mass Spectrometry of Carbohydrates*, **43**, 36.

DOKKUM VAN W (1995), 'Tolerance for galacto-oligosaccharides in humans', *Borculo Domo Ingredients: Studies on Elix'or*.

DOMBO M, YAMAMOTO H and NAKAJIMA H (1997), 'Production, health benefits and applications of galacto-oligosaccharides', *New Technologies For Healthy Foods &Nutraceuticals*, 143–156.

DONALD A S R and FEENEY J (1988), 'Separation of human milk oligosaccharides by recycling chromatography. First isolation of lacto-*N*-neo-difucohexaose II and 3´-galactosyllactose from this source', *Carbohydr Res*, **178**, 79–91.

DUA V K, GOSO K, DUBE V E and BUSH C A (1985), 'Characterisation of lacto-*N*-hexaose and two fucosylated derivatives from human milk by high-performance liquid chromatography and proton NMR spectroscopy', *J Chromatogr*, **328**, 259–269.

DUGUID J P and GILLIES R R (1957), 'Fimbriae and adhesive properties of dysentery bacilli', *J Pathol Bacteriol*, **74**, 397–411.

EGGE H and PETER-KATALINIC J (1987), 'Fast atom bombardment mass spectrometry for structural elucidation of glycoconjugates', *Mass Spectrom Rev*, **6**, 331–394.

EGGE H, PETER-KATALINIC J, KARAS M and STAHL B (1991), 'The use of fast atom bombardment and laser desorption mass spectrometry in the analysis of complex carbohydrates', *Pure Appl Chem*, **63**, 491–498.

EL RASSI Z (Ed.) (1995), *Carbohydrate Analyses: High Performance Liquid Chromatography and Capillary Electrophoresis, J. Chrom Library* no. 58, Amsterdam, Elsevier.

EMÖDY L, CARLSSON A, LJUNGH A and WADSTRÖM T (1988), 'Mannose-resistant haemagglutination by *Campylobacter pylori*', *Scand J Infect Dis*, **20**, 353–354.

ENGFER M B, STAHL B, FINKE B, SAWATZKI G and DANIEL H (2000), 'Human milk oligosaccharides are resistant to enzymatic hydrolysis in the upper gastrointestinal tract' *Am. J. Clin. Nutr*, **71**, 1589–1596.

ERNST B, HART G W and SINAY P (Eds) (2000), '*Carbohydrates in Chemistry and Biology*' Wiley, Weinheim.

EVANS D G, EVANS D J, MOULDS J J and GRAHAM D Y (1988), *N*-Acetylneuraminyllactose-binding fibrillar hemagglutination of *Campylobacter pylori*: a putative colonisation factor antigen', *Infect Immun*, **56**, 2896–2906.

FIÈVRE S, WIERUSZESKI J M, MICHALSKI J C, LEMOINE J, MONTREUIL J and STRECKER G (1991), 'Primary structure of a trisialylated oligosaccharide from human milk', *Biochem Biophys Res Commun*, **177**, 720–725.

FINKE B (2000), 'Isolierung und Charakterisierung von Oligosacchariden aus humanen und tierischen Milchen', PhD thesis, Univ. Giessen, Shaker Verlag Aachen, Germany.

FINKE B, STAHL B, PFENNINGER F, KARAS M, DANIEL H and SAWATZKI G (1999), 'Analysis of high molecular weight oligosaccharides from human milk by liquid chromatography and MALDI-MS', *Anal Chem*, **71**, 3755–3762.

FINKE B, STAHL B, PRITSCHET M, FACIUS D, WOLFGANG J and BOEHM G (2002), 'Preparative continuous annular chromatography (P-CAC) enable the large scale fractionation of fructans', *J Agric Food Chem*, **50**, 4743–4748.

FRANK A (1998), 'Prebiotics stimulate calcium absorption: a review', *Milchwissenschaft*, **53**, 427–429.

GIBSON G R (1999), 'Dietary modulation of the human gut microflora using the prebiotics oligofructose and inulin', *J Nutr*, **129**, 1438S–1441S.

GIBSON G R and ROBERFROID M B (1995), 'Dietary modulation of the human colonic micro-biota: introducing the concept of prebiotics', *J Nutr*, **125**, 1401–1412.

GIBSON G R, BEATTY E R, WANG X and CUMMINGS J H (1995), 'Selective stimulation of bifidobacteria in the human colon by oligofructose and inulin', *Gastroenterology*, **108**, 975–982.

GINSBURG V, ZOPF D A, YAMASHITA K and KOBATA A (1976), 'Oligosaccharides of human milk. Isolation of a new pentasaccharide, lacto-*N*-fucopentaose V', *Arch Biochem Biophys*, **175**, 565–568.

GNOTH M J., KUNZ C, KINNE-SAFFRAN E and RUDLOFF S (2000), 'Human milk oligosaccharides are minimally digested *in vitro*', *J Nutr*, **130**, 3014–3020.

GRIMMONPREZ L and MONTREUIL J (1968), 'Etude physico-chemique de six nouveaux oligosaccharides isolés du lait de femme', *Bull Soc Chim Biol*, **50**, 843–855.

GRÖNBERG G, LIPNIUNAS P, LUNDGREN T, ERLANSSON K, LINDH F and NILSSON B (1989), 'Isolation of monosialylated oligosaccharides from human milk and structural analysis of three new compounds', *Carbohydr Res*, **191**, 261–278.

GRÖNBERG G, LIPNIUNAS P, LUNDGREN T, LINDH F and NILSSON B (1992), 'Structural analysis of five new monosialylated oligosaccharides from human milk', *Arch Biochem Biophys*, **296**, 597–610.

GUERARDEL Y, MORELLE W, PLANCKE Y, LEMOINE J and STRECKER G (1999), 'Structural analysis of three sulfated oligosaccharides isolated from human milk', *Carbohydr Res*, **320**, 230–238.

GUESRY P R, BODANSKI H, TOMSIT E and AESCHLIMANN J M (2000), 'Effect of 3 doses of fructo-oligosaccharides in infants', *J Pediatr Gastroenterol Nutr*, **31**, S252.

GUGGENBICHLER J P, DE BETTIGNIES-DUTZ A, MEISSNER P, SCHELLMOSER S and JURENITSCH

J (1997), 'Acidic oligosaccharides from natural sources block adherence of *Escherichia coli* on uroepithelial cells', *Pharm Pharmacol Lett*, **7**, 35–38.

GYÖRGY P, NORRIS R F and ROSE C S (1954), 'A variant of *Lactobacillus bifidus* requiring a special growth factor', *Arch Biochem Biophys*, **48**, 193–201.

HAATAJA S, TIKKANEN K, LIUKKONEN J, FRANCOIS-GERARD C and FINNE J (1993), 'Characterisation of a novel bacterial adhesion specificity of *Streptococcus suis* recognising blood group P receptor oligosaccharides', *J Biol Chem*, **268**, 4311–4317.

HAEUW-FIEVRE S, WIERUSZESKI J M, PLANCKE Y, MICHALSKI J C and MONTREUIL J (1993), 'Primary structure of human milk octa-, dodeca- and tridecasaccharides determined by a combination of ^{1}H-NMR spectroscopy and fast-atom-bombardment mass spectrometry', *Eur J Biochem*, **215**, 361–371.

HARVEY D (1999), 'Matrix assisted laser desorption/ionisation mass spectrometry of carbohydrates', *Mass Spectrom Rev*, **18**, 349–451.

HAW J F (1992), 'Nuclear-magnetic-resonance spectroscopy', *Anal Chem*, **64**, 243R–254R.

HEUVEL VAN DEN E G H M, MUYS T, DOKKUM VAN W and SCHAAFSMA G (1999), 'Oligofructose stimulates calcium absorption in adolescents', *Am J Clin Nutr*, **69**, 544–548.

HEUVEL VAN DEN E G J M, SCHAAFSMA G, MUYS T and DOKKUM VAN W (1998), 'Nondigestible oligosaccharides do not interfere with calcium and nonheme-iron absorption in young, healthy men', *Am J Clin Nutr*, **67**, 445–451.

HEUVEL VAN DEN E G J M, SCHOTERMAN M H C and MUIJS T (2000), Transgalactooligosaccharides stimulate calcium absorption in postmenopausal women', *J Nutr*, **130**, 2938–2942.

HIRMO S, KELM S, SCHAUER R, NILSSON B and WADSTRÖM T (1996), 'Adhesion of *Helicobacter pylori* strains to α-2,3-linked sialic acids', *Glycoconjugate J*, **13**, 1005–1011.

HOEBREGS H, BALIS P, DEVRIES J, V EEKELEN J, FARNELL, P, GRAY K, GOEDHUYS B, HERMANS M, HEROFF J, VAN LEUWEN M, LI B W, MARTIN D, PIETERS M, QUEMENER B, ROOMANS H, SLAGHEK T, THIBAULT J F, VAN DER WAAL, W and DEWITT D (1997), 'Fructans in foods and food products, ion-exchange chromatographic method: collaborative study', *J AOAC Intern*, **80**, 1029–1037.

HOLMGREN J, SVENNERHOLM A M and LINDBLAD M (1983), 'Receptor-like glycocompounds in human milk that inhibit classical and El Tor *Vibrio cholerae* cell adherence (hemagglutination)', *Infect Immun*, **40**, 563–569.

IDOTA T, KAWAKAMI H, MURAKAMI Y and SUGAWARA M (1995), 'Inhibition of cholera toxin by human milk fractions and sialyllactose', *Biosc Biotech Biochem*, **59**, 417–419.

ITO M, DEGUCHI Y, MIYAMORI A, MATSUMOTO K, KIKUCHI H, MATSUMOTO K, KOBAYASHI Y, YAJIMA T and KAN T (1990), 'Effects of administration of galactooligosaccharides on the human faecal microflora, stool weight and abdominal sensation', *Microb Ecol Health Dis*, **3**, 285–292.

ITO M, DEGUCHI Y, MATSUMOTO K, KIMURA M, ONODERA N and YAJIMA T (1993), 'Influence of galactooligosaccharides on the human fecal microflora', *J Nutr Sci Vitaminol*, **39**, 635–640.

ITO M, KIMURA M, DEGUCHI Y, MIYAMORI-WATABE A, YAJIMA T and KAN T (1993), Effects of transgalactosylated disaccharides on the human intestinal microflora and their metabolism', *J Nutr Sci Vitaminol*, **39**, 279–88.

IUB-IUPAC JOINT COMMISSION ON BIOCHEMICAL NOMENCLATURE (JCBN) (1980), 'Abbreviated terminology of oligosaccharide chains', Recommendations, *J Biol Chem*, **257**, 334.

IUPAC/IUBMB (1997), Joint Commission on Biochemical Nomenclature (JCBN) Nomenclature of Carbohydrates, Recommendations 1996 *Carbohydr Res*, **297**(12), 9.

JACKSON P (1994), In: *Methods Enzymol.*, W J Lennarz and G W Hart (Eds), *Acad. Press*, San Diego, **230**, 250–264.

JENKINS D J A, KENDALL C W C and VUKSAN V (1999), 'Inulin, oligofructose and intestinal function', *J Nutr*, **129**,1431S–1433S.

JENSEN R G (1995), *Handbook of Milk Composition*, San Diego, Academic Press.

KALLIOMÄKI M, SALMINEN S, ARVILOMMI H, KERO P, KOSKINEN P and ISOLAURI E (2001), 'Probiotics in primary prevention of atopic diseases: a randomised placebo-controlled trial', *Lancet*, **357**, 1067–1079.

KARAS M, BAHR U and DULCKS T (2000), 'Nano-electrospray ionisation mass spectrometry: addressing analytical problems beyond routine', *Fresenius J Anal Chem*, **366**, 669–676.

KINSELLA J E and TAYLOR S L (Eds) (1995), 'Commercial β-galactosidase' *in Food Processing Advances in Food and Nutrition Research*, San Diego, Academic Press.

KITAGAWA H, NAKADA H, NUMATA Y, KUROSAKA A, FUKUI S, FUNAKOSHI I, KAWASAKI T, SHIMADA I, INAGAKI F and YAMASHINA I (1990), 'Occurrence of tetra- and pentasaccharides with the sialyl-Le^a structure in human milk', *J Biol Chem*, **265**, 4859–4862.

KITAGAWA H, NAKADA H, FUKUI S, FUNAKOSHI I, KAWASAKI T and YAMASHINA I (1991), 'Novel oligosaccharides with the sialyl-Le^a structure in human milk', *Biochem*, **30**, 2869–2876.

KITAGAWA H, NAKADA H, FUKUI S, FUNAKOSHI I, KAWASAKI T, YAMASHINA I, TATE S I and INAGAKI F (1993), 'Novel oligosaccharides with the sialyl-Le^a structure in human milk', *J Biochem*, **114**, 504–508.

KLEESSEN B, SYKURA B, ZUNFT H J and BLAUT M (1997), 'Effects of inulin and lactose on faecal microflora, microbial activity, and bowel habit in elderly constipated persons', *Am J Clin Nutr*, **65**, 1397–1402.

KOBATA A and GINSBURG V (1969), 'Oligosaccharides of human milk. Isolation and characterisation of a new pentasaccharide, lacto-*N*-fucopentaose III', *J Biol Chem*, **244**, 5496–5502.

KOBATA A and GINSBURG V (1972a), 'Oligosaccharides of human milk. Isolation and characterisation of a new hexasaccharide, Lacto-*N*-hexaose', *J Biol Chem*, **247**, 1525–1529.

KOBATA A and GINSBURG V (1972b), 'Oligosaccharides of human milk. Isolation and characterisation of a new hexasaccharide, Lacto-*N*-neohexaose', *Arch Biochem Biophys*, **150**, 273–281.

KUHN R (1959), 'Biochemie der Rezeptoren und Resistenzfaktoren. Von der Widerstandsfähigkeit der Lebewesen gegen Einwirkung der Umwelt', *Naturwissenschaften*, **46**, 43–50.

KUHN R and BAER H H (1956a), 'Die Konstitution der Lacto-*N*-tetraose', *Chem Ber*, **89**, 504–511.

KUHN R and BROSSMER R (1956), 'Über *O*-Acetyl-lactaminsäure-lactose aus Kuh-Colostrum und ihre Spaltbarkeit durch Influenza-Virus', *Chem Ber*, **89**, 2013–2025.

KUHN R and BROSSMER R (1959), 'Über das durch Viren der Influenza-Gruppe spalt-bare Trisaccharid der Milch', *Chem Ber*, **92**, 1667–1671.

KUHN R GAUHE A (1958); 'Über die Lacto-difuco-tetraose der Frauenmilch', *Justus Liebigs Ann Chem*, **611**, 249–253.

KUHN R and GAUHE A (1960), 'Über ein kristallisiertes, Le^a-aktives Hexasaccharid aus Frauenmilch', *Chem Ber*, **93**, 647–651.

KUHN R and GAUHE A (1962), 'Die Konstitution der Lacto-*N*-neotetraose', *Chem Ber*, **95**, 518–522.

KUHN R and GAUHE A (1965), 'Bestimmung der Bindungsstellen von Sialinsäure-resten in Oligosacchariden mit Hilfe von Perjodat', *Chem Ber*, **98**, 395–413.

KUHN R, BAER H H and GAUHE A (1956), 'Kristallisation und Konstitutionsermittlung der Lacto-*N*-Fucopentaose I', *Chem Ber*, **89**, 2514–2523.

KUHN R, BAER H H and GAUHE A (1958), 'Die Konstitution der Lacto-*N*-fucopen-taose II', *Chem Ber*, **91**, 364.

KUNZ C, RUDLOFF S, SCHAD W and BRAUN D (1999), 'Lactose-derived oligosaccharides in milk of the elephants–comparison to human milk', *Brit J Nutr*, **82**, 391–399.

KUNZ C, RUDLOFF S, BAIER W, KLEIN N and STROBEL S (2000), 'Oligosaccharides in human milk: structural, functional and metabolic aspects', *Ann Rev Nutr*, **20**, 699–722.

KUNZ C, RUDLOFF S and MICHAELSEN K F (2002), 'Concentration of human milk oligosaccharides during late lactation', *FASEB J*, **15**, A986.

KUO C, TAKAHASHI N, SWANSON A F, OZEKI Y and HAKOMORI S (1996), 'An *N*-linked high-mannose type oligosaccharide, expressed at the outer membrane protein of *Chlamydia trachomatis*, mediates attachment and infectivity of the microorganism to HeLa cells', *J Clin Invest*, **98**, 2813–2818.

LEE Y C (1990), 'High-performance anion-exchange chromatography for carbohydrate analysis', *Anal Biochem*, **189**, 151–162.

LEE Y C (1996), 'Carbohydrate analyses with high-performance anion-exchange chromatography', *J Chromatogr*, **720**, 137–149.

LENNARZ W J and HART G W (Eds) (1994), *Methods in Enzymology: Techniques in Glycobiology*, 230, San Diego, Academic Press.

LIDESTRI M, AGOSTI M, MARINI A and BOEHM G (2002), 'Prebiotics might stimulate calcium absorption in preterm infants', *Acta Paediatr* (suppl.); in press.

LIO S, LING Y and TSAI C E (1994), 'Biotechnically synthesised oligosaccharides and polydextrose reduce constipation and putrefactive metabolites in the human', *J Clin Nutr Soc*, **19**, 221–232.

LIUKKONEN J, HAATAJA S, TIKKANEN K, KELM S and FINNE J (1992), 'Identification of *N*-acetylneuraminyl α-2,3 poly-N-acetyllactosamine glycans as the receptors of sialic acid binding *Streptococcus suis* strains', *J Biol Chem*, **267**, 21105–21111.

LOO J V, CUMMINGS J, DELZENNE N, ENGLYST H, FRANK A, HOPKINS M, KOK N, MACFARLANE G, NEWTON D, QUINGLEY M, ROBERFROID M, VLIET T and HEUVEL E (1999), 'Functional food properties of non-digestible oligosaccharides: a consensus report from the ENDO project (DGXII AIRII-CT94-1095)', *B J Nutr*, **81**, 121–132.

LOVELESS R W and FEIZI G (1989), 'Sialo-oligosaccharide receptor for *Mycoplasma pneumoniae* and related oligosaccharides of poly-*N*-acetyllactosamine series are polarised at the cilia and apical-microvillar domains of the ciliated cells in human bronchial epithelium', *Infect Immun*, **57**, 1285–1289.

MARTIN M J, MARTIN-SOSA S, GARCIA-PARDO L A and HUESO P (2001), 'Distribution of bovine milk sialoglycoconjugatesduring lactation' *J Diary Sci*, **84**, 995–1000.

MENNE E, GUGGENBUHL N and ROBERFROID M (2000), 'Fn-type chicory inulin hydrolysate has a prebiotic effect in humans', *J Nutr*, **130**, 1197–1199.

MEYER P D, TUNGLAND B C, CAUSEY J L and SLAVIN J L (2001), '*In-vitro* und *in-vivo* – Effekte von Inulin auf das Immunsystem', *Ernährungs-Umschau*, **48**, 13–16.

MILLAR M R, LINTON C J, CADE A, GLANCY D, HALL M and JALAL H (1996), 'Application of 16S rRNA gene PCR to study bowel flora of preterm infants with and without necrotising enterocolitis', *J Clin Microbiol*, **34**, 2506–2510.

MILLER-PODRAZA H, MILH M A, BERGSTRÖM J and KARLSSON K A (1996), 'Recognition of glycoconjugates by *Helicobacter pylori*: an apparently high-affinity binding of human polyglycosylceramides, a second sialic acid-based specificity', *Glycoconjugate J*, **13**, 453–460.

MITSUOKA T, HIDAKA H and EIDA T (1987), 'Effect of fructo-oligosaccharides on intestinal microflora', *Die Nahrung*, **31**, 427–36.

MOCK K K, DAVEY M and COTTRELL J S (1991), 'The analysis of underivatised oligosaccharides by matrix-assisted laser desorption mass spectrometry', *Biochem Biophys Res Commun*, **177**, 644–651.

MONNIG C A and KENNEDY R T (1994), 'Capillary electrophoresis', *Anal Chem*, **66**, 280R–314R.

MONTREUIL J (1956), 'Structure de deux triholosides isolés du lait de femme', *C R Acad Sci*, **242**, 192–193.

MORO E (1908), 'Karottensuppe bei Ernährungsstörungen der Säuglinge', *Münchener Medizinische Wochenzeitschrift*, **31**, 1637–1640.

MORO G, MINOLI I, MOSCA M, FANARO S, JELINEK J, STAHL B and BOEHM G (2002), 'Dosage-related bifidogenic effects of galacto- and fructooligosaccharides in formula-fed term infants', *J Pediatr Gastroenterol Nutr*, **34**, 291–295.

MOURICOUT M, PETIT J M, CARIAS J R and JULIEN R (1990),'Glycoprotein glycans that inhibit adhesion of *Escherichia coli* mediated by K99 fimbriae : treatment of experimental colibacillosis', *Infect Immun*, **58**, 98–106.

MURRAY P A, LEVINE M J, TABAK L A and REDDY M S (1982), 'Specificity of salivary–bacterial interactions: II. Evidence for a lectin on *Streptococcus sanguis* with specificity for a Neu5Acα2,3Galß1,3GalNAc sequence', *Biochem Biophys Res Comm*, **106**, 390–396.

NAKAKUKI T (Ed) (1994), *'Oligosaccharides Production, Properties and Applications'*, Switzerland, Gordon and Breach.

NAKAMURA T, URASHIMA T, NAKAGAWA M and SAITO T (1998), 'Sialyllactose occurs as free lactones in ovine colostrum', *Biochim Biophys Acta*, **1381**, 286–292.

NAUGHTON P J, MIKKELSEN L L and JENSEN B B (2001), 'Effects of non-digestible oligosaccharides on *Salmonella enterica* serovar *Typhimurium* and non-pathogenic *Escherichia coli* in the pig small intestine *in vitro*', *Appl Environ Microbio*, **67**(8), 3391–3395.

NEWBURG D S (1999), 'Human milk oligosaccharides that inhibit pathogens' *Curr Med Chem*, **6**, 117–127.

NEWBURG D S and NEUBAUER S H (1995) 'Carbohydrates in milk'. In: *Handbook of Milk Composition*, San Diego, Academic Press.

NÖHLE U and SCHAUER R (1981),' Uptake metabolism and excretion of orally and intravenously administered, ^{14}C- and ^{3}H-labelled *N*-acetylneuraminic acid mixture in the mouse and rat', *Hoppe Seylers Z Physiol Chem*, **362**, 1495–14506.

OHTA A, OHTSUKI M, HOSONO A, ADACHI T, HARA H and SAKATA T (1998), 'Dietary fructooligosaccharides prevent osteopenia after gastrectomy in rats', *J Nutr*, **128**, 106–110.

OHTSUKA K, BENNO Y, ENDO K, UEDA H, OZAMU O, UCHIDA T and MITSUOKA T (1989), 'Effects of 4´galactosyllactose intake on human fecal microflora', *Bifidus*, **2**, 143–49.

OUADIA A, KARAMANOS Y and JULIEN R (1992), 'Detection of the ganglioside *N*-glycolyl-neuraminyl-lactosyl-ceramide by biotinylated *Escherichia coli* K99 lectin', *Glycoconjugate J*, **9**, 21–26.

OUWEHAND A C, ISOLAURI E, HE F, HASHIMOTO H, BENNO Y and SALMINEN S (2001), 'Differences in bifidobacterium flora composition in allergic and healthy infants', *J Allergy Clin Immunol*, **108**, 144–145.

PARKINNEN J and FINNE J (1987), 'Isolation of sialyl oligosaccharides and sialyl oligosaccharides phosphates from bovine colostrum and human urine', *Methods Enzymol*, **138**, 289–300.

PARKINNEN J, FINNE J, ACHTMAN M, VÄISÄNEN V and KORHONEN T K (1983), '*Escherichia coli* strains binding neuraminyl α2-3 galactosides', *Biochem Biophys Research Comm*, **111**, 456–461.

PFENNINGER A, KARAS M, FINKE B and STAHL B (2002a), 'Structural analysis of underivatised human milk oligosaccharides in the negative ion mode by nano-electrospray MS^n', Part I Methodology', *J Am. Mass Spectrom*, 2002, **13**, 1331–1340.

PFENNINGER A, KARAS M, FINKE B and STAHL B (2002b), 'Structural analysis of underivatised human milk oligosaccharides in the negative ion mode by nano-electrospray MS^n', Part II Application to isomeric mixtures', *J Am. Mass Spectrom*, 2002, **13**, 1341–1348.

PFENNINGER A, CHAN S Y, KARAS M, FINKE B, STAHL B and COSTELLO C E (2002c), 'Mass spectral analysis of compositions for neutral, high molecular weight oligosaccharides from human milk', *Glycobiology*; sent for publication.

POLONOVSKI M and LESPAGNOL A (1933), 'Nouvelles acquisitions sur les composes glucidiques du lait de femme', *Bull Soc Chim Biol*, **15**, 320–349.

RAMPHAL R, CARNOY C, FIEVRE S, MICHALSKI J C, HOUDRET N, LAMBLIN G, STRECKER G and ROUSSEL P (1991), '*Pseudomonas aeruginosa* recognises carbohydrate chains containing type 1 (Galβ1-3GlcNAc) or type 2 (Galβ1-4GlcNAc) disaccharide units', *Infect Immun*, **59**, 700–704.

RIVERO-URGELL, M and SANTAMARIA-ORLEANS A (2001), 'Oligosaccharides: application in infant food', *Early Human Development*, **65** (Suppl), S43–52.

ROBERFROID M B and DELZENNE N M (1998), 'Dietary fructans', *Ann Rev Nutr*, **18**, 117–143.

ROBERFROID M and SLAVIN J (2000), 'Nondigestible oligosaccharides', *Crit Rev Food Sci Nutr*, **40**, 461–480.

ROGERS G N, HERRLER G, PAULSON J C and KLENK H D (1986), 'Influenza C virus uses 9-O-acetyl-*N*-acetylneuraminic acid as high affinity receptor determinant for attach-ment to cells, *J Biol Chem*, **261**, 5947–5951.

RUDLOFF S, STEFAN C, POHLENTZ G and KUNZ C (2002), 'Detection of ligands for selectins in the oligosaccharide fraction of human milk', *Eur. J Nutr*, **41**, 85–92.

SAAVEDRA J, TSCHERNIA, MOORE N, ABI-HANNA A, COLETTA F, EMENHISER C and YOLKEN R (1999), 'Gastro-intestinal function in infants consuming a weaning food supplemented with oligofructose, a prebiotic', *J Pediatr Gastroenterol Nutr*, **29**, 95.

SABHARWAL H, NILSSON B, GRÖNBERG G, CHESTER M A, DAKOUR J, SJÖBLAD S and LUNDBLAD A (1988a), 'Oligosaccharides from faeces of preterm infants fed on breast milk', *Arch Biochem Biophys*, **265**, 390–406.

SABHARWAL H, NILSSON B, CHESTER M A, LINDH F, GRÖNBERG G, SJÖBLAD S and LUNDBLAD A (1988b), 'Oligosaccharides from faeces of a blood-group B, breast-fed infant', *Carbohydr Res*, **178**, 145–154.

SAITO T, ITOH T and ADACHI S (1984), 'Presence of two neutral disaccharides containing N-acetylhexosamine in bovine colostrum as free forms', *Biochim Biophys Acta*, **801**, 147–150.

SAITO T, ITOH T and ADACHI S (1987), 'Chemical structural of three neutral trisaccharides isolated in free forms from bovine colostrum', *Carbohydr Res*, **165**, 43–51.

SARNEY D B, HALE C, FRANKEL G and VULFSON E N (2000), 'A novel approach to the recovery of biological active oligosaccharides from milk using a combination of enzymatic treatment and nanofiltration', *Biotechnol Bioeng*, **69**, 461–467.

SCHAAFSMA G, MEULING W J A, DOKKUM W VAN and BOULEY C (1998), 'Effects of a milk product, fermented by Lactobacillus acidophilus and with fructo-oligosaccharides added, on blood lipids in male volunteers', *Eur J Clin Nutr*, **52**, 436–40.

SCHENKMAN S and EICHINGER D (1993), '*Trypanosoma cruzi* trans sialidase and cell invasion', *Parasitol Today*, **9**, 218–222.

SCHNEIR M L and RAFELSON M E (1966), 'Isolation of two structural isomers of *N*-acetylneuraminyllactose from bovine colostrum', *Biochim Biophys Acta*, **130**, 1–11.

SCHOLZ-AHRENS K, SCHAAFSMA G, VAN DEN HEUVEL E G M H and SCHREZENMEIR J (2001), 'Effect of prebiotics on mineral metabolism', *Am J Clin Nutr*, **73**, 459S–464S.

SEGAL E (1996), 'Inhibitors of *Candida albicans* adhesion to prevent candidiasis' In: 'Toward anti-adhesion therapy for microbial diseases', (I Kahane and I Ofek, Eds), *Advances in Experimental Medicine and Biology*, **408**, New York, Plenum.

SHERMA J (1994), 'Planar chromatography', *Anal Chem*, **66**, 67R–83R.

SMITH D F, PRIETO P A, MCCRUMB D K and WANG W C (1987), 'A novel sialylfucopentaose in human milk', *J Biol Chem*, **262**, 12040–12047.

SNOW BRAND MILK PRODUCTS CO., Technical Research Institute (1995), 'A new oligosac-charide; 4′-galactosyl-lactose in human milk', Annual Meeting at Hokkaido University.

SOUTHGATE D A T (1995), 'Dietary fiber analysis RSC Food analysis monographs' P S Belton (Ed), The Royal Society of Chemistry, Cambridge.

SPENGLER B, KIRSCH D and KAUFMANN R (1991), 'Metastable decay of peptides and proteins in matrix-assisted laser-desorption mass spectrometry', *Rapid Comm. Mass Spectrom*, **5**, 198–202.

SPILLER G A (2001), *CRC Handbook of Dietary Fiber in Human Nutrition*, Boca Raton, CRC Press.

STAHL B, STEUP M, KARAS M and HILLENKAMP F (1991), 'Analysis of neutral oligosaccharides by matrix assisted laser desorption/ionisation mass spectrometry' *Anal Chem*, **63**, 1463–1466.

STAHL B, THURL S, ZENG J, KARAS M, HILLENKAMP F, STEUP M and SAWATZKI G (1994), 'Oligosaccharides from human milk as revealed by matrix-assisted laser desorption/ionisation mass spectrometry', *Anal Biochem*, **223**, 218–226.

STONE A L, MELTON D J and LEWIS M S (1998), 'Structure function relations of heparin mimetic sulphated xylan oligosaccharied: inhibition of human immunodeficiency virus-1 infectivity *in vitro*', *Glycoconjugate J*, **15**, 697–712.

STRECKER G, FIÈVRE S, WIERUSZESKI J M, MICHALSKI J C and MONTREUIL J (1992), 'Primary structure of four human milk octa-, nona- and undeca-saccharides estab-lished by ^{1}H- and ^{13}C-nuclear magnetic resonance spectroscopy', *Carbohydr Res*, **226**, 1–14.

STRECKER G, WIERUSZESKI J M, MICHALSKI J C and MONTREUIL J (1988), 'Structure of a new nonasaccharide isolated from human milk: VI^{2}Fuc, V^{4}Fuc, III^{3}Fuc-p-Lacto-*N*-hexaose', *Glycoconjugate J*, **5**, 385–396.

STRÖMBERG N, MARKLUND B I, LUND B, ILVER D, HAMERS A, GAASTRA W, KARLSSON K A and NORMARK S (1990), 'Host-specificity of uropathogenic *Escherichia coli* depends on differences in binding specifity to Galα1-4Gal-containing isoreceptors', *EMBO J*, **9**, 2001–2010.

SVENSSON L (1992), 'Group-C rotavirus requires sialic acid for erythrocyte and cell receptor binding', *J Virol*, **66**, 5582–5585.

TACHIBANA Y, YAMASHITA K and KOBATA A (1978), 'Oligosaccharides of human milk: structural studies of di- and trifucosyl derivatives of Lacto-*N*-octaose and Lacto-*N*-neooctaose', *Arch Biochem Biophys*, **188**, 83–89.

TANAKA R and MATSUMOTO K (1998), 'Recent progress on prebiotics in Japan, including galactooligosaccharides', *Bull IDF*, **336**, 21–27.

TANAKA R, TAKAYAMA H, MOROTOMI M, KUROSHIMA T, UEYAMA S, MATSUMOTO K, KURODA and A MUTAI M (1983), 'Effects of administration of TOS and bifidobacterium breve 4006 on the human faecal flora', *Bifidobacteria Microflora*, **2**, 17–24.

TEURI U and KORPELA R (1998), 'Galacto-oligosaccharides relieve constipation in elderly people', *Ann Nutr Metab*, **42**, 319–27.

TEURI U I, KORPELA R, SAXELIN M, MONTONEN L and SALMINEN S J (1998), 'Increased faecal frequency and gastrointestinal symptoms following ingestion of galacto-oligosaccharide-containing yogurt', *J Nutr Sci Vitaminol*, **44**, 465–71.

TERRAZAS L I, WALSH K L, PISKORSKA D, MCGUIRE E and HARN D A JR (2001), 'The schistosome oligosaccharide lacto-N-neotetraose expands Gr1(+) cells that secrete anti-inflammatory cytokines and inhibit proliferation of naive CD4(+) cells: a potential mechanism for immune polarisation in helminth infections', *J Immunol*, *1*, **167**(9):5294–5303.

THURL S, HENKER J, SIEGEL M, TOVAR K and SAWATZKI G (1997), 'Detection of four human milk groups with respect to Lewis blood group dependent oligosaccharides. *Glycoconjugate J*, **14**, 795–799.

THURL S, MÜLLER-WERNER B and SAWATZKI G (1996), 'Quantification of individual oligosaccharide compounds from human milk by gel chromatography', *Anal Biochem*, **235**, 202–206.

TISSIER H C R (1899), 'La vécation chromophile d'Escherich et la Bacterium coli', *Soc Biol*, **51**, 943–945.

URASHIMA T, SAITO T, NAKAMURA T and MESSER M (2001), 'Oligosaccharides of milk and colostrum in non-human mammals' *Glycoconjugate J*, **18**, 357–371

URASHIMA T, MURATA S and NAKAMURA T (1997b), 'Structural determination of mono-sialyl trisaccharides obtained from caprine colostrum', *Comp Biochem Physiol*, **116**, 431–435.

URASHIMA T, NAKAMURA T and SAITO T (1997c), 'Biological significance of milk oligo-saccharides – homology and heterogeneity of milk oligosaccharides among mammalian species', *Milk Science*, **46**, 211–220.

Oligosaccharides 243

URASHIMA T, BUBB W A, MESSER M, TSUJI Y and TANEDA Y (1994), 'Studies of the neutral trisaccharides of goat (*Capra hircus*) colostrum and of the one- and two-dimensional ^{1}H and ^{13}C NMR spectra of 6'-N-acetylglucosaminyllactose', *Carbohydr Res*, **262**, 173–184.

URASHIMA T, SAITO T and KIMURA T (1991a), 'Chemical structures of three neutral oligosaccharides obtained from horse (Thoroughbred) colostrum', *Comp Biochem Physiol B*, **100**, 177–183.

URASHIMA T, SAITO T, OHMISYA K and SHIMAZAKI K (1991b), 'Structural determination of three neutral oligosaccharides in bovine (Holstein–Friesian) colostrum, including the novel trisaccharide; GalNAcα1-3Galß1-4Glc', *Biochim Biophys Acta*, **1073**, 225–229.

URASHIMA T, SAITO T, NISHIMURA J and ARIGA H (1989), 'New galactosyllactose containing α-glycosidid linkage isolated from ovine (*Booroola dorset*) colostrum', *Biochim Biophys Acta*, **992**, 375–378.

URASHIMA T, SAKAMOTO T, ARIGA H and SAITO T (1989b), 'Structure determination of three neutral oligosaccharides obtained from horse colostrum', *Carbohydr Res*, **194**, 280–287.

VARKI A, CUMMINGS R, ESKO J, FREEZE H, HART G and MATH J (Eds) (1999), Essentials of Glycobiology, New York, Cold Spring Harbor Laboratory Press.

VEH R W, MICHALSKI J C, CORFIELD A P, SANDER-WEWER M, GIES D and SCHAUER R (1981), 'New chromatographic system for the rapid analysis and preparation of colostrum sialyloligosaccharides', *J Chromatogr*, **212**, 313–322.

VELUPILLAI P and HARN D A (1994), 'Oligosaccharide-specific induction of interleukin 10 production by B220+ cells from schistosome-infected mice: a mechanism for regulation of CD4+ T-cell subsets', *Proc Natl Acad Sci*, **91**, 18–22.

VIVERGE D, GRIMMONPREZ L and SOLERE M (1997), 'Chemical characterisation of sialyl oligosaccharides isolated from goat (*Capra hircus*) milk', *Biochim Biophys Acta*, **1336**, 157–164.

VLIEGENTHART J F G, DORLAND L and VAN HALBEEK H (1983), 'High-resolution, ^{1}H-nuclear magnetic resonance spectroscopy as a tool in the structural analysis of carbohydrates related to glycoproteins', *Adv Carbohydr Chem Biochem*, **41**, 209–374.

WEIS W, BROWN J H, CUSACK S, PAULSON J C, SKEHEL J J and WILEY D C (1988), 'Structure of the influenza virus haemagglutinin complexed with its receptor, sialic acid', *Nature*, **333**, 426–431.

WIERUSZESKI J M, CHEKKOR A, BOUQUELOT S, MONTREUIL J, STRECKER G, PETER-KATALINIC J and EGGE H (1985), 'Structure of two new oligosaccharides isolated from human milk: Sialylated lacto-*N*-fucopentaoses I and II', *Carbohydr Res*, **137**, 127–138.

XANTHOU M (1998), 'Immune protection of human milk', *Biol Neonate*, **74**, 121–133.

YAHIRO M, NISHIKAWA I, MURAKAMI Y, YOSHIDA H and AHIKO K (1982), 'Studies on application of galactosyl lactose for infant formula II. Changes of faecal characteristics on infants fed galactosyl lactose', Reports of Research Laboratory, Snow Brand Milk Products Co. Ltd., No. 78, 27–32.

YAMASHITA K, TACHIBANA Y and KOBATA A (1977), 'Structural studies of two new octasaccharides, difucosyl derivatives of para-lacto-N-neohexaose', *J Biol Chem*, **252**, 5408–5411.

YAMASHITA K, TACHIBANA Y and KOBATA A (1976a), 'Oligosaccharides of human milk. Isolation and characterisation of three new disialylfucosyl hexasaccharides', *Arch Biochem Biophys*, **174**, 582–591.

YAMASHITA K, TACHIBANA Y and KOBATA A (1976b), 'Oligosaccharides of human milk. Isolation and characterisation of two new nonasaccharides, monofucosyllacto-*N*-octaose and monofucosyllacto-N-neooctaose', *Biochemistry*, **15**, 3950–3955.

YAMASHITA K and KOBATA A (1974), 'Oligosaccharides of human milk. Isolation and characterisation of a new trisaccharide, 6'-galactosyllactose', *Arch Biochem Biophys*, **161**, 164–170.

ZOPF D and ROTH S (1996), 'Oligosaccharide anti-infective agents', *Lancet*, **347**, 1017–1021.

10

Lactic acid bacteria (LAB) in functional dairy products

R. Fondén, Arla Foods ICS, Sweden, M. Saarela, J. Mättö and
T. Mattila-Sandholm, VTT Biotechnology, Finland

10.1 Introduction

When our ancestors developed the knowledge to use milk, they also learned how
to use lactic acid bacteria (LAB) in dairy products even though they did not know
about the existence of microorganisms. When milk was left at room temperature,
it was spontaneously fermented and was used as such or after dilution with water.
As an alternative it could be concentrated by drainage of whey, resulting in a new
product, which could be stored for several months: the first cheese. Owing to this
long period of consumption, such products were found to be safe and healthy
(Fleichman and Weigmann, 1932). By trial and error, the practice of inoculating
milk with a small amount of previously fermented milk was developed. Heating
the milk before inoculation increased the quality of the product and it became
possible to develop a range of specific, traditional fermented milk products (Table
10.1).

About two hundred years ago cheese and butter production became industrialised. Traditional products were standardised and slightly modified to adjust them
to industrial production. Starters were grown in manufacturing plants, but, as the
size of the operations grew, the importance of standardised starters increased. In
the late 19th century companies specialised in the starter production were founded.
In the beginning the starters were still undefined, containing an unknown mixture
of different strains of the suitable species. Later on, starters were analysed and
defined, allowing them to be composed of specific strains with properties selected
to give a standard product. In general, defined starters are composed of some
specific strains with less species variety than undefined starters, and do not usually
contain yeast.

Table 10.1 Fermented milks produced by traditional starters

Name	Origin	LAB and other microorganisms	Reference
Dickmilch	Northern Europe	*Lactococcus* spp., *Leuconostoc* spp., yeasts	Fleichman and Weigman (1932)
Kefir	Northern Caucasus	As Dickmilch + *Lactobacillus kefiri, Lactobacillus parakefiri, Lactobacillus kefirigranum*	Kurmann *et al.* (1992)
Kumiss	Asian Russia, Mongolia, Tibet	*Lactobacillus delbrueckii,* yeasts	Kurmann *et al.* (1992)
Skyr	Nordic countries	*Lactococcus* spp., *Lactobacillus casei,* yeasts	Orla-Jensen and Sadler (1940)
Tätmjölk	Nordic countries (northern parts)	*Lactococcus* spp. (including ropy variants)	Larsson (1988)
Yoghurt	South-eastern Europe, Asia	*L. delbrueckii* subsp. *bulgaricus, Staphylococcus thermophilus*	Burke (1938)

LAB ferment the lactose of milk into lactic acid and flavour compounds such as acetaldehyde, carbon dioxide and diacetyl (Marshall, 1986). At the low pH of fermented milk, lactic acid and flavour compounds inhibit the growth of most bacteria present in the unfermented milk, resulting in a safe and nutritious product. Fermented milks, especially yoghurt, are believed to be both safe and healthy. This idea originated from the publicity following the work of Metchnikoff (1907). He suggested that, because of the LAB yoghurt contained, it had a favourable impact on longevity due to reduced formation of poisonous products in the colon. Later research suggested that the traditional LAB present in yoghurt were not such an important part of the intestinal microbiota (Rettger, 1915). However, LAB in yoghurt may be of importance to lactose malabsorbers as they tolerate yoghurt better than milk (Kolars *et al.*, 1984; Marteau *et al.*, 1990). Yoghurt may even have an impact on the fermentation pattern in the colon by increasing production of short-chain fatty acids (Rizkalla *et al.*, 2000).

Following the original work of Metchnikoff, fermented milks with intestinal LAB were developed. In the beginning products with intestinal LAB *Lactobacillus acidophilus, Lactobacillus casei* and *Bifidobacterium bifidum* were used (Table 10.2). In the USA the first milk fermented by *L. acidophilus* was developed based on the work of Rettger (1915). This product was produced by inoculation of sterilised milk using *L. acidophilus* as the single strain responsible for the fermentation (Burke, 1938). In Japan a fermented milk drink, exclusively fermented with *L. casei* Shirota, was developed during 1930s. Later intestinal strains were used in combination with supporter cultures of LAB (Fleichman and Weigmann, 1932; Kurmann *et al.*, 1992). In addition to *L. acidophilus* and *L. casei* bifidobacteria were also introduced alone, or together with *L. acidophilus*, into the products. In all of these products the number of the intestinal LAB might be quite variable, from less than 10^6 up to 10^8 cfu (colony forming units)/g.

Table 10.2   First fermented milks with intestinal LAB

Product	Intestinal bacteria	First documented	Reference
Acidophilus	*L. acidophilus*	1922	Burke (1938)
Yakult	*L. casei* Shirota	1935	
Bifidus	*B. bifidum/B. longum*	1968	Schuhler-Mayloth *et al.* (1968)

10.2   Production of dairy products using LAB

LAB used in the production of modern fermented milks are mainly the same as used traditionally. Traditional LAB used in yoghurt are *Lactobacillus delbrueckii* subsp. *bulgaricus* and *Streptococcus thermophilus*. By reducing the numbers of lactobacilli and by selecting new strains, post-acidification occurring in the traditional yoghurts has been reduced and novel, less acidic products have been developed. In Scandinavian countries, in particular, milk is often fermented with mixed mesophilic starters *Lactococcus* spp. and *Leuconostoc* spp. instead of yoghurt starters (Kurmann *et al.*, 1992).

Production of fermented milk starts with normal milk or milk fortified with minerals, vitamins or milk solids and standardised to the required fat and protein content (Fig. 10.1). Milk is homogenised, heat-treated (usually at 90 °C for 5 min), and chilled to the fermentation temperature, which is 32–43 °C for yoghurt and around 20 °C for *filmjölk*, the modern product similar to *dickmilch*. Milk is then inoculated with starter culture so that the numbers of LAB are about 10^7 cfu/g. Most starters used today are added in concentrated and pellet-frozen form directly to the product milk at a level of roughly 0.01%. As an alternative, a bulk starter can be produced in the dairy plant. This is still the most common way to add starters during cheese production.

During fermentation, LAB ferment lactose and thereby lower the pH. In milk fermentation the pH is allowed to drop to 4.3–4.5. Fermentation time varies between 4 and 20 hours depending on the temperature used. Fermentation takes place either in the package or in the fermentation vat. After fermentation, milk is chilled and stored, resulting in a product with a shelf-life of between two and four weeks. The final level of LAB in the products is about 10^9 cfu/g. The process is optimised to give a product with as low whey separation as possible and with a limited post-acidification. More extensive information on the production of fermented milks is given by Cogan and Accolas (1996), Tamine and Robinson (1999) and IDF (1988, 1992).

Several different LAB are used in the production of cheese. In cheese both water activity and oxygen tension are low, which limits the growth of most bacteria. However, some microorganisms are still able to grow in cheese and influence its maturation. Examples are LAB combined with white and blue moulds, propionic acid bacteria and *Micrococcus* spp. Because of the fairly high

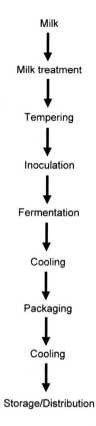

Milk

Milk treatment

Tempering

Inoculation

Fermentation

Cooling

Packaging

Cooling

Storage/Distribution

Fig. 10.1   Main process steps in the production of fermented milks.

process temperatures, LAB used to ferment lactose during cheese making are either mesophilic or thermophilic (such as the ones used in the production of filmjölk or yoghurt). Besides starter LAB, other LAB are usually involved in the maturation of the cheese. These include *L. casei, Lactobacillus curvatus, L. delbrueckii* subsp. *lactis, Lactobacillus helveticus, Lactobacillus paracasei, Lacto-bacillus rhamnosus, Lactobacillus sakei* subsp. *sakei, Lactobacillus salivarius, Lactobacillus xylosus* and *Lactobacillus zeae*. These LAB might be added but they more often originate either from raw milk or whey. Compared to the modern production of fermented milks, cheese production is less aseptic and thus it still resembles original production conditions (IDF, 2002).

The cheese-making process starts with the treatment of the cheese milk. The fat content is standardised by separation and the milk is heat-treated at 72 °C for 15 seconds (several traditional cheeses are produced from raw milk without any heat treatment). Milk is chilled to 30 °C and rennet and starter are added. After coagulation, coagulum is broken and cheese curd heated to temperatures between 40 and 50 °C depending on the type of cheese to be produced. Cheese curd and whey are separated, the curd is pressed, salt is added, and the cheese is matured at different time/temperature conditions. In the final cheese, LAB are present in the order of 10^6–10^8 cfu/g. These are usually other LAB than the ones responsible for the original lactose fermentation. More extensive information on the microbiology and production of cheese is available in Fox (1993) and Law (1999).

10.3   Dairy products with probiotic LAB

Most dairy products with probiotic LAB are fermented milks. Milk is fermented either only with probiotic LAB or with the assistance of a supporter culture (Saxelin et al., 1999; Svensson, 1999). The process steps are the same as in the production of standard fermented milks (Lourens-Hatting and Viljoen, 2001). In the production of fermented milk containing probiotics, the selection of probiotic strains and supporter cultures, if used, is very important. In addition to the health aspect of probiotic LAB, technological issues, such as suitability for growth and processing into a concentrated culture, growth and survival in the fermented milk, and impact on sensoric properties, have also to be considered (Saxelin et al., 1999; Mattila-Sandholm et al., 2002).

In the production of ordinary fermented milks, starters produce acid shortly after inoculation and thus inhibit the growth of those contaminating vegetative cells and spores surviving the heat treatment at 90 °C. Use of a supporter culture together with a probiotic culture results in a similar reduction in spores and contaminants. In contrast, when exclusively probiotic bacteria are used as starters, acidification can be slower and growth of contaminating bacteria has to be controlled by other means. Possible solutions, alone or in combination, are to increase the acidification rate by using growth-promoting substances, to add high numbers of starter bacteria, or to select probiotic bacteria stimulating the growth of each other (Fondén et al., 2000; Lourens-Hatting and Viljoen, 2001). A reduction of contaminating bacteria can also be achieved by using UHT-milk and an extended disinfection of the processing equipment. This procedure has been used successfully in the production of a fermented acidophilus milk. Yeast-extract is first added to the milk as a growth stimulant, then milk is UHT-treated and roughly 10^8 cfu/g L. acidophilus are added, resulting in a product with high levels of the probiotic bacteria and a shelf-life of two weeks (Fondén, 1989).

When the problem of contamination is solved by the use of a supporter culture in the production of probiotic products, the viability of various LAB cultures needs to be considered. The most important issue is that all the LAB used must be able to grow together without inhibiting each other (Fondén et al., 2000). As reviewed by Lourens-Hatting and Viljoen (2001), the survival of some probiotic bacteria is influenced by the ability of the supporter culture to eliminate oxygen and produce low post-acidification. The inoculation rate of the added LAB should, in combination with optimal incubation conditions, result in a proper level of probiotic LAB in the final product, which should be at least 10^6 cfu/g (Hawley *et al.*, 1959; Gilliland, 1989). Probiotic LAB should preferably be added to milk before fermentation since addition afterwards results in limited survival (Hull *et al.*, 1984). A suitable final pH is ensured both by using strains with low post-acidification and by using proper cooling conditions. Taking into account the above factors, it has been demonstrated that fermented milks with selected probiotic LAB can be stored for several weeks with a minimal loss of viability (Saxelin *et al.*, 1999).

Today, most available dairy products with probiotic bacteria are fermented milks such as yoghurt. However, other milk products such as ordinary 'sweet' milk, cheese, ice cream and milk powder can also be used as carriers of probiotic LAB. One of the main advantages of using non-fermented milk products as carriers of probiotics is the absence of fermentation end-products. Organic acids and flavour compounds have a negative impact on the survival of probiotic LAB. In the production of sweet milk, probiotic bacteria are added after the heat treatment of milk. The numbers of added probiotic LAB and the shelf-life of the sweet milk product are limited by the acid fermentation that occurs during cold storage. The maximum number of probiotic LAB that can be added to sweet milk depends on the storage temperature and time as well as on the properties of the strain used. Sweet acidophilus milk is an example of a sweet milk product with added probiotic *L. acidophilus* (Speck, 1975; Young and Nelson, 1978).

Ice cream is another suitable carrier for probiotic bacteria as the storage at a low temperature results in longer survival. Survival seems to depend on pH, since survival is better in ice cream mixes of neutral pH than in ice cream similar to frozen yoghurt. Products with at least 10^7 cfu/g and with a shelf life of up to eight months have been produced (Hekmat and McMahon, 1992; Hagen and Narvhus, 1999; Davidson *et al.*, 2000; Alampreese *et al.*, 2002).

Milk powders with probiotic bacteria have a long shelf-life compared with products with high water activity. It has been demonstrated that most probiotic bacteria can be produced as freeze-dried cultures with 10^{12} cfu/g with good storage stability even at +25 °C (Saxelin *et al.*, 1999). These cultures can be mixed with milk powder or similar ingredients to make dried food products with suitable concentrations of probiotic bacteria. As an alternative, a concentrated probiotic culture may be added into concentrated milk and spray-dried. However, numbers of viable probiotic bacteria are lower and the storage stability of spray-dried products is poorer than in freeze-dried products (Gardiner *et al.*, 2000).

Cheese provides LAB conditions that assist both bacterial growth and survival. Many of the LAB found in cheese belong to the same species as probiotic bacteria, including *L. casei*, *L. paracasei*, *L. rhamnosus* and *Lactobacillus plantarum* (Lindberg *et al.*, 1996; Gardiner *et al.*, 1998). An *L. paracasei* strain, *Lactobacillus* F19, originating from human GI microbiota and used as a probiotic bacterium, has even been isolated from cheese produced from milk not inoculated with this strain (Björneholm and Fondén, 2002). The suitability of cheese as a carrier of probiotic LAB has been shown both for mature cheeses and fresh cheeses (Gomes *et al.*, 1995; Gardiner *et al.*, 1998; Vinderola *et al.*, 2000).

10.4   The health benefits of probiotic LAB

Owing to their perceived health benefits, lactobacilli and bifidobacteria have been increasingly included in yoghurts and fermented milks during the past two decades (Daly and Davis, 1998). A major development in the functional food sector has emerged from the use of probiotic bacteria and prebiotic carbohydrates that enhance health-promoting microbiota in the intestine. There is growing evidence to support the concept that the maintenance of healthy gut microbiota may provide protection against various gastrointestinal disorders including gastrointestinal (GI) infections, inflammatory bowel diseases and even cancer (Haenel and Bendig, 1975; Mitsuoka, 1982; Salminen *et al.*, 1998).

The human large intestine has a rich and dynamic microbiota consisting of at least 400–500 bacterial species (Berg, 1996). Maintaining a balanced microbial ecosystem is essential for the normal functions of the GI tract, especially in preventing infections and stimulating the host's immune response. Several factors including stress, antibiotic treatments and other medications can alter the GI tract microbiota, predisposing the host to various diseases (Salminen *et al.*, 1995; Schaafsma, 1995). Overprescription and misuse of antibiotics has led to a situation where increasing numbers of pathogens are becoming resistant to antibiotics (Austin *et al.*, 1999; Robredo *et al.*, 2000). The World Health Organization (WHO) has indicated that alternative disease control strategies, such as the use of probiotics – live microbial preparations with documented health benefits for consumers by maintaining or improving their intestinal microbiota balance (Fuller, 1989) – may be needed in the future in the prevention and treatment of certain infections (Bengmark, 1998).

To benefit the health of the consumer, a probiotic bacterium has to reach its target site (gut) alive. It should also have good technological properties so that it can be manufactured and added into foods without losing viability and functionality, or creating unpleasant flavours or textures. It should survive passage through the upper GI tract and be able to function in the gut environment. Feeding trials with different probiotic strains have shown that the probiotic strain usually disappears from the GI tract within a couple of weeks after the ingestion has been discontinued (Fukushima *et al.*, 1998; Johansson *et al.*, 1998; Donnet-Hughes *et al.*, 1999; Alander *et al.*, 1999, 2001). Obtaining a benefit from the probiotic consumption necessitates regular ingestion of probiotic products (daily dosage usually between 10^9–10^{11} cfu).

Most bacteria with probiotic properties belong to the genera *Lactobacillus* and *Bifidobacterium*, which are common but non-dominant members of the indigenous microbiota of the human GI tract (Sghir *et al.*, 2000; Walter *et al.*, 2001). The probiotic potential of various *Lactobacillus* and *Bifidobacterium* strains has been discussed in numerous reviews and includes well-documented management of intestinal disorders such as lactose intolerance, infant gastroenteritis and rotavirus-associated diarrhoea, antibiotic-associated intestinal symptoms (mainly diarrhoea), and food allergy in babies (Salminen *et al.*, 1998; Isolauri *et al.*, 1999, 2001; Marteau *et al.*, 2001; Kaur *et al.*, 2002). These disorders and diseases are associated

Table 10.3   Probiotic health effects. Note that not all effects can be attributed to the same probiotic strain

Established effects	Potential effects
Alleviation of lactose intolerance symptoms	Alleviation of IBD symptoms
Treatment of viral (rotavirus) diarrhoea	Alleviation of IBS symptoms
Treatment of infant gastroenteritis	Improvement of constipation
Treatment of antibiotic associated diarrhoea	Antimutagenic/anticarcinogenic activity
Alleviation of atopic dermatitis symptoms in children	Treatment of candidal and bacterial vaginitis
Positive effects on superficial bladder cancer and cervical cancer	Cholesterol and blood pressure lowering
Modulation of intestinal microbiota	Eradication of multidrug-resistant microbes
Immune modulation	Infection control
Lowering biomarkers (harmful faecal enzymes)	Prevention of transmission of AIDS and sexually transmitted diseases

with intestinal microbiota imbalance and increased gut permeability (Salminen *et al.*, 1996a,b). In addition to these beneficial effects on disturbed intestinal microbiota, probiotics can modulate immune responses, lower biomarkers such as harmful faecal enzyme activities, and show positive effects against superficial bladder cancer and cervical cancer (McFarland, 2000). Other potential areas of probiotic nutritional management include alleviation of inflammatory bowel disease (IBD) and irritable bowel syndrome (IBS) symptoms, mucosal vaccines and immunomodulation, infection control and eradication of multidrug-resistant microbes, treatment of candidal vaginitis, prevention of transmission of AIDS and sexually transmitted diseases, lowering cholesterol and blood, and antimutagenic/ anticarcinogenic activity (Table 10.3) (Alvarez-Olmos and Oberhelman, 2001; Kopp-Hoolihan, 2001; Marteau *et al.*, 2001; Kaur *et al.*, 2002). The most widely studied probiotic strains in human clinical trials include *L. rhamnosus* GG (by far the most extensively studied strain), *Bifidobacterium lactis* Bb-12, *L. casei* Shirota, *L. plantarum* DSM9843 (299V), *Lactobacillus reuteri* (BioGaia Biologics) and *Lactobacillus johnsonii* LJ-1 (Saarela *et al.*, 2000).

When assessing the health-promoting effects of probiotics, it is important to keep in mind that all probiotic strains are different. Even strains representing the same species usually have different properties. Several attributes of probiotic bacteria should be considered before clinical trials are performed. These include:

- origin (preferably human);
- safety;
- viability/activity in delivery vehicles (e.g. food);
- resistance to acid and bile;
- adherence to gut epithelial tissue;
- ability to persist in the GI tract;
- production of antimicrobial substances;
- ability to stimulate host's immune response;
- ability to influence metabolic activities (Salminen *et al.*, 1996b).

Earlier clinical studies on probiotic efficacy generally suffered from ambiguities in probiotic product identification, stability and dosage, and from the small number of test subjects and controls recruited. Considerable efforts have been made to correct these deficiencies, and evidence on the health-promoting effects of probiotics is currently accumulating from randomised and placebo-controlled double-blind studies. Today, plenty of evidence exists on the positive effects of probiotics on human health. However, this has usually been demonstrated in diseased human populations only (Salminen et al., 1998). Thus there is an urgent need for evidence for probiotic health benefits in average (generally healthy) populations.

10.5   Enhancing the viability and stability of LAB

Traditional starters need to be viable and metabolically active during fermentation to be able to grow and produce the desired sensory compounds during processing, whereas their viability after fermentation is not necessary. In contrast, probiotics are by definition live bacteria and the product should contain a suitable level of the specific probiotic strain(s) throughout storage. Although several studies have shown that non-viable probiotics can have beneficial effects, such as immune modulation and carcinogen binding (Ouwehand and Salminen, 1998), good viability is generally considered a prerequisite for optimal functionality (Saarela et al., 2000). The quality assurance of probiotic products is also usually based on techniques detecting viable cells. Examination of existing products has revealed deficiencies in cell numbers and labelling of some probiotic products on the market (Hamilton-Miller and Shah, 2001; Temmerman et al., 2002), which demonstrates the necessity for optimising the parameters affecting the viability of probiotics.

Several factors such as strain characteristics, food matrix, temperature, pH and accompanying microbes affect the viability of a probiotic strain (Fig. 10.2). The ability to tolerate environmental stresses varies greatly between different species and even between strains within a species (Schmidt and Zink, 2000). The stability of the probiotic strain in the product can be enhanced in several ways. The packaging materials used and the storage conditions of the final product are important for the quality of products. The parameters used in the fermentation and downstream processing affect the stability of probiotics in the product. Their viability can be enhanced by optimising the growth medium and the cryo-protectant, for example. Encapsulation techniques using resistant starch or alginate as a protective material have been developed to enhance the stability of probiotics (Shah, 2000; Sultana et al., 2000; Mattila-Sandholm et al., 2002). Encapsulation has increased the viability of probiotics in novel product types such as frozen yoghurt. Combinations of prebiotics (fermentable non-digestible carbohydrates) and probiotics are called synbiotics (Roberfroid, 1998). Prebiotics may have favourable impact on the stability of selected probiotic(s). However, as the concept of synbiotic is relatively new, there are only few specific studies on the interaction between pro- and prebiotics.

The culture technique, based on reproduction of bacterial cells on agar plates,

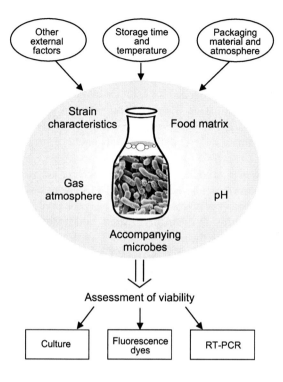

Fig. 10.2 Factors affecting the viability and stability of LAB and methods for assessment of viability.

is the traditional method used for quality assurance of probiotic products. In products containing a high background of accompanying microbes and/or a mixture of several probiotic strains, culture on selective media followed by identification of each strain by accurate molecular methods is needed to assess viability of the probiotic strain(s) (Björneholm *et al.*, 2002). Beside culture, fluorescent dyes have commonly been applied in studies on bacterial viability. A technique called reverse transcription polymerase chain reaction (RT-PCR) also has the potential to detect live and active cells.

Fluorescent dyes can be applied in the detection of viable, damaged and dead bacterial cells in a sample. Live/dead bacteria stained with a suitable fluorophore are detected with epifluorescence microscopy, fluorometer, or flow cytometer, which also allows rapid enumeration of the cells. Typically, a dual approach including staining of viable cells with one dye followed by counterstaining of dead cells with another stain to obtain the total cell number is used (Breeuwer and Abee, 2000). Fluorescent probes used in the viability assessment of lactic acid bacteria include nucleic acid probes such as propidium iodide (PI), TOTO-1 (excluded from intact cells), SYTO9, DAPI (which stain both viable and non-viable cells, and are used together with other probes), and physiological indicators such as 2'7'-bis-(2 carboxyethyl)-5(and-6) carboxyfluorescein (BCECF), carboxyfluorescein diacetate (CFDA), N-(fluorescein thio-ureanyl)-glutamate (FTUG), and bisoxonol (BOX) (Molenaar et al., 1992; Karwoski et al., 1995; Glaasker et al., 1996; Ueckert et al., 1995, 1997, 1998; Auty et al., 2001; Bunthof et al., 2001; Bunthof and Abee, 2002). The time needed for analysis of the viability of cells is substantially shorter with fluorescent dyes (1–2 hours) than with culture (2–5 days) (Breeuwer and Abee, 2000). However, the commonly used fluorophores are non-specific and they cannot differentiate bacterial genera or species.

In RT-PCR the target of PCR amplification is messenger RNA (mRNA), which is rapidly destroyed in dead bacterial cells by RNase activity. In RT-PCR the mRNA is first transcribed into copy DNA (cDNA), which is then amplified by PCR. In lactic acid bacterial research RT-PCR has been used, for example, in nisin gene functionality studies in *Lactococcus lactis* and in detecting inducible nitric oxidase synthase mRNA in the probiotic strain *L. rhamnosus* GG (Moschetti et al., 1996; Korhonen et al., 2001). Although RT-PCR is well suited for studying viable and active bacterial cells in laboratory conditions, there is a drawback in using it for viability studies in products. Under processing conditions where bacterial RNA remains fairly intact but bacterial RNase is destroyed or inactivated, a misleading result of the bacterial viability may be obtained, since under these circumstances the dead bacterial cells may amplify the numbers of cells identified in viability studies of the final product.

10.6   Enhancing the functionality of LAB

Reviews covering the functionality of probiotic LAB have mostly been limited to selection of suitable strains and to the viability of the LAB in the product (Saarela et al., 2000; Dunne et al., 2001; O'Sullivan, 2001). However, the composition of the probiotic product matrix and the status of the LAB during ingestion are also important factors contributing to the functionality of probiotic LAB. A good example of this is the fact that lactose malabsorbers tolerate milk better when it is included in a meal instead of consumed separately. Although most lactose malabsorbers are able to tolerate several grams of lactose when consumed in milk, an increased intake of lactose results in flatulence, abdominal pain and sometimes even diarrhoea (Hertzler et al., 1996; Suarez et al., 1995; Vesa et al., 1996). Fermented milk products including yoghurt and milk with *L. acidophilus* and *L. casei* are usually tolerated better than plain milk (Kolars et al., 1984). It has been suggested that the activity of β-galactosidase in combination with a slower gastric emptying of the fermented milks results in increased tolerance (Marteau et al., 1990; Lin et al., 1991). Furthermore, by slowly increasing daily intake of lactose, the tolerated dose also increases. During the adaptation process the activity of the GI microbiota is changed as hydrogen production decreases and bacterial β-galactosidase activity increases (Johnson et al., 1993; Fondén, 2001). Enhancing the usefulness of LAB-containing dairy products to lactose malabsorbers requires evaluation of not only the selection of a LAB strain producing β-galactosidase, but also the impact of the product on gastric emptying rate and the influence of the probiotic LAB on the intestinal fermentation of lactose during the adaptation process.

Most attention in product development is given to the alleviation of lactose intolerance symptoms even if other non-absorbable, fermentable carbohydrates can also cause identical symptoms. Fructans, fructose, lactulose, sorbitol and even some starches are examples of such carbohydrates (Rumessen et al., 1998; Lebenthal-Bendor et al., 2001). As in lactose intolerance, including LAB able to ferment the specific carbohydrates could reduce the symptoms in products with a slow gastric emptying effect. The GI microbiota also plays a role here illustrated in individuals with a disturbed microbiota (due to intake of antibiotics), showing increased sensitivities to fructans (Orrhage et al., 2000).

Since probiotic LAB should be able to survive the conditions in the GI tract, probiotic strains are often selected based on their resistance to low pH and to a high concentration of bile acids (Dunne et al., 2001). According to in vitro models, milk components are in general protective against harmful conditions of the GI tract (Conway et al., 1987; Charteris et al., 1998). However, fermented milks can increase the exposure of probiotic cells to a low pH and high concentrations of undissociated organic acids. Other properties of the product, e.g. viscosity, which influences gastric emptying time, are also important factors contributing to the exposure of probiotic cells to a low pH (Marteau et al., 1997). Low preformed acid and high buffering capacity may explain the better survival of probiotic LAB in sweet milk and cheese compared to yoghurt (Pettersson et al., 1983; Gardiner et al., 1999). Similarly, exposure to bile acids depends on the fat content of the product. Low-fat dairy products will result in a lower excretion of bile acids (Cummings et al., 1978).

LAB are capable of displaying adaptive responses to stressful conditions, increasing their adaptability to intestinal conditions (Lemay et al., 2000). Acid tolerance in LAB can be triggered by allowing the final pH of the growth medium to drop to 4.5 during fermentation (Lorca and Valdez, 2001). In addition, bile and acid resistance can be triggered by exposing LAB cells to sublethal temperatures or NaCl concentrations. Utilising the stress response of probiotic LAB provides further possibilities of increasing their functionality (Kim et al., 2001). LAB can also be protected from the harsh conditions in the upper part of the GI tract by physical barriers such as enterocoating or microencapsulation. The high survival of probiotic LAB in cheese may depend on a similar barrier effect, since during the digestion process small pieces of cheese are formed, protecting LAB cells from the environment (Gardiner et al., 1999).

Adhesion to intestinal cells and mucus is one of the important functional properties of probiotic LAB. Adhesion is usually improved by milk components (Conway et al., 1987). This effect may be due to the presence of divalent ions in milk, since calcium and magnesium have been shown to favour adhesion (Chauviere et al., 1992). In contrast, some free fatty acids (mainly polyunsaturated) and milk proteins reduce adhesion (Kankaanpää et al., 2001; Ouwehand et al., 2001). The impact of dairy products on GI transit and adhesion of probiotic LAB is clearly a quite complex phenomenon, necessitating the performance of *in vivo* trials with the specific LAB-containing product.

10.7   Future trends

Today's consumers are increasingly attentive to the relationship between diet and well-being. Not only are they aware of the positive effects of certain foods or food compounds on health, there is also an increased preoccupation about the incidence of toxic food components, food-related disease and gastrointestinal disorders. From these trends, a clear need emerges to meet the current demand for safe and healthy food of high quality, preferably produced by traditional methods and from high-quality raw materials, and minimally treated with unnatural or technical means in order to retain the original character and unique organoleptic and sensory qualities of these foods. Functional starter cultures will lead to improved organoleptic, sensory and nutritional properties of the food on the one hand and the inhibition of food-borne pathogens on the other.

Emerging scientific developments will result in fermented foods with strong identities and enhanced quality and safety, that are natural and healthy, and with increased appeal to the consumer. Functional starter cultures are starter cultures that possess at least one inherent functional property. This may be a preservative, organoleptic, technological, health-promoting or nutritional function. Probiotics will remain an important functional ingredient, and novel strains will be identified and foods will be developed to fulfil the needs of specific consumer groups. New non-digestible carbohydrates will be developed with controlled fermentation rates, and the use of prebiotics to modulate the microbiota composition, as well as specific dietary fibres to enhance microbiota's performance and gut function, will be developed.

There is a clear trend towards the use of multiple, defined starter cultures to improve control of industrial fermentation processes. Examples are found in the dairy industry, and promising results are in the pipeline for the processing of cereals. Future possibilities of producing new ingredients for nutritionally optimised foods, which promote consumer health through the microbial reactions, are listed below:

- Selection and improvement of functional starter cultures.
- Effect of food fermentation conditions on the microbes' functionality.
- Predictive modelling and simulation of microbes' behaviour and functionality.
- Targeted screening and selection of functional starter cultures, microbiological analysis, phylogenetic positioning.
- Ecology and technology studies of mixed starter cultures of LAB.
- Effect of food properties on microbial functionality.
- Mechanisms of action of probiotics in the GI tract and biomarkers for their assessment.

- Development of probiotics and bioactive compounds to prevent GI diseases, GI infections and allergies.
- Ensurance of the stability of probiotics also in new types of food applications by developing feasible technologies (e.g. process and material development for microencapsulation).
- Genomics-based analysis of the functionality of food microbes.
- Metabolomics- and physionomics-based analysis of the functionality of food microbes.
- Metabolic engineering and flux analysis for improved functionalities of novel starter cultures.
- Consumers' acceptability of novel functional starter cultures.

10.8   Sources of further information and advice

Reviews published in the Supplements of *British Journal of Nutrition* (2000 Suppl 2) and *American Journal of Clinical Nutrition* (2001, 73, 6) give an update of knowledge available in the late 1990s. Newly published books edited by Tannock (2002) and Brandi, Biavati and Biasco (2002) give further information.

IDF, the International Dairy Federation, is the organisation of the international dairy world. References are made to several bulletins published by the organisation (IDF 1988,1992, 2002). The activities of the organisation can be followed by reading the home page of IDF at www.fil-idf.org.

In order to follow the development of LAB and dairy science, the two main journals of special interest that should be followed are *International Dairy Journal* and *Journal of Dairy Science*. To follow the development on the market and questions related to the interface between R&D and marketing the journal *New Nutrition Business* is highly recommended.

10.9   References

ALAMPREESE C, FOSCHINO R, ROSSI M, POMPEI C and SAVANI L (2002), 'Survival of *Lactobacillus johnsonii* La1 and influence of its addition in retail-manufactured ice cream produced with different sugar and fat concentrations', *Int Dairy J*, **12**(2), 201–208.

ALANDER M, SATOKARI R, KORPELA R, SAXELIN M, VILPPONEN-SALMELA T, MATTILA-SANDHOLM T and VON WRIGHT A (1999), 'Persistence of colonization of human colonic mucosa by a probiotic strain, *Lactobacillus rhamnosus* GG, after oral consumption', *Appl Environ Microbiol*, **65**, 351–354.

ALANDER M, MÄTTÖ J, KNEIFEL W, JOHANSSON M, KÖGLER B, CRITTENDEN R, MATTILA-SANDHOLM T and SAARELA M (2001), 'Effect of galacto-oligosaccharide supplementation on human faecal microflora and on survival and persistence of *Bifidobacterium lactis* Bb-12 in the gastrointestinal tract', *Int Dairy J*, **11**, 817–825.

ALVAREZ-OLMOS M I and OBERHELMAN R A (2001), 'Probiotic agents and infectious diseases: a modern perspective on a traditional therapy', *Clin Infect Dis*, **32**, 1567–1576.

AUSTIN D J, KRISTINSSON K G and ANDERSON R M (1999), 'The relationship between the volume of antimicrobial consumption in human communities and the frequency of resistance', *PNAS*, **96**, 1152–1156.

AUTY M A E, GARDINER G E, MCBREARTY S J, O'SULLIVAN E O, MULVIHILL D M, COLLINS J K, FITZGERALD G F, STANTON C and ROSS R P (2001), 'Direct *in situ* assessment of bacteria in probiotic dairy products using viability staining in conjunction with confocal scanning laser microscopy', *Appl Environ Microbiol*, **67**, 420–425.
BENGMARK S (1998), 'Ecological control of the gastrointestinal tract. The role of probiotic bacteria', *Gut*, **42**, 2–7.
BERG R D (1996), 'The indigenous gastrointestinal microflora', *Trends Microbiol*, **4**, 430–435.
BJÖRNEHOLM S and FONDÉN R (2002), '*Lactobacillus* F19 – closing the broken circle' *Microb Ecol Health Dis*, Suppl (3), 3.
BJÖRNEHOLM S, EKLÖW A, SAARELA M and MÄTTÖ J (2002), 'Enumeration and identification of *Lactobacillus paracasei* subsp. *paracasei* F19', *Microbiol Ecol Health Dis*, Suppl 3, 7–13.
BRANDI G, BIAVATI B and BIASCO G (2002), *Bifidobacteria: Microbiological Aspects and Probiotic Potentialities*, Heidelberg, Springer-Verlag.
BREEUWER P and ABEE T (2000), 'Assessment of viability of microorganisms employing fluorescence techniques', *Int J Food Microbiol*, **55**, 193–200.
BUNTHOF C J and ABEE T (2002), 'Development of a flow cytometric method to analyze subpopulations of bacteria in probiotic products and dairy starters', *Appl Environ Microbiol*, **68**, 2934–2942.
BUNTHOF C J, BLOEMEN K, BREEUWER P, ROMBOUTS F M and ABEE T (2001), 'Flow cytometric assessment of viability of lactic acid bacteria', *Appl Environ Microbiol*, **67**, 2326–2335.
BURKE A (1938), *Practical Manufacture of Cultured Milks and Kindred Products*, Milwaukee, Olsen Publishing Co.
CHARTERIS W P, KELLY P M, MORELLI L and COLLINS J K (1998), 'Development and application of an *in vitro* methodology to determine the transit tolerance of potentially probiotic *Lactobacillus* and *Bifidobacterium* species in the upper human gastrointestinal tract', *J Appl Microbiol*, **84**(5), 759–768.
CHAUVIERE G, COCONNIER M H, KERNEIS S, FOURNIAT J and SERVIN A L (1992), 'Adhesion of human *Lactobacillus acidophilus* strain LB to human enterocyte-like Caco-2 cells', *J Gen Microbiol*, **138**(8),1689–1696.
COGAN T M ACCOLAS J P (1996), *Dairy Starter Cultures*, Weiheim, VCH Verlagsgesellschaft.
CONWAY P L, GORBACH S L and GOLDIN B R (1987), 'Survival of lactic acid bacteria in the human stomach and adhesion to intestinal cells', *J Dairy Sci*, **70**(1), 1–12.
CUMMINGS J H, WIGGINS H S, JENKINS D J, HOUSTON H, JIVRAJ T, DRASAR B S and HILL M J (1978), 'Influence of diets high and low in animal fat on bowel habit, gastrointestinal transit time, fecal microflora, bile acid, and fat excretion', *J Clin Invest*, **61**(4), 953–963.
DALY C and DAVIS R (1998), 'The biotechnology of lactic acid bacteria with emphasis on applications in food safety and human health', *Agric. Food Sci Finland*, **7**, 251–265.
DAVIDSON R H, DUNCAN S E, HACKNEY C R, EIGEL W N and BOLING J W (2000), 'Probiotic culture survival and implications in fermented frozen yoghurt characteristics', *J Dairy Sci*, **83**(4), 666–673.
DONNET-HUGHES A, ROCHAT F, SERRANT P, AESCHLIMANN J M and SCHIFFRIN J E (1999), 'Modulation of nonspecific mechanisms of defence by lactic acid bacteria: effective dose', *J Dairy Sci*, **82**, 863–869.
DUNNE C, O'MAHONY L, MURPHY L, THORNTON G, MORRISSEY D, O'HALLORAN S, FEENEY M, FLYNN S, FITZGERALD G, DALY C, KIELY B, O'SULLIVAN G C, SHANAHAN F and COLLINS J K (2001), '*In vitro* selection criteria for probiotic bacteria of human origin: correlation with *in vivo* findings', *Am J Clin Nutr*, **73**(2 Suppl), 386S–392S.
FLEICHMAN W and WEIGMANN H (1932), 'Die Sauermilcharten' in *Lehrbuch der Milchwirtschaft*, 7 Auflage, Berlin, Paul Parey, 369–386.
FONDÉN R (1989) 'Development of new fermented milk products', *Food Techn Int Europe*, **1989**(1), 158–160.

FONDÉN R (2001), 'Adaptation to lactose in lactose malabsorbers – importance of the intestinal microflora', *Scand J Nutr*, **45**(4), 174.

FONDÉN R, GRENOV B, RENIERO R, SAXELIN M and BIRKELAND S E (2000), 'Industrial panel statements: technological aspect'. In: M Alander and T. Mattila-Sandholm, *Functional Foods for EU-health in 2000, Fourth Workshop, FAIR CT96-1028, PROBDEMO, VTT Symposium, Rovaniemi, Finland*. VTT, Helsinki, Vol. 198, 43–50.

FOX P F (1993), *Cheese: Chemistry, Physics and Microbiology*, 2nd edn, London, Chapman & Hall.

FUKUSHIMA Y, KAWATA Y, HARA H, TERADA A and MITSUOKA T (1998), 'Effect of a probiotic formula on intestinal immunoglobulin A production in healthy children', *Int J Food Microbiol*, **42**, 39–44.

FULLER, R (1989), 'Probiotics in man and animals', *J Appl Bacteriol*, **66**, 365–378.

GARDINER G, ROSS R P, COLLINS J K, FITZGERALD G and STANTON C (1998), 'Development of a probiotic cheddar cheese containing human-derived *Lactobacillus paracasei* strains', *Appl Environ Microbiol*, **64**(6), 2192–2199.

GARDINER G, STANTON C, LYNCH P B, COLLINS J K, FITZGERALD G and ROSS R P (1999), 'Evaluation of cheddar cheese as a food carrier for delivery of a probiotic strain to the gastrointestinal tract', *J Dairy Sci*, **82**(7),1379–1387.

GARDINER G E, O'SULLIVAN E, KELLY J, AUTY M A E, FITZGERALD G F, COLLINS J K, ROSS R P and STANTON C (2000), 'Comparitive survival rates of human-derived probiotic *Lactobacillus paracasei* and *L. salivarius* strains during heat treatment and spray drying', *Appl Envir Microbiol*, **66**(6), 2605–2612.

GILLILAND S E (1989), 'Acidophilus milk products: a review of potential benefits to the consumers', *J Dairy Sci*, **72**(10), 2483–2494.

GLAASKER E, KONINGS W N and POOLMAN B (1996), 'The application of pH-sensitive fluorescent dyes in lactic acid bacteria reveals distinct extrusion systems for unmodified and conjugated dyes', *Mol Membr Biol*, **13**, 173–181.

GOMES A M P, MALCATA F A, KLAVER M and GRANDE H G (1995), 'Incorporation and survival of *Bifidobacterium* sp. strain Bo and *Lactobacillus acidophilus* strain Ki in a cheese product', *Neth Milk Dairy J*, **49**(2), 71–95.

HAENEL H and BENDIG J (1975) 'Intestinal flora in health and disease', *Progr Food Nutr Sci*, **1**, 21–64.

HAGEN M and NARVHUS A (1999), 'Production of ice cream containing probiotic bacteria', *Milchwissenschaft*, **54**(3), 265–268.

HAMILTON-MILLER J M T and SHAH S (2001), 'Deficiencies in microbiological quality and labelling of probiotic supplements', *Int J Food Microbiol*, **72**, 175–176.

HAWLEY H B, SHEPHERD P A and WHEATER D M (1959), 'Factors affecting the implantation of lactobacilli in the intestine', *J Appl Bact*, **22**, 360.

HEKMAT S and MCMAHON D J (1992), 'Survival of *Lactobacillus acidophilus* and *Bifidobacterium bifidum* in ice cream for use as a probiotic food', *J Dairy Sci*, **75**(6), 1415–1422.

HERTZLER S R, HUYNH B C and SAVAIANO D A (1996), 'How much lactose is low lactose?', *J Am Diet Assoc*, **96**(3), 243–246.

HULL R R, ROBERTS A V and MAYES J J (1984), 'Survival of *Lactobacillus acidophilus* in yoghurt', *Austral J Dairy Technol*, **39**(4), 164–166.

IDF (1988), *Fermented Milks – Science and Technology*, IDF Bulletin, **227**, Brussels, IDF.

IDF (1992), *New Technologies for Fermented Milks*, IDF Bulletin, **277**, Brussels, IDF.

IDF (2002), *Health benefits and safety evaluation of certain food components*, IDF Bulletin, **377**, Brussels, IDF.

ISOLAURI E, SALMINEN S and MATTILA-SANDHOLM T (1999), 'New functional foods in the treatment of food allergy', *Ann Med*, **31**, 299–302.

ISOLAURI E, SUTAS Y, KANKAANPÄÄ P, ARVILOMMI H and SALMINEN S (2001), 'Probiotics: effects on immunity', *Am J Clin Nutr*, **73** (suppl), 444S–450S.

JOHANSSON M-L, NOBAEK S, BERGGREN A, NYMAN M, BJÖRCK I, AHRNE S, JEPPSSON B and MOLIN G (1998), 'Survival of *Lactobacillus plantarum* DSM 9843 (299v), and effect on the short-chain fatty acid content of faeces after ingestion of a rose-hip drink with fermented oats', *Int J Food Microbiol*, **42**, 29–38.

JOHNSON A O, SEMENYA J G, BUCHOWSKI M S, ENWONWU C O and SCRIMSHAW N S (1993), 'Adaptation of lactose maldigesters to continued milk intakes', *Am J Clin Nutr*, **58**(6), 879–881.

KANKAANPÄÄ P E, SALMINEN S J, ISOLAURI E and LEE Y K (2001), 'The influence of polyunsaturated fatty acids on probiotic growth and adhesion', *FEMS Microbiol Lett*, **194**(2), 149–53.

KARWOSKI M, VENELAMPI O, LINKO P and MATTILA-SANDHOLM T (1995), 'A staining procedure for viability assessment of starter culture cells', *Food Microbiol*, **12**, 21–29.

KAUR I P, CHOPRA K and SAINI A (2002), 'Probiotics: potential pharmaceutical applications', *Eur J Pharmac Sci*, **15**,1–9.

KIM W S, PERL L, PARK J H, TANDIANUS J E and DUNN M W (2001), 'Assessment of stress response of the probiotic *Lactobacillus acidophilus*', *Curr Microbiol*, **43**(5),346–350.

KOLARS J C, LEVITT M D, AOUJI M and SAVIANO D A (1984), 'Yogurt – an autodigesting source of lactose', *New Engl J Med*, **310**(1), 1–3.

KOPP-HOOLIHAN L (2001), 'Prophylactic and therapeutic uses of probiotics: a review', *J Am Diet Assoc*, **101**, 229–238.

KORHONEN R, KORPELA R, SAXELIN M, MÄKI M, KANKAANRANTA H and MOILANEN E (2001), 'Induction of nitric oxide synthesis by probiotic *Lactobacillus rhamnosus* GG in J774 macrophages and human T84 intestinal epithelial cells', *Inflammation*, **25**, 223–232.

KURMANN J, RAŠI J and KROGER, M (1992), *Encyclopedia of Fermented Fresh Milk Products*, New York, AVI.

LARSSON I (1988), *Tätmjölk, tätgräs, surmjölk och skyr*, Sthlm Studies Scand Philology, New Series 18, Stockholm, Almqvist and Wicksell (incl. English summary).

LAW B A (1999), *Technology of Cheesemaking*. Boca Raton, CRC Press.

LEBENTHAL-BENDOR Y, THEUER R C, LEBENTHAL A, TABI I and LEBENTHAL E (2001), 'Malabsorption of modified food starch (acetylated distarch phosphate) in normal infants and in 8–24-month-old toddlers with non-specific diarrhea, as influenced by sorbitol and fructose', *Acta Paediatr*, **90**(12), 1368–1372.

LEMAY M-J, RODRIGUE N, CARIÉPY C and SAUCIER L (2000), 'Adaptation of *Lactobacillus alimentarius* to environmental stresses', *Int J Food Microbiol*, **55**, 249–253.

LIN M Y, SAVAIANO D and HARLANDER S (1991), 'Influence of nonfermented dairy products containing bacterial starter cultures on lactose maldigestion in humans', *J Dairy Sci*, **74**(1), 87–95.

LINDBERG A, CHRISTIANSSON A, RUKKE E, EKLUND T and MOLIN G (1996), 'Bacterial flora of Norwegian and Swedish semi-hard cheeses after ripening, with special references to *Lactobacillus*', *Neth Milk Dairy J*, **50**(4), 563–572.

LORCA G L and VALDEZ G F (2001), 'A low-pH-inducible, stationary-phase acid tolerance response in *Lactobacillus acidophilus* CRL 639', *Curr Microbiol*, **42**(1), 21–25.

LOURENS-HATTINGH A and VILJOEN B C (2001), 'Review: yoghurt as probiotic carrier food', *Int Dairy J*, **11**(1), 1–17.

MARSHALL V (1986), 'The microflora and production of fermented milks'. In: M R Adams, *Micro-organisms in the Production of Food*, Progress in Industrial Microbiol, Elsevier, Amsterdam, **23**, 1–44.

MARTEAU P, FLOURIE B and POCHART P (1990), 'Effect of the microbial lactase (EC 3.2.1.2.3) activity in yoghurt on the intestinal absorption of lactose: an *in vivo* study in lactase-deficient humans', *Br J Nutr*, **64**(1), 71–79.

MARTEAU P, MINEKUS M, HAVENAAR R and HUIS IN'T VELD J H (1997), 'Survival of lactic acid bacteria in a dynamic model of the stomach and small intestine: validation and the effects of bile', *J Dairy Sci*, **80**(6), 1031–1037.

MARTEAU P R, DE VRESE M, CELLIER C J and SCHREZENMEIR J (2001), 'Protection from gastrointestinal diseases with the use of probiotics', *Am J Clin Nutr*, **73** (2 suppl), 430S–436S.

MATTILA-SANDHOLM T, MYLLÄRINEN P, CRITTENDEN R, MOGENSEN G, FONDÉN R and SAARELA M (2002), 'Technological challenges for future probiotic foods', *Int Dairy J*, **12**(2), 173–182.

MCFARLAND L V (2000), 'A review of the evidence of health claims for biotherapeutic agents', *Microbial Ecol Health Dis*, **12**(1), 65–76.

METCHNIKOFF E (1907), *Essaises Optimiste*, Paris, Maloine.

MITSUOKA T (1982), 'Recent trends in research on intestinal flora', *Bifidobacteria Microflora*, **1**, 3–24.

MOLENAAR D, BOHUIS H, ABEE T, POOLMAN B and KONINGS W N (1992), 'The efflux of a fluorescent probe is catalyzed by an ATP-driven extrusion system in *Lactococcus lactis*', *J Bacteriol*, **174**, 3118–3124.

MOSCHETTI G, VILLANI F, BLAIOTTA G, BALDINELLI B and COPPOLA S (1996), 'Presence of non-functional nisin genes in *Lactococcus lactis* subsp. *lactis* isolated from natural starters', *FEMS Microbiol Lett*, **145**(1), 27–32.

ORLA-JENSEN S and SADLER W (1940), 'Bakteriologische untersuchungen über das isländische Sauermilchpräparat', *Zentralblatt f Bakt II Abt*, **102**, 260.

ORRHAGE K, SJOSTEDT S and NORD C E (2000), 'Effect of supplements with lactic acid bacteria and oligofructose on the intestinal microflora during administration of cefpodoxime proxetil', *J Antimicrob Chemother*, **46**(4), 603–612.

O'SULLIVAN D J (2001), 'Screening of intestinal microflora for effective probiotic bacteria', *J Agric Food Chem*, **49**(4), 1751–1760.

OUWEHAND A C and SALMINEN S J (1998), 'The health effects of cultured milk products with viable and non-viable bacteria', *Int Dairy J*, **8**, 749–758.

OUWEHAND A C, TUOMOLA E M, TÖLKKÖ S and SALMINEN S (2001), 'Assessment of adhesion properties of novel probiotic strains to human intestinal mucus', *Int J Food Microbiol*, **64**(1–2), 19–26.

PETTERSSON L, GRAF W and SEWELIN U (1983), 'Survival of *Lactobacillus acidophilus* NCDO 1748 in the human gastrointestinal tract. 2: Ability to pass the stomach and intestine *in vivo*', in *Nutrition and the Intestinal Microflora*, XV Symposium of the Swedish Nutrition Foundation, Almqvist & Wicksell International, 123–125.

RETTGER L (1915), 'The influence of milk feeding on mortality and growth and on the characteristics of the intestinal flora', *J Exp Med*, **21**, 365–388.

RIZKALLA S W, LUO J, KABIR M, CHEVALIER A, PACHER N and SLAMA G (2000), 'Chronic consumption of fresh but not heated yogurt improves breath-hydrogen status and short-chain fatty acid profiles: a controlled study in healthy men with or without lactose maldigestion', *Am J Clin Nutr*, **72**(6), 1474–1479.

ROBERFROID M B (1998), 'Prebiotics and synbiotics: concepts and nutritional properties', *Br J Nutr*, **80**, S197–S202.

ROBREDO B, SINGH K V, BAQUERO R, MURRAY B E and TORRES C (2000), 'Vancomycin-resistant enterococci isolated from animals and food', *Int J Food Microbiol*, **54**, 197–204.

RUMESSEN J and GUDMAND-HOYER E (1998), 'Fructans of chicory: intestinal transport and fermentation of different chain lengths and relation to fructose and sorbitol malabsorption', *Am J Clin Nutr*, **68**(2), 357–64.

SAARELA M, MOGENSEN G, FONDEN R, MÄTTÖ J and MATTILA-SANDHOLM T (2000), 'Probiotic bacteria: safety, functional and technological properties', *J Biotechnol*, **84**(3), 197–215.

SALMINEN, S, ISOLAURI, E and ONNELA T (1995), 'Gut microflora in health and disease', *Chemotherapy*, **41**, 5–15.

SALMINEN S, ISOLAURI E and SALMINEN E (1996a), 'Probiotics and stabilisation of the gut mucosal barrier', *Asia Pacific J Clin Nutr*, **5**, 53–56.

SALMINEN S, VON WRIGHT A, LAINE M, VUOPIO-VARKILA J, KORHONEN T and MATTILA-SANDHOLM T (1996b), 'Development of selection criteria for probiotic strains to assess their potential in functional foods: a Nordic and European approach', *Biosci Microbiol*, **15**, 61–67.

SALMINEN S, OUWEHANDA A C and ISOLAURI E (1998), 'Clinical applications of probiotic bacteria', *Int Dairy J*, **8**, 563–572.

SAXELIN M, GRENOV B, SVENSSON U, FONDEN R, RENIERO R and MATTILA-SANDHOLM T (1999), 'The technology of probiotics', *Trends Food Sci Technol*, **10**(12), 387–392.

SCHAAFSMA G (1995), 'Application of lactic acid bacteria in novel foods from a nutritional perspective´, in G Novel and J-F Le Querler, *Lactic Acid Bacteria: Actes du Colloque LACTIC 94*. Caen, Presses Universitaires de Caen, p. 85–93.

SCHMIDT G and ZINK R (2000), 'Basic features of the stress response in three species of bifidobacteria: *B. longum, B. adolescentis*, and *B. breve*', *Int J Food Microbiol*, **55**(1–3), 41–45.

SCHUHLER-MAYLOTH R, RUPPERT A and MULLER F (1968), 'Die Mikroorganismen der Bifidusgrupp (syn. *Lactobacillus bifidus*) 2. Mittelung: Die technologie der Bifiduskultur im milchverarbeitenden Betrieb', *Milchwissenschaft*, **23**(9), 554–558.

SGHIR A, GRAMET G, SUAU A, ROCHET V, POCHART P and DORE J (2000), 'Quantification of bacterial groups within human fecal flora by oligonucleotide probe hybridization', *Appl Environ Microbiol*, **66**, 2263–2266.

SHAH N P (2000), 'Probiotic bacteria: selective enumeration and survival in dairy foods', *J Dairy Sci*, **83**, 894–907.

SPECK M L (1975), 'Market outlook for acidophilus food products', *Cultured Dairy Products J*, **10**(4), 8–10.

SUAREZ F L, SAVAIANO D A and LEVITT M D (1995), 'A comparison of symptoms after the consumption of milk or lactose-hydrolyzed milk by people with self-reported severe lactose intolerance', *New Engl J Med*, **333**(1),1–4.

SULTANA K, GODWARD G, REYNOLDS N, ARUMUGASWAMY R, PEIRIS P and KAILASPATHY K (2000), 'Encapsulation of probiotic bacteria with alginate-starch and evaluation of survival in simulated gastrointestinal conditions and in yoghurt', *Int J Food Microbiol*, **62**, 47–55.

SVENSSON U (1999), 'Industrial perspectives' in Tannock GW, *Probiotics: A critical review*, Wymondham, Horizon Scientific Press, 57–64.

TAMIME A Y and ROBINSON R K (1999), *Yoghurt: Science and Technology*, 2nd edn, Cambridge, Woodhead Publishing Ltd.

TANNOCK G W (2002), *Probiotics and Prebiotics:Where Are We Going?*, Wymondham, Caister Academic Press.

TEMMERMAN R, HUYS G, POT B and SWINGS J (2002), 'Quality analysis and label correctness of commercial probiotic products', *Innov Food Technol*, **14**, 72–73.

UECKERT J, BREEUWER P, ABEE T, STEPHENS P, NEBE VON CARON G and TER STEEG P F (1995), 'Flow cytometry applications in physiological study and detection of foodborne microorganisms', *Int J Food Microbiol*, **28**, 317–326.

UECKERT J, NEBE VON CARON G, BOS A P TER STEEG P F (1997), 'Flow cytometric analysis of *Lactobacillus plantarum* to monitor lag times, cell division and injury', *Lett Appl Microbiol*, **25**, 295–299.

UECKERT J E, TER STEEG P F and COOTE P J (1998), 'Synergistic antibacterial action of heat combination with nisin and magainin II amide', *J Appl Microbiol*, **85**, 487–494.

WALTER J, HERTEL C, TANNOCK G W, LIS C M, MUNRO K and HAMMES W P (2001), 'Detection of *Lactobacillus, Pediococcus, Leuconostoc*, and *Weissella* species in human feces using group-specific PCR primers and denaturing gradient gel electrophoresis', *Appl Environ Microbiol*, **67**, 2578–2585.

VESA T H, KORPELA R A and SAHI T (1996), 'Tolerance to small amounts of lactose in lactose maldigesters', *Am J Clin Nutr*, **64**(2),1 97–201.

VINDEROLA C G, PROSELLO W, GHIBERTO T D and REINHEIMER J A (2000), 'Viability of probiotic (*Bifidobacterium, Lactobacillus acidophilus* and *Lactobacillus casei*) and nonprobiotic microflora in Argentinian Fresco cheese', *J Dairy Sci*, **83**(9), 1905–1911.

YOUNG C K and NELSON F E (1978), 'Survival of *Lactobacillus acidophilus* in sweet acidophilus milk during refrigerated storage', *J Food Protec*, **41**(4), 248–250.

11

Conjugated linoleic acid (CLA) as a functional ingredient

S. Gnädig, Y. Xue, O. Berdeaux, J.M. Chardigny and J-L. Sebedio,
Institut National de la Recherche Agronomique, France

11.1   Introduction

During the 1980s Michael Pariza and his colleagues found that a fraction isolated from grilled minced beef could inhibit carcinogenesis (Pariza and Hargraves, 1985). This fraction was shown to contain isomers of octadecadienoic acid having two conjugated double bonds. These fatty acids have been referred to since as conjugated linoleic acid or CLA. CLA are usually found in dairy products and meat from ruminants (Sébédio *et al.*, 1999). Out of the many positional and geometrical isomers of CLA the 9*cis*,11*trans* octadecadienoic acid (or 9*c*,11*t*-18:2) predominates in these food products. Following the discovery of Pariza and co-workers, much research was initiated. Particular attention was given to the development of suitable analytical techniques to determine the isomer mixtures present in foods as earlier studies were not able to provide any details about the isomer mixtures utilized. Methods to synthesize pure isomers, especially those present in commercial mixtures previously prepared, mainly the 9*c*,11*t*- and the 10*t*,12*c*-18:2 are now available. Recent research now gives us better knowledge of the precise role of two isomers so far studied. This review will describe the progress made in CLA research, especially in the field of analysis, synthesis and biological effects on animals and humans. Further research needs will also be discussed.

11.2   Natural sources of CLA

As already mentioned, the major sources of naturally occurring CLA are foods containing ruminant fat while some products derived from plant oils also contain small amounts of CLA. Milk and dairy products have been shown to contain the

Table 11.1   CLA contents in foods, g/100 g of
total fatty acids (Gnädig, 1996)

	CLA content
Butter	0.63–2.02
Milk	0.46–1.78
Beef	0.67–0.99
Lamb	1.62–2.02
Fish	0.04–0.28
Yoghurt	0.43–1.12
Cheese	0.50–1.70
Pork	0.15
Turkey	0.96
Plant oils	n.d.

highest amounts of CLA. The CLA contents of various normally consumed foods are presented in Table 11.1.

Among different meat products, meat from ruminants shows higher contents of CLA than meat of non-ruminant origin. Highest CLA amounts were found in lamb (Banni *et al.*, 1996). For seafood and poultry, except for turkey, only low CLA contents were reported (Chin *et al.*, 1992). For dairy products, CLA contents up to 30.0 mg/g fat were reported (O'Shea *et al.*, 1998). A study by Lavillonnière *et al.* (1998) revealed variations of the CLA content in cheeses between 5.3 and 15.8 mg/g fat. About 75–90% of total CLA in ruminant fat is $9c,11t$-18:2 as the natural CLA isomer, also called rumenic acid (Kramer *et al.*, 1998).

Plant oils or margarine contain only small amounts of CLA (0.1–0.5 mg/g fat) (Chin *et al.*, 1992). CLAs are formed as a result of industrial processing – oil refining processes (mainly deodorization) and catalytic process of hydrogenation to produce margarine (Jung and Ha, 1999). Otherwise, CLA appear in oils due to high-temperature treatments. As an example, sunflower oil after its use as frying oil contained CLA amounts up to 0.5 g CLA/100 g of oil (Juaneda *et al.*, 2001). Further analysis of the CLA composition in different oil and margarine samples showed that, among total occurring CLA, the all-*trans* isomers were formed favourably and less than 20% of total CLA was the $9c,11t$-18:2 (Chin *et al.*, 1992, Juaneda *et al.*, 2001, Jung *et al.*, 2001). Processed, canned or infant foods showed CLA amounts comparable to those of unprocessed foods (Chin *et al.*, 1992). CLA in foods such as chocolates, pastries and bakeries originated predominantly from dairy fat, and the CLA level in the final foodstuff reflects the proportion of dairy fat in the preparation (Fritsche and Steinhart, 1998b).

The natural occurring CLA in milk is formed using two different pathways, which are summarized in Fig. 11.1. A first pathway is the bioconversion of polyunsaturated fatty acids (PUFA) in the rumen. The ingested dietary unsaturated fatty acids, e.g. linoleic acid, are metabolized by enzymes of different bacteria present in the rumen (Harfoot and Hazelwood, 1988). Various *trans* fatty acids appear along this biohydrogenation pathway as intermediates. $9c,11t$-18:2 is

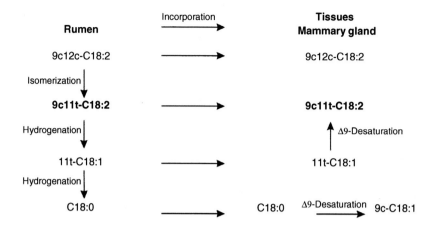

Fig. 11.1   Pathways of CLA biosynthesis (9c,11t-18:2).

formed during this bacterial transformation. Kepler and Tove (1967) isolated a linoleate isomerase (EC 5.2.1.5) from the rumen bacterium *Butyrivibrio fibrisolvens*, which is responsible for the isomerization of linoleic acid into 9c,11t-18:2 in a first step. Then, the double bond in position Δ9 is hydrogenated to form vaccenic acid (11t-18:1). In the last step vaccenic acid is hydrogenated into stearic acid. It was suggested that this last step is rate limiting. The intermediates (9c,11t-18:2 and vaccenic acid) accumulate and are absorbed in the intestine and incorporated into different tissues. It was recently demonstrated that vaccenic acid also leads to the formation of 9c,11t-18:2. It is Δ9 desaturated in adipose tissue and in the mammary gland of the lactating cow to produce 9c,11t-18:2 (Griinari *et al.*, 2000). The endogenous synthesis in the mammary gland was reported to be very important, as about 60% of CLA in milk fat is formed via this pathway in the lactating cow. As far as the second most important CLA isomer in milk fat is concerned, the 7t,9c-18:2, Corl *et al.* (2002) have recently demonstrated that it is only formed via the Δ9 desaturation pathway of 9t-18:1.

Various factors are known to influence the CLA content in milk, such as the food of the ruminant, the season, the animal breeding type, the number of times the animal has lactated and the current stage of lactation (Jahreis *et al.*, 1997; Fritsche and Steinhart, 1998b; Sébédio *et al.*, 1999). Stanton *et al.* (1997) showed that a high number of lactations led to high CLA concentrations in milk fat. An interaction between the lactation stage and the CLA level was not examined, as the milk sampling in this study excluded the beginning and the end of the lactation period. But it seems that the lactation stage can also affect the CLA content in milk, as body fat stores of the cow are mobilized at the beginning of lactation (Jahreis *et al.*, 1999).

The CLA concentration in milk is season dependent. All ruminants showed a decreased CLA level during winter, with the lowest CLA level found in March, and inversely an increase was observed during summer, which is positively correlated to the grazing period. Indeed, pasture in spring time contains higher amounts of PUFAs, which caused a higher bacterial biohydrogenation in the rumen (Jahreis *et al.*, 1997). The relationship between pasture feeding and CLA level was also described by Dhiman *et al.* (1996). The CLA level increased from 8.4 to 22.7 mg/g milk fat when the pasture level was increased in the diet. To show the influence of different diets on the CLA content in milk, an experiment was carried out to compare three different types of farm management applying three different feeding methods. Cows were fed either with maize silage throughout the year (indoor group), with grazing in summer and maize silage in winter (conventional group) or with grazing during summer and clover–alfalfa–grass silage in winter (ecological group). The highest amounts of CLA in milk were measured in the ecological group, whereas smallest CLA contents were found in the indoor-maize silage group (Jahreis *et al.*, 1996). The application of ecological feeding conditions rich in pasture would result in the production of milk naturally high in CLA (Jiang *et al.*, 1996).

Various experiments were carried out to increase the CLA content in milk, by modulating the dietary regimen of the cows. Milk from cows offered a diet supplemented with oils rich in PUFAs had up to fivefold higher CLA levels. Enrichment with sunflower, linseed or rapeseed oil also caused an increase in CLA levels (Kelly *et al.*, 1998). A positive correlation was found between the content of linoleic acid in the oil and the increase of CLA in the milk. The same effects were obtained by feeding with extruded oilseeds (Dhiman *et al.*, 1999). Feeding fish oils which contained large amounts of PUFAs, mainly eicosapentaneoic acid and docosahexaneoic acid, resulted in the same effects. CLA content increased significantly with the formation of biohydrogenation products in the rumen (Donovan *et al.*, 2000; Jones *et al.*, 2000). A recent study of Chouinard and coworkers (2001) investigated the effects of different dietary fat supplements on CLA content in milk. The use of canola, soybean and linseed oil increased CLA content in the milk fat by three- to fivefold compared with a control diet. The influence of processing methods to produce soybeans as dietary supplement in the cows' diet was also examined. It was found that the feeding of extruded and roasted soybeans resulted in two- to threefold higher contents of CLA in milk fat compared to a diet containing raw ground soybeans.

Direct feeding of CLA in the regimen and the abomasal infusion of CLA was tested as another possibility to enrich CLA in cows' milk (Chouinard *et al.*, 1999). It was found that direct CLA supplementation reduced milk fat yield, suggesting a decrease of *de novo* fatty acid synthesis (Loor and Herbein, 1998). Moreover, the amount of CLA that induced a substantial reduction of milk fat synthesis was found to be low. As an example, infusion with 0.10 g/day of $10t,12c$-18:2 (0.05% of diet) resulted in a 44% reduction in milk fat yield (Baumgard *et al.*, 2001). It was demonstrated that only the $10t,12c$-18:2 inhibited milk fat synthesis, whereas the $9c,11t$-18:2 had no effect. The application of this method to enrich CLA in milk seems not to be worth pursuing because of the disadvantageous reduction of milk fat resulting in the reduction of total CLA in milk.

11.3   Commercial production of CLA

11.3.1   Alkali isomerization

Isomerization of linoleic acid ($9c,12c$-18:2) under strong alkali conditions (Chipault and Hawkins, 1959) produces complex CLA mixtures containing two predominant isomers. $9c,11t$-18:2 (43–45%) and $10t,12c$-18:2 (43–45%). These two major isomers are accompanied by small amounts of other CLA isomers with double bonds in 8,10 or 11,13 positions. In addition to *cis–trans* and *trans–cis* isomers, all-*cis* and all-*trans* CLA isomers are formed (Christie, 1997). By varying the solvent used (ethylene glycol, glycerol, propylene glycol, *tert*-butanol, water, dimethyl sulphoside (DMSO), *N,N*-dimethyl formamide (DMF), the catalysts (lithium, sodium or potassium hydroxide) or the reaction vessels, it is possible to obtain various CLA mixtures possessing different properties and an enrichment of single isomers (Reaney *et al.*, 1999). Alkali isomerization of linoleic acid is used for a commercial production of CLA. Reaney *et al.* (1999) have reviewed the literature on commercial production of CLA.

11.3.2   Dehydration of hydroxy fatty acid

Dehydration of hydroxy fatty acid has been used to prepare a number of conjugated fatty acids. For example, dehydration of ricinoleic acid is the best method to prepare pure $9c,11t$-18:2. Ricinoleic acid (12-hydroxy-octadec-9Z-enoic acid) is a relative inexpensive starting material, easily isolated from castor oil by countercurrent distribution (Tassignon *et al.*, 1994; Berdeaux *et al.*, 1997).

As a first step, the hydroxy group was modified to become a better leaving group and methyl ricinoleate was transformed into a mesylate. Methyl 12-mesyloxyoleate can undergo competitive elimination or substitution depending on the experimental conditions. Gunstone and Said (1971) showed that the elimination was the dominant reaction and conjugated and non-conjugated methyl octadecadienoates were the major reaction products with an appropriate base. In particular, heating for 12 hours with the polycyclic base 1,8-diazo-bicyclo[5.4.0]undec-5-ene (DBU) or 1.5-diazobicyclo[4.3.0]non-5-ene (DBN) gave 100% of elimination and mainly the $9c,11t$-octadecadienoate isomer; a reaction time of four hours may be sufficient (Bascetta *et al.*, 1984). Berdeaux *et al.* (1997) adapted this method using DBU as base to prepare 50–60 g batches of $9c,11t$-18:2 from ricinoleate in overall yield of >70%. Although it is an efficient reaction, the necessity of using expensive elimination reagents such as DBU increases the production cost and makes this synthesis pathway uneconomical for a commercial synthesis (Reaney *et al.*, 1999).

Schneider *et al.* (1964) developed a convenient method for preparing $9t,11t$-18:2 from ricinelaic acid (12-hydroxy-octadec-9E-enoic acid) which was heated at 235 °C under vacuum to form a polyester. Pyrolysis of this polyester and simultaneous distillation of the products gave $9t,11t$-18:2 in a low yield (35%). More recently, demesylation of mesyloxy derivative of methyl ricinelaidate with DBU furnished $9t,11t$-18:2 in 76% yield (Lie Ken Jie *et al.*, 1997). Methyl $9t,11t$-

18:2 was also prepared from 9,12-dihydroxy-10*t*-octadecenoate using chlorotrimethylsilane and sodium in acetonitrile in 72% yield (Lie Ken Jie and Wong, 1992). Yurawecz *et al.* (1994) converted allylic hydroxy oleate to a mixture of CLA isomers by an acid-catalyzed methylation using HCl/methanol or BF$_3$/ methanol. They obtained a mixture of 9*c*,11*t*-, 9*t*,11*t*-, 7*t*,9*c*- and 7*t*,9*t*-18:2 isomers in a good yield (72–78%). This method was used to prepare mixture of CLA isomers as analytical standard (Yurawecz *et al.*, 1994).

11.3.3 Reduction of acetylenic bonds

Santalbic acid (octadec-11*E*-en-9-ynoic acid) isolated from *Santalum album* seed oil or prepared from ricinoleic acid (Lie Ken Jie *et al.*, 1996), is a useful precursor to synthesize 9*c*,11*t*-18:2. Zinc in aqueous *n*-propanol was highly selective in the reduction of the acetylenic bond to the corresponding Z-olefinic bond. Using this procedure, Lie Ken Jie *et al.* (1997) prepared 9*c*,11*t*-18:2 in the pure form from methyl santalbate in 82% yield. Following the same procedure, they prepared pure 9*c*,11*c*-18:2 isomer from its Z isomer (octadec-11Z-en-9-ynoic acid). Lindlar catalyst and H$_2$ have also been used to reduce the acetylenic bond of methyl santalbate into the *cis*-bond (Smith *et al.*, 1991). Moreover, the Lindlar catalyzed reduction of methyl santalbate has been used in presence of deuterium gas to produce 9*c*,11*t*-18:2-9,10d$_2$ in yield of 65–70% (Adlof, 1999). Iron in 15% *n*-propanol solution has also been used to reduce acetylenic bond of methyl santalbate and 9*c*,11*t*-18:2 was obtained in 85% yield (Morris *et al.*, 1972). Morris *et al.* (1972) prepared some *cis/cis* CLA isomers (6*c*-/7*c*,9*c*-/8*c*,10*c*-/9*c*,11*c*-18:2) by reduction of the corresponding synthesized conjugated methyl octadecadiynoate with iron in aqueous propanol in 56–77% yield.

11.3.4 Multiple step syntheses

Pure CLA isomers can also be produced by total stereoselective multiple step syntheses, in small quantities (up to 1 g), the isomeric purity can be higher than 98%. Stereoselective synthesis guarantees the exact chemical structure of the final molecule and limits the formation of by-products due to several purification procedures performed during the synthesis. Moreover, these syntheses should allow the preparation of labeled molecules of CLA with deuterium, ^{13}C or ^{14}C. Recently Adlof (1997) described a multistep synthesis of deuterium-labeled 9*c*,11*t*- and 9*t*,11*t*-18:2 with isotopic and chemical purities >95% for each geometrical isomers, in 12% overall yield. The two isomers were synthesized as a mixture via a Wittig coupling reaction. The two isomers were readily separated by a combination of reverse-phase and silver resin chromatography.

Loreau *et al.* (2001) described the stereoselective multistep syntheses of 9*c*,11*t*, 10*t*,12*c* and 10*c*,12*c*-[1-^{14}C] conjugated linoleic acid isomers with high radio-chemical and isomeric purities (>98%) (Fig. 11.2). They described the stereo-controlled preparation of (Z,E)- or (Z,Z)-heptadecadienyl bromo precursors of the CLA isomers by using the sequential substitution of (*E*)- or (*Z*)-1,2-dichloro-

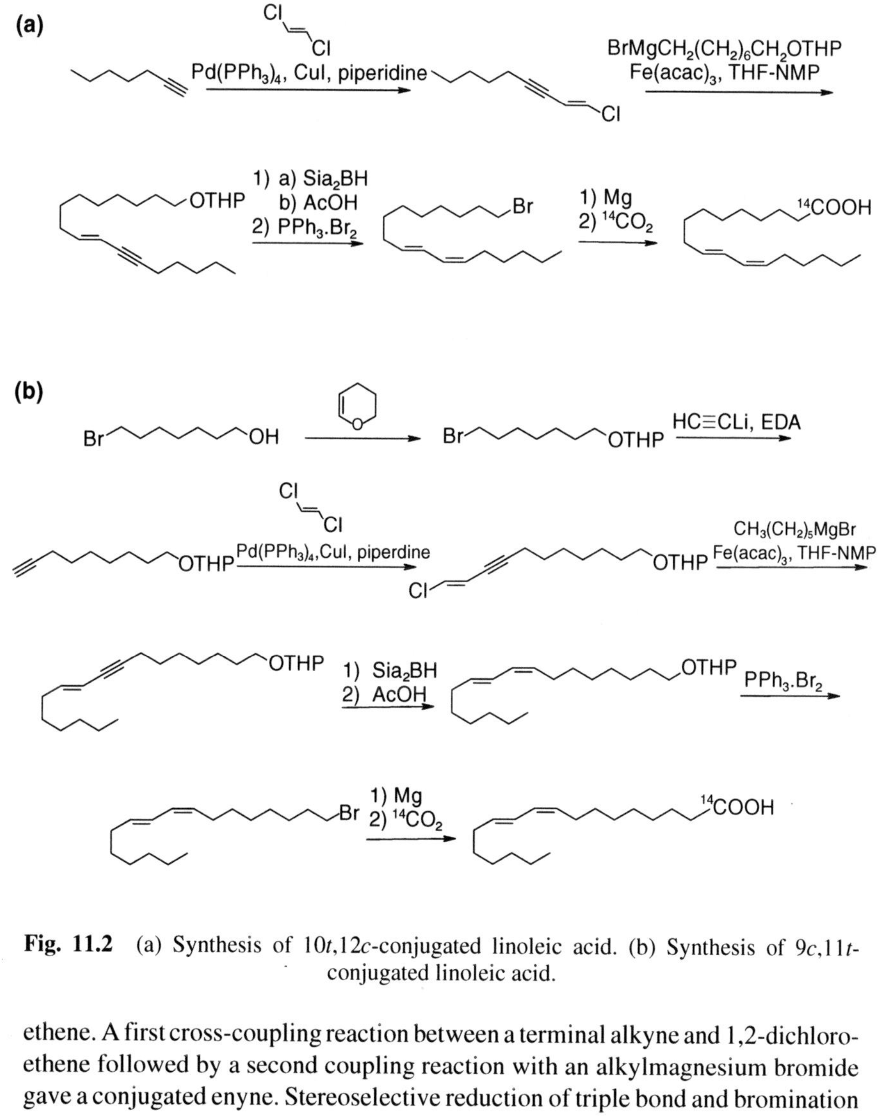

(a)

(b)

Fig. 11.2   (a) Synthesis of 10*t*,12*c*-conjugated linoleic acid. (b) Synthesis of 9*c*,11*t*-conjugated linoleic acid.

ethene. A first cross-coupling reaction between a terminal alkyne and 1,2-dichloro-ethene followed by a second coupling reaction with an alkylmagnesium bromide gave a conjugated enyne. Stereoselective reduction of triple bond and bromination afforded the expected bromo precursor. After formation of the heptadecadienyl-magnesium bromide, (9,11)- and (10,12)-CLA isomers were obtained with ^{14}C carbon dioxide.

Stereoselective synthesis procedures for 9*c*,11*c*-, 9*t*,11*t*-, 10*t*,12*c*, 7*t*,9*c*- and 9*c*,11*t*-18:2 using comparable pathways have been published by Lehmann (2001). All isomers were synthesized from commercial reagents via acetylenic-coupling reactions, and stereoselective hydrogenation of the first double bond in *E* or *Z* configuration. The second double bond was formed selectively in *E* or *Z* configuration by a Wittig reaction. All synthesized CLA isomers showed a good isomeric purity without any further purification.

Some long chain metabolites of 9c,11t-18:2 were prepared stereoselectively by total synthesis. Labelle *et al.* (1990) described syntheses of 5c,8c,11c,13t-20:4. The final product was obtained through two highly stereoselective Wittig reaction under *cis*-olefination condition, the first one between 6-(*t*-butylsiloxy)-3Z-buten-1-yltriphenylphosphonium iodide obtained in six steps from commercial 3-butyn-1-ol, and methyl 5-oxopentanoate and the second one between (Z,Z) (methyl 5,8-undecadienoate)-11-yl) triphenylphosphonium iodide with commercial (*E*) 2-nonenal to give the final molecule. However, in order to introduce the conjugated double bond system with specific stereochemistry *cis* at the newly formed double bond, lithium hexamethyldisilazide (LiHMDS) was used as a weak base in the presence of hexamethyl phosphoramide (HMPA) in tetrahydrofuran (THF) at −78 °C in the second Wittig reaction, to avoid isomerization of the existing *trans* double bond. The conjugated C20:4 was obtained in 15 steps with high isomeric purity. The same authors prepared 5c,8c,11c,13t,15t-20:5 following the same protocol as described for 5c,8c,11c,13t-20:4 but using (*E,E*)-2,4-nonadienal as the aldehyde in the second Wittig condensation. However, this synthesis does not allow the preparation of the corresponding labelled analogue.

Gnädig *et al.* (2001) developed efficient and highly stereoselective multi-step syntheses of the 6c,9c,11t-18:3 and the 8c,11c,13t-20:3 and their [1-^{14}C] radiolabelled analogues. The synthesis pathway for the 8c,11c,13t-20:3 is presented in Fig. 11.3. In a first step the synthesis involves a highly stereoselective Wittig reaction between 3-(*t*-butyldiphenylsilyloxy)propanal and 7-(2-tetra-hydropyranyloxy)heptanylphosphonium salt, which gave (3Z)-1-(*t*-butyldiphenyl silyloxy)-10-(2-tetrahydropyranyloxy)dec-3-ene. Then the *t*-butyldiphenylsilyl derivative was deprotected selectively and the obtained alcohol function was converted via a bromide into a phosphonium salt. The second stereoselective Wittig condensation between the phosphonium salt and commercial (2E)-non-2-enal under *cis*-olefinic conditions using LiHMDS as base afforded the (7Z,10Z,12E)-1-(2-tetrahydropyranyloxy)nonadeca-7,10,12-triene in a very high isomeric purity. The intermediary product was bromated and transformed by reaction with Mg in a Grignard reagent, which was one-carbon elongated by unlabelled or labelled carbon dioxide to obtain the 8c,11c,13t-20:3 in good isomeric purity and high radiochemical purity for its [1-^{14}C]-radiolabelled analogue. The 6c,9c,11t-18:3 was synthesized in a similar way by using 5-(2-tetrahydro-pyranyloxy)pentanylphosphonium salt in place of 7-(2-tetrahydropyranyloxy)-heptanylphosphonium salt. Other reactions were unchanged and products were obtained in similar yields. As 8c,11c,13t-20:3, the 6c,9c,11t-18:3 was obtained in a very good purity and its [1-^{14}C]-radiolabelled analogue in a high radiochemical purity.

Georger and Hudson (1988) prepared parinaric acid (9t,11t,13t,15t-18:4) labeled with deuterium atoms at all vinyl positions by using the Wittig reaction to couple diene phosphorane with α,β-unsaturated aldehyde-ester. The preparation of each component included the stereoselective reduction of a substituted propynoic ester with lithium aluminium deuteride introducing the *trans* double bond as well as most of the deuterium atoms in one step in high purity.

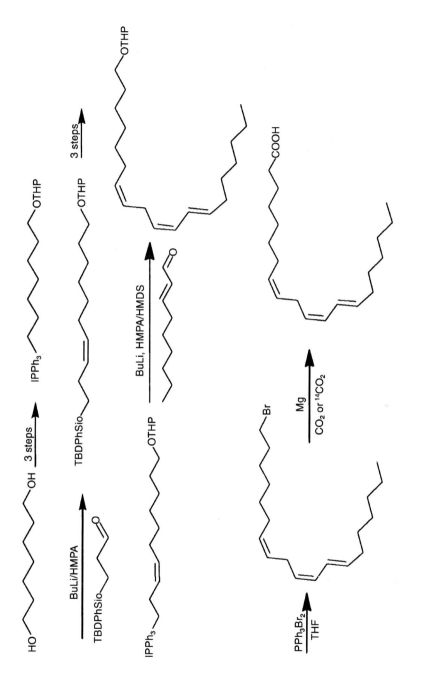

Fig. 11.3 Synthesis of 8c,11c,13t-20:3.

11.3.5 Biochemical syntheses

At this time, the use of biosynthetic method utilizing specific enzymes isolated from algal or bacterial sources to produce CLA isomers is limited to the preparation of 9c,11t-18:2 isomer. First, 9c,11t-18:2 isomer was prepared from linoleic acid using the linoleate isomerase enzyme isolated from the rumen bacterium *Butyrivibrio fibrisolvens* (Kepler and Tove, 1967). Using the same linoleate isomerase enzyme from *B. fibrisolvens*, the same authors prepared also ^{14}C- and ^{2}H-labeled 9c,11t-18:2 from linoleic acid (Kepler and Tove; 1967, Kepler *et al.*, 1966). Recently, a microbial synthesis of 9c,11t-18:2 from linoleic acid using a culture of *Lactobacillus* sp. has been described (Reaney *et al.*, 1999). A total CLA level of 7 mg/g cells was produced containing >96% 9c,11t-18:2 isomer.

11.4   Analytic methods

As described before, many positional and geometrical CLA isomers are present in natural and synthetically produced materials. Therefore, high-quality analytical methods are needed to determine the exact isomer composition of the complex CLA mixtures in synthetic CLA, foodstuff or biological matrices. Great care has to be taken with the analysis of CLAs as they are unstable and very sensitive to isomerization (Shantha *et al.*, 1993). Therefore, detected CLA levels could be incorrect because of the use of unsuitable analytical methods in the past (Hamilton, 2001). It has been suggested that any data reported more than five years ago should be critically regarded (Christie, cited in Hamilton, 2001).

To avoid any problems, the CLA-containing substance has to be transformed into fatty acid methyl esters (FAME) using mild methylation methods. Acid catalysts or high temperatures lead to isomerization, where *cis–trans* and *trans–cis* isomers are converted into *trans–trans* isomers, and to the formation of methoxy artefacts (Kramer *et al.*, 1997; Park *et al.*, 2001b). Also methoxy artefacts appeared using the methylation with trimethylsilyl-diazomethane, an often recommended weak methylation method (Christie *et al.*, 2001). Usually sodium methoxide is used as a weak base.

CLA contents are currently determined using gas chromatography (GC) coupled to a flame ionization detector (FID). To obtain a good separation of the CLA isomers, high polar capillary columns 50 and 100 m in length are used. Although a good separation of the CLA isomers can be reached, some of the positional isomers, e.g. 7t,9c-18:2 and 9c,11t-18:2, co-elute and the samples have to be analyzed using other techniques (Rickert and Steinhart, 2001). The utilization of silver ion (Ag$^+$)-high-performance liquid chromatography (HPLC) improved the resolution of the CLA isomers, since two to six columns were coupled in series. The CLA isomers are separated into three groups depending on the configuration of the double bounds, in *trans–trans, cis–trans/trans–cis and cis–cis* respectively. Furthermore Ag$^+$-HPLC led to a better separation of the CLA with less coelution compared to GC (Sehat *et al.*, 1999).

Exact structural determination of single CLA isomers is possible using different spectroscopic methods. The analysis is often done using GC coupled with mass spectrometry (MS). Therefore, the FAME are transformed into a 4-methyl-1,2,4-triazoline-3,5-dione (MTAD) adduct or a 4,4-dimethyloxazoline derivative (DMOX) (Dobson and Christie, 1996; Dobson, 1997). The use of these derivatives in GC–MS permits the localization of the double bond position in the fatty acid molecule. To get information about the configuration of the double bond, GC–Fourier-transformed infra red spectroscopy (FTIR) is applied. The IR-spectra for conjugated *trans–trans* and *cis–trans/trans–cis* isomers exhibited absorption at 990 cm^{-1} and characteristic bands at 988 and 949 cm^{-1}, respectively. However, differentiation between a *cis–trans* and *trans–cis* diene is not possible through FTIR (Mossoba *et al.*, 1999). The positions and relative intensities of the =C–H stretch bands are highly characteristic and could discriminate between *cis–trans/ trans–cis* (3020 and 3002 cm^{-1}), *cis–cis* (3037 and 3005 cm^{-1}) and *trans–trans* (3017 cm^{-1}) conjugated double bond systems (Fritsche *et al.*, 1997).

11.5   The influence of processing on the CLA content of dairy products

The relationship between the CLA content and different parameters of food processing in dairy products is a controversial area. Because of their chemical structure containing a conjugated double bond system, CLAs have been seen as more sensitive to oxidation or to isomerization during heat treatment than linoleic acid. Research has been undertaken to determine the influence of heat treatments, such as grilling, cooking and frying, on CLA content, to test the stability in food products and to avoid a CLA decrease by oxidative damage. Home-made food preparation had no influence on CLA content. Shantha *et al.* (1994) demonstrated that frying, baking, broiling or microwaving did not change the CLA content in grilled hamburger beef patties. These findings were supported by a recent study, which described that only intensive heating for 15 min, using temperatures higher than 200 °C, led to isomerization of CLA in milk, whereas moderate heating had no effect. Heating at 225 °C for 15 min decreased 9c,11t-18:2 from 1.7% to 1.1% of total fatty acids (Precht *et al.*, 1999).

The influence of storage on the CLA content was also tested in butter, yoghurt and sour cream. No modification of the CLA content was detected (Shantha *et al.*, 1995). Dairy products often undergo a microbial fermentation during processing. The use of different fermentation cultures, processing temperatures or ripening periods could affect the CLA level in the final foodstuff (Fritsche and Steinhart, 1998a). These suggestions were confirmed by the findings of Jiang *et al.* (1998), who reported the ability of two strains of *Propionibacterium* spp. to produce CLA in culture by conversion of linoleic acid into 9c,11t-18:2 (Jiang *et al.*, 1998). More recently, six further *Lactobacillus* spp. were identified as able to convert linoleic acid into CLA (T.Y. Lin *et al.*, 1999a; Lin, 2000). The use of these strains could be one way of enriching yoghurts and cheese in CLA during fermentation.

As cheese processing involves bacterial fermentation, various studies have been carried out on this subject. Only minor changes in the CLA contents of different cheddar cheese varieties were found (3.20 mg/g fat *vs* 3.55 mg/g fat) (H. Lin *et al.*, 1999). Another study, published by a Swedish group (Jiang *et al.*, 1997) tested various hard cheeses. They observed the same CLA amounts in raw material and in the final product. Other processing parameters during cheese fabrication were investigated to see if these were able to modify CLA content. Changes in milling and pH values or the addition of butylhydroxyanisol (BHA) or amino acids (lysine, tyrosine) led to a small decrease of the CLA content in the final cheese (2.70 mg/g fat *vs* 2.19 mg/g fat by addition of tyrosine and reduced pH) (Lin *et al.*, 1998).

Various studies have been undertaken to determine the CLA content in cheese spread using different processing conditions. Tests compared processing under atmospheric conditions and under nitrogen as protective gas. It was shown that processing under atmospheric conditions at temperatures of 80–90 °C could increase the CLA content in the product. This effect was amplified by the addition of whey proteins as hydrogen donors or antioxidants as butylhydroxytoluene (BHT). Also the use of iron as Fe^{2+}/Fe^{3+} ascorbate as additive seems to be useful to increase the CLA level (Shantha *et al.*, 1992, 1995).

In contrast, the addition of antioxidants (ascorbate or whey protein concentrates) acting as hydrogen donors to beef patties did not influence the CLA content of the samples (Shantha and Decker, 1995). The controversial results concerning the relationship of antioxidants and hydrogen donors with the CLA content could be related to the foodstuff itself and the presentation of the fatty acids in the food matrix. However, the changes in CLA content, reported during processing, are much less important than the seasonal differences in milk, so that the CLA content in the processed food largely depends on the CLA content in the original milk.

11.6   Functional benefits of CLA: cancer

In the early 1980s Pariza and Hargraves (1985) reported that an extract of grilled ground beef exhibited mutagenesis inhibitory activity in experiments using the AMES test. The active molecules were later identified as CLA (Ha *et al.*, 1987). Since that time, researchers have paid more attention to CLA, and there is now growing evidence of the multisite and multistage anticarcinogensis activity of CLA. CLA could inhibit many chemical-induced cancers of the mammary, stomach, colon and skin, and a variety of tumour cell lines (including breast/mammary cancer, colorectal cancer, prostate cancer, lung adenocarcinoma, malignant melonoma, neuroglioma) demonstrated inhibition by CLA. Moreover, CLA may inhibit carnogenesis at each of its major stages, such as initiation, promotion (post-initiation), progression and metastasis (Pariza, 1997; Pariza *et al.*, 1999).

Though the mechanisms through which CLA inhibits tumourigenesis are not completely established, they seem to fall into the following categories:

• an antioxidant mechanism involving lipid peroxidation production toxic to cancer cells

- changing the fatty acid profile in tumour cells and interfering with eicosanoid metabolism
- reducing the formation of carcinogen–DNA adducts
- influencing the growth and development of certain types of mammary cells by interfering with the estrogen response system, inducing apoptosis and regulating gene expression such as c-*myc*, stearoyl-CoA desaturase, PPARα, (peroxisome proliferator-activated receptor) cyclin A, B1, D1 cyclin-dependent kinase inhibitors (CDKI) (P16 and P21) (Kritchevsky, 2000).

In the following sections, we try to summarize CLA multisite and multistage anticarcinogenesis and to explain all these possible mechanisms.

11.7   Multisite anticarcinogenesis

11.7.1   *In vivo* – mammary/breast cancer model

Although CLA demonstrates multisite anticarcinogenesis, research has concentrated on inhibition by CLA of mammary cancer because CLA has shown higher retention in adipose tissue (Sugano *et al.*, 1997). The retention of conjugated linoleic acid in the mammary gland may be associated with tumour inhibition. Ip *et al.* (1997) examined the influence of dietary CLA on tumour development in the rat after the use of dimethylbenz(a)anthracene (DMBA) as carcinogen administrated in a single dose. One per cent CLA (from Nu-Chek Prep, Inc: 43.3% *c*9,*t*11- and *t*9,*c*11-CLA; 45.3% *t*10,*c*12; 1.9% *c*9,*c*11-CLA; 1.4% *c*10,*c*12-CLA; 2.6% *t*9,*t*11- and *t*10,*t*12-CLA; 4.4% linoleic acid) was given in the diet for 4 and 8 weeks, and continuously after the exposure to the carcinogen. It was shown that dietary CLA for 4 and 8 weeks did not prevent the occurrence of mammary carcinoma. However, significant tumour inhibition was observed in rats fed with a CLA diet for the entire duration of the experiment. This demonstrated that the physiological accumulation of CLA in the mammary gland contributed to cancer protection in the tumour post-initiation stage. Analysis of fatty acids in mammary gland showed that more CLA was incorporated into mammary tissue neutral lipids than in mammary tissue phospholipids. Moreover, CLA levels in neutral lipids disappeared as CLA was removed from the diet, which resulted in the subsequent occurrence of new tumours (Ip *et al.*, 1997). However, five weeks of short-term CLA feeding, from weaning to the time of carcinogen administration (7,12-dimethylbenz(a)anthrane and methylnitrosourea) at 50 days of age, significantly protect against subsequent tumour occurrence, which is related to maturation of the mammary gland to the adult stage (Ip *et al.*, 1994, 1995).

CLA at dietary concentrations that are close to the levels consumed by humans showed a significant anticarcinogenic activity in experiments using the rat mammary cancer model. Diets containing 0.5, 1 or 1.5% CLA isomer mixture could reduce the total tumour numbers by 32, 56 and 60%, respectively (Ip *et al.*, 1991). A further study showed that dietary CLA between 0.05 and 0.5% caused a dose-dependent inhibition of mammary tumourigenesis where the tumour yield was reduced to 22, 36, 50 and 58% by feeding 0.05, 0.1, 0.25 and 0.5% CLA in the diet. No further beneficial effect was observed by increasing the CLA level in the diet to 1% (Ip et al., 1994). Moreover, the inhibition by CLA of mammary cancer is independent of the level or type of fat in diet. To study the influence of the level or type of fat in diet on CLA tumour inhibition, a fat blend (10, 13.3, 16.7 or 20% by weight in the diet) or 20% fat diet containing either corn oil or lard was used. The results showed that the inhibition by 1% CLA on tumour size was not influenced by the level or type of fat in the diet (Ip et al., 1996).

The main sources of CLA are dairy products and foods derived from ruminants. Therefore, they are in the form of triacylglycerols. It is necessary to carry out the animal experiment using CLA as triacylglycerols. The working group of Ip was the first to compare the effects of CLA as free fatty acids and CLA as triacylglycerol on the development of mammary carcinogenesis using the rat methylnitrosourea (MNU) model. They reported that the anticancer activities of both molecules (triacylglycerols and free fatty acids) were identical (Ip et al., 1995). Subsequently, CLA-enriched butter fat (CLA as triacylglyerols in natural form) was used in an animal experiment and it was shown that natural CLA possessed the biological action in cancer inhibition (Ip et al., 1999a; O'Shea et al., 2000). In addition, Shirai's group recently reported that conjugated fatty acids derived from safflower oil (CFA-S) containing large amounts of CLA may retard the development of 2-amino-1-methyl-6-phenylimidazo(4,5-b) pyridine (PhIP)-induced mammary tumours (Kimoto et al., 2001; Futakuchi et al., 2002).

11.7.2 *In vivo* – other cancer models
In other animal models, CLA proved to be equally effective in tumour inhibition. The benzo(a)pyrene-induced mouse forestomach model was firstly used to study the effect of CLA on carcinogenesis in animal. Ha et al. (1990) gave a synthetic CLA isomer mixture (c9,t11-, t10,c12-, t9, t11- and t10,t12-CLA) to female ICR mice treated with benzo(a)pyrene. Dietary CLA significantly reduced tumour incidence (up to 30%) as well as tumour multiplicity compared with the control group. In mouse skin tumour promotion elicited by 12-O-tetradecanoyl phorbol-13-acetate (TPA), mice fed diets with 1.0–1.5% CLA developed significantly fewer tumours than those fed a diet without CLA (Belury et al., 1996). Using a rat colon cancer model and different carcinogens such as 2-amino-3-methylimidazo (4,5-f) quinoline (IQ), heterocyclic amine, 1,2-dimethylhydrasine, azoxymethane (AOM), researchers demonstrated that CLA significantly decreased the number of colonic aberrant crypt foci (ACF) and inhibited colon tumour incidence (Liew et al., 1995; Park et al., 2001a; Xu and Dashwood, 1999; Kohno et al., 2002).

11.7.3 *In vitro* model
Tumour cell lines are easily available and the treatment is less complicated than that in animal model. A number of tumour cell lines were used to study the CLA anticarcinogenesis. Shultz et al. (1992) firstly confirmed that CLA isomer mixtures

(43% $c9,t11$- and $t9,c11$-, 41.5% $t10,c12$-, 1.6% $c9,c11$-, 2.4% $c10,c12$-, 11.2% $t9,t11$- and $t10,t12$-) significantly reduced human (malignant melanoma M21-HPB, colorectal HT-29, breast MCF-7) cancer cells proliferation by 18–100%. Subsequently, much research attempted to explain CLA anticarcinogenesis mechanism using different human cancer cells, mainly human mammary cancer cell MCF-7. For example, the relation between CLA, linoleic acid, cyclooxygenase and lipoxygenase inhibitor, estrogen reponse system, antioxidant enzyme defense responses and liperoxidation, cell signal transduction, arachidonic acid distribution, stearoyl-CoA desaturase activity were studied in MCF-7 breast cancer cell (O'Shea *et al.*, 1998, 1999; Cunningham *et al.*, 1997; Durgam and Fernandes, 1997; Park *et al.*, 2000; Choi *et al.*, 2002). Besides breast cancer MCF-7 cells, human breast (MDA-MB-231 and MDA-MB-468) (Visonneau *et al.*, 1997), gastric (SGC-7901) (Liu *et al.*, 2002), colorectal (HT-29, MIP-101 and SW480), prostate (PC-3 and DU-145) (Palombo *et al.*, 2002), hepatoma (HepG2) (Igarashi and Miyazawa, 2001), and three different lung adenocarcinoma (A-427, SK-LU-1, A549) cancer cell lines were used in CLA research. However, CLA showed no influence on human glioblastoma cell line (A-172) growth (Schonberg and Krokan, 1995).

11.8   Multistage anticarcinogenesis

11.8.1   Initiation and promotion (post-initiation)
For mammary tumour initiation research, the exposure of CLA started during the early post-weaning and pubertal period before carcinogen administration, which was sufficient to reduce subsequent tumourigenesis induced by different carcinogens such as MNU and DMBA; for mammary tumour promotion (post-initiation), CLA feeding was given after carcinogen administration, a continuous intake (20 weeks) of CLA was required for inhibition of carcinogenesis, but 4 or 8 weeks of CLA feeding did not show any effect on the tumour inhibition (Ip *et al.*, 1994, 1995, 1997). In mouse skin tumour promotion elicited by TPA, Belury *et al.* (1996) also confirmed the fact that CLA inhibited the tumour promotion.

In addition, DNA-adduct, as one of the biological makers of tumour initiation, was a hot topic for CLA research. Generally, CLA can inhibit different mutagen–DNA adduct formation in different organs. For example, Schut reported that CLA treatment caused inhibition of adduct formation in the liver, lung, large intestine and kidney in CFD1 mouse given IQ. IQ–DNA adduct formation in the kidneys of females was reduced by up to 95.2% (Schut and Zu, 1993). The F344 rat given CLA also had lower IQ–DNA and PhIP–DNA adduct in various organs including the mammary gland and the colon (Schut *et al.*, 1997). Recently, Futakuchi *et al.* (2002) confirmed that CFA-S (conjugated fatty acids derived from safflower oil; mainly CLA) could also suppress PhIP–DNA adduct formation in the epithelial cells of mammary gland. But Josyula *et al.* (1998) found that dietary CLA did not absolutely decrease DNA adduct. They reported that CLA increased IQ–DNA adduct formation by 30.5% on day 8 in the colon of female Fisher 344 rats given

AIN-76 diet with 4% CLA and treated with IQ or PhIP after two weeks and had no effect on adduct levels in liver or white blood cells (Josyula *et al.*, 1998; Josyula and Schut, 1998).

11.8.2   Progression and metastasis

To date, there have been few reports assessing the effects of CLA on the progression or metastasis of established tumour. In the chemical-induced tumour model, the progression or metastasis stage is difficult to confirm. Fortunately, the growth and spread of the transplantable tumours in animal provide an ideal way to assess the effect of CLA on tumour progression and metastasis because of the high similarity to human cancer progression and metastasis *in vivo*. In one paper, CB-17 *scid/scid* mice (severe combined immunodeficient) were fed 1% CLA (42.3% *c*9,*t*11- and 44.2% *t*10,*c*12-) diets starting two weeks before MDA-MB468 transplantation throughout the whole of experiment. It was found that CLA not only inhibited tumour growth, but also abrogated the spread of breast cancer cells to lung, peripheral blood and bone marrow (Visonneau *et al.*, 1997). Hubbard *et al.* (2000) also reported that 0.1, 0.5 and 1.0% CLA diets significantly retarded the growth rate of transplantable mice mammary tumour cell line 4526 in *BALB/cfC3H* mouse in a dose-dependent manner and decreased the numbers of pulmonary nodules. The same conclusion was drawn in the human prostatic cancer cell line DU-145 transplantable model in SCID mice (Cesano *et al.*, 1998). However, Wong *et al.* (1997) reported that 0.1, 0.3 or 0.9% CLA diets had no obvious effect on the growth of the established, aggressive mammary tumour. Further research should be pursued.

11.9   Mechanisms of CLA anticarcinogenesis

11.9.1   Antioxidant mechanism

To explain its anticarcinogenic action, it was firstly suggested that CLAs may have an antioxidative activity because of their chemical structure containing a conjugated double bond system. Ha *et al.* (1990) described that lowest concentrations of CLA showed a high antioxidative action which was more potent than that of α-tocopherol. Leung's research was also in agreement with the above conclusion (Leung and Liu, 2000). However, higher concentrations of CLA seem to show prooxidative activity, which suggests that there is a balance between antioxidation and prooxidation of CLA.

11.9.2   Cytotoxicity of lipid peroxidation production to cancer cells

Two research groups directed by Shultz and O'Shea suggested that the cytotoxicity of lipid peroxidation induced by CLA plays an important role in anticarcinogenesis (O'Shea *et al.*, 1999; Shultz *et al.*, 1992). Shultz *et al.* (1992) treated MCF-7 breast cancer cells with 1.78-7.14×10^{-5} CLA, and found that treated MCF-7 cancer cells showed significantly growth inhibition in a dose- and time-dependent manner. Further studies confirmed that cytostatic and cytotoxic effects of CLA were more pronounced (8–81%) than LA at similar LA and CLA concentrations (Shultz *et al.*, 1992). In O'Shea's group, when MCF-7 and SW480 cells were treated with CLA amounts greater or equal than 15 ppm for eight days or more, the amount of the lipid peroxidation product was significantly increased and the cytotoxicity to tumour cell induced by CLA is related to the extent of lipid peroxidation product (O'Shea *et al.*, 1999). Moreover, it was found that bovine milk fat enriched with CLA could decrease the MCF-7 cell numbers up to 90% and with the increasing CLA concentration between 16.9 and 22.6 ppm in milk fat lipid peroxidation increased 15-fold (O'Shea *et al.*, 1999). But the cytotoxicity of lipid peroxidation induced by CLA may be only one part of a CLA anticarcinogenesis mechanism. Schonberg and Krokan (1995) reported that in the lung adenocarcinoma cell lines, 40 μM CLA significantly increased the lipid peroxidation level by twice as much as 40 μM LA. However, 30 μM vitamin E could completely abolish the lipid peroxidation and the growth inhibition of CLA was only partially restored, which indicates that a more complex mechanism may explain the anticarcinogenic activities (Schonberg and Krokan, 1995).

11.9.3 Influencing growth and development of certain types of mammary cells

Interestingly, CLAs may influence the morphological status of the mammary gland, by which mammary cancer risk was reduced. It was shown that the feeding of 0.5 or 1% CLA to rats from weaning to about seven weeks of age in the normal diet reduced the density of terminal end buds (TEB) and lobuloalveolar buds. TEB and lobuloalveolar buds are the target sites for the development of mammary cancer, so their decrease may reduce the cancer risk. Moreover, mammary cells showed lower proliferative activity in the sites of TEB and lobuloalveolar buds (Banni *et al.*, 1999; Thompson *et al.*, 1997).

The same team also tried to determine that a high CLA butter had similar biological activities as the CLA isomer mixture. They reported that, in rats given butter fat CLA during the pubescent mammary gland development period, mammary epithelial mass was reduced by 22%, the size of the TEB population was decreased by 30%, the proliferation of TEB cells were suppressed by 30% and mammary tumour yield was inhibited by 50% (Ip *et al.*, 1999). In general, it might be more important for mammary tumour chemoprevention that CLA or triacylglycerols CLA is given to the animal during the development of mammary gland.

11.9.4 Interfering in the estrogen response system

One paper reported that CLA selectively inhibited proliferation of estrogen receptor (ER) positive MCF-7 cells as compared with ER negative MDA-MB-231 cells and a higher percentage of MCF-7 cells treated with CLA remained in the G0/G1 phase of cell cycle compared with control and those treated with linoleic acid

(Durgam and Fernandes, 1997). That CLA might inhibit MCF-7 cell growth by interfering with the hormone-regulated mitogenic pathway needs to be further investigated.

11.9.5   Inducing apoptosis

Tumour chemoprevention can be carried out by two pathways: one is to inhibit the proliferation of tumour cells and the other is to induce the apoptosis of tumour cells. Ip *et al.* (1999) reported that CLA isomer mixture inhibited the growth of mammary epithelial cell organoids (MEO) by a reduction in DNA synthesis and an induction of normal mammary epithelial cells apoptosis in primary culture. They suggest that CLA-induced apopotosis may play an important role in reducing the epithelial carcinogen-susceptible target cell population. Moreover, they found that CLA increased the chromatin condensation and induced DNA ladder in a rat mammary tumour cell line. This supports the speculation that CLA may induce tumour cell apoptosis. CLA inhibited the formation of premalignant lesions in the rat mammary gland by increasing apoptosis and inducing the expression of bcl-2 in the lesions, but showed no influence on the level of bak or bax in these lesions, and the apoptosis and expression of apoptosis-related protein in normal mammary gland alveole or terminal end buds (Ip *et al.*, 2000). This may be an indication that CLA are more sensitive to premalignancy. Recently, Park *et al.* (2001a) also made the same conclusion in colonic mucosa of 1,2-dimethylhydrazine-treated rat.

11.9.6   Regulating the gene expression

We have discussed that CLA could induce the expression of bcl-2 protein to increasing the apoptosis occurrence. In addition, Durgam and Fernandes (1997) reported that CLA inhibited the expression of c-myc in MCF-7 cell. Using a human hepatoblastoma cell line (HepG-2), it was shown that $t10,c12$-CLA did not influence SCD gene transcription, mRNA and protein levels, but could decrease the SCD activity $c9,t11$-CLA showed no influence on the SCD gene expression and activity (Choi *et al.*, 2001). Subsequently, it was found that $c9,t11$- and $t10,c12$-CLA did not repress SCD mRNA in MDA-MB231 and MCF-7, but could decrease SCD protein in MDA-MB231 and SCD activity in MCF-7 (Choi *et al.*, 2002). In addition, Belury and co-workers found that CLA are ligands and activators of PPARα, suggesting that CLA might exert their protective role against skin tumour promotion by activating PPARα (Moya-Camarena and Belury, 1999; Moya-Camarena *et al.*, 1999a; Thuillier *et al.*, 2000). Recently, Liu *et al.* (2002) reported that $c9,t11$-CLA could decrease the expression of cyclin A, B(1), D(1) and increase the expression of P16(ink4a) and P21(cip/waf1), cyclin-dependent kinase inhibitors (CDKI) in human gastric cancer cell line (SGC-7901).

11.9.7   Changing fatty acid profile and interfering eicosanoid metabolism

CLAs are positional and structural isomers of linoleic acid and probably enjoy the same enzyme for chain desaturation and elongation. It was shown that 1% CLA diet could depress the linoleic acid metabolites including C18:3, C20:3 and C20:4. Moreover a significant drop of C20:4 was observed, which may directly influence the release of archidonic acid metabolites (prostaglandins and leucotrienes) (Banni *et al.*, 1999). Miller *et al.* (2001) reported that the growth inhibition of CLA isomer mixture or *c*9,*t*11-CLA may be induced by changes in AA distribution among cellular lipids and an altered prostaglandin profile. They found CLA isomer mixture or *c*9,*t*11-CLA could increase AA uptake into the monoglyceride fraction of MCF-7 cells and into the triacylglycerol fraction of SW480 with reduced uptake into phospholipids of SW480; unlike phosphatidylethanolamine, the uptake of AA into phosphatidylcholine was decreased by the *c*9,*t*11-CLA isomer in both cell lines. In addition, CLA isomer mixture and the *c*9,*t*11-CLA could also decrease the PGE_2 levels and increase $PGF_{2\alpha}$. Igarashi and Miyazawa (2001) also suggested that the growth inhibitory effect of CLA on HepG2 is due to changes in fatty acid metabolism.

11.10   Functional benefits of CLA: lipid and protein metabolism

CLA is recognized as having some modulating effects on energy and protein metabolism (Fitch-Haumann, 1996). Studies have been reported in animals, including rodents (mice, rats, hamsters) as well as pigs, and a few human trials have also been carried out. In rodents, the first report demonstrating that CLA (fed as a mixture of isomers as free fatty acids) may alter the body composition was published in 1997 (Park *et al.*, 1997). In this paper, Park *et al.* showed that feeding ICR mice with CLA (0.5% in the diet at the expense of corn oil) induced a decrease in body fat content which was partly balanced by an increase in the body protein content as well as body water content. In this early study, the effects of CLA were studied on both male and female mice. A similar trend was observed in males and females, but the extent of the decrease of body fat was slightly higher for females (Table 11.2).

Table 11.2 Empty carcass weight (ECW), fat and protein content of male and female ICR mice fed a diet containing 0.5% of CLA for 32 d (males) or 28 d (females)

	ECW (g)	Fat (%)	Protein (%)
Males			
Control	32.4 ± 1.1	10.13 ± 1.17	17.76 ± 0.30
CLA	32.2 ± 0.8	4.34 ± 0.40*	18.58 ± 0.14*
Females			
Control	25.0 ± 0.9	18.68 ± 3.08	17.67 ± 0.61
CLA	23.1 ± 1.0	7.47 ± 0.59*	20.09 ± 0.24*

* Significantly different from control.

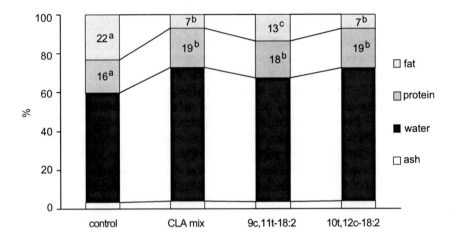

Fig. 11.4   Body composition of female ICR mice after 4 weeks (Park *et al.*, 1999).

Two years later, the same authors showed that these effects on body composition were mainly due to the dietary intake of the *t*10,*c*12 CLA isomer. Again, the effects were studied in both sexes and some differences were observed, but the general trend was the same (Fig. 11.4). In both males and females, the major body fat reduction was obtained when the animals were fed a CLA mixture containing more 10*t*,12*c* isomer. However, the effects obtained after feeding the 9*c*11*t* isomer were also significant, but to a lesser extent (Park *et al.*, 1999b).

Using another strain (AKR/J mice) fed a high-fat or a low-fat diet, West *et al.* (1998) showed a reduced energy intake along the six-week experimental CLA feeding period. It was associated with a significant decrease of body fat, as illustrated by the reduction of inguinal, epididymal, retroperitoneal and mesenteric adipose depot weights. Even if high-fat fed mice were fattier, the reduction of the weight of the fat pads had the same effect. Compared with the previous studies by Park *et al.* (1999b), West *et al.* reported that CLA decreased the carcass lipid content, but also decreased the body protein content (West *et al.*, 1998). To our knowledge, it is the only study reporting such an effect on body protein.

Like Park *et al.* (1999a), Gavino *et al.* (2000) reported a decrease in body weight gain in hamsters fed a CLA mixture, but such effect was not obtained with pure 9*c*,11*t* isomer. Again, it was suggested that the 10*t*,12*c*-18:2 was the active isomer on this parameter and it was suggested that the effect may be due to a lower fat mass accumulation.

Besides these studies carried out with only one dose of CLA in the diet, DeLany *et al.* (1999) studied the dose response of male AKR/J mice fed CLA (0.25–1%) for up to 12 weeks. CLA treatment decreased the body fat content (25.1% in 1% CLA fed mice *versus* 31.4% in the control animals). This decrease in body fat content was observed when CLA was present at least at 0.50% in the diet. As reported by others (see above), the body fat loss was partly balanced by a significant increase of protein content and was not associated with any major effect on food intake. In order to study the time course of the body composition modifications, a kinetic study was carried out (Park et al., 1999a). The authors suggested that CLA firstly increases the protein content and that the decrease in body fat occurs with a slight delay.

Whereas most of the data have been obtained in mice, a few studies have been carried out on other rodents such as rats. Stangl (2000) fed rats submitted to food restriction with 3% (by weight) of a CLA mixture at the expense of sunflower oil. CLA-fed animals gained 11% less weight than control animals and they presented less body fat. The author concluded that CLA acts as a strong repartitioning agent in the Sprague–Dawley male rat by modulating the body composition. Recently, Sisk et al. (2001) showed that feeding CLA reduced the fat pad weight in lean rats but the opposite effect was obtained in obese Zucker rats. This difference in response was associated with an effect on fat cell size.

Besides the studies carried out in various rodents, pigs have also been studied, since the data on rodents suggested that feed conversion and carcass composition would be beneficially affected by CLA treatments. The first study on pigs was published in 1997 by Dugan et al. (1997) who reported that feeding CLA with 2% of a CLA mixture in the diet induced a reduction of deposition of subcutaneous fat associated with an increased lean mass gain. The effect on feed conversion was borderline significant ($P = 0.06$). Later, Ostrowska et al. (1999) studied the effects of increasing CLA intake in female pigs (up to 10 g/kg of diet of a 55% pure CLA mixture). Whereas CLA did not alter weight gain and food intake, feed efficiency was improved by CLA supplementation. Feeding CLA also decreased fat deposition and improved the lean tissue deposition in finisher pigs. The decrease in fat deposition was dose-dependent in the range of CLA supplementation studied.

More recently, Wiegand et al. (2001) showed that a commercial CLA mixture (details of isomers not available) significantly improved the feed efficiency from 330 g/pkg food to 350 g/pkg food. Rib fat was also significantly decreased. These effects were obtained with only 0.75% of CLA in the pig diet.

Taken altogether, these data in pigs strongly suggest that CLA can be considered as a repartitioning agent which is also able to improve feed efficiency in finisher pigs. The effect on feed efficiency was not observed in growing pigs (Ramsay et al., 2001) except in the study reported by Dugan et al. (1997). Using a more complex CLA mixture, Bee (2000) studied the effects of CLA on pregnant and lactating sows. CLA did not modified the pig weight at birth or at weaning. The fatty acid profile of milk and back fat of sows was modified, but the effect on the body composition of the newborn has not been reported.

As overweight and obese subjects are more and more numerous in Western countries, CLA has been suggested to be a good candidate for modulating some imbalance occurring during these situations. However, compared with animal studies, very few trials have been reported (and published as a full paper) in humans. Blankson et al. (2000) showed that feeding CLA (mixture of isomers, up to 6.8 g per day for days) induced a decrease in body fat mass associated with an increase in lean mass. However, a dose-dependent relationship was difficult to establish, as no significant effect was observed with an intermediate daily intake (5.1 g/d), which is not clearly understood. On the other hand, the study reported by Zambell et al. (2000) did not show any effect on body composition on overweight women supplemented for 64 days with a CLA mixture. But Thom et al. (2001) reported that CLA given as a Tonalin® mixture induced a significant decrease in body fat after three months of CLA intake.

This discrepancy has to be discussed considering the difference in the population studied as well as the CLA mixture used. In the first case, the volunteers were men and women with a body mass index (BMI) from 25 to 35, whereas the second study was carried out only with women presenting a BMI between 22 and 23 (non-overweight subjects). Moreover, the CLA mixture used differed, which means that the intake of individual isomers differed. As the CLA isomer profile was also different, it is difficult to conclude, as some antagonism or synergy between isomers cannot be excluded when taking into account the present knowledge.

The effects of CLA on body composition have to be taken with care before building a conclusion. The effects seem to be species-dependent and the mechanism of action are not fully understood. It seems that CLA may alter fat deposition and also modifies protein metabolism, but further studies are mandatory to have a good knowledge of CLA action on body composition. However, growing animals seem to be more responsive than adult organisms.

11.11   The process of CLA metabolism

11.11.1   In animals

As mentioned before, it has often been suggested that CLA may interact with the arachidonic acid metabolism perhaps by competing with linoleic acid. Conjugated C18:3, C20:3 and C20:4 fatty acids, as corresponding metabolites, were identified in animals fed CLA. In a first study Sébédio et al. (1997) detected conjugated C20:3 and C20:4 fatty acids in the liver and adipose tissue. Male Wistar rats were fed a fat-free diet for two weeks and afterwards force-fed with a CLA mixture containing mainly $c9,t11$-18:2 and $t10,c12$-18:2 for six days while maintaining the fat-free diet. Conjugated C20:3 and C20:4 were identified as respective long chain fatty acid metabolites of $c9,t11$-18:2 and $t10,c12$-18:2. A second experiment was carried out by Banni et al. (1999) under normal nutritional conditions. Female Sprague–Dawley rats were fed a standard diet enriched with CLA mixture for one month. They found conjugated C18:3 and C20:3 in liver and mammary tissue, but no conjugated C20:4 was detected. Recent data on this subject using a CLA-enriched butter fat, which contained the natural isomer composition (mainly $c9,t11$-18:2), described the formation of conjugated C18:3, C20:3 and C20:4 (Banni et al., 2001). The incorporation of the metabolites of CLA were found primarily in triacylglycerols. Furthermore, the level of conjugated C20:3 fatty acid

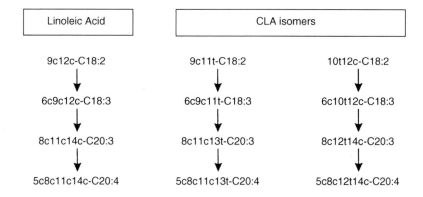

Fig. 11.5 Hypothetical metabolic pathway of 9*c*,11*t*-18:2 and 10*t*,12*c*-18:2 compared to linoleic acid.

in the liver was about four times higher than those of conjugated C18:3 and C20:4. Similar results were discussed in a feeding study on rats using pure CLA isomers (Sébédio *et al.*, 2001). For *c*9,*t*11-18:2 more conjugated C20:3 than C18:3 were found in the liver, whereas the inverse effect was reported for 10*t*,12*c*-18:2. These *in vivo* results were confirmed by an *in vitro* Δ6-desaturase study carried out on rat liver microsomes using radiolabeled CLA in isomeric mixture as substrate (Belury and Kempa-Steczko, 1997). CLAs were converted into C18:3 fatty acids, suggesting that the metabolites were the respective Δ6-desaturation products of CLA.

These various results reinforce the hypothesis that CLA could be metabolized *in vivo* into long chain PUFA using the same pathway as linoleic acid. The hypothetical metabolic pathway of CLA into conjugated C20:4 is shown in Fig. 11.5.

11.11.2   In humans

As other fatty acids, CLA metabolism is shared between three major pathways, incorporation in tissue lipid classes (acylation), oxidation and desaturation/ elongation. Again, few studies have been carried out on these aspects in humans. Cawood *et al.* (1983) showed that CLAs are incorporated in various human fluids, including bile and serum. Breast adipose tissue has also been reported to contain some CLA (Chajes *et al.*, 2002), but extensive data on CLA incorporation in human tissues are not available.

We studied the content of CLA isomers and their metabolites in plasma samples collected during two intervention studies. The first study was the *Trans*Line study (Sébédio *et al.*, 2000). Plasma lipids have been extracted according to Moilanen and Nikkari (1981) and further methylated. Reverse phase HPLC fractionation allowed us to analyse two fractions by GC: the first one contained CLA isomers, as well as other 18:2 fatty acids and 2:3 fatty acids, and the second one contained 18:3 and 20:4 moieties. All the samples contained CLA isomers and also some conjugated 20:3 fatty acids, suggesting that humans, like rodents (see above), are able to metabolize dietary rumenic acid (the major CLA isomer present in normal diet) into conjugated 20:3.

During the other intervention study, the subjects received 3 g per day of one of the following CLA isomers: rumenic acid (i.e. *c9,t*11 isomer) and *t*10,*c*12 isomer. These CLA were incorporated as triacylglycerols in a dairy drinkable preparation. The plasma lipids were analysed using a similar procedure than described above for the samples from the *Trans*Line study. The results showed that the incorporation of CLAs in plasma phospholipids ranged from 0.30 to 1.54% of total fatty acids for rumenic acid and from 0.23 to 1.14% for the other isomer.

11.11.3   Influence on PPAR

The mechanisms for the beneficial effects of CLAs are not known. It has been hypothesized that their biological effects as anticarcinogenic agents and the influence on body composition may be related to an interaction with PPAR (Meadus *et al.*, 2002). It has been shown that CLA can act as a ligand for some of the PPAR family. Moreoever, experiments on Sprague–Dawley rats and SENCAR mice showed that CLA could activate PPAR α and γ (Belury *et al.*, 1997; Moya-Camarena *et al.*, 1999a, b). Hepatic mRNA and protein levels of several enzymes, known to be linked to peroxisome proliferation (acyl coenzyme A, cytochrome P450 4A1, fatty acid binding protein), were found at elevated levels. Using a human cell line only a binding to the PPAR was observed. Two recent publications investigated the influence of CLA on PPAR α and γ activation related to body composition. The first one, using PPAR α null mice and wild-type mice, showed that body weight gain was similar in the two mice models. Compared with the control group the weight gain in the CLA groups were decreased. This suggested that the influence of CLA on body composition is independent of PPAR α (Peters *et al.*, 2001). The second publication examined a relationship between body composition and PPAR γ in the male pig. It was shown that the CLA treatment increased PPAR γ but not PPAR α. Furthermore higher levels of adipocyte fatty acid binding protein were detected in the CLA group, suggesting a promotion of intramuscular fat content (Meadus *et al.*, 2002). Regarding these differences in the publications published so far, the influence of CLA on PPAR activation is inconclusive at this moment and has to be further examined (Moya-Camarena and Belury, 1999).

11.12   Functional benefits of CLA: atherosclerosis

Recent *in vivo* studies have shown that CLA significantly decreased total and low-density lipoprotein (LDL) cholesterol in rabbits and protected against arterial lipid accumulation (Lee *et al.*, 1994). A further study on the same animal model showed a 30% decrease of atherosclerotic lesions (Kritchevsky *et al.*, 2000). Using a hamster model, only 10*t*12*c*- and not 9*c*,11*t*-18:2 decreased triacylglycerols, total cholesterol and non-HDL-cholesterol (Gavino *et al.*, 2000; de Deckere *et al.*,

1999). In the rat model (adult male Sprague–Dawley rats) the feeding of high CLA diets containing 3 and 5% of CLA mixture in the diet, reductions of LDL and HDL cholesterol were detected (Stangl, 2000). In contrast, Munday *et al.* (1999) reported an increase of aortic fatty streaks in C57BL/6 mice, but the serum triacylglycerol level was reduced and the HDL to total cholesterol ratio was increased. A six-week study on female swine showed pro-atherosclerotic effects after feeding a CLA mixture. Very low-density lipoprotein (VLDL) and LDL cholesterol were increased by CLA. Moreover, the LDL to HDL cholesterol ratio was significantly up-regulated (Stangl *et al.*, 1999). The influence of CLA on atherosclerosis seems to be controversial and the influence of the animal model on the observed effects should be further investigated.

Heart disease induced by atherosclerosis is identified as the first case of human mortality. Therefore, the possible relationship between CLA and atherosclerosis in humans has a large research interest. A human study was carried out to look at effects of CLA treatment on plasma lipoproteins and tissue fatty acid composition. In a 63-day study healthy normolipidemic women were supplemented with CLA. Blood cholesterol and lipoprotein levels were not altered by the CLA treatment. $9c,11t$-18:2 was incorporated in plasma at only 4.23% of ingested CLA. Contrary to animal studies, short-term nutritional trials with CLA in humans did not change lipoprotein levels and seem to be ineffective in the prevention of atherosclerosis (Benito *et al.*, 2001a).

CLAs were also described to induce an antithrombotic effect. The effects of several CLA isomers on human platelet aggregation were examined *in vitro* (Truitt *et al.*, 1999). A CLA mixture was tested in comparison to linoleic acid using platelet aggregation agents as arachidonic acid, collagen and thrombin. CLA was effective to inhibit platelet aggregation. To investigate the influence of CLA on platelet cyclooxygenase and lipoxygenase activities, formation of $[^{14}C]$-thromboxane B_2 (TXB_2) and $[^{14}C]$-12hydroxyeicosatetranoic acid (HETE) was measured. All tested CLA-isomers inhibited TXB_2, while the HETE level was unchanged. An *in vivo* experiment on platelet function was carried out in humans. However, the daily intake of 3.9 g of CLA for 63 days had no influence on *in vitro* platelet aggregation. The authors concluded that short-term consumption of CLA did not exhibit antithrombotic properties in humans (Benito *et al.*, 2001b).

11.13   Functional benefits of CLA: immune function

Some studies indicate that CLA could protect against immune-induced cachexia (growth suppression or weight loss). Usually, an enhanced immunological function leads to a decreased growth. Feeding CLA protected against these catabolic effects of such an immune stimulation (Cook and Pariza, 1998). Miller *et al.* (1994) described how dietary CLA in mice prevent endotoxin-induced growth suppression compared to a control group. The authors hypothesized a relationship between CLA and possible changes in the interleukin-1 level. The capacity of CLA to change interleukin production was later observed by Hayek *et al.* (1999).

CLA may also influence the allergic reaction pathways, by alteration of serum immunoglobulins (Ig). IgA, IgG and IgM were increased, whereas IgE was decreased (Sugano et al., 1998). IgA, IgG and IgM are implicated in the defense metabolism against virus or bacteria. As IgE is related to allergic reactions, its diminution is desired.

The influence of CLA on the immunological function may have implications for human health. A study in humans examined possible changes in immune status by testing the number of circulating white blood cells, granulocytes, monocytes, lymphocytes and lymphocyte proliferation. No changes in the investigated parameters were observed (Kelley et al., 2000). In another experiment the specific antibody production after hepatitis B vaccination was tested depending on a CLA treatment. The antibody formation after CLA-treatment (9c,11t-18:2/10t,12c-18:2; 50/50) was significantly higher (Mohede et al., 2001), suggesting that CLA could be able to enhance immune function in humans.

11.14   Functional benefits of CLA: diabetes

The discussion concerning antidiabetic effects of CLA is controversial. An experiment using Zucker diabetic fatty rats showed a normalization of impaired glucose tolerance and improved hyperinsulinemia induced by CLA (Houseknecht et al., 1998). To get information about the active isomer, a second experiment was carried out with the same animal model using CLA mixture and a 9c,11t-18:2 enriched butter (90.5% 9c,11t-18:2 of total CLA). Only the CLA mixture, not 9c,11t-18:2, reduced improved glucose tolerance. Furthermore, an improved insulin action in muscle was observed (Ryder et al., 2001).

A recent study showed inverse effects. C57BL/6J mice fed a CLA-enriched diet developed a state resembling lipoatrophic diabetes, with a marked insulin resistance (hyperinsulinemia) (Tsuboyama-Kasaoka et al., 2000). The hyperinsulinemia was reversible by continuous infusion with leptin, suggesting leptin may be antagonistic to the CLA-induced effects.

Only a few studies have been carried out on possible antidiabetic effects of CLA using different animal models. One reason for the contrary results reported above could be a difference in the metabolic pathways of the two animal species.

Type II diabetes (diabetes induced by insulin resistance) represents one of the diseases of the affluent society and at the moment very little is known about the effect of CLA on insulin resistance in humans. A nutritional study was carried out on overweight middle-aged men with insulin resistance syndrome. One group received a CLA mixture and the other one the 10t,12c-18:2 (Riserus and Vessby, 2001). Compared with the control group, glucose metabolism remained unchanged for the CLA mixture trial. The treatment with pure 10t,12c-18:2 increased the insulin resistance and glycemia. It can be concluded that the two different CLA treatments showed diverging effects. The influence of CLA on type II diabetes is not clear, therefore more results concerning this subject are needed.

11.15 Conclusion and future trends

The increase of CLA levels in food or the creation of functional foods containing high amounts of CLA may be beneficial for human health because of their biological activities discussed above. Current data exist predominantly from experiments on animals, where CLA showed high anticarcinogenic effects and produced strong evidence for the influence on body composition in the growing animal. Moreover CLA may have beneficial effects for cardiovascular diseases, as they showed a lipid lowering potential. Currently, experiments on humans have investtigated mainly the influence on body composition and the results obtained remain controversial. This may be related to different subjects in the studies and the use of CLA mixtures of variable isomer compositions. Furthermore, all human studies have been short-term experiments.

More experiments using isolated isomers to evaluate the effects of each CLA isomer, and long-term intervention studies, are needed. In addition, the toxicological risks have to be evaluated, as little data are available on safety aspects of a CLA supplementation.

11.16 References

ADLOF R (1997), Preparation of methyl *cis*-9,*trans*-11- and *trans*-9,trans-11-octadecadienoate-17,17,18,18-d4, two of the isomers of conjugated linoleic acid, *Chem Phys Lipids*, **88**, 107–112.

ADLOF R (1999), The lindlar-catalyzed reduction of methyl santalbate: a facile preparation of methyl 9-*cis*,11-*trans*-octadecadienoate-9,10-d2, *J Am Oil Chemists Soc*, **76**, 301–304.

BANNI S, CARTA G, CONTINI M S, ANGIONI E, DEIANA M, DESSI M A, MELIS M P and CORONGIU F P (1996), Characterization of conjugated diene fatty acids in milk, dairy products, and lamb tissues, *J Nutr Biochem*, **7**, 150–155.

BANNI S, ANGIONI E, CASU V, MELIS M P, CARTA G, CORONGIU F P, THOMPSON H and IP C (1999), Decrease in linoleic acid metabolites as a potential mechanism in cancer risk reduction by conjugated linoleic acid, *Carcinogenesis*, **20**, 1019–1024.

BANNI S, CARTA G, ANGIONI E, MURRU E, SCANU P, MELIS M P, BAUMAN D E, FISCHER S M and IP C (2001), Distribution of conjugated linoleic acid and metabolites in different lipid fractions in the rat liver, *J Lipid Res*, **42**, 1056–1061.

BASCETTA E, GUNSTONE F D and SCRMGEOUR C M (1984), Synthesis, characterisation, and transformation of a lipid cyclic peroxide, *J Chem Soc Perkin Trans I*, 2199–2205.

BAUMGARD L H, SANGSTER J K and BAUMAN D E (2001), Milk fat synthesis in dairy cows is progressively reduced by increasing supplemental amounts of *trans*-10,*cis*-12 conjugated linoleic acid (CLA), *J Nutr*, **131**, 1764–1769.

BEE G (2000), Dietary conjugated linoleic acids alter adipose tissue and milk lipids of pregnant and lactating sows, *J Nutr*, **130**, 2292–2298.

BELURY M A and KEMPA-STECZKO A (1997), Conjugated linoleic acid modulates hepatic lipid composition in mice, *Lipids*, **32**, 199–204.

BELURY M A, NICKEL K P, BIRD C E and WU Y (1996), Dietary conjugated linoleic acid modulation of phorbol ester skin tumor promotion, *Nutr Cancer*, **26**, 149–157.

BELURY M A, MOYA-CAMARENA S Y, LIU K L and HEUVEL J P V (1997), Dietary conjugated linoleic acid induces peroxisome-specific enzyme accumulation and ornithine decarboxylase activity in mouse liver, *J Nutr Biochem*, **8**, 579–584.

BENITO P, NELSON G J, KELLEY D S, BARTOLINI G, SCHMIDT P C and SIMON V (2001a), The effect of conjugated linoleic acid on plasma lipoproteins and tissue fatty acid composition in humans, *Lipids*, **36**, 229–236.

BENITO P, NELSON G J, KELLEY D S, BARTOLINI G, SCHMIDT P C and SIMON V (2001b), The effect of conjugated linoleic acid on platelet function, platelet fatty acid composition, and blood coagulation in humans, *Lipids*, **36**, 221–227.

BERDEAUX O, CHRISTIE W W, GUNSTONE F D and SÉBÉDIO J L (1997), Large-scale synthesis of methyl *cis*-9,*trans*-11-octadecadienoate from methyl ricinoleate, *J Am Oil Chem Soc*, **74**, 1011–1015.

BLANKSON H, STAKKESTAD J A, FAGERTUN H, THOM E, WADSTEIN J and GUDMUNDSEN O (2000), Conjugated linoleic acid reduces body fat mass in overweight and obese humans, *J Nutr*, **130**, 2943–2948.

CAWOOD P, WICKENS D G, IVERSEN S A, BRAGANZA J M and DORMANDY T L (1983), The nature of diene conjugation in human serum, bile and duodenal juice, *FEBS Lett*, **162**, 239–243.

CESANO A, VISONNEAU S, SCIMECA J A, KRITCHEVSKY D and SANTOLI D (1998), Opposite effects of linoleic acid and conjugated linoleic acid on human prostatic cancer in SCID mice, *Anticancer Res*, **18**, 1429–1434.

CHAJES V, LAVILLONNIERE F, FERRARI P, JOURDAN M L, PINAULT M, MAILLARD V, SEBEDIO J L and BOUGNOUX P (2002), Conjugated linoleic acid content in breast adipose tissue is not associated with the relative risk of breast cancer in a population of French patients, *Cancer Epidemiol Biomarkers Prev*, **11**, 672–673.

CHIN S F, LIU W, STORKSON J M, HA Y L and PARIZA M W (1992), Dietary sources of conjugated dienoic isomers of linoleic acid, a newly recognized class of anticarcinogens, *J Food Comp Anal*, **5**, 185–197.

CHIPAULT J R and HAWKINS J M (1959), The determination of conjugated *cis–trans* and *trans–trans* methyl octadecadienoates by infrared spectrometry, *J Am Oil Chem Soc*, **36**, 535–539.

CHOI Y, PARK Y, PARIZA M W and NTAMBI J M (2001), Regulation of stearoyl-CoA desaturase activity by the *trans*-10,*cis*-12 isomer of conjugated linoleic acid in HepG2 cells, *Biochem Biophys Res Commun*, **284**, 689–693.

CHOI Y, PARK Y, STORKSON J M, PARIZA M W and NTAMBI J M (2002), Inhibition of stearoyl-CoA desaturase activity by the *cis*-9,*trans*-11 isomer and the *trans*-10,*cis*-12 isomer of conjugated linoleic acid in MDA-MB-231 and MCF-7 human breast cancer cells, *Biochem Biophys Res Commun*, **294**, 785–790.

CHOUINARD P Y, CORNEAU L, SAEBO A and BAUMAN D E (1999), Milk yield and composition during abomasal infusion of conjugated linoleic acids in dairy cows, *J Dairy Sci*, **82**, 2737–2745.

CHOUINARD P Y, CORNEAU L, BUTLER W R, CHILLIARD Y, DRACKLEY J K and BAUMAN D E (2001), Effect of dietary lipid source on conjugated linoleic acid concentrations in milk fat, *J Dairy Sci*, **84**, 680–690.

CHRISTIE W W (1997), Isomers in commercial samples of conjugated linoleic acid, *J Am Oil Chem Soc*, **74**, 1231.

CHRISTIE W W (2001), in R J Hamilton, 'CLA is focus for SCI seminar', *INFORM*, **12**, 629–633.

CHRISTIE W W, SÉBÉDIO J L and JUANEDA P (2001), A practical guide to the analysis of conjugated linoleic acid, *INFORM*, **12**, 147–152.

COOK M E and PARIZA M (1998), The role of conjugated linoleic acid (CLA) in health, *Int Dairy J*, **8**, 459–462.

CORL B, BAUMGARD L, GRIINARI J, DELMONTE P, MOREHOUSE K, YURAWECZ M and BAUMAN D (2002), *trans*-7,*cis*9 CLA is endogenously synthesized by D-9-desaturation in lactating dairy cows, AOCS annual meeting, Montreal.

CUNNINGHAM D C, HARRISON L Y and SCHULTZ T D (1997), Proliferative responses of normal human mammary and MCF-7 breast cancer cells to linoleic acid, conjugated linoleic acid and eicosanoid synthesis inhibitors in culture, *Anticancer Research*, **17**, 197–204.

DE DECKERE E A M, VAN AMELSVOORT J M M, MCNEILL G P and JONES P (1999), Effects of conjugated linoleic acid (CLA) isomers on lipid levels and peroxisome proliferation in the hamster, *Br J Nutr*, **82**, 309–317.

DELANY J P, BLOHM F, TRUETT A A, SCIMECA J A and WEST D B (1999), Conjugated linoleic acid rapidly reduces body fat content in mice without affecting energy intake, *Am J Physiol Reg Integrative Comp Physiol*, **45**, R1172–R1179.

DHIMAN T R, ANAND G R, SATTER L D and PARIZA M W (1996), Conjugated linoleic acid content of milk from cows fed different diets, *J Dairy Sci*, **79 (Suppl. 1)**, 35.

DHIMAN T R, HELMINK E D, MCMAHON D J, FIFE R L and PARIZA M W (1999), Conjugated linoleic acid content of milk and cheese from cows fed extruded oilseeds, *J Dairy Sci*, **82**, 412–419.

DOBSON G (1997), Identification of conjugated fatty acids by gas chromatography-mass spectrometry of 4-methyl-1,2,4-triazoline-3,5-dione adducts, *J Am Oil Chem Soc*, **75**, 137–142.

DOBSON G and CHRISTIE W W (1996), Structural analysis of fatty acids by mass spectrometry of picolinyl esters and dimethyloxazoline derivatives, *Trends in Analytical Chemistry*, **15**, 130–137.

DONOVAN D C, SCHINGOETHE D J, BAER R J, RYALI J, HIPPEN A R and FRANKLIN S T (2000), Influence of dietary fish oil on conjugated linoleic acid and other fatty acids in milk fat from lactating dairy cows, *J Dairy Sci*, **83**, 2620–2628.

DUGAN M E R, AALHUS, J L, SCHAEFER A L and KRAMER J K G (1997), The effect of conjugated linoleic acid on fat to lean repartitioning and feed conversion in pigs, *Can J Anim Sci*, **77**, 723–725.

DURGAM V R and FERNANDES G (1997), The growth inhibitory effect of conjugated linoleic acid on MCF-7 cells is related to estrogen response system, *Cancer Lett*, **116**, 121–130.

FITCH-HAUMANN B (1996), Conjugated linoleic acid offers research promise, *INFORM*, **7**, 152–159.

FRITSCHE J and STEINHART H (1998a), Analysis, occurrence, and physiological properties of *trans* fatty acids (TFA) with particular emphasis on conjugated linoleic acid isomers (CLA) – a review, *Fett/Lipid*, **100**, 190–210.

FRITSCHE J and STEINHART H (1998b), Amounts of conjugated linoleic acid (CLA) in german foods and evaluation of daily intake, *Z Lebensm Unters Forsch A*, **206**, 77–82.

FRITSCHE J, MOSSOBA M M, YURAWECZ M P, ROACH J A G, SEHAT N, KU Y and STEINHART H (1997), Conjugated linoleic acid (CLA) isomers in human adipose tissue, *Z Lebensm Unters Forsch A*, **205**, 415–418.

FUTAKUCHI M, CHENG J L, HIROSE M, KIMOTO N, CHO Y M, IWATA T, KASAI M, TOKUDOME S and SHIRAI T (2002), Inhibition of conjugated fatty acids derived from safflower or perilla oil of induction and development of mammary tumors in rats induced by 2-amino-1-methyl-6-phenylimidazo[4,5-b]pyridine (PhIP), *Cancer Lett*, **178**, 131–139.

GAVINO V C, GAVINO G, LEBLANC M J and TUCHWEBER B (2000), An isomeric mixture of conjugated linoleic acids but not pure *cis*-9,*trans*-11-octadecadienoic acid affects body weight gain and plasma lipids in hamsters, *J Nutr*, **130**, 27–29.

GEORGER M M and HUDSON B S (1988), Synthesis of all-*trans*-parinaric acid-d8 specifically deuteriated at all vinyl positions, *J Org Chem*, **53**, 3148–3153.

GNÄDIG S (1996), Konjugierte Linolsäureisomere in Lebensmitteln, humanem Fettgewebe und humanem Blutplasma, Diplomarbeit, Fachbereich Chemie, Inst. für Biochemie und Lebensmittelchemie, Abt. Lebensmittelchemie, Universität Hamburg, Hamburg.

GNÄDIG S, BERDEAUX O, LOREAU O, NOEL J P and SEBEDIO J L (2001), Synthesis of (6Z,9Z,11E)-octadecatrienoic and (8Z,11Z,13E)-eicosatrienoic acids and their [1-(14)C]-radiolabeled analogs, *Chem Phys Lipids*, **112**, 121–135.

GRIINARI J M, CORI B A, LACY S H, CHOUINARD P Y, NURMELA K V V and BAUMAN D E (2000), Conjugated linoleic acid is synthesized endogenously in lactating dairy cows by Delta(9)-desaturase, *J Nutr*, **130**, 2285–2291.

GUNSTONE F D and SAID A I (1971), Methyl 12-mesyloxyoleate as a source of cyclopropane esters and of conjugated octadecadienoates, *Chem Phys Lipids*, **7**, 121–134.

HA Y L , GRIMM N K and PARIZA M W (1987), Anticarcinogens from fried ground beef: heat-altered derivatives of linoleic acid, *Carcinogenes*, **8**, 1881–1887.

HA, Y L, STORKSON J and PARIZA M W (1990), Inhibition of benzo(a)pyrene-induced mouse forestomach neoplasia by conjugated dienoic derivatives of linoleic acid, *Cancer Res*, **50**, 1097–1101.

HAMILTON R J (2001), CLA is focus for SCI seminar, *INFORM*, **12**, 629–633.

harfoot c g and hazelwood g p (1988), Lipid metabolism in the rumen, in P Hobson, *The Rumen Microbial Ecosystem*, London, Elsevier Science Publishers, 285–322.

HAYEK M G, HAN S N, WU D Y, WATKINS B A, MEYDANI M, DORSEY J L, SMITH D E and MEYDANI S N (1999), Dietary conjugated linoleic acid influences the immune response of young and old C57BL/6NCrlBR mice, *J Nutr*, **129**, 32–38.

HOUSEKNECHT K L, VANDENHEUVEL J P, MOYACAMARENA S Y, PORTOCARRERO C P, PECK L W, NICKEL K P and BELURY M A (1998), Dietary conjugated linoleic acid normalizes impaired glucose tolerance in the Zucker diabetic fatty fa/fa rat, *Biochem Biophys Res Comm*, **244**, 678–682.

HUBBARD N E, LIM D, SUMMERS L and ERICKSON K L (2000), Reduction of murine mammary tumor metastasis by conjugated linoleic acid, *Cancer Lett*, **150**, 93–100.

IGARASHI M and MIYAZAWA T (2001), The growth inhibitory effect of conjugated linoleic acid on a human hepatoma cell line, HepG2, is induced by a change in fatty acid metabolism, but not the facilitation of lipid peroxidation in the cells, *Biochim Biophys Acta*, **1530**, 162–171.

IP C, CHIN S F, SCIMECA J A and PARIZA M W (1991), Mammary cancer prevention by conjugated dienoic derivative of linoleic acid, *Cancer Res*, **51**, 6118–6124.

IP C, SINGH M, THOMPSON H J and SCIMECA J A (1994), Conjugated linoleic acid suppresses mammary carcinogenesis and proliferative activity of the mammary gland in the rat, *Cancer Res*, **54**, 1212–1215.

IP C, SCIMECA J A and THOMPSON H (1995), Effect of timing and duration of dietary conjugated linoleic acid on mammary cancer prevention, *Nutr Cancer*, **24**, 241–247.

IP C, BRIGGS S P, HAEGELE A D, THOMPSON H J, STORKSON J and SCIMECA J A (1996), The efficacy of conjugated linoleic acid in mammary cancer prevention is independent of the level or type of fat in the diet, *Carcinogenesis*, **17**, 1045–1050.

IP C, JIANG C, THOMPSON H J and SCIMECA J A (1997), Retention of conjugated linoleic acid in the mammary gland is associated with tumor inhibition during the post-initiation phase of carcinogenesis, *Carcinogenesis*, **18**, 755–759.

IP C, BANNI S, ANGIONI E, CARTA G, MCGINLEY J, THOMPSON H, BARBANO D and BAUMAN D (1999), Conjugated linoleic acid-enriched butter fat alters mammary gland morphogenesis and reduces cancer risk in rats, *J Nutr*, **129**, 2135–2142.

IP C, IP M, LOFTUS T, SHOEMAKER S and SHEA-EATON W (2000), Induction of apoptosis by conjugated linoleic acid in cultured mammary tumor cells and premalignant lesions of the rat mammary gland, *Cancer Epidemiol Biomarkers Prevent*, **9**, 689–696.

JAHREIS G, FRITSCHE J and STEINHART H (1996), Monthly variation of milk composition with special regard to fatty acids depending on season and farm management systems – convential versus ecological, *Fett/Lipid*, **98**, 356–359.

JAHREIS G, FRITSCHE J and STEINHART H (1997), Conjugated linoleic acid in milk fat – high variation depending on production system, *Nutr Res*, **17**, 1479–1484.

JAHREIS G, FRITSCHE J and KRAFT J (1999), Species-dependent, seasonal, and dietary variation of conjugated linoleic acid in milk. In: M P Yurawecz, M M Mossoba, J K G Kramer, M W Pariza and G J Nelson (Eds), *Advances in Conjugated Linoleic Acid Research*, Vol. 1, Champaign, IL, AOCS Press, 215–225.

JIANG J, BJOERCK L, FONDÉN R and EMANUELSON M (1996), Occurrence of conjugated *cis*-9,*trans*-11-octadecadienoic acid in bovine milk: effects of feed and dietary regimen, *J Dairy Sci*, **79**, 438–445.

JIANG J, BJORCK L and FONDEN R (1997), Conjugated linoleic acid in Swedish dairy products with special reference to the manufacture of hard cheese, *Int Dairy J*, **7**, 863–867.

JIANG J, BJORCK L and FONDEN R (1998), Production of conjugated linoleic acid by dairy starter cultures, *J Appl Microbiol*, **85**, 95–102.

JONES D F, WEISS W P and PALMQUIST D L (2000), Short communication: influence of dietary tallow and fish oil on milk fat composition, *J Dairy Sci*, **83**, 2024–2026.

JOSYULA S and SCHUT H A (1998), Effects of dietary conjugated linoleic acid on DNA adduct formation of PhIP and IQ after bolus administration to female F344 rats, *Nutr Cancer*, **32**, 139–145.

JOSYULA S, HE Y H, RUCH R J and SCHUT H A (1998(, Inhibition of DNA adduct formation of PhIP in female F344 rats by dietary conjugated linoleic acid, *Nutr Cancer*, **32**, 132–138.

JUANEDA P, CORDIER O, GREGOIRE S and SEBEDIO J L (2001), Conjugated linoleic acid (CLA) isomers in heat-treated vegetable oils, *OCL*, **8**, 94–97.

JUNG M O, YOON S H and JUNG M Y (2001), Effects of temperature and agitation rate on the formation of conjugated linoleic acids in soybean oil during hydrogenation process, *J Agric Food Chem*, **49**, 3010–3016.

JUNG M Y and HA Y L (1999), Conjugated linoleic acid isomers in partially hydrogenated soybean oil obtained during nonselective and selective hydrogenation processes, *J Agric Food Chem*, **47**, 704–708.

KELLEY D S, TAYLOR P C, RUDOLPH I L, BENITO P, NELSON G J, MACKEY B E and ERICKSON K L (2000), Dietary conjugated linoleic acid did not alter immune status in young healthy woman, *Lipids*, **35**, 1065–1071.

KELLY M L, BERRY J R, DWYER D A, GRIINARI J M, CHOUINARD P Y, VANAMBURGH M E and BAUMAN D E (1998), Dietary fatty acid sources affect conjugated linoleic acid concentrations in milk from lactating dairy cows, *J Nutr*, **128**, 881–885.

KEPLER C R and TOVE S B (1967), Biohydrogenation of unsaturated fatty acids, *J Biol Chem*, **242**, 5686–5692.

KEPLER C R, HIRONS K P, MCNEIL J J and TOVE S B (1966), Intermediates and products of the biohydrogenation of linoleic acid by *Butyrivibrio fibrisolvens*, *J Biol Chem*, **241**, 1350–1354.

KIMOTO N, HIROSE M, FUTAKUCHI M, IWATA T, KASAI M and SHIRAI T (2001), Site-dependent modulating effects of conjugated fatty acids from safflower oil in a rat two-stage carcinogenesis model in female Sprague–Dawley rats, *Cancer Lett*, **168**, 15–21.

KOHNO H, SUZUKI R, NOGUCHI R, HOSOKAWA M, MIYASHITA K and TANAKA T (2002), Dietary conjugated linoleic acid inhibits azoxymethane-induced colonic aberrant crypt foci in rats, *Jpn J Cancer Res*, **93**, 133–142.

KRAMER J K C, FELLNER V, DUGAN M E R, SAUER F D, MOSSOBA M M and YURAWECZ M P (1997), Evaluating acid and base catalysts in the methylation of milk and rumen fatty acids with special emphasis on conjugated dienes and total *trans* fatty acids, *Lipids*, **32**, 1219–1228.

KRAMER J K C, PARODI P W, JENSEN R G, MOSSOBA M M, YURAWECZ M P and ADLOF R O (1998), Rumenic acid: a proposed common name for the major conjugated linoleic acid isomer found in natural products, *Lipids*, **33**, 835.

KRITCHEVSKY D (2000), Antimutagenic and some other effects of conjugated linoleic acid, *Br J Nutr*, **83**, 459–465.

KRITCHEVSKY D, TEPPER S A, WRIGHT S, TSO P and CZARNECKI S K (2000), Influence of conjugated linoleic acid (CLA) on establishment and progression of atherosclerosis in rabbits, *J Am Coll Nutr*, **19**, 472S–477S.

LABELLE M, FALGUEYRET J P, RIENDEAU D and ROKACH J (1990), Synthesis of two analogues of arachidonic acid and their reactions with 12-lipoxygenase, *Tetrahedron*, **46**, 6301–6310.

LAVILLONNIÈRE F, MARTIN J C, BOUGNOUX P and SÉBÉDIO J L (1998), Analysis of conjugated linoleic acid isomers and content in French cheeses, *J Am Oil Chem Soc*, **75**, 343–352.

LEE K N, KRITCHEVSKY D and PARIZA M W (1994), Conjugated linoleic acid and atherosclerosis in rabbits, *Atherosclerosis*, **108**, 19–25.

LEHMANN L (2001), Identifizierung und stereoselektive Synthesen ungesättigter Signalstoffe,

Doktorarbeit, Fachbereich Chemie, Inst. für Organische Chemie, Universität Hamburg, Hamburg.

LEUNG Y H and LIU R H (2000), *trans*-10,*cis*-12-Conjugated linoleic acid isomer exhibits stronger oxyradical scavenging capacity than *cis*-9,*trans*-11-conjugated linoleic acid isomer, *J Agric Food Chem*, **48**, 5469–5475.

LIE KEN JIE M S F and WONG K P (1992), Dehydration reactions involving methyl 9,12-dihydroxy-10-octadecenoate, *Chem Phys Lipids*, **62**, 177–183.

LIE KEN JIE M S F, KHYSAR PASHA M and AHMAD F (1996), Ultrasound-assisted synthesis of santalbic acid and a study of triacylglycerol species in *Santalum album* (Linn.) seed oil, *Lipids*, **31**, 1083–1089.

LIE KEN JIE M S F, KHYSAR PASHA M and SHAHIN ALAM M (1997), Synthesis and nuclear magnetic resonance properties of all geometrical isomers of conjugated linoleic acids, *Lipids*, **32**, 1041–1044.

LIEW C, SCHUT H A J, CHIN S F, PARIZA M W and DASHWOOD R H (1995), Protection of conjugated linoleic acid against 2-amino 3-methylimido[4,5-F] quinoline-induce colon carcinogenesis in the F344 rat: a study of inhibitory mechanisms, *Carcinogenesis*, **16**, 3037–3044.

LIN H, BOYLSTON T D, LUEDECKE L O and SHULTZ T D (1998), Factors affecting the conjugated linoleic acid content of cheddar cheese, *J Agric Food Chem*, **46**, 801–807.

LIN H, BOYLSTON T D, LUEDECKE L O and SHULTZ T D (1999b), Conjugated linoleic acid content of cheddar-type cheeses as affected by processing, *J Food Sci*, **64**, 874–878.

LIN T Y (2000), Conjugated linoleic acid concentration as affected by lactic cultures and additives, *Food Chem*, **69**, 27–31.

LIN T Y, LIN C W and LEE C H (1999a), Conjugated linoleic acid concentration as affected by lactic cultures and added linoleic acid, *Food Chem*, **67**, 1–5.

LIU J R, LI B X, CHEN B Q, HAN X H, XUE Y B, YANG Y M, ZHENG Y M and LIU R H (2002), Effect of *cis*-9, *trans*-11-conjugated linoleic acid on cell cycle of gastric adenocarcinoma cell line (SGC-7901), *World J Gastroenterol*, **8**, 224–229.

LOOR J J and HERBEIN J H (1998), Exogenous conjugated linoleic acid isomers reduce bovine milk fat concentration and yield by inhibiting *de novo* fatty acid synthesis, *J Nutr*, **128**, 2411–2419.

LOREAU O, MARET A, CHARDIGNY J M, SÉBÉDIO J L and NOEL J P (2001), Sequential substitution of 1,2-dichloro-ethene: a convenient stereoselective route to (9Z,11E)-, (10E,12Z)- and (10Z,12Z)-[1-14C] conjugated linoleic acid isomers, *Chem Phys Lipids*, **110**, 57–67.

MEADUS W J, MACINNIS R and DUGAN M E (2002), Prolonged dietary treatment with conjugated linoleic acid stimulates porcine muscle peroxisome proliferator activated receptor gamma and glutamine-fructose aminotransferase gene expression *in vivo*, *J Mol Endocrinol*, **28**, 79–86.

MILLER A, STANTON C and DEVERY R (2001), Modulation of arachidonic acid distribution by conjugated linoleic acid isomers and linoleic acid in MCF-7 and SW480 cancer cells, *Lipids*, **36**, 1161–1168.

MILLER C C, PARK Y M W P and COOK M E (1994), Feeding conjugated linoleic acid to animals partially overcomes catabolic responses due to endotoxin injection, *Biochem Biophys Res Commun*, **198**, 1107–1112.

MOHEDE I, ALBERS R, VAN DER WIELEN R, BRINK L and DOROVSKA-TARAN V (2001), Immuno-modulation: CLA stimulates antigen specific antibody production in humans, 1st International Conference on Conjugated Linoleic Acid (CLA), Ålesund, Norway.

MOILANEN T and NIKKARI T (1981), The effect of storage on the fatty acid composition of human serum, *Clin Chim Acta*, **114**, 111–116.

MORRIS S G, MAGIDMAN P and HERB S F (1972), Hydrogenation of triple bonds to double bonds in conjugated methyl octadecadiynoate and methyl santalbate, *J Am Oil Chem Soc*, **49**, 505–507.

MOSSOBA M M, KRAMER J K G, YURAWECZ M P, SEHAT N, ROACH J A G, EULITZ K, FRITSCHE

J, DUGAN M E R and KU Y (1999), Impact of novel methodologies on the analysis of conjugated linoleic acid (CLA). Implications of CLA feeding studies, *Fett/Lipid*, **101**, 235–243.

MOYA-CAMARENA S Y and BELURY M A (1999), Species differences in the metabolism and regulation of gene expression by conjugated linoleic acid, *Nutr Rev*, **57**, 336–340.

MOYA-CAMARENA S Y, VANDEN HEUVEL J P and BELURY M A (1999a), Conjugated linoleic acid activates peroxisome proliferator-activated receptor α and β subtypes but does not induce hepatic peroxisome proliferation in Sprague–Dawley rats, *Biochim Biophys Acta*, **1436**, 331–342.

MOYA-CAMARENA S Y, VANDEN HEUVEL J P, BLANCHARD S G, LEESNITZER L A and BELURY M A (1999b), Conjugated linoleic acid is a potent naturally occurring ligand and activator of PPARα, *J Lipid Res*, **40**, 1426–1433.

MUNDAY J S, THOMPSON K G and JAMES K A C (1999), Dietary conjugated linoleic acids promote fatty streak formation in the C57BL/6 mouse atherosclerosis model, *Br J Nutr*, **81**, 251–255.

O'SHEA M, LAWLESS F, STANTON C and DEVERY R (1998), Conjugated linoleic acid in bovine milk fat: a food-based approach to cancer chemoprevention, *Trends Food Sci Technol*, **9**, 192–196.

O'SHEA M, STANTON C and DEVERY R (1999), Antioxidant enzyme defence responses of human MCF-7 and SW480 cancer cells to conjugated linoleic acid. *Anticancer Res*, **19**, 1953–1959.

O'SHEA M, DEVERY R, LAWLESS F, MURPHY J and STANTON C (2000), Milk fat conjugated linoleic acid inhibits growth of human mammary MCF-7 cancer cells, *Anticancer Res*, **20**, 3591–3602.

OSTROWSKA E, MURALITHARAN M, CROSS R F, BAUMAN D E and DUNSHEA F R (1999), Dietary conjugated linoleic acids increase lean tissue and decrease fat deposition in growing pigs, *J Nutr*, **129**, 2037–2042.

PALOMBO J D, GANGULY A, BISTRIAN B R and MENARD M P (2002), The antiproliferative effects of biologically active isomers of conjugated linoleic acid on human colorectal and prostatic cancer cells, *Cancer Lett*, **177**, 163–172.

PARIZA M W (1997), Conjugated linoleic acid, a newly recognised nutrient, *Chem Ind*, 464–466.

PARIZA M W and HARGRAVES W A (1985), A beef-derived mutagenesis modulator inhibits initiation of mouse epidermal tumors by 7,12-dimethylbenz[a]anthracene, *Carcinogenesis*, **6**, 591–593.

PARIZA M, PARK Y and COOK M (1999), Conjugated linoleic acid and the control of cancer and obesity, *Toxicol Sci*, **52** (Suppl), 107–110.

PARK H S, RYU J H, HA Y L and PARK J H (2001a), Dietary conjugated linoleic acid (CLA) induces apoptosis of colonic mucosa in 1,2-dimethylhydrazine-treated rats: a possible mechanism of the anticarcinogenic effect by CLA, *Br J Nutr*, **86**, 549–555.

PARK Y, ALBRIGHT K J, LIU W, STORKSON J M, COOK M E and PARIZA M W (1997), Effect of conjugated linoleic acid on body composition in mice, *Lipids*, **32**, 853–858.

PARK Y, ALBRIGHT K J, STORKSON J M, LIU W, COOK M E and PARIZA M W (1999a), Changes in body composition in mice during feeding and withdrawal of conjugated linoleic acid, *Lipids*, **34**, 243–248.

PARK Y, STORKSON J M, ALBRIGHT K J, LIU W and PARIZA M W (1999b), Evidence that the *trans*-10,*cis*-12 isomer of conjugated linoleic acid induces body composition changes in mice, *Lipids*, **34**, 235–241.

PARK Y, ALLEN K and SHULTZ T (2000), Modulation of MCF-7 breast cancer cell signal transduction by linoleic acid and conjugated linoleic acid in culture, *Anticancer Res*, **20**, 669–676.

PARK Y, ALBRIGHT K J, CAI Z Y and PARIZA M W (2001b), Comparison of methylation procedures for conjugated linoleic acid and artifact formation by commercial (trimethylsilyl)diazomethane, *J. Agric. Food Chem.*, **49**, 1158–1164.

PETERS J M, PARK Y, GONZALEZ F J and PARIZA M W (2001), Influence of conjugated linoleic acid on body composition and target gene expression in peroxisome proliferator-activated receptor alpha-null mice, *Biochim Biophys Acta*, **1533**, 233–242.

PRECHT D, MOLKENTIN J and VAHLENDICK M (1999), Influence of the heating temperature on the fat composition of milk fat with the emphasis on *cis-/trans*-isomerization, *Nahrung/Food*, **43**, 25–33.

RAMSAY T G, EVOCK-CLOVER C M, STEELE N C and AZAIN M J (2001), Dietary conjugated linoleic acid alters fatty acid composition of pig skeletal muscle and fat, *J Anim Sci*, **79**, 2152–2161.

REANEY M J T, LIU Y-D and WESTCOTT N D (1999), Commercial production of conjugated linoleic acid. In: M P Yurawecz, M M Mossoba, J K G Kramer, M W Pariza and G J Nelson (Eds), *Advances in Conjuged Linuleic Acid Research*, Vol. 1, Champaign, IL, AOCS Press, 39–54.

RICKERT R and STEINHART H (2001), Bedeutung, Analytik sowie Vorkommen von konjugierten Linolsäureisomeren (CLA) in Lebensmitteln, *Ernährungs-Umschau*, **48**, 4–7.

RISERUS U and VESSBY B (2001), Diverging effects of CLA isomers on insulin resistance and lipid metabolsim in obese men with metabolic syndrome, 1st International Conference on Conjugated Linoleic Acid (CLA), Ålesund, Norway.

YYDER J W, PORTOCARRERO C P, SONG X M, CUI L, YU M, COMBATSIARIS T, GALUSKA D, BAUMAN D E, BARBANO D M, CHARRON M J, ZIERATH J R and HOUSEKNECHT K L (2001), Isomer-specific antidiabetic properties of conjugated linoleic acid – improved glucose tolerance, skeletal muscle insulin action, and UCP-2 gene expression, *Diabetes*, **50**, 1149–1157.

SCHNEIDER W J, GAST L E and TEETER H M (1964), A convenient laboratory method for preparing *trans,trans*-9,11-octadecadienoic acid, *J Am Oil Chem So*, **41**, 605–606.

SCHONBERG S and KROKAN H E (1995), The inhibitory effect of conjugated dienoic derivatives (CLA) of linoleic acid on the growth of human cell lines is in part due to increased lipid peroxidation, *Anticancer Res*, **15**, 1241–1246.

SCHUT H A and ZU H X (1993), Application of the 32P-postlabelling assay to the inhibition of 2-amino-3-methylimidazo[4,5-f]quinoline (IQ)–DNA adduct formation by dietary fatty acids, *IARC Sci Publ*, 181–188.

SCHUT H A, CUMMINGS D A, SMALE M H, JOSYULA S and FRIESEN M D (1997), DNA adducts of heterocyclic amines: formation, removal and inhibition by dietary components, *Mutat Res*, **376**, 185–194.

SEBEDIO J L, JUANEDA P, DOBSON G, RAMILISON I, MARTIN J C, CHARDIGNY J M and CHRISTIE W W (1997), Metabolites of conjugated isomers of linoleic acid (CLA) in the rat, *Biochim Biophys Acta*, **1345**, 5–10.

SEBEDIO J L, GNÄDIG S and CHARDIGNY J M (1999), Recent advances in conjugated linoleic acid, *Curr Opin Clin Nutr Metab Care*, **2**, 499–506.

SEBEDIO J L, VERMUNT S H F, CHARDIGNY J M, BEAUFRERE B, MENSINK R P, ARMSTRONG R A, CHRISTIE W W, NIEMELA J, HENON G and RIEMERSMA R A (2000), The effect of dietary *trans* α-linoleic acid on plasma lipids and platelet fatty acid composition: the *Trans*Line study, *Eur J Clin Nutr*, **54**, 104–113.

SEBEDIO J L, ANGIONI E, CHARDIGNY J M, GREGOIRE S, JUANEDA P and BERDEAUX O (2001), The effect of conjugated linoleic acid isomers on fatty acid profiles of liver and adipose tissues and their conversion to isomers of 16:2 and 18:3 conjugated fatty acids in rats, *Lipids*, **36**, 575–582.

SEHAT N, RICKERT R, MOSSOBA M M, KRAMER J K G, YURAWECZ M P, ROACH J A G, ADLOF R O, MOREHOUSE K M, FRITSCHE J, EULITZ K D, STEINHART H and KU Y (1999), Improved separation of conjugated fatty acid methyl esters by silver ion high-performance liquid chromatography, *Lipids*, **34**, 407–413.

SHANTHA N C, DECKER E A and USTUNOL Z (1992), Conjugated linoleic acid concentration in processed cheese, *J Am Oil Chem Soc*, **69**, 425–428.

SHANTHA N C, DECKER E A and HENNIG B (1993), Comparison of methylation methods for the quantitation of conjugated linoleic acid isomers, *J Assoc Off Am Chem Int*, **76**, 644–649.

SHANTHA N C, CRUM A D and DECKER E A (1994), Evaluation of conjugated linoleic acid concentrations in cooked beef, *J Agric Food Chem*, **42**, 1757–1760.

SHANTHA N C, RAM L N, O'LEARY J, HICKS C L and DECKER E A (1995), Conjugated linoleic acid concentrations in dairy products as affected by processing and storage, *J Food Res*, **60**, 695–720.

SHANTHA D C and DECKER E A (1995), Conjugated linoleic acid concentrations in cooked beef containing antioxidants and hydrogen donors, *J Food Lipids*, **2**, 57–64.

SHULTZ T D, CHEW B P, SEAMAN W R and LUEDECKE L O (1992), Inhibitory effect of conjuagetd dienoic derivatives of linoleic acid and ß-carotene on the *in vitro* growth of human cancer cells, *Cancer Lett*, **63**, 125–133.

SISK M B, HAUSMAN D B, MARTIN R J and AZAIN M J (2001), Dietary conjugated linoleic acid reduces adiposity in lean but not obese Zucker rats, *J Nutr*, **131**, 1668–1674.

SMITH G N, TAJ M and BRAGANZA J M (1991), On the identification of a conjugated diene component of duodenal bile as 9Z,11E-octadecadienoic acid, *Free Radic BiolMed*, **10**, 13–21.

STANGL G I (2000), High dietary levels of conjugated linoleic acid mixture alter hepatic glycerophospholipid class profile and cholesterol-carrying serum lipoproteins of rats, *J Nutr Biochem*, **11**, 184–191.

STANGL G I, MÜLLER H and KIRCHGESSNER M (1999), Conjugated linoleic acid effects on circulating hormones, metabolites and lipoproteins, and its proportion in fasting serum and erythrocyte membranes of swine, *Eur J Nutr*, **38**, 271–277.

STANTON C, LAWLESS F, KJELLMER G, HARRINGTON D, DEVERY R, CONNOLLY J F and MURPHY J (1997), Dietary influences on bovine milk *cis-9,trans*-11-conjugated linoleic acid content, *J Food Sci*, **62**, 1083–1086.

SUGANO M, TSUJITA A, YAMASAKI K, IKEDA I and KRITCHEVSKY D (1997), Lymphatic recovery, tissue distribution, and metabolic effects of conjugated linoleic acid in rats, *J Nutr Biochem*, **8**, 38–43.

SUGANO M, TSUJITA A, YAMASAKI M, NOGUCHI M and YAMADA K (1998), Conjugated linoleic acid modulates tissue levels of chemical mediators and immunoglobulins in rats, *Lipids*, **33**, 521–527.

TASSIGNON P, DE WAARD P, DE RIJK T, TOURNOIS H, DE WIT D and DE BUYCK L (1994), An efficient countercureent distribution method for large scale isolation of dimorphecolic acid methyl ester, *Chem Phys Lipids*, **71**, 187–196.

THOM E, WADSTEIN J and GUDMUSEN O (2001), Conjugated linoleic acid reduces body fat in healthy exercising humans, *J Int Med Res*, **29**, 392–396.

THOMPSON H, ZHU Z J, BANNI S, DARCY K, LOFTUS T and IP C (1997), Morphological and biochemical status of the mammary gland as influenced by conjugated linoleic acid: implication for a reduction in mammary cancer risk, *Cancer Res*, **57**, 5067–5072.

THUILLIER P, ANCHIRAICO G J, NICKEL K P, MALDVE R E, GIMENEZ-CONTI I, MUGA S J, LIU K L, FISCHER S M and BELURY M A (2000), Activators of peroxisome proliferator-activated receptor-alpha partially inhibit mouse skin tumor promotion, *Mol Carcinog*, **29**, 134–142.

TRUITT A, MCNEILL G and VANDERHOEK J Y (1999), Antiplatelet effects of conjugated linoleic acid isomers, *Biochim Biophys Acta*, **1438**, 239–246.

TSUBOYAMA-KASAOKA N, TAKAHASHI M, TANEMURA K, KIM H J, TSUYOSHI T, OKUYAMA H, KASAI M, IKEMOTO S and EZAKI O (2000), Conjugated linoleic acid supplementation reduces adipose tissue by apoptosis and develops lipodystrophy in mice, *Diabetes*, **49**, 1534–1542.

VISONNEAU S, CESANO A, TEPPER S A, SCIMECA J A, SANTOLI D and KRITCHEVSKY D (1997), Conjugated linoleic acid suppresses the growth of human breast adenocarcinoma cells in SCID mice, *Anticancer Res*, **17**, 969–973.

WEST D B, DELANY J P, CAMET P M, BLOHM F, TRUETT A A and SCIMECA J (1998), Effects of conjugated linoleic acid on body fat and energy metabolism in the mouse, *Am J Physiol*, **275**, R667–R672.

WIEGAND B R, PARRISH F C, SWAN J E, LARSEN S T and BAAS T J (2001), Conjugated linoleic acid improves feed efficiency, decrease subcutaneous fat, and improves certain aspects of meat quality in stress-genotype pigs, *J Anim Sci*, **79**, 2187–2195.

WONG M W, CHEW B P, WONG T S, HOSICK H L, BOYLSTON T D and SHULTZ T D (1997), Effects of dietary conjugated linoleic acid on lymphocyte function and growth of mammary tumors in mice, *Anticancer Res*, **17**, 987–994.

XU M and DASHWOOD R (1999), Chemoprevention studies of heterocyclic amine-induced colon carcinogenesis, *Cancer Lett*, **143**, 179–183.

YURAWECZ M P, HOOD J K, ROACH J A G, MOSSOBA M, DANIELS D H, KU Y, PARIZA M and CHIN S F (1994), Conversion of allylic hydroxy oleate to conjugated linoleic acid and methoxy oleate by acid catalysed methylation procedures, *J Am Oil Chem Soc*, **71**, 1149–1155.

ZAMBELL K L, KEIM N L, VAN LOAN M D, GALE B, BENITO P, KELLEY D S and NELSON G J (2000), Conjugated linoleic acid supplementation in humans: effects on body composition and energy expenditure, *Lipids*, **35**, 777–782.

Part III

Product development

12

Enhancing the functionality of prebiotics and probiotics

R. Rastall, The University of Reading, UK

12.1 Introduction

Most functional food products are designed to provide the consumer with some kind of health benefit above and beyond mere nutrition. In the case of probiotic and prebiotic functional foods, this involves the manipulation of the gut flora. Currently, however, in the EU a specific health claim cannot be made for a food product (Gibson *et al.*, 2000). It could be argued, however, that modulation of the gut flora using probiotics and prebiotics is a functional aspect of the gut microecology.

An EU consensus document (the gastrointestinal tract group of the EU Concerted Action on Functional Food Science in Europe, FUFOSE) discusses the established and possible effects of prebiotics (Salminen *et al.*, 1998). The established effects include non-digestibility and low energy value, a stool bulking effect and modulation of the gut flora, promoting bifidobacteria and repressing clostridia.

Postulated areas for development in the future might include prevention of intestinal disorders (e.g. ulcerative colitis, irritable bowel syndrome) and gastrointestinal infections including diarrhoea, modulation of the immune response, prevention of colon carcinogenesis, reduction in serum triacylglycerols and cholesterol and the improvement of bioavailability of minerals such as calcium and magnesium. This chapter discusses how prebiotics and probiotics might be developed in order bring about some of these health benefits.

12.2 The functional enhancement of prebiotics

There are several recognised prebiotic oligosaccharides in use around the world (Playne and Crittenden, 1996; Crittenden and Playne, 1996). In Europe and the

Table 12.1   Some desirable attributes in functionally enhanced prebiotics

Desirable attribute in prebiotic	Properties of oligosaccharides
Active at low dosage and lack of side effects	Highly selectively and efficiently metabolised by 'beneficial' bacteria but not by gas producers, putrefactive organisms, etc.
Persistence through the colon	Controlled molecular weight distribution
Protection against colon cancer	Stimulate butyrate production in the colon
Enhance the barrier effect against pathogens	Structural basis unknown
Inhibit adhesion of pathogens	Possess receptor sequence
Targeting at specific probiotics	Selectively metabolised by restricted species of *Lactobacillus* and/or *Bifidobacterium*

USA, the fructans (inulin and fructooligosaccharides, FOS) are the market leaders followed by the galactooligosaccharides. In Japan the list is much more extensive, including isomaltooligosaccharides, soybean oligosaccharides, xylooligo-saccharides, gentiooligosaccharides and lactosucrose. Many of these prebiotics have not yet been subjected to rigorous testing in human volunteers using modern molecular microbiological techniques. In many cases we still know little about their strengths and weaknesses relative to other prebiotics and little about their food functionality.

Despite our relatively poor knowledge of the structure–function relationships in prebiotic oligosaccharides, it is possible to identify several desirable attributes in a functionally enhanced prebiotic. We can also apply an understanding of carbohydrate chemistry and of the roles of carbohydrates in foods to identify the properties needed in the oligosaccharide to achieve these attributes (Table 12.1). No single molecule will possess all of these properties, but each is a practical target for prebiotic development. These properties will now be explored in more detail.

12.2.1   Lack of side effects and activity at low dosage

It is well documented that, at excessive dosage, fructans can cause problems with gas production, leading to intestinal bloating and discomfort (Stone-Dorshow and Levitt, 1987). This gas is not produced as a result of metabolism by bifidobacteria or lactobacilli but results from the action of other microorganisms in the gut (Gibson and Macfarlane, 1995). While we do not know enough about the relation-ship between structure and resultant fermentation behaviour to rationally design low-gas-producing prebiotics, there are pronounced differences between the current oligosaccharides. Rycroft *et al.* (2001) compared the properties of a range of commercial oligosaccharides from Europe and Japan and found a large variation in gas evolution rates when used as substrates in batch culture with human faecal inocula. Fructans resulted in the highest gas liberation levels while galacto-oligosaccharides and isomaltooligosaccharides resulted in the least gas. Metabolism of the prebiotic by the target microorganisms rather than other members of the colonic flora is also likely to reduce the effective dosage required for a prebiotic effect.

12.2.2   Persistence of the prebiotic effect to distal regions of the colon

Many chronic gut disorders arise in the distal regions of the colon. Of particular note is colonic cancer (Rowland, 1992). Much research has suggested that prebiotics are protective against development of colon cancer (Rowland and Tanaka, 1993; Bouhnik *et al.*, 1996; Buddington *et al.*, 1996; Reddy *et al.*, 1997; Hylla *et al.*, 1998). In order for a prebiotic to have a protective effect, however, it is necessary for the prebiotics to reach the distal colon and exert selective influence on the microflora in that region.

One approach to achieving this persistence to the distal colon is the regulation of molecular weight distributions. Most of the current prebiotics are low molecular weight with the exception of inulin (Playne and Crittenden, 1996). The current paradigm for prebiotic action is the possession of cell-associated bacterial glycosidases by probiotics, which degrade the carbohydrate prior to metabolism of the resultant monosaccharides (O'Sullivan, 1996). Longer carbohydrate chains will thus be metabolised more slowly than short ones (Rycroft *et al*, 2001). This idea has led to industrial forms of inulin/FOS mixtures with controlled chain length distributions ('Synergie II' manufactured by Orafti, Tienen, Belgium). In such products, the long chain inulin exerts a prebiotic effect in more distal colonic regions than the FOS with lower molecular weight, which is more rapidly fermented in the proximal colon. This approach should lead to the development of more persistent prebiotics from other polysaccharide systems. It is generally found that polysaccharides are not selectively fermented in the colon (Cummings and Macfarlane, 1991) whereas low molecular weight oligosaccharides frequently are (Crittenden, 1999). In any polysaccharide system, selectivity is likely to decrease with increasing molecular weight. For this reason, development of a persistent prebiotics will probably prove to be a compromise between persistence and prebiotic activity.

The manufacturing technology to achieve a controlled chain length distribution in a mixture of oligosaccharides is being developed (*see* Section 12.5.1). Plant-derived inulin is intrinsically limited in its degree of polymerisation (DP), restricting scope for increasing persistence. Other polysaccharides such as laevan might be a better substrate for controlled hydrolysis. Laevan is a very high molecular weight polysaccharide (Han, 1990) and it is likely that it would take longer to be metabolised throughout the colon.

12.2.3   Protection against colon cancer

Colon cancer is the second most prevalent cancer in humans (Gibson and Macfarlane, 1994). The colonic microflora has long been suspected of having a role in the aetiology of colon cancer (Rowland, 1988). It is believed, for instance that cancers of the large intestine originate 100 times more frequently than in the small intestine (Morotomi *et al.*, 1990). Many species of colonic bacteria produce genotoxic enzymes and produce carcinogens, and tumour promoters from a variety of food components (Gibson and Macfarlane, 1994). It is believed that prebiotics may help protect against cancer by either the production of butyrate or by modulating the colonic flora away from protein and lipid metabolism towards the more benign carbohydrate metabolism.

Butyrate is believed to be protective against colon cancer as it is known to stimulate apoptosis in colonic cancer cell lines and it is the preferred fuel for healthy colonocytes (Prasad, 1980; Kim *et al.*, 1982). While it is probably desirable to have high levels of butyrate generation in the distal colon, it must be borne in mind that lactobacilli and bifidobacteria do not produce butyrate. The principal butyrate organisms that are believed to produce butyrate in the colon are the clostridia and eubacteria (Cummings and Macfarlane, 1991). An obvious development target for prebiotics would be an oligosaccharide that stimulated eubacteria but not clostridia. Some prebiotics have been found to result in the generation of butyrate (Olano-Martin *et al.*, 2000; Rycroft *et al.*, 2001), although the microflora changes responsible are still uncertain as most such studies do not enumerate the changes in eubacteria.

An alternative approach would be to develop prebiotics that shift the metabolism of closridia and bacteroides towards saccharolysis rather than proteolysis or lipolysis. Clostridia are believed to convert dietary lipids such as phosphatidylcholine into carcinogens such as diacylglycerol (DAG, Morotomi *et al.*, 1990). DAG is believed to be a protein kinase C activator, resulting in stimulation of mucosal cell proliferation. We currently have an inadequate knowledge of the effects of various prebiotics upon cancer biomarkers in the human colon and more research is urgently needed in this area.

12.2.4   The barrier effect against gastrointestinal pathogens

A very fertile area for the development of functionally enhanced prebiotics is the enhancement of the barrier effect of the normal gut microflora against pathogens. Dietary manipulation of this barrier to invading pathogens could make an important contribution to food safety. Bifidobacteria and lactobacilli produce a range of anti-microbial compounds from bacteriocins, active against related species, to more broad spectrum agents active against a range of Gram-positive and Gram-negative pathogens such as salmonellae, campylobacters and *Escherichia coli* (Anand *et al.*, 1985; Gibson and Wang, 1994; Araya-Kojima *et al.*, 1995). This anti-pathogen effect can be reproducibly seen in gut model systems and is likely to operate in the human gut.

In addition to the production of specific anti-microbial agents, fermentation of prebiotics results in the production of short-chain fatty acids (SCFA; Gibson and Roberfroid, 1995). The resultant reduction in colonic pH inhibits growth and activity of pathogens. There will also be increased competition with other organisms for nutrients and for adhesion receptors on the gut wall (Araya-Kojima *et al.*, 1995). The regulation of anti-microbial production by means of substrates in the

Table 12.2   Receptor oligosaccharide sequences for selected gastrointestinal pathogens and toxins

Oligosaccharide sequence	Pathogen/toxin
Galα4Gal	*E. coli* (P-piliated), Vero cytotoxin
Galβ4GlcNAcβ3Gal	*S. pneumoniae*
GalNAcβ4Gal	*P. aeruginosa, H. influenzae, S. aureus, K. pneumoniae*
Sialic acids	*E. coli* (S-fimbriated)
Galα3Galβ4GlcNAc	*C. difficile* toxin A
Fucose	*V. cholerae*
GlcNAc	*E. coli, V. cholerae*
Mannose	*E. coli, K. aerogenes, Salmonella* spp. (Type 1-fimbriated)

gut is imperfectly understood. In particular there is a surprising lack of information about the effects of different prebiotics on the expression levels and activity spectrum of anti-microbial agents produced by indigenous probiotics. This would seem to be an obvious target for prebiotic development.

12.2.5   Anti-adhesive activities against pathogens and toxins

An interesting development to the prebiotic concept is the modulation of the colonic flora using an anti-adhesive strategy. Many gastrointestinal pathogens recognise oligosaccharide sequences carried on glycoproteins and glycolipids on the gut wall as receptors (Table 12.2; Karlsson, 1989; Finlay and Falkow, 1989). There is currently much interest in the development of pharmaceutical agents based around this concept and there are several preparations in clinical trials. The potential for the incorporation of receptor oligosaccharides into foods, however, has not been fully explored, largely due to the high cost of oligosaccharides. Recent advances in the biotechnological manufacture (see section 12.5) of complex oligosaccharides, however, are likely to lead to significant cost reductions in the future.

12.3   Targeted prebiotics

A very exciting area for the future development of enhanced prebiotics is that of targeted prebiotics. Prebiotics could conceivably be developed that will display selectivity towards particular species of probiotic. Currently, prebiotics are defined as acting at the level of bifidobacteria and lactobacilli (Gibson and Roberfroid, 1995). Virtually all of the available data on the prebiotic properties of oligosaccharides describes microflora changes at the genus level. This is due to the challenges of accurately identifying the colonic microflora using conventional microbiological techniques. It would, however, be highly desirable to develop prebiotics that are targeted at particular species of *Bifidobacterium* and *Lactobacillus*. Such targeted prebiotics would have many applications.

12.3.1   Synbiotics with defined health benefits
Commercial probiotics are highly selected to have useful properties such as resistance to acid and bile and survival upon freeze-drying (Svensson, 1999; Lee *et al.*, 1999). In addition, many of them have been developed for particular health benefits such as immune stimulation or anti-pathogen activity. It would be expected that synbiotic versions of these strains made with targeted prebiotics would display better survival and colonisation in the gut (Bielecka *et al.*, 2002).

12.3.2   Infant formulae and weaning foods
The gut flora of breast-fed infants is generally dominated by bifidobacteria, while the gut flora of a formula-fed infant resembles that of an adult, with a wider range of microbial groups present (Cooperstock and Zedd, 1983; Benno *et al.*, 1984). It is believed that this bifidobacterial flora leads to health benefits for the infant, including a decreased susceptibility to microbial infections. It would be of interest to synthesise prebiotics with high selectivity towards those bifidobacteria that are present in the guts of breast-fed infants as the basis of novel infant foods.

12.3.3   Functional foods for healthy ageing
Above the age of 55–60 years, bifidobacterial counts in the colon decrease markedly compared with those of young adults (Mitsuoka, 1990; Kleessen *et al.*, 1997). It is conceivable that this decrease in bifidobacterial numbers leads to reduced resistance to gastrointestinal infections. It is well known that elderly people tend to suffer more such infections. The development of functional foods for the elderly is the subject of the EU-funded *Crownalife* project, part of the *ProEUhealth* cluster. The project will characterise the elderly gut flora across Europe, will run a synbiotic intervention trial and also develop prebiotics targeted at the bifidobacteria present in elderly people. The aim of the targeted prebiotic development is increased resistance to gastrointestinal pathogens such as *E. coli*, *Salmonella* sp. and *Campylobacter jejuni*.

12.3.4   The development of targeted prebiotics
Targeted prebiotics might be developed through one of two approaches. Screening a wide range of oligosaccharide structures for their prebiotic properties will give information on their selectivity towards particular species. In order to achieve the necessary structural diversity, economical manufacturing technologies for complex oligosaccharides are required.

An alternative approach might be to use enzymes expressed by probiotic bacteria as synthetic catalysts. Such enzymes will produce oligosaccharide mixtures, which may be more readily metabolised by the producing organism, resulting in higher selectivity. Early results using pure culture testing are encouraging. Novel β-galacto-oligosaccharide mixtures were synthesised from lactose using β-galactosidases from a range of probiotics (Rabiu *et al.*, 2001). The growth of each of the probiotics was measured on each of the synthetic mixtures in pure culture (Fig. 12.1). Several of the probiotics tested displayed higher growth rates on the oligosaccharide mixtures resulting from their own enzymes relative to other bacteria on the same oligosaccharide mixture. To date, these products have not been tested in mixed culture owing to the problems of identification. These problems will be solved by the use of species-specific oligonucleotide probes.

12.4   Current manufacturing technologies for prebiotics

Current prebiotics are either extracted from plants or manufactured from cheap, readily available sources, generally by means of enzymatic hydrolysis or synthesis reactions (Playne and Crittenden, 1996).

12.4.1   Extraction from plants
Currently, both inulin (De Leenheer, 1994) and soybean oligosaccharides (Koga *et al.*, 1993) are extracted from their corresponding plants. Inulin is extracted from chicory by a modification of the sugar beet process (De Leenheer, 1994), with Orafti in Tienen, Belgium, being Europe's largest manufacturer. Soybean oligosaccharides are extracted from soybean whey (Koga *et al.*, 1993) by the Calpis Food Industry Co. in Japan.

12.4.2   Polysaccharide hydrolysis
Fructooligosaccharides and xylooligosaccharides are both manufactured by hydrolysis of their parent polysaccharides, although fructooligosaccharides are also manufactured by synthesis from sucrose (*see* Section 12.4.3). Fungal inulinase is used to hydrolyse chicory inulin to oligosaccharides with low monosaccharide contents (De Leenheer, 1994). Inulin contains polysaccharide chains terminating in either a reducing fructose residue or a non-reducing glucose residue. Consequently, FOS produced from inulin have reducing activity. Xylooligo-saccharides, principally xylobiose, are manufactured by the enzymatic hydrolysis corn cob xylan (Koga and Fujikawa, 1993). Final products are processed using membrane technology to remove xylose and high molecular weight materials.

12.4.3   Enzymatic synthesis
Galactooligosaccharides, lactosucrose, isomaltooligosaccharides and some fructooligosaccharides are manufactured by enzymic glycosyl transfer reactions from cheap sugars such as sucrose and lactose or from starch (Playne and Crittenden, 1996).

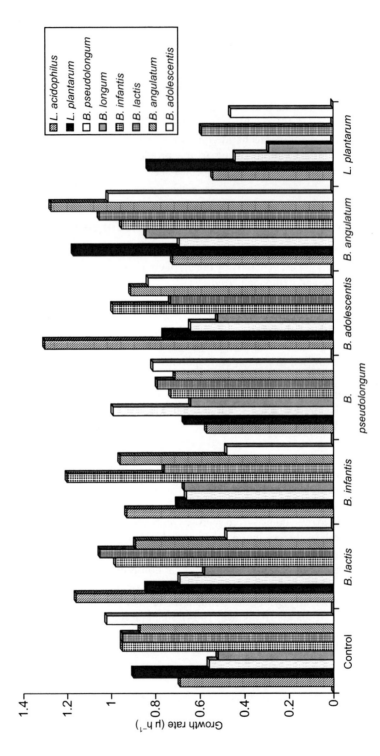

Fig. 12.1   Growth of selected probiotic strains of galactooligosaccharide mixtures generated by probiotic β-galactosidases. The control is a commercial galactooligosaccharide mixture. In most cases, probiotics display higher growth rates on their homologous oligosaccharide mixture.

Fructooligosaccharides are made from sucrose by Meiji Seika in Japan and by Beghin-Meiji Industries in Europe (Playne and Crittenden, 1996). All of the sucrose-derived FOS terminate in a non-reducing glucose residue. Ion-exchange chromatography can be used to remove glucose and sucrose (Kono, 1993). Galactooligosaccharides are manufactured from lactose using β-galactosidase (Matsumoto, 1993). At high lactose concentration the enzyme forms galactooligosaccharides (Schoterman and Timmermans, 2000). The products are complex mixtures of di- to pentasaccharides principally containing β1→3 and β1→6 linkages. Galactooligosaccharides are made by Yakult Honsha and Snow Brand in Japan and by Borculo Domo Ingredients in Europe.

Lactosucrose is made using a similar approach, this time using β-fructofuranosidase to transfer fructose from sucrose to the reducing glucosyl group on lactose (Kitahata and Fujita, 1993). Commercial production is centred in Japan by Hayashibara Shoji Inc. and Ensuiko Sugar Refining Co. Isomaltooligosaccharides (IMO) are manufactured by a more complex route (Yatake, 1993). Starch is first hydrolysed to maltooligosaccharides with α-amylase and pullulanase. The α1→4 linked maltooligosaccharides are then converted into α1→6 linked isomaltooligosaccharides by α-glucosidase. The commercial product manufactured by Showa Sangyu Co. in Japan consists of IMO of varying molecular weight together with panose.

12.5   Emerging manufacturing technologies for second generation prebiotics

If the full potential of enhanced prebiotics is to be realised, new technological innovations will be required. The challenge, as ever, for biotechnologists is to achieve the manufacturing technologies at economically viable costs. Two areas of development are being explored in laboratories in Europe at the current time.

12.5.1   Controlled polysaccharide hydrolysis

As discussed above, polysaccharide hydrolysis is a commercial manufacturing approach to prebiotics. Several of the functional enhancements discussed above, however, will demand a more controlled partial hydrolysis in order to achieve control over the molecular weight distribution of the products. Controlled partial hydrolysis has been investigated with dextran and pectins. Mountzouris *et al.* (1998, 2001) utilised *endo*-dextranase in an enzyme membrane reactor to make controlled IMO. By controlling factors such as residence time and the ratio of enzyme to substrate, dextran could be converted into different IMO preparations with average molecular weights varying from trisaccharide up to 12,000 Da. The fractions displayed good prebiotic fermentation in an *in vitro* three-stage model of the human gut (Olano-Martin *et al.*, 2000). Selective fermentation extended to the third vessel of this system, which models the distal region of the gut (Macfarlane *et al.*, 1992, 1998).

Enzyme membrane reactors have also been used for the controlled depolymerisation of pectins (Olano-Martin *et al.*, 2001). In this case, low molecular weight products were produced. Such reactor systems show great promise but are in need of further research and development before they become a practical tool for oligosaccharide manufacture. This approach might, in future be extended to other polysaccharides. Bacterial extracellular polysaccharides (EPS), for example, display huge structural variation and complexity and can often be manufactured economically by fermentation processes (Sutherland, 1975; Morin, 1998). They frequently contain rare monosaccharide residues in a wide range of linkage configurations. An interesting source of enzymes to degrade such polysaccharides can be isolated from bacteriophage. These enzymes have been used in sequencing studies of polysaccharides (Geyer *et al.*, 1983; Stirm, 1994; Sutherland, 1999) but not in a commercial manufacturing technology. This approach could extend our collection of oligosaccharide structures for screening as candidate probiotics with enhanced specificity at the species level.

12.5.2 Enzymatic synthesis of oligosaccharides

The enzymatic synthesis of oligosaccharides can be achieved through several routes (Rastall and Bucke, 1992; Kren and Thiem, 1997; Crout and Vic, 1998). Many of these have not yet reached industrial implementation, although they are subjects of active research programmes.

One class of enzymes with obvious potential for the industrial synthesis of novel prebiotics is the glycosyltransferases. These can be divided into the so-called Leloir enzymes (Rastall and Bucke, 1992), which use sugar nucleotides as glycosyl donors, and the sucrases, which use sucrose. The sucrases can be used to generate novel oligosaccharides (Robyt, 1995). Dextransucrase has been used in the manufacture of novel α-glucooligosaccharides that are selectively fermented by desirable microorganisms on the skin (Dols *et al.*, 1999) and the gut (Djouzi *et al.*, 1995; Djouzi and Andrieux 1997). These oligosaccharides have been formulated into cosmetics but have not yet been fully evaluated as prebiotic food ingredients.

Glycosidase enzymes can also be used to synthesise novel oligosaccharides under appropriate conditions (Rastall and Bucke, 1992; Kren and Thiem, 1997; Crout and Vic, 1998). Vic *et al.* (1997), for example synthesised the trisaccharide receptor for *Clostridium difficile* Toxin A using glycosidases in kinetically controlled syntheses. Suwasono and Rastall (1996) synthesised the receptor active Manα1→2Man di- and trisaccharides using α1,2-mannosidase from *Aspergillus phoenicis*. These oligosaccharides are receptors for type-1 fimbriated enteric microorganisms and may have potential for development as anti-adhesive agents.

Glycosidases may also be used to manufacture novel oligosaccharides as candidate prebiotics. Many glycosidases can recognise a range of sugars as acceptors, building up heterooligosaccharides. α-Mannosidase enzymes have been used to attach a mannose residue onto several disaccharide acceptors to manufacture novel trisaccharides (Rastall *et al.*, 1992; Smith *et al.*, 1997; Suwasono and Rastall, 1998). The prebiotic potential of such oligosaccharides is now under investigation.

12.6   The functional enhancement of probiotics

Although it is not too difficult to conceive of desirable enhancements to probiotic bacterial strains, the technology required to achieve these enhancements is complex. Genetic manipulation tools for probiotics have been developed (Kullen and Klaenhammer, 1999), but these have not yet been widely applied to the enhancement of probiotics destined for human food applications.

It is possible to define the characteristics that one would want in a probiotic strain (Shah, 2001) and this is covered in more depth elsewhere in this volume. One such enhancement might be the metabolism of specific prebiotics.

12.6.1   Metabolism of specific prebiotics

It is possible to achieve stable, functional expression of heterologous enzyme genes in *Lactobacillus* species. This raises the possibility of generating probiotics that can metabolise particular carbohydrates that most lactobacilli can not. Certain prebiotic oligosaccharides such as inulin, for instance, are not particularly well fermented by many lactobacilli. There would be an obvious advantage in the development of synbiotics, which can metabolise unusual oligosaccharides.

To date, no such probiotics destined for human food applications have been developed. Several genetically modified *Lactobacillus* species expressing glycanase genes have been constructed as enhanced silage inocula. Most work has focused on *Lactobacillus plantarum* strain Lp80, a common silage inoculum strain. Engineered forms have been made with good expression levels of α-amylase (Schierlink, *et al*, 1989; Fitzsimons *et al.*, 1994) cellulases (Schierlink *et al.*, 1989, 1990), xylanases (Schierlink *et al.*, 1990), laevanases (Hols *et al.*, 1994; Wanker *et al.*, 1995) and chitinase (Brurberg *et al.*, 1994). The laevanase gene has also been expressed in *L. casei* (Wanker *et al.*, 1995) and the chitinase gene was expressed in *L. lactis* (Brurberg *et al.*, 1994). The engineered lactobacilli displayed enhanced growth and lactic fermentation on complex biomass polysaccharides with no need for added fermentable carbohydrates or enzymes. Of particular interest is the expression of the laevanase gene from *Bacilllus subtilis*, as this enzyme enabled the recombinant strain to grow on laevan and on inulin (Wanker *et al.*, 1995).

12.7   Conclusion and future trends

Prebiotic oligosaccharides are set to make a further impact on the area of functional foods. Current forms are efficient at stimulating bifidobacteria; however, the next generation of prebiotics will entail multiple functionality. Biotechnological procedures are increasingly being applied to the area and developments with novel foods are proceeding quickly. It is clear that for the further development of this field reliable testing methods are needed. One example is through the application of molecular procedures to improve our knowledge of the gut microflora, its interactions with the human host and its role in maintenance of health. Advances in these areas will ultimately provide us with very sophisticated dietary tools with which to manipulate this important ecosystem and improve public health.

12.8   References

ANAND S K, SRINIVASAN R A and RAO L K (1985), 'Antibacterial activity associated with *Bifidobacterium bifidum* – II', *Cult Dairy Prod J*, **2**, 21–23.

ARAYA-KOJIMA A, YAESHIMA T, ISHIBASHI N, SHIMAMUR S and HAYASAWA H (1995), 'Inhibitory effects of *Bifidobacterium longum* BB536 on harmful intestinal bacteria', *Bifidobacteria Microflora*, **14**, 59–66.

BENNO Y, SAWADA K and MITSUOKA T (1984), 'The intestinal microflora of infants: composition of fecal flora in breast-fed and bottle-fed infants', *Micro Immun*, **28**, 975–986.

BIELECKA M, BIEDRZYCKA E and MAJKOWSKA A (2002), 'Selection of probiotics and prebiotics for synbiotics and confirmation of their *in vivo* effectiveness', *Food Res Int*, **35**, 125–131.

BOUHNIK Y, FLOURIE B, RIOTTOT M, BISETTI N, GAILING M F and GUIBERT A (1996), 'Effects of fructo-oligosaccharides ingestion on faecal bifidobacteria and selected metabolic indexes of colon carcinogenesis in healthy humans', *Nutr Cancer*, **26**, 21–29.

BRURBERG M B, HAANDRIKMAN A J, LEENHOUTS K J, VENEMA G and NES I F (1994), 'Expression of a chitinase gene from *Serratia marcescens* in *Lactococcus lactis* and *Lactobacillus plantarum*', *Appl Micro Biotechnol*, **42**, 108–115.

BUDDINGTON R K, WILLIAMS C H, CHEN S-C and WITHERLY S A (1996), 'Dietary supplementation of neosugar alters the fecal flora and decreases activities of some reductive enzymes in human subjects', *Am J Clin Nutr*, **63**, 709–716.

COOPERSTOCK M S and ZEDD A J (1983), 'Intestinal flora of infants', in D J Hentges, *Human Intestinal Microflora in Health and Disease*, London, Academic Press, 79–99.

CRITTENDEN R G (1999), 'Prebiotics', in G Tannock, *Probiotics: A Critical Review*, Wymondham, Horizon Scientific Press, 141–156.

CRITTENDEN R G and PLAYNE M J (1996), 'Production, properties and applications of food-grade oligosaccharides', *Trends Food Sci Tech*, **7**, 353–361.

CROUT D H G and VIC G (1998), 'Glycosidases and glycosyl transferases in glycoside and oligosaccharide synthesis', *Curr Opin Chem Biol*, **2**, 98–111.

CUMMINGS J H and MACFARLANE G T (1991), 'The control and consequences of bacterial fermentation in the human colon', *J Appl Bact*, **70**, 443–459.

DE LEENHEER L (1994), 'Production and use of inulin: industrial reality with a promising future', in H van Bekkum, H Roper and A G J Voragen, *Carbohydrates as Organic Raw Materials III*, Weinheim, VCH, 67–92.

DJOUZI Z and ANDRIEUX C (1997), 'Compared effects of three oligosaccharides on metabolism of intestinal microflora in rats inoculated with a human faecal flora', *Br J Nutr*, **78**, 313–324.

DJOUZI Z, ANDRIEUX C, PELENC V, SOMARRIBA S, POPOT F, PAUL F, MONSAN P and SZYLIT O (1995), 'Degradation and fermentation of α-gluco-oligosaccharides by bacterial strains from human colon: *in vitro* and *in vivo* studies in gnotobiotic rats', *J Appl Bact*, **79**, 117–127.

DOLS M, MONCHOIS V, REMAUD-SIMEON M, WILLEMOT R-M and MONSAN P F (1999), 'The production of α1→2-terminated glucooligosaccharides', in C Bucke, *Carbohydrate*

Biotechnology Protocols, Methods in Biotechnology, Vol. 10, Totowa, NJ, Humana Press, 129–139.

FINLAY B B and FALKOW S (1989), 'Common themes in microbial pathogenicity', *Microbiol Rev*, **53**, 210–230.

FITZSIMONS A, HOLS P, JORE J, LEER R J, O'CONNELL M and DELCOUR J (1994), 'Development of an amylolytic *Lactobacillus plantarum* silage strain expressing the *Lactobacillus amylovorus* alpha-amylase gene', *Appl Env Microbiol*, **60**, 3529–3535.

GEYER H, HIMMELSPACH K, KWIATKOWSKI B, SCHLECHT S and STIRM S (1993), 'Degradation of bacterial surface carbohydrates by virus associated enzymes', *Pure Appl Chem*, **55**, 637–653.

GIBSON G R and MACFARLANE G T (1994), 'Intestinal bacteria and disease', in S A W Gibson, *Human Health – The Contribution of Microorganisms*, London, Springer-Verlag, 53–62.

GIBSON G R and MACFARLANE G T (1995), *Human Colonic Bacteria: Role in Nutrition, Physiology and Pathology*, Boca Raton, CRC Press.

GIBSON G R and ROBERFROID M B (1995), 'Dietary modulation of the human colonic microbiota: introducing the concept of prebiotics', *J Nutr*, **125**, 1401–1412.

GIBSON G R and WANG X (1994), 'Regulatory effects of bifidobacteria on the growth of other colonic bacteria', *J Appl Bact*, **77**, 412–420.

GIBSON G R, BERRY OTTAWAY P and RASTALL R A (2000), *Prebiotics: New Developments in Functional Foods*, Oxford, Chandos Publishing.

HAN Y W (1990), 'Microbial Levan', in S L Neidleman and A I Laskin, *Advances in Applied Microbiology*, Vol. 35, London, Academic Press, 171–194.

HOLS P, FERAIN T, GARMYN D, BERNARD N and DELCOUR J (1994), 'Use of homologous expression-secretion signals and vector-free stable chromosomal integration in engineering of *Lactobacilllus plantarum* for alpha-amylase and laevanase expression', *Appl Env Microbiol*, **60**, 1401–1413.

HYLLA S, GOSTNER A, DUSEL G, ANGER H, BARTRAM H-P, CHRISTL S U *et al.* (1998), 'Effects of resistant starch on the colon in healthy volunteers: possible implications for cancer prevention', *Am J Clin Nutr*, **67**, 136–142.

KARLSSON K-A (1989), 'Animal glycosphingolipids as membrane attachment sites for bacteria', *Ann Rev Biochem*, **58**, 309–350.

KIM Y S, TSAO D, MORITA A and BELLA A (1982), 'Effect of sodium butyrate and three human colorectal adenocarcinoma cell lines in culture', *Falk Symposium*, **31**, 317–323.

KITAHATA S and FUJITA K (1993), 'Xylsucrose, isomaltosucrose and lactosucrose'. In: *Oligosaccharides. Production, Properties and Applications*, Japanese Technology Review, Vol. 3, Switzerland, Gordon & Breach, 158–174.

KLEESSEN B, SYKURA B, ZUNFT H-J and BLAUT M (1997), 'Effects of inulin and lactose on fecal microflora, microbial activity and bowel habit in elderly constipated persons', *Am J Clin Nut*, **65**, 1397–1402.

KOGA K and FUJIKAWA S (1993), 'Xylooligosaccharides'. In: *Oligosaccharides. Production, Properties and Applications*, Japanese Technology Review, Vol. 3, Switzerland, Gordon & Breach, 130–143.

KOGA Y, SHIBUTA T and O'BRIEN R (1993), 'Soybean oligosaccharides'. In: *Oligosaccharides. Production, Properties and Applications*, Japanese Technology Review, Vol. 3, Switzerland, Gordon & Breach, 90–106.

KONO T (1993), 'Fructooligosaccharides'. In: *Oligosaccharides. Production, Properties and Applications*, Japanese Technology Review, Vol. 3, Switzerland, Gordon & Breach, 50–78.

KREN V and THIEM J (1997), 'Glycosylation employing bio-systems: from enzymes to whole cells', *Chem Soc Rev*, **26**, 463–473.

KULLEN M J and KLAENHAMMER T (1999), 'Genetic modification of lactobacilli and bifidobacteria', in G Tannock, *Probiotics: A Critical Review*, Wymondham, Horizon Scientific Press, 65–83.

LEE Y-K, NOMOTO K, SALMINEN S and GORBACH S L (1999), *Handbook of Probiotics*, New York, Wiley.

MACFARLANE G T, GIBSON G R and CUMMINGS J H (1992), 'Comparison of fermentation reactions in different regions of the colon', *J Appl Bact*, **72**, 57–64.

MACFARLANE G T, MACFARLANE S and GIBSON G R (1998), 'Validation of a three-stage compound continuous culture system for investigating the effect of retention time on the ecology and metabolism of bacteria in the human colonic microbiota', *Micro Ecol*, **35**, 180–187.

MATSUMOTO K (1993), 'Galactooligosaccharides'. In: *Oligosaccharides. Production, Properties and Applications*, Japanese Technology Review, Vol. 3, Switzerland, Gordon & Breach, 90–106.

MITSUOKA T (1990), 'Bifidobacteria and their role in human health', *J Ind Microbiol*, **6**, 263–268.

MORIN A (1998), 'Screening of polysaccharide producing organisms, factors influencing the production and recovery of microbial polysaccharides', in S Dimitriu, *Polysaccharides: Structural Diversity and Functional Versatility*, New York, Marcel Dekker Inc.

MOROTOMI M, GUILLEM J G, LOGERFO P and WEINSTEN I B (1990), 'Production of diacylglycerol, an activator of protein kinase C by human intestinal microflora', *Cancer Res*, **50**, 3595–3599.

MOUNTZOURIS K C, GILMOUR S G, GRANDISON A S and RASTALL R A (1998), 'Modelling of oligodextran production in an ultrafiltration stirred cell membrane reactor', *Enz Microbial Technol*, **24**, 75–85.

MOUNTZOURIS K C, GILMOUR S G and RASTALL R A (2001), 'Continuous production of oligodextrans via controlled hydrolysis of dextran in an enzyme membrane reactor', *J Food Sci*, **67**, 1767–1771.

OLANO-MARTIN E, MOUNTZOURIS K C, GIBSON G R and RASTALL R A (2001), 'Continuous production of oligosaccharides from pectin in an enzyme membrane reactor', *J Food Sci*, **66**, 966–971.

OLANO-MARTIN E, MOUNTZOURIS K C, GIBSON G R and RASTALL R A (2000), '*In vitro* fermentability of dextran, oligodextran and maltodextrin by human gut bacteria', *Brit J Nutr*, **83**, 247–255.

O'SULLIVAN M G (1996), 'Metabolism of bifidogenic factors – an overview', *Bull Int Dairy Fed*, **313**, 23–30.

PLAYNE M J and CRITTENDEN R (1996), 'Commercially available oligosaccharides', *Bull Int Dairy Fed*, **313**, 10–22.

PRASAD K N (1980), 'Butyric acid: a small fatty acid with diverse biological functions', *Life Sci*, **27**, 1351–1358.

RABIU B A, JAY A J, GIBSON G R and RASTALL R A (2001), 'Synthesis and fermentation properties of novel galacto-oligosaccharides by beta-galactosidases from bifidobacterium species', *Appl Env Microbiol*, **67**, 2526–2530.

RASTALL R A and BUCKE C (1992), 'Enzymatic synthesis of oligosaccharides', *Biotechnol Gen Eng Rev*, **10**, 253–281.

RASTALL R A, REES N H, WAIT R, ADLARD M W and BUCKE C (1992), 'Alpha-mannosidase-catalysed synthesis of novel manno-, lyxo-, and heteromanno-oligosaccharides: a comparison of kinetically and thermodynamically mediated approaches', *Enz Micro Technol*, **14**, 53–57.

REDDY B S, HAMID R and RAO C V (1997), 'Effect of dietary oligofructose and inulin on colonic preneoplastic aberrant crypt foci inhibition', *Carcinogenesis*, **18**, 1371–1374.

ROBYT J (1995), 'Mechanisms in the glucansucrase synthesis of polysaccharides and oligosaccharides from sucrose', *Adv Carbohydr Chem*, **51**, 133–168.

ROWLAND I R (1988), *Role of the Gut Flora in Toxicity and Cancer*, London, Academic Press.

ROWLAND I R (1992), 'Metabolic interactions in the gut', in R Fuller, *Probiotics: the Scientific Basis*, Andover, Chapman and Hall, 29–53.

ROWLAND I R and TANAKA R (1993), 'The effects of transgalactosylated oligosaccharides on gut flora metabolism in rats associated with a human faecal microflora', *J Appl Bact*, **74**, 667–674.

RYCROFT C E, JONES M R, GIBSON G R and RASTALL R A (2001), 'A comparative *in vitro* evaluation of the fermentation properties of prebiotic oligosaccharides', *J Appl Microbiol*, **91**, 878–887.

SALMINEN S, BOULEY C, BOUTRON-RUAULT M-C, CUMMINGS J H, FRANCK A, GIBSON G R, ISOLAURI E, MOREAU M-C, ROBERFROID M and ROWLAND I R (1998), 'Functional food science and gastrointestinal physiology and function', *Br J Nutr*, **80**, S147–S171.

SCHIERLINK T, MAHILLON J, JOOS P, DHAESE P and MICHIELS F (1989), 'Integration and expression of alpha-amylase and endoglucanase genes in the *Lactobacillus plantarum* chromosome', *Appl Env Microbiol*, **55**, 2130–2137.

SCHIERLINK T, MEUTTER J D, ARNAUT G, JOSS H, CLAEYSSENS M and MICHIELS F (1990), 'Cloning and expression of cellulase and xylanase genes in *Lactobacillus plantarum*', *Appl Microbiol Biotechnol*, **33**, 534–541.

SCHOTERMAN H C and TIMMERMANS H J A R (2000), 'Galacto-oligosaccharides', in G R Gibson and F Angus, *Prebiotics and Probiotics*, LFRA Ingredients Handbook, Leatherhead, Leatherhead Food RA Publishing, 19–46.

SHAH N P (2001), 'Functional foods from probiotics and prebiotics', *Food Tech*, **51**, 46–53.

SMITH N K, GILMOUR S G and RASTALL R A (1997), 'Statistical optimisation of enzymatic synthesis of derivatives of trehalose and sucrose', *Enz Microbial Technol*, **21**, 349.

STIRM S (1994), 'Examination of the repeating units of bacterial exopolysaccharides', in J N BeMiller, D J Manners and R J Sturgeon, *Methods in Carbohydrate Chemistry*, Vol., X, New York, John Wiley.

STONE-DORSHOW T and LEVITT M D (1987), 'Gaseous response to ingestion of a poorly absorbed fructo-oligosaccharide', *Am J Clin Nutr*, **46**, 61–65.

SUTHERLAND I (1975), *Surface Carbohydrates of the Prokaryotic Cell*, New York, Academic Press.

SUTHERLAND I (1999), 'Polysaccharases for microbial exopolysaccharides', *Carbohydr Polym*, **38**, 319–328.

SUWASONO S and RASTALL R A (1996), 'A highly regioselective synthesis of mannobiose and mannotriose by reverse hydrolysis using specific 1,2-alpha-mannosidase from *Aspergillus phoenicis*', *Biotechnol Lett*, **18**, 851–856.

SUWASONO S and RASTALL R A (1998), 'Enzymatic synthesis of manno- and heteromanno-oligosaccharides using α-mannosidases: a comparative study of linkage-specific and non-linkage-specific enzymes", *J Chem Technol Biotechnol*, **73**, 37–42.

SVENSSON U (1999), "Industrial perspectives", in G Tannock, *Probiotics: A Critical Review*, Wymondham, Horizon Scientific Press, 57–64.

VIC G, TRAN C H, SCIGELOVA M and CROUT D H G (1997), "Glycosidase-catalysed synthesis of oligosaccharides: a one step synthesis of lactosamine and of the linear B type 2 trisaccharide α-D-Gal-(1→3)-β-D-Gal-(1→4)-β-D-GlcNAcSEt involved in the hyperacute rejection response in xenotransplantation from pigs to man and as the specific receptor for toxin A from *Clostridium difficile*', *Chem Commun*, 169–170.

WANKER E, LEER R J, POUWELS P H and SCHWAB H (1995), 'Expression of *Bacillus subtilis* laevanase gene in *Lactobacillus plantarum* and *Lactobacillus casei*', *Appl Microbiol Biotechnol*, **43**, 297–303.

YATAKE T (1993), 'Anomalously linked oligosaccharides'. In: '*Oligosaccharides. Production, Properties and Applications*', Japanese Technology Review, Vol. 3, Switzerland, Gordon & Breach, 79–89.

13

Safety evaluation of probiotics

A.C. Ouwehand and S. Salminen, University of Turku, Finland

13.1   Introduction

Most probiotics belong to the genera *Lactobacillus* or *Bifidobacterium*. The chapter will therefore mainly consider the safety issues associated with members of these genera, although species of other genera are also used as probiotics and similar safety considerations apply to these. Many criteria are used in the selection and use of probiotics. The two of greatest importance are efficacy and safety. Probiotic products have been shown to provide a variety of documented health benefits to all consumers (Salminen *et al.*, 1998). The safety record of probiotics is also good, indicating no or negligible risks. However, in order to be able to continue to ensure the safety of probiotics and dairy starters, safety assessments are necessary.

Lactobacilli and bifidobacteria are generally regarded as safe. This safety status is based on a number of empirical observations:

- Lactobacilli, in particular dairy starters, have a long history of safe use.
- Lactobacilli and bifidobacteria are extremely rarely associated with disease.
- Bifidobacteria and to a lesser extent lactobacilli are major members of the normal human intestinal microflora.

Considering the above, what are the safety concerns for probiotics?

Most probiotics do not have a particularly long history of safe use. The longest is probably exemplified by certain strains belonging to the *Lactobacillus acidophilus* group and *Lactobacillus casei* Shirota, which has been on the market for over 60 years. This is, however, short compared with the history of safe use of many dairy-associated lactic acid bacteria such as *Lactobacillus delbrueckii* subsp. *bulgaricus*, *Streptococcus thermophilus* and *Lactoccus lactis*.

Table 13.1   Suggested potentially predisposing factors for *Lactobacillus* and probiotic-associated infections

Predisposing factor	Remark	Reference
Dental manipulation	Risk for endocarditis	Harty *et al.* (1993)
Abnormal heart valve	Risk for endocarditis	Gallemore *et al.* (1995); Penot *et al.* (1998)
Indwelling catheter	Risk for septicaemia	Hennequin *et al.* (2000)
Colonoscopy	Actual causal relation is uncertain	Avlami *et al.* (2001)
Compromised immune system	Caused by AIDS or transplantation	Horwitch *et al.* (1995)
Traumatic injury	e.g. Surgery	Husni *et al.* (1997)
Diabetes		Jones *et al.* (2000)
Poor nutritional status		Husni *et al.* (1997)
Peritoneal inflammation		Husni *et al.* (1997)
Treatment with vancomycin	Many lactobacilli have intrinsic resistance	Chomarat and Espinouse (1991)

Dairy starters are not specifically adapted to survive in the gastrointestinal (GI) tract as probiotics are. The survival and transient colonisation of the GI tract require different safety considerations.

With the advances in medical care, an increasing part of the population is considered to be immunocompromised at some time of their life. A reduced immune function has been suggested to be a potential risk factor for *Lactobacillus*-associated infections, as in fact for infections by other microbes as well (*see* Table 13.1). Finally, critical voices suggest there is an increase in *Lactobacillus* infections (Antony, 2000) or that lactobacilli may be emerging pathogens (Fruchart *et al.*, 1997). It is therefore important to provide a sound scientific basis for the safety status of probiotic lactobacilli and bifidobacteria and to develop criteria for future safety assessment of new and novel probiotic strains.

13.2   Key safety issues

13.2.1   The microbe
All microbes that have, in principle, the ability to grow under the conditions that prevail in the human body can theoretically colonise this habitat and can thus be potential pathogens. This includes the majority of the mesophilic bacteria, and lactobacilli and bifidobacteria among others. However, lactobacilli and bifidobacteria are, in general, not even considered to be potential pathogens. The frequency of bacteraemia caused by these organisms is extremely low; for lactobacilli 0.1–0.24% (Gasser 1994; Saxelin *et al.*, 1996a). For bifidobacteria no such data are available since their involvement in bloodstream infections is even more rare (Hata *et al.*, 1988), which is interesting considering that bifidobacteria are among the main members of the culturable normal intestinal microflora. The lactobacilli isolated

from bacteraemia cases have not been identified as dairy starters or probiotics (Saxelin *et al.*, 1996a). Since lactobacilli and bifidobacteria have a natural association with human mucosal surfaces of the mouth, GI tract and genitourinary tract, the lactobacilli involved in the bacteraemia therefore probably originate from the patient's own microflora (Wang *et al.*, 1996). However, two cases have been reported where probiotic lactobacilli may have been involved in the aetiology of disease: a liver abscess caused by an *L. rhamnosus* indistinguishable from *L. rhamnosus* GG (Rautio *et al.*, 1999) and a case of endocarditis possibly associated with a probiotic *L. rhamnosus* (Mackay *et al.*, 1999). For the latter, however, the identification is relatively uncertain, as only unconventional methods were used.

Despite these observations the general safety of lactobacilli and bifidobacteria in general and of probiotics in particular appears to be good considering the large amounts of these bacteria consumed by people worldwide. This is, in fact, a far better safety record than even for water, with on average one reported fatal intoxication per year (Chen and Huang, 1995). The above two cases were not fatal; *Lactobacillus* infections are themselves usually not fatal and can be relatively easily treated once diagnosed. However, owing to their extremely low occurrence, recognition, identification and diagnosis by clinicians may sometimes be difficult.

13.2.2 The host

Because the human body, from the microbe's point of view, is a habitat rich in nutrients, many defence mechanisms have evolved to resist colonisation by microbes. Only a relatively small number of microbes have adapted themselves to overcome these defence mechanisms, i.e. the 'true' pathogens. Under normal conditions the defence barriers function well and so non-pathogenic, and in fact also many pathogenic microorganisms, do not gain access into the human body. However, under adverse conditions such as disease and trauma, the defence barriers break down and pathogens, and even generally non-pathogenic organisms such as lactobacilli and bifidobacteria, may enter into the human body (Falkow, 1997). A number of potentially predisposing factors for *Lactobacillus*-associated infections have been suggested (Table 13.1). Indeed, most *Lactobacillus* and *Bifidobacterium* infections have occurred in subjects with severe underlying diseases (Gasser, 1994; Husni *et al.*, 1997; Saxelin *et al.*, 1996a).

Although a compromised immune system appears to be an obvious risk factor, it is important to note that probiotics can also significantly contribute to the protection of, in particular, immunocompromised subjects (Bengmark and Jeppson 1995). This has also clearly been shown in animal experiments using lethally irradiated mice (Dong *et al.*, 1987) and congenital immune-deficient mice (Wagner *et al.*, 1997b). In both cases, probiotics provided significant protection for the host. However, a higher mortality among offspring of some of the congenital immune-deficient mice colonised exclusively with *Lactobacillus* GG and *Lactobacillus reuteri* was observed (Wagner *et al.*, 1997a). This clearly indicates that, in the case of *Lactobacillus*-associated disease, the health status of the host plays a significant role. The health status is likely to be even more important than the properties of the lactobacilli.

Table 13.2 Relative safety of probiotics and starters. Modified after Salminen *et al.* (2002)

Safe	Uncertain	Risk
Bifidobacterium	*Bacillus***	*Peptostreptococcus*
*Carnobacterium**	*Bifidobacterium dentium*	*Streptococcus*
Lactobacillus	*Enterococcus***	
Lactococcus	*Escherichia coli***	
Leuconostoc	*Lactococcus garvieae**	
Oenococcus	*Vagococcus*	
Pediococcus		
Propionibacterium (classical)		
Saccharomyces cerevisiae		
Streptococcus thermophilus		
Weisella		

*Some strains are pathogenic for fish or may have been associated with mastitis.
**Some strains with a history of safe use are known.

13.3 Identifying probiotic strains

Proper identification of a probiotic strain and deposition in an international culture collection for future reference are of major importance for industrial applications since competition in the probiotic market is strong and the probiotic that has been invested in should be protected. Species identification alone may not be sufficient and the probiotic should be characterised to strain level. Apart from protection of commercial interests, proper identification may also indicate safety and technical applicability. Most current probiotics belong to a limited number of species of, mainly, the genera *Bifidobacterium* and *Lactobacillus*; other genera are used as probiotics, albeit to a lesser extent (Table 13.2).

Most species in the genera *Bifidobacterium* and *Lactobacillus* are considered safe. There is some concern that *L. rhamnosus* may be more often associated with bacteraemia than other *Lactobacillus* species (Adams and Marteau, 1995; Berufs-genossenschaft der chemischen Industrie, 1998). However, since *L. rhamnosus* is among the most common *Lactobacillus* species in the human intestine (Ahrné *et al.*, 1998; Apostolou *et al.*, 2001), this may just be a result of their prevalence in the intestine. Among bifidobacteria, only *B. dentium*, *B. denticolens* and *B. inopinatum* are of potential concern as they have been found to be associated with human caries (Crociani *et al.*, 1996; Becker *et al.*, 2002).

Probiotics from the genus *Enterococcus* are a less clear-cut case. They are mainly *Enterococcus faecium* and, to a lesser extent, *Enterococcus faecalis*. Although enterococci are common members of the normal microflora of the GI tract, are naturally present in certain types of cheese and sausages (Herranz *et al.*, 2001) and are being used as probiotics (Agerholm-Larsen *et al.*, 2000), particularly in veterinary use, their safety status remains uncertain (Franz *et al.*, 1999). Several species of the genus *Enterococcus* are pathogens; enterococci are also a common cause of nosocomial infections and have a widespread ability to transfer genetic material, e.g. antibiotic resistance (Wade, 1997). From this it is obvious that it is extremely important to be able to differentiate between probiotic or starter strains of enterococci and clinical strains.

Because of changes in taxonomy, identification to strain level and use of a strain designation is important as illustrated by the probiotic *Lactobacillus* GG. The strain was patented as *L. acidophilus* (Gorbach and Goldin, 1989), subsequently reclassified as *L. casei* (Saxelin, 1997) and, because of revision of the species *L. casei*, it is now *L. rhamnosus* (Alander *et al.*, 1997). Similar changes in strain taxonomy have been performed for other probiotic organisms as well and are likely to happen in the future with the continuing development of taxonomy.

13.4   Potential risk factors: acute toxicity

For generally non-pathogenic microorganisms such as lactobacilli and bifido-bacteria it is difficult to identify potential virulence factors. Therefore, potential risk factors have been proposed that may be of significance when assessing the safety of probiotics. These potential risk factors are based on the knowledge on virulence factors of 'true' pathogens: their relevance for lactobacilli and bifido-bacteria remains to be determined (Table 13.3).

As far as enterococci are concerned, their virulence is still not well understood, making distinction between safe probiotic strains and clinical isolates difficult. However, the presence of haemolysin, hyaluronidase, gelatinase and aggregation substance and their ability to transfer antibiotic resistance may be considered putative virulence factors (Franz *et al.*, 1999). Assessment of their safety is therefore of particular importance.

One of the most direct ways to assess the safety of a probiotic is to determine the LD_{50}. This gives general information on the potential toxicity, or absence of it. Acute and chronic toxicity tests have been performed for a number of probiotic strains. The LD_{50} for mice was found to be high for all tested strains, for some even more than 50 g/kg body mass (Table 13.4). Despite the high exposure, no translocation outside the intestine was observed (Zhou *et al.*, 2000).

13.5   Potential risk factors: microbial metabolism

Many of the suggested risk factors are related to the metabolism of the microbe. Hyaluronidase and gelatinase activity may damage the extracellular matrix proteins, thereby causing direct tissue damage. Not much is known about this enzymatic activity in lactobacilli and bifidobacteria. However, it is a relatively common property of enterococci (Franz *et al.*, 1999). Degradation of intestinal mucus is a normal phenomenon in the caecum and colon whereby several microorganisms are involved. However, excessive degradation may cause damage to the intestinal mucosal barrier function. The ability of probiotics to degrade intestinal mucus has been investigated for several strains *in vitro*. So far, no degradation has been

Table 13.3 Suggested potential risk factors of probiotic lactobacilli and bifidobacteria

Related to	Property	Remark

Microbial metabolism	Hyaluronidase activity
	Gelatinase activity
	DNAse activity
	Mucus degradation
	D-Lactic acid formation
	Aminoacid decarboxylase activity
	Bile salt deconjugation
	Bile salt dehydroxylase activity
	Phosphatidylinositol-specific
	    phospholipase C
		Of particular importance for
		    Enterococcus sp.
		
		Compromises mucosal barrier
		D-Lactic acidosis
		Formation of biogenic amines
		Formation of primary bile salts
		Formation of secondary bile salts
		Translocation

Microbial properties	Capsule formation Transfer of genetic material	Resistance to phagocytosis Acquisition of virulence genes

Binding	Adhesion to intestinal mucosa
	Adhesion to extracellular
	    matrix proteins
	Binding of essential nutrients or
	    therapeutic compounds
		Translocation
		Translocation

Blood	Resistance to complement-mediated killing Resistance to phagocytosis Haemolysis Haemagglutination Platelet aggregation	Bacteraemia Bacteraemia Anaemia, oedema Thrombosis, endocarditis
Immunology	Cell wall composition Modulation of the immune response	Arthritis Inflammation, immune suppression

Table 13.4 Acute toxicity of probiotic bacteria, combined from Momose *et al.* (1979), Donohue and Salminen (1996), Takahashi and Onoue (1999) and Zhou *et al.* (2000)

Probiotic strain	LD_{50} (g/kg body mass)
Enterococcus faecium AD1050*	>6.60
*Streptococcus equinus**	>6.39
Lactobacillus fermentum AD0002*	>6.62
Lactobacillus salivarius AD0001*	>6.47
Lactobacillus acidophilus HN017	>50
Lactobacillus rhamnosus GG (ATCC 53103)	>6.00
Lactobacillus rhamnosus HN001	>50
Lactobacillus helveticus	>6.00
Lactobacillus delbrueckii subsp. *bulgaricus*	>6.00
Lactobacillus casei Shirota	>2.00
Bifidobacterium longum	25
Bifidobacterium longum BB-536	>50 (0.52 via intraperitoneal route)
Bifidobacterium lactis HN019	>50

* Non-viable, heat-inactivated, preparations.

observed for any of the tested strains (Ruseler-van Embden *et al.*, 1995; Zhou *et al.*, 2001). Also in human studies, no change in excretion of mucus was observed upon consumption of probiotics (Ouwehand *et al.*, 2002), suggesting that the tested probotics (*L. reuteri*, *L. rhamnosus* Lc705 and *Propionibacterium freudenreichii* JS) do not take part in the degradation of intestinal mucus.

D-Lactic acidosis is associated with short bowel syndrome and other forms of malabsorption. The reduced intestinal capacity for carbohydrate absorption and the oral consumption of easily fermentable carbohydrates cause an increased delivery of carbohydrates in the colon. This, together with an over-growth of endogenous lactobacilli, that by fermentation produce D-lactic acid, form the basis of D-lactic acidosis. Human tissue contains D-2-hydroxy acid dehydrogenase which converts D-lactate to pyruvate. This suggests that D-lactic acidosis only occurs when absorption exceeds its metabolism. During active over-growth, lactobacilli may make up 40–60% of the total faecal flora (Bongaerts *et al.*, 1997; Kaneko *et al.*, 1997). Endogenous *Lactobacillus* species such as *L. acidophilus*, *Lactobacillus fermentum* and *Lactobacillus delbrueckii* subsp. *lactis* have been implicated in the aetiology (Bongaerts *et al.*, 1997; Kaneko *et al.*, 1997).

Although dietary lactobacilli have been suggested to be involved in D-lactic acidosis (Perlmutter *et al.*, 1983), no attempt has been made to compare the D-lactic acid-producing stool isolates with the dietary isolates. Therefore, the causality remains speculative. Nevertheless, the observations may indicate that consumption of D-lactate-producing microbes is not desirable in this patient group. Antibiotic treatment of these patients has been observed to be an additional potential factor in the aetiology of D-lactic acidosis (Coronado *et al.*, 1995). Interestingly, in one such case a L-lactic acid-producing probiotic (*L. rhamnosus* GG) was found to reverse the state, after correction of the diet and alternative antibiotic treatment had failed (Gavazzi *et al.*, 2001).

Biogenic amines have been implicated in some rare cases of food poisoning from the consumption of fermented foods containing high amounts of these substances. Non-starter lactic acid bacteria have been suggested to be mainly responsible for biogenic amine production in cheese (González de Liano *et al.*, 1998). Formation of histamine through the decarboxylation of histidine appears to be performed mainly by *L. buchneri* in cheese (Sumner *et al.*, 1985) and *Lactobacillus hilgardii* in wine (Arena and Manca de Nadra, 2001), although also *Lactobacillus helveticus* and *Lc. lactis* have been found to produce small quantities of histamine. *Lactobacillus brevis* and several *Leuconostoc* and *Lactococcus* species are tyrosine-decarboxylating, leading to the formation of tyramine (González de Liano *et al.*, 1998). These findings suggest that microbes with no, or a low, ability to produce biogenic amines should be selected. Alternatively, organisms that do have the ability to produce biogenic amines should only be used in products not containing precursors for the production of amines and in addition have a low residual activity in the intestine.

Bile salts are the water-soluble end-products of the cholesterol metabolism of the liver and play an important role in digestion, in particular of lipids. Bile salts are actively absorbed from the terminal ileum and are subsequently resecreted in bile, forming an enterohepatic cycle. Because bile acids are part of the cholesterol metabolism, they have often been targeted as a potential means of reducing serum cholestrol levels, with varying success (Rowland, 1999). The first step in microbial bile salt metabolism is deconjugation through hydrolysis of the glycine/taurine moiety from the side chain of the steroid core. These deconjugated, primary, bile acids are less effective in solubilisation of dietary lipids. Too much and too early deconjugation, in the upper small intestine, may disturb lipid digestion and the uptake of fat-soluble vitamins. Primary bile acids can subsequently be dehydroxylated to yield secondary bile acids. The latter are most hydrophobic and toxic to hepatocytes and the gastric and intestinal mucosa, and have been suggested to be cancer promoters and to be involved in the formation of gal stones (de Boever and Verstreate, 1999).

Many *Bifidobacterium* and *Lactobacillus* species are able to deconjugate bile salts (Tanaka *et al.*, 1999). However, little is known about whether they also have 7α-dehydroxylase activity and can generate secondary bile acids. Because of the delicate balance in the bile circulation, care should be taken not to excessively increase the deconjugation of bile acids, especially in the upper small intestine. Considering the detrimental properties of secondary bile acids, no increase in 7α-dehydroxylase activity can be accepted anywhere in the intestine. Potential probiotics and starters should not exhibit this property.

Phosphatidylinositol-specific phospholipase C (PI-SPC) has been implicated as an important virulence factor involved in translocation. Several lactic acid bacteria were observed to possess this activity. A PI-SPC positive *L. rhamnosus* was found to translocate in Balb/C mice, while a PI-SPC negative *L. rhamnosus* did not (Rodriguez *et al.*, 2001). Considering the fact that only two strains were tested, it seems premature to consider this a risk factor. However, further research is certainly warranted.

13.6   Potential risk factors: microbial properties and binding

13.6.1   Microbial properties
Some suggested virulence factors are related to specific intrinsic properties of lactobacilli and bifidobacteria, e.g. the ability to produce a capsule or the ability to transfer genetic material.

The ability to form a capsule has been implicated with an increased resistance to antibiotic treatment and phagocytosis (Sims, 1964). Certain fermented dairy products do indeed rely on the formation of capsules in order to obtain their characteristic texture (Roginski, 2002). However, the property of capsule formation was found to be rare among *L. rhamnosus* isolates from clinical, food and faecal origin (Baumgartner *et al.*, 1998), suggesting that this potential virulence factor is probably not a relevant safety concern.

With the exception of enterococci, the ability to transfer genetic material is low in lactic acid bacteria. Also, the presence of antibiotic resistance plasmids and transposable genetic elements is relatively rare among lactobacilli and bifidobacteria (von Wright and Sibakov, 1998). However, it does occur and plasmids carrying resistance for erythromycin and related macrolides have been observed (Teuber *et al.*, 1999). Thus, although the spread of antibiotic resistance through lactobacilli is far less common than with enterococci, it cannot be ruled out. As far as intrinsic antibiotic (vancomycin) resistance is concerned, this is due to differences in the composition of the microbial cell wall and has not been found to be transferable (Tynkkynen *et al.*, 1998). It should therefore not be considered a risk factor *per se*. However, since vancomycin is often used as a 'last resort' antibiotic in patients with severe diseases, this may lead to over-growth of endogenous lactobacilli. This, in combination with the general poor health of the patient may increase the risk for *Lactobacillus* translocation. If this does occur, there are many other antibiotics with which to successfully treat a *Lactobacillus* infection.

13.6.2 Binding

Adhesion to the intestinal mucosa is one of the main selection criteria for probiotics. This is thought to be important for prolonged transient colonisation, modulation of the immune system and competitive exclusion of pathogens (Ouwehand *et al.*, 1999). However, adhesion to the intestinal mucosa is also the first step in pathogenesis (Finlay and Falkow, 1997). The question therefore arises whether this selection also leads to the selection of potentially pathogenic organisms. In a study comparing *Lactobacillus* bacteraemia isolates with current probiotics and dairy starters, the bacteraemia isolates were found to adhere significantly better to intestinal mucus than the probiotic strains. When split to species level, bacteraemia *L. rhamnosus* isolates were found to adhere significantly better than faecal *L. rhamnosus* isolates. Although this would suggest that clinical isolates indeed adhere better than faecal and probiotic strains, the variation between the isolates in each group was considerable, suggesting that other factors must have been involved in the translocation of the bacteraemia isolates as well (Apostolou *et al.*, 2001). Considering these findings and the suggested importance of adhesion for probiotic effects, this needs further attention with a larger number of isolates. Strong adhesion has also been suggested to be a risk factor for platelet aggregation (see below) (Harty *et al.*, 1994). However, in a comparative study with bacteraemia *Lactobacillus* isolates, no correlation could be observed. In fact, the strains causing platelet aggregation exhibited a low level of adhesion, indicating that adhesion is not associated with platelet aggregation (Kirjavainen *et al.*, 1999).

Extracellular matrix (ECM) proteins are typically exposed in wound tissue. Pathogenic microorganisms therefore often have affinity for these proteins, which serve as receptors for invading microbes. Lactobacilli have been observed to bind to ECM proteins (Styriak *et al.*, 2001). This binding has been observed to be beneficial to the host: damaged gastric mucosa was found to heal quicker after colonisation with lactobacilli, probably through exclusion of potential pathogens (Elliott *et al.*, 1998). This suggests that binding of lactobacilli to the ECM is more a benefit to the host than a risk.

Many lactic acid bacteria have been observed to be able to absorb environmental toxins; mycotoxins (Peltonen *et al.*, 2000), heterocyclic amines (Sreekumar and Hosono, 1998) and possibly even heavy metals. Reducing the bioavailability of these compounds is obviously a desirable trait. However, it is important that such organisms do not also bind to therapeutic compounds or essential nutrients as reported recently (Turbic *et al.*, 2002).

13.7   Other potential risk factors

13.7.1   Survival of blood-related defence mechanisms
Blood contains many different defence systems to eliminate invading microbes. Resistance to these mechanisms is important for successful infection by a pathogen. The complement system in blood can opsonise bacteria and facilitate their phagocytosis by leucocytes. The activated complement system can also form a complex that kills bacteria. Baumgartner and co-workers (1998) observed that *L. rhamnosus* strains were, in general, resistant to serum and in some cases even grew in it. No significant difference was observed in resistance between clinical, faecal or dairy isolates. This indicates that serum resistance is likely to be an intrinsic property that may have little relevance for the safety. The serum resistance of other species of *Lactobacillus* and *Bifidobacterium*, however, needs to be determined.

Upon phagocytosis, peripheral blood mononucleocytes produce a burst of reactive oxygen species, which will kill and digest the phagocytosed particle. The ability to avoid the induction of such a respiratory burst may enhance the survival of a translocated pathogen and hence the risk for infection. This has, to date, not been investigated for lactobacilli or bifidobacteria.

Haemolysis is a common virulence factor among pathogens. It serves mainly to make iron available to the microbe and causes anaemia and oedema to the host. Lactobacilli have been observed to exhibit α-haemolysis towards sheep blood (Baumgartner *et al.*, 1998; Horwitch *et al.*, 1995). However, the relevance of this is uncertain for human safety. In addition, haemolysis was found to be equally common among clinical, faecal and dairy *L. rhamnosus* isolates. Since it is not a particular trait of the clinical isolates, it may not be a relevant safety concern.

Haemagglutination has been observed to be a rare phenomenon among lactobacilli (Baumgartner *et al.*, 1998) and is therefore not likely to be a risk factor. Spontaneous aggregation of platelets leads to the formation of trombi, oedema and has been implicated in infective endocarditis. Probiotics have not been observed to cause platelet aggregation and it was found to be limited among clinical *Lactobacillus* isolates as well (Table 13.5). The relevance for probiotic safety is therefore uncertain.

13.7.2   Immunological effects
In health, the immune system is in a delicate balance between inflammation and unresponsiveness. Disturbance of this balance may have dramatic consequences. Many components of the microbial cell and in particular whole probiotics have

Table 13.5 Induction of platelet aggregation by lactic acid bacteria of different origins, combined from Korpela *et al.* (1997), Kirjavainen *et al.* (1999) and unpublished results

Origin	Species	Platelet aggregation
Probiotic	*Lactobacillus rhamnosus* GG	nd*
	Lactobacillus johnsonii La1	nd
	Lactobacillus paracasei F19	nd
	Lactobacillus crispatus M247	nd
	Lactobacillus crispatus MU5	nd
	Lactobacillus salivarius UCC118	nd
	Bifidobacterium lactis Bb12	nd
Bacteraemia	*Lactobacillus rhamnosus* T4813	94.5%
	Lactobacillus zea T8320	nd
	Lactobacillus rhamnosus T4846	44.2%
	Lactobacillus casei T9393	87.5%
	Lactobacillus casei T7709	nd
	Lactobacillus curvatus T3300	nd
	Lactobacillus rhamnosus T15756	nd
Other	*Lactobacillus rhamnosus* ATCC7469	nd
	Lactobacillus rhamnosus T5080	nd
	Enterococcus faecium T2L6	nd

nd = not detected.

been observed to influence the immune system. This is therefore an important point in the safety consideration for probiotics.

Selected probiotic strains have been observed to selectively up-regulate the immune system in subjects with reduced immune response (Majamaa *et al.*, 1995; Gill *et al.*, 2001). Interestingly, the same probiotics also seem to be able to reduce an unnecessary strong and potentially damaging immune response (Majamaa and Isolauri, 1997; Pelto *et al.*, 1998). However, probiotics do not appear to affect the immune response of healthy subjects (Spanhaak *et al.*, 1998; Campbell *et al.*, 2000). It therefore appears that probiotics have the ability to let the immune response return to its normal range in the dynamic balance. No reports are available indicating, for example, increases in allergic reactions or immune deficiency upon the use of probiotics, and there are no known negative effects from the modulation of the immune system by probiotics. Disturbance of the immune system therefore seems more a theoretical concern than a true health risk.

Most probiotics are Gram-positive organisms. The cell wall of these organisms consists of a large amount of peptidoglycan, which is a powerful immune stimulator (Stewart-Tull, 1980). Several studies have shown that cell wall fragments from lactobacilli and bifidobacteria are arthritogenic in a rat model (Severijnen *et al.*, 1989). However, a predisposing factor is needed: an increased bowel permeability (Phillips, 1989). In subjects with a reduced intestinal mucosal barrier function, this could pose a risk. However, treatment of patients with juvenile chronic arthritis with lactobacilli caused a reduction in the faecal urease activity, which is thought to cause tissue damage through the production of ammonia. No adverse effects were observed in the patients (Malin *et al.*, 1997). Thus, if probiotics do have an effect, it appears to be beneficial rather then detrimental.

13.8   Post-marketing surveillance

The manufacturer of a probiotic product has the final responsibility for its safety. Once a probiotic product has been brought onto the market, surveillance for any cases of disease related to the probiotic is important in order to be able to continue to ascertain the safety of the probiotic. Any *Lactobacillus* or *Bifidobacterium* isolates associated with disease should be compared to the probiotic strain of the same species (Saxelin *et al.*, 1996b). This illustrates that accurate taxonomic identification is essential. Especially in case reports on *Lactobacillus-* or *Bifidobacterium*-associated infections, proper identification is important; unfortunately this is often not the case.

Proper storage and established availability of any clinical *Lactobacillus* or *Bifidobacterium* isolates for future research are essential. This guarantees the safety of the current probiotics on the market and these isolates may, at the same time, provide information on the potential risk factors of these genera (Apostolou *et al.*, 2001).

13.9   Safety issues for new generation probiotics

New probiotic microorganisms and novel probiotic products are likely to be introduced to the market. These will require different safety assessment procedures, which have been established, e.g. in the European Novel Foods Directive and in the US Premarketing Approval Clearances.

13.9.1 Novel probiotic species

Most of the current probiotics belong to the genera *Lactobacillus* and *Bifidobacterium*. However, some probiotics belong to other genera (Table 13.2), which in most cases do not have a history of safe use. For some of these organisms other, possibly more strict, safety assessments may be necessary. It is likely that a number of novel probiotics will represent other genera of lactic acid bacteria, *Lactococcus*, *Leuconostoc*, etc., or they may be of other food-related genera, for example, *Propionibacterium*. However, most interesting will be probiotics from outside this traditional field as e.g. *Oxalobacter formigenes*, which by metabolising oxalate may reduce the risk for kidney stones (Sidhu *et al.*, 2001). Such probiotics will set new standards for the safety assessment.

13.9.2   Genetically modified probiotics

Genetically modified organisms (GMO) have a low consumer acceptance, at least in Europe. It is therefore unlikely that GMO versions of current probiotic products will enter the market any time soon. However, in clinical applications GMO probiotics may be more readily accepted and some potential applications in this area are under investigation. GMO probiotics could be used to deliver antigens for vaccines. This would provide a safer method of vaccination than the use of attenuated pathogens (Mercenier *et al.*, 2000).

A GMO, *Lactococcus lactis*, has been observed to be able to produce IL-10 in the mouse intestine (Steidler *et al.*, 2000). This may provide new treatment strategies for inflammatory bowel disease, and similar applications may be useful for other diseases. The safety of such organisms that are able to produce these very powerful bioactive substances is of major importance as excess production of these substances in a healthy individual may be detrimental. However, in general the safety assessment and regulations for the use of GMOs, including lactobacilli, is strict and may provide sufficient protection. The details of regulations concerning GMOs fall outside the scope of this chapter.

13.9.3   Non-viable probiotics

Although probiotics are generally defined as live microorganisms, there is some indication that non-viable probiotics may also have beneficial health effects (Ouwehand and Salminen 1998). Non-viable probiotics would have clear practical and economic benefits: longer shelf-life, no need for cooled transportation and storage, etc. They may also have a safety advantage. Although the risk for infection due to probiotic *Lactobacillus* or *Bifidobacterium* preparations is extremely small, non-viable probiotics would not have this risk at all. This is likely to provide mainly a theoretical advantage, though non-viable probiotics could be considered safer when used in extremely high-risk immunosuppressed patients.

13.9.4   Novel applications

The main application for probiotics is their use in foods, aiming at affecting the composition or activity of the intestinal microflora or directly affecting the function of the intestine. However, the probiotic principle should be expected to work in any part of the body that has a normal microflora. With the exception of the urogenital tract, extra-intestinal applications of probiotics have received little attention. Such probiotic preparations would clearly need different safety requirements.

13.10   The safety of animal probiotics

Probiotics are also commonly used in veterinary applications (Nousiainen and Setälä, 1998). Despite the fact that the probiotics will be used for animals, more strict safety assessment may be necessary than for humans. They should prove to be safe for both animals and humans. In the case of probiotics for farm animals, the probiotics may enter the food chain (Nikoskelainen *et al.*, 2001). Because of the often intimate relation between a pet and its owner, spread of the probiotic from animal to human is possible. Because enterococci are commonly used in probiotic preparations for animals, this may be reason for some concern.

The safety requirements for animals are different from those for humans; *Lacto-coccus garvieae* has been found to be associated with mastitis in cows (Collins *et al.*, 1983) and septicaemia in fish (Schmidtke and Carson 1999). Although *Lc. garvieae* has been found to be associated with disease in humans, this has so far been very limited and underlying diseases have been implicated (Fefer *et al.*, 1998). Its true pathogenesis for humans remains therefore to be determined (Aguirre and Collins 1993). Another *Lactococcus* species that has been found to be involved in disease of fish is *Lc. piscium* (Williams *et al.*, 1990). This species has not been found to be associated with disease in humans.

13.11   The current regulatory context

13.11.1   Dietary supplements
The dietary supplement market for probiotics is large and growing in many countries. A multitude of bacterial genera and species is used and include many different *Lactobacillus* and *Bifidobacterium* species. In addition to these, many of these supplements also contain enterococci, yeasts and sometimes even *Escherichia coli*, *Bacillus* and *Clostridium* strains. The safety of these organisms has to be assessed separately.

13.11.2   Safety assessment in the United States
In the United States, the Food and Drug Administration (FDA) controls the safety of foods and dietary supplements. Probiotics have been sold in the US as components of conventional foods or as dietary supplements. The safety of traditional lactic starter bacteria has been accepted. However, new and less traditional strains of microbes have to be more carefully assessed prior to distribution to consumers with potentially compromised health. In the US, Title 21 of the Code of Federal Regulations (21 CFR) and the FDA Office of Premarket Approval lists microorganisms that are approved food additives or which enjoy Generally Recognized as Safe (GRAS) status. 'GRAS' status is always considered only for a specified use. Thus, for instance, microbes themselves are not considered GRAS, but their traditional use in dairy foods is.

Based on the regulation, a substance that will be added to food is subject to premarket approval by the FDA unless its use is considered GRAS by qualified experts. In addition, a substance that is used in accordance with a sanction granted prior to 6 September 1958 is not subject to premarket approval. On 17 April 1997, the FDA issued a proposed rule (The GRAS proposal 62 FR 18938) that would establish a notification procedure whereby any person (or interested company) may notify the FDA of a determination by that person that a particular use of a substance is GRAS for specific use.

Currently, it can be understood that harmless lactic acid-producing bacteria, including *L. acidophilus* and other lactic acid bacteria (specifically *S. thermophilus*,

Lc. lactis subsp. *cremoris*, *Lc. lactis* subsp. *lactis*, *L. delbrueckii* subsp. *bulgaricus*, *L. delbrueckii* subsp. *lactis*, *L. fermentum* and three *Leuconostoc* species) can be used as additional ingredients in specified standardised foods such as cultured milk (including yoghurt and buttermilk), sour cream, cottage cheese and yoghurt, provided that the mandatory cultures of *L. bulgaricus* and *S. thermophilus* are also used in yoghurt. Many organisms used in foods are not included on the FDA list, but the absence of a species or a strain does not constitute a safety issue. In a more specific manner, the FDA has also approved a GRAS petition by Nestlé US concerning the use of *Bifidobacterium lactis* Bb12 and *Streptococcus thermophilus* Th4 in infant formula and foods for infants and children. Based on the published information by the FDA, other strains of *Lactobacillus*, *Bifidobacterium* and *Streptococcus* have been withdrawn from the petition. No additional lactic acid bacteria or currently used probiotics have been petitioned or approved at this time (April 2002).

Regarding the safety of enterococci, the picture is less clear-cut. Foods containing enterococci are safely consumed on a regular basis. However, safety reports seem to agree that enterococci pose a greater threat to health than other lactic acid bacteria. According to the US food regulations, such food substances that were used prior to 1958 do not need approval procedures.

13.11.3   Safety assessment in the European Union

In the European Union foods that were not available to a significant part of the population in one of the member states before 1998 are considered novel foods; this includes also probiotics and starters. The purpose of the EC Novel Foods Regulation (258/97) is to ensure the free movement of novel foods, while protecting the interests of consumers, especially with respect to safety, health and information. New probiotic or starter strains that are substantially equivalent to existing conventional organisms concerning their composition, nutritional value, metabolism, intended use and level of undesirable substances contained in them are not regarded as novel foods. Otherwise, however, the strains will need to be assessed according to the novel food regulations and procedures.

13.11.4   Safety assessment in Germany

In Germany, an occupational foundation of the chemical industry has established an expert group to assess the safety of microbiology and biotechnology. This expert group has also assessed the safety of microbes and published a list containing the classification of bacteria used by different industries including food and feed industries (Berufsgenossenschaft der chemischen Industrie, 1998). The classification divides bacteria into four risk groups. Almost all lactic acid bacteria and bifidobacteria belong to risk group 1, indicating that there are no known risks in the handling of these bacteria by humans or animals. Some lactic acid bacteria, such as *L. rhamnosus* wild types and *L. uli* and *L. catenaformis*, belong to risk group 2, indicating that caution should be taken when handling these bacteria.

Similar classification applies to most lactic acid bacteria and bifidobacteria; among the bifidobacteria, *B. dentium* is classified in risk group 2. In the *Enterococcus* group, a significant part of the species belong to risk group 2, associating these bacteria with a low but existing risk of infection at the workplace. Also, many of the class 2 enterococci are scheduled for further reassessment. Basically, the German classification applies to occupational exposure, but the general safety focus may be applied in a wider perspective in some cases.

13.11.5   Safety assessment in Japan

In Japan probiotics and other functional foods fall under the Food for Specified Health Uses (FOSHU) regulation. FOSHU regulations do not specifically define the safety aspects for probiotic microbes but for functional foods in general. In order for a product to be approved as FOSHU, companies need to go through an application process that typically takes about one year to complete. Applications are reviewed by local prefecture authorities and the Ministry of Health and Welfare. Applications must include scientific documentation demonstrating the medical or nutritional basis for a health claim, the basis for the recommended dose of the functional ingredient, information demonstrating the safety of the ingredient, information on physical and chemical characteristics, relevant test methods, and a compositional analysis. This information is expected to be accompanied by scientific papers in the fields of medicine and/or nutrition that substantiate the health claim.

13.11.6   International Dairy Federation and dairy starter and probiotic strains

Within the International Dairy Federation (IDF) an expert action team has been preparing a position document on the properties of starter cultures and probiotics used by the dairy industry. The IDF work is proceeding in collaboration with the European Food and Feed Culture Association and the aim is to produce a list of cultures with the safe history of use in dairy foods. Also, a document on the use prior to 1998, the introduction of the European Union Novel Foods Directive, is being prepared and the worldwide position document on this subject is under review.

13.12   Conclusion and future trends

The current safety record of food starter cultures and probiotics appears to be excellent. However, future development of probiotics, and also industrial starter culture and fermentation culture organisms, requires stringent guidelines for safety assessment of such organisms. Therefore, constant surveillance of food microorganisms and their appearance in clinical situations is extremely important. Methods for both the premarket safety assessment and post-marketing surveillance of human populations should be developed and refined to guarantee the safety of future products in all populations including subjects with immune suppression. This will enable the future use of microbes and microbial fermentations for a widening area in food technology and in functional and clinical food areas.

13.13   Sources for further information and advice

ADAMS M R and NOUT M J R (eds) (2001), *Fermentation and Food Safety*, Aspen Publishers, Inc., Gaithersburg, 290 pp.

ELMER G W, MCFARLAND L V and SURAWICZ C M (eds) (1999), *Biotherapeutic Agents and Infectious Diseases*, Humana Press, Inc., Totowa, 316 pp.

HANSON L Å and YOLKEN R H (eds) (1999), *Probiotics, Other Nutritional Factors, and Intestinal Microflora*, Lippincott-Raven Publishers, Philadelphia, 306 pp.

LEE Y-K, NOMOTO K, SALMINEN S and GORBACH S L (1999), *Handbook of Probiotics*, John Wiley and Sons, Inc., New York, 211 pp.

ROGINSKI H, FUQUAY J W and FOX P F (eds) (2002), *Encyclopedia of Dairy Sciences*, Academic Press, New York.

SALMINEN S, OUWEHAND A C and VON WRIGHT A (eds) (2003), *Lactic Acid Bacteria: Microbiology and Functional Aspects*, Marcel Dekker, New York.

13.14   References

ADAMS M R and MARTEAU P (1995), 'On the safety of lactic acid bacteria from food', *Int J Food Microbiol*, **27**, 263–264.

AGERHOLM-LARSEN L, RABEN A, HAULRIK N, HNASEN A S, MANDERS M and ASTRUP A (2000), 'Effect of 8 week intake of probiotic milk products on risk factors for cardiovascular diseases', *Eur J Clin Nutr*, **54**, 288–297.

AGUIRRE M and COLLINS M D (1993), 'Lactic acid bacteria and human clinical infection', *J Appl Bact*, **75**, 95–107.

AHRNÉ S, NOBAEK S, JEPPSON B, ADLERBERTH I, WOLD A E and MOLIN G (1998), 'The normal *Lactobacillus* flora of healthy human rectal and oral mucosa', *J Appl Microbiol*, **85**, 88–94.

ALANDER M, KORPELA R, SAXELIN M, VILPPONEN-SALMELA T, MATTILA-SANDHOLM T and VON WRIGHT A (1997), 'Recovery of *Lactobacillus rhamnosus* GG from human colonic biopsies', *Lett Appl Microbiol*, **24**, 361–364.

ANTONY S J (2000), 'Lactobacillemia: an emerging cause of infection in both the immunocompromised and the immunocompetent host', *J Natl Med Assoc*, **92**, 83–86.

APOSTOLOU E, KIRJAVAINEN P V, SAXELIN M, RAUTELIN H, VALTONEN V, SALMINEN S J and OUWEHAND A C (2001), 'Good adhesion properties of probiotics: a potential risk for bacteremia?', *FEMS Immunol Med Microbiol*, **31**, 35–39.

ARENA M E and MANCA DE NADRA M C (2001), 'Biogenic amine production by *Lactobacillus*', *J Appl Microbiol*, **90**, 158–162.

AVLAMI A, KORDOSSIS T, VRIZIDIS N and SIPSAS N V (2001), '*Lactobacillus rhamnosus* endocarditis complicating colonoscopy', *J Infect*, **42**, 1–2.

BAUMGARTNER A, KUEFFER M, SIMMEN A and GRAND M (1998), 'Relatedness of *Lactobacillus rhamnosus* strains isolated from clinical specimens and such from food-stuffs, humans and technology', *Lebensm Wiss Technol*, **31**, 489–494.

BECKER M R, PASTER B J, LEYS E J, MOESCHBERG M L, KENYON S G, GALVIN J L, BOCHES S K, DEWIRST F E and GRIFFEN A L (2002), 'Molecular analysis of bacterial species associated with childhood caries', *J Clin Microbiol*, **40**, 1001–1009.

BENGMARK S and JEPPSON B (1995), 'Gastrointestinal surface protection and mucosa reconditioning', *J Parent Ent Nutr*, **19**, 410–415.

BERUFSGENOSSENSCHAFT DER CHEMISCHEN INDUSTRIE (1998), Sichere Biotechnologie. Eingruppierung biologischer Agenzien: Bakterien. Merkblatt B 006. Heidelberg, Jedermannn Verlag Dr Otto Pfeffer oHG.

BONGAERTS G P A, TOLBOOM J J M, NABER A H J, SPERL W J K, SEVERIJNEN R S V M, BAKKEREN J A J M and WILLEMS J L (1997), 'Role of bacteria in the pathogenesis of short bowel syndrome-associated D-lactic acidemia', *Microb Pathogen*, **22**, 285–293.

CAMPBELL C G, CHEW B P, LUEDECKE L O and SHULTZ T D (2000), 'Yogurt consumption does not enhance immune function in healthy premenopausal women', *Nutr Canc*, **37**, 27–35.

CHEN X and HUANG G (1995), 'Autopsy case report of a rare acute iatrogenic water intoxication with a review of the literature', *Forensic Sci Int*, **76**, 27–34.

CHOMARAT M and ESPINOUSE D (1991), '*Lactobacillus rhamnosus* septicemia in patients with prolonged aplasia receiving ceftazidime-vancomycin', *Eur J Clin Microbiol Infect Dis*, **10**, 44.

COLLINS M D, FARROW J A E, PHILLIPS B A and KANDLER O (1983), '*Streptococcus garvieae* sp nov and *Streptococcus plantarum* sp nov' *J Gen Microbiol*, **129**, 3427–3431.

CORONADO B E, OPAL S M and YOBURN D C (1995), 'Antibiotic induced D-lactic acidosis', *Ann Int Med*, **122**, 839–842.

CROCIANI F, BIAVATI B, ALESSANDRINI A, CHIARINI C and SCARDOVI V (1996), '*Bifidobacterium inopinatum* sp. nov. and *Bifidobacterium denticolens* sp. nov., two new species isolated from human dental caries', *Int J Syst Bacteriol*, **46**, 564–71.

DE BOEVER P and VERSTREATE W (1999), 'Bile salt deconjugation by *Lactobacillus plantarum* 80 and its implication for bacterial toxicity', *J Appl Microbiol*, **87**, 345–352.

DONG M-Y, CHANG, T-W and GORBACH, SL (1987), 'Effects of feeding *Lactobacillus* GG on lethal irradiation in mice', *Diagn Microbiol Infect Dis*, **7**, 1–7.

DONOHUE D C and SALMINEN S (1996), 'Safety of probiotic bacteria', *Asia Pacific J Clin Nutr*, **4**, 25–28.

ELLIOTT S N, BURET A, MCKNIGHT W, MILLER M J S and WALLACE J L (1998), 'Bacteria rapidly colonize and modulate healing of gastric ulcers in rats', *Am J Physiol*, **275**, G425–G432.

FALKOW S (1997), 'What is a pathogen?', *ASM News*, **63**, 359–365.

FEFER J J, RATZAN K R, SHARP S E and SAIZ E (1998), '*Lactococcus garvieae* endocarditis: report of a case and review of the literature', *Diagn Microbiol Infect Dis*, **32**, 127–130.

FINLAY B B and FALKOW S (1997), 'Common themes in microbial pathogenicity revisited', *Microbiol Mol Biol Rev*, **61**, 136–169.

FRANZ C M A P, HOLZAPFEL W H and STILES M E (1999), 'Enterococci at the crossroads of food safety', *Int J Food Microbiol*, **47**, 1–24.

FRUCHART C, SALAH A, GRAY C, MARTIN E, STAMATOULLAS A, BONMARCHAND G, LEMELAND J F and TILLY H (1997), '*Lactobacillus* species as emerging pathogens in neutropenic patients', *Eur J Clin Microbiol Infect Dis*, **16**, 681–684.

GALLEMORE G H, MOHON R T and FERGUSON D A (1995) '*Lactobacillus fermentum* endocarditis involving a native matral valve', *J Tenn Med Assoc*, **88**, 306–308.

GASSER F (1994), 'Safety of lactic acid bacteria and their occurrence in human clinical infections', *Bull Inst Pasteur*, **92**, 45–67.

GAVAZZI C, STACCHIOTTI S, CAVALLETTI R and LODI R (2001), 'Confusion after antibiotics', *Lancet*, **357**, 1410.

GILL H, RUTHERFORD K J, CROSS M L and GOPAL P K (2001), 'Enhancement of immunity in the elderly by dietary supplementation with the probiotic *Bifidobacterium lactis* HN019', *Am J Clin Nutr*, **74**, 833–839.

GONZÁLEZ DE LIANO D, CUESTA P and RODRÍGUEZ A (1998), 'Biogenic amine production by wild lactococcal and leuconostoc strains', *Lett Appl Microbiol*, **26**, 270–274.

GORBACH S L and GOLDIN B R (1989) '*Lactobacillus* strains and methods of selection', US Patent: no. 4,839,281.

HARTY D W S, PATRIAKIS M and KNOX K W (1993), 'Identification of *Lactobacillus* strains isolated from patients with infective endocarditis and comparison of their surface-associated properties with those of other strains of the same species', *Microb Ecol Health Dis*, **6**, 191–201.

HARTY D W S, OAKEY H J, PATRIKAKIS M, HUME E B H and KNOX K W (1994), 'Pathogenic potential of lactobacilli', *Int J Food Microbiol*, **24**, 179–189.

HATA D, YOSHIDA A, OHKUBO H, MOCHIZUKI Y, HOSOKI Y, TANAKA R and AZUMA R (1988), 'Meningitis caused by *Bifidobacterium* in an infant', *Ped Infect Dis J*, **7**, 669–671.

HENNEQUIN C, KAUFFMANN-LACROIX C, JOBERT A, VIARD T, RIQOUR C, JACQUEMIN J L and BERCHE P (2000), 'Possible role of catheters in *Saccharomyces boulardii* fungemia', *Eur J Clin Microbiol Infect Dis*, **19**, 16–20.

HERRANZ C, CASAUS P, MUKHOPADHYAY S, MARTÍNEZ J M, RODRÍGUEZ J M, NES I F, HERNÁNDEZ P E and CINTAS L M (2001), '*Enterococcus faecium* P21: a strain occurring naturally in dry-fermented sausages producing the class II bacteriocins enterocin A and enterocin B', *Food Microbiol*, **18**, 115–131.

HORWITCH C A, FURSETH H A, LARSON A M, JONES T L, OLLIFFE J F and SPACH D H (1995), 'Lactobacillemia in three patients with AIDS', *Clin Infect Dis*, **21**, 1460–1462.

HUSNI R N, GORDON S M, WASHINGTON J A and LONGWORTH D L (1997), '*Lactobacillus* bacteremia and endocarditis: review of 45 cases', *Clin Infect Dis*, **25**, 1048–1055.

JONES R, ELLBOGEN M and DUNLAY R (2000), '*Lactobacillus* allograft pyelonephritis and bacteremia', *Nephron*, **86**, 502.

KANEKO T, BANDO Y, KURIHARA H, SATOMI K, NONOYAMA K and MATSUURA N (1997), 'Fecal microflora in a patient with short-bowel syndrome and identification of dominant lactobacilli', *J Clin Microbiol*, **35**, 3181–3185.

KIRJAVAINEN P V, TUOMOLA E M, CRITTENDEN R G, OUWEHAND A C, HARTY D W S, MORRIS L F, RAUTELIN H, PLAYNE M J, DONOHUE D C and SALMINEN S J (1999), 'In vitro adhesion and platelet aggregation properties of bacteremia-associated lactobacilli', *Infect Immun*, **67**, 2653–2655.

KORPELA R, MOILAEN E, SAXELIN M and VAPAATALO H (1997), '*Lactobacillus rhamnosus* GG (ATCC 53103) and platelet aggregation *in vitro*', *Int J Food Microbiol*, **37**, 83–86.

MACKAY A D, TAYLOR M B, KIBBLER C C and HAMILTON-MILLER J M T (1999), '*Lactobacillus* endocarditis caused by a probiotic organism', *Clin Microbiol Infect*, **5**, 290–292.

MAJAMAA H and ISOLAURI E (1997), 'Probiotics: A novel approach in the management of food allergy', *J Allergy Clin Immunol*, **99**, 179–185.

MAJAMAA H, ISOLAURI E, SAXELIN M and VESKARI T (1995), 'Lactic acid bacteria in the treatment on acute rotavirus gastroenteritis', *J Ped Gastroenterol Nutr*, **20**, 333–338.

MALIN M, VERRONEN P, KORHONEN H, SYVÄOJA E-L, SALMINEN S, MYKKÄNEN H, ARVILOMMI H, EEROLA E and ISOLAURI E (1997), 'Dietary therapy with *Lactobacillus* GG, bovine colostrum or bovine immune colostrum in patients with juvenile chronic arthritis: evaluation of effect on gut defence mechanisms', *Inflammopharmacology*, **5**, 219–236.

MERCENIER A, MULLER-ALOUF H and GRANGETTE C (2000), 'Lactic acid bacteria as live vaccines', *Curr Issues Mol Biol*, **2**, 17–25.

MOMOSE H, IGARASHI M, ERA T, FUKUDA Y, YAMADA M and OGASA K (1979), 'Toxicological studies on *Bifidobacterium longum* BB-536', *Ouyou Yakuri*, **17**, 881–887.

NIKOSKELAINEN S, SALMINEN S, BYLUND G and OUWEHAND A C (2001), 'Characterization of the properties of human- and dairy-derived probiotics for prevention of infectious diseases in fish', *Appl Environ Microbiol*, **67**, 2430–2435.

NOUSIAINEN J and SETÄLÄ J (1998),'Lactic acid bacteria as animal probiotics'. In: S Salminen and A von Wright, *Lactic Acid Bacteria: Microbiology and Functional Aspects*, New York: Marcel Dekker, Inc, 437–473.

OUWEHAND A C, KIRJAVAINEN P V, GRÖNLUND M-M, ISOLAURI E and SALMINEN S J (1999), 'Adhesion of probiotic micro-organisms to intestinal mucus', *Int Dairy J*, **9**, 623–630.

OUWEHAND A C and SALMINEN S J (1998), 'The health effects of viable and non-viable cultured milks', *Int Dairy J*, **8**, 749–758.

OUWEHAND A C, LAGSTRÖM H, SUOMALAINEN T and SALMINEN S (2002), 'Effect of probiotics on constipation, fecal azoreductase activity and fecal mucin content in the elderly', *Ann Nutr Metab*, **46**, 159–162.

PELTO L, ISOLAURI E, LILIUS E-M, NUUTILA J and SALMINEN S (1998), 'Probiotic bacteria downregulate the milk-induced inflammatory response in milk-hypersensitve subjects but have a immunostimulatory effect in healthy subjects', *Clin Exp Allergy*, **28**, 1474–1479.

PELTONEN K D, EL-NEZAMI H S, SALMINEN S J and AHOKAS J T (2000), 'Binding of aflatoxin B$_1$ by probiotic bacteria', *J Sci Food Agric*, **80**, 1942–1945.

PENOT J P, LAGRANGE P, DARODES N, PLOY M C, VIOLET T, VIROT P, MANSOUR L and BESAID J (1998), '*Lactobacillus acidophilus* endocarditis', *Presse Med*, **27**, 1009–1012.

PERLMUTTER D H, BOYLE J T, CAMPOS J M, EGLER J M and WATKINS J B (1983), 'D-Lactic acidosis in children: an unusual metabolic complication of small bowel resection', *J Pediatr*, **102**, 234–238.

PHILLIPS P E (1989), 'How do bacteria cause chronic arthritis?' *J Reumathol*, **16**, 1017–1019.

RAUTIO M, JOUSIMIES-SOMER H, KAUMA H, PIETARINEN I, SAXELIN M, TYNKKYNEN S and KOSKELA M (1999), 'Liver abscess due to a *Lactobacillus rhamnosus* strain indistinguishable from *L rhamnosus* strain GG', *Clin Infect Dis*, **28**, 1159–1160.

RODRIGUEZ A V, BAIGORI M D, ALVAREZ S, CASTRO G R and OLIVER G (2001), 'Phosphatidylinositol-specific phospholipase C activity in *Lactobacillus rhamnosus* with capacity to translocate', *FEMS Microbiol Lett*, **204**, 33–38.

ROGINSKI H (2002), 'Fermented milks of northern Europe' in Roginski H, Fuquay J W and Fox P F *Encyclopedia of Dairy Sciences*, New York, Academic Press, *in press.*

ROWLAND I (1999), 'Probiotics and benefits to human health – the evidence in favour', *Environ Microbiol*, **1**, 375–382.

RUSELER-VAN EMBDEN J G H, VAN LIESHOUT L M C and MARTEAU P (1995), 'No degradation of intestinal mucus glycoproteins by *Lactobacillus casei* strain GG', *Microecol Ther*, **25**, 304–309.

SALMINEN S, BOULEY C, BOUTRON-RUAULT M-C, CUMMINGS J H, FRANCK A, GIBSON G R, ISOLAURI E, MOREAU M-C, ROBERFROID M and ROWLAND I (1998), 'Functional food science and gastrointestinal physiology and function', *Br J Nutr*, **80**, S147–S171.

SALMINEN S J, VON WRIGHT A J, OUWEHAND A C and HOLZAPFEL W H (2001), 'Safety assessment of probiotics and starters', in Adams M R and Nout M J R, *Fermentation and Food Safety*, Gaithersburg: Aspen Publishers, Inc, 239–251.

SAXELIN M (1997), '*Lactobacillus* GG – a human probiotic strain with thorough clinical documentation', *Food Rev Int*, **13**, 293–313.

SAXELIN M, CHUANG N-H, CHASSY B, RAUTELIN H, MÄKELÄ P H, SALMINEN S and GORBACH S L (1996a), 'Lactobacilli and bacteremia in southern Finland, 1989–1992', *Clin INFECT DIS*, **22**, 564–566.

SAXELIN M, RAUTELIN H, SALMINEN S and MÄKELÄ P H (1996b), 'Safety of commercial products with viable *Lactobacillus* strains', *Infect Dis Clin Pract*, **5**, 331–335.

SCHMIDTKE L M and CARSON J (1999), 'Induction, characterisation and pathogenicity in rainbow trout *Oncorhynchus mykiss* (Walbaum) of *Lactococcus garvieae* L-forms', *Vet Microbiol*, **69**, 287–300.

SEVERIJNEN A J, VAN KLEEF R, HAZENBERG M P and VAN DE MERWE J P (1989), 'Cell wall fragments from major residents of the human intestinal flora induce chronic arthritis in rats', *J Reumathol*, **16**, 1061–1068.

SIDHU H, ALLISON M J, MAY CHOW J, CLARK A and PECK A B (2001), 'Rapid reversal of hyperoxaluria in a rat model after probiotic administration of *Oxalobacter formigens*', *J Urol*, **166**, 1487–1491.

SIMS W (1964), 'A pathogenic Lactobacillus', *J Path Bact*, **87**, 99–105.

SPANHAAK S, HAVENAAR R and SCHAAFSMA G (1998) 'The effect of consumption of milk fermented by *Lactobacillus casei* strain Shirota on the intestinal microflora and immune parameters in humans', *Eur J Clin Nutr*, **42**1, 65–72

SREEKUMAR O and HOSONO A (1998), 'The heterocyclic amine binding receptors of *Lactobacillus gasseri* cells', *Mut Res*, **421**, 65–72.

STEIDLER L, HANS W, SCHOTTE L, NEIRYNCK S, OBERMEIER F, FALK W, FIERS W and REMAUT
E (2000), 'Treatment of murine colitis by *Lactococcus lactis* secreting interleukin-10',
Science, **289**, 1352–1355.

STEWART-TULL D E S (1980), 'The immunological activities of bacterial peptidoglycans',
Ann Rev Microbiol, **34**, 311–340.

STYRIAK I, ZATKOVIC B and MARSALKOVA S (2001), 'Binding of extracellular matrix proteins
by lactobacilli', *Folia Microbiol*, **46**, 83–85.

SUMNER S S, SPECKHARD M W, SOMERS E B and TAYLOR S L (1985), 'Isolation of histamine-
producing *Lactobacillus buchneri* from Swiss cheese implicated in a food poisoning
outbreak', *Appl Environ Microbiol*, **50**, 1094–1096.

TAKAHASHI M and ONOUE M (1999), 'Safety of *Lactobacillus casei* strain Shirota', in Yakult
Central Institute for Microbiological Research *Lactobacillus casei Strain Shirota –
Intestinal Flora and Human Health*, Tokyo: Yakult Honsha Co, Ltd, 263–281.

TANAKA H, DOESBURG K, IWASAKI T and MIERAU I (1999), 'Screening of lactic acid bacteria
for bile salt hydrolase activity', *Int Dairy J*, **82**, 2530–2535.

TEUBER M, MEILE L and SCHWARZ F (1999), 'Acquired antibiotic resistance in lactic acid
bacteria from food', *Antonie van Leeuwenhoek*, **76**, 115–137.

TURBIC A, AHOKAS J T and HASKARD C A (2002), 'Selective *in vitro* binding of dietary
mutagens, individually or in combination, by lactic acid bacteria', *Food Add Contam*, **19**,
144–152.

TYNKKYNEN S, SINGH K V and VARMANEN P (1998), 'Vancomycin resistance factor of
Lactobacillus rhamnosus GG in relation to enterococcal vancomycin resistance (*van*)
genes', *Int J Food Microbiol*, **41**, 195–204.

WADE J J (1997), '*Enterococcus faecium* in hospitals', *Eur J Clin Microbiol Infect Dis*, **16**,
113–119.

WAGNER R D, PIERSON C, WARNER T, DOHNALEK M, FARMER J, ROBERTS L, HILTY M and
BALISH E (1997a), 'Biotheraputic effects of probiotic bacteria on candidiasis in
immunodeficient mice', *Infect Immunol*, **65**, 4165–4172.

WAGNER R D, WARNER T, ROBERTS L, FARMER J and BALISH E (1997b), 'Colonization of
congenitally immunodeficient mice with probiotic bacteria', *Infect Immunol*, **65**, 3345–
3351.

WANG X, ANDERSSON R, SOLTESZ V, LEVEAU P and IHSE I (1996), 'Gut origin sepsis,
macrophage function, and oxygen extraction associated with acute pancreatitis in the rat',
World J Surg, **20**, 299–308.

WILLIAMS A M, FRYER J L and COLLINS M D (1990), '*Lactococcus piscium* sp nov: a new
Lactococcus species from salmonid fish', *FEMS Microbiol Lett*, **68**, 109–114.

VON WRIGHT A and SIBAKOV M (1998), 'Genetic modification of lactic acid bacteria'. In:
Salminen S and von Wright A, *Lactic Acid Bacteria: Microbiology and Functional
Aspects*, New York: Marcel Dekker, 161–210.

ZHOU J S, SHU Q, RUTHERFORD K J, PRASAD J, GOPAL P K and GILL H (2000), 'Acute oral
toxicity and bacterial translocation studies on potentially probiotic strains of lactic acid
bacteria', *Food Chem Toxicol*, **38**, 153–161.

ZHOU J S, GOPAL P K and GILL H S (2001), 'Potential probiotic lactic acid bacteria
Lactobacillus rhamnosus (HN001), *Lactobacillus acidophilus* (HN017) and *Bifidobacterium lactis* (HN019) do not degrade gastric mucin *in vitro*', *Int J Food Microbiol*,
63, 81–90.

14

Clinical trials

P. Marteau, Paris V University, France

14.1 Introduction

The establishment of the efficacy and tolerance of functional dairy product (as well as that of other functional foods) relies on clinical trials. Various kinds of studies are usually performed during the development or after commercialisation of food products and the 'evidence pyramid' is usually used to illustrate the power of these different types of study to establish evidence for the beneficial effects (Fig. 14.1). Research ideas (base of the pyramid) are usually first tested in laboratory models, then in animals, and finally in humans. The main pitfall of the clinical studies is the risk of biases, which can alter the results. Randomised controlled trials (RCTs) are the last step to evaluate the effectiveness and efficacy of a product. Their methodology aims at reducing the potential for bias and allowing comparison between the product and a control (no treatment or another product). This chapter considers the general and specific aspects of clinical trials on functional dairy products in humans. The general aspects (which are shared with all clinical trials for foods or drugs) are the needs for a good protocol, respect of ethics rules, proper statistics and publication of the results. Interpretation of the trials should follow the rules of evidence-based medicine. One of the specific aspects of studies with dairy products is the difficulty of choosing a placebo or a control, especially when blinded protocols are needed.

14.2 Setting up a clinical trial: protocols

The major points that should be stated in a clinical trial protocol are shown in Table 14.1.

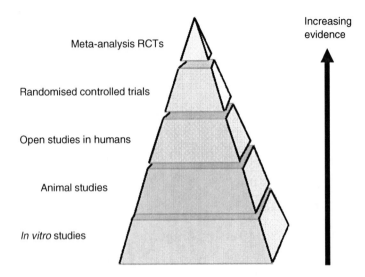

Fig. 14.1 The 'pyramid of evidence' (adapted from http://www.hsl.unc.edu/lm/ebin/question.htm).

Table 14.1 Topics that are usually stated in a clinical trial protocol

- Title – any amendment(s) and date(s).
- Sponsor – investigator(s), clinical laboratories and other institutions involved.
- Name and description of the product(s).
- Summary of findings from previous studies relevant to the trial (and references).
- Summary of the known and potential risks and benefits, if any, to human subjects.
- Description of the route of administration, dosage, dosage regimen, and treatment period(s).
- A statement that the trial will be conducted in compliance with the protocol, good clinical practice (GCP) and the applicable regulatory requirement(s).
- Inclusion criteria – exclusion criteria – withdrawal criteria.
- Description of the objectives and the purpose of the trial.
- A specific statement of the primary end-points and the secondary end-points.
- A description of the type/design of trial to be conducted (e.g. double-blind, placebo-controlled, parallel design) and a schematic diagram of trial design, procedures and stages.
- Measures taken to minimise/avoid bias.
- Maintenance of trial treatment randomisation codes and procedures for breaking codes.
- Treatment(s) permitted and not permitted before and/or during the trial.
- Procedures for monitoring compliance.
- Specification of the efficacy parameters.
- Methods and timing for assessing, recording and analysing safety parameters.
- Procedures for eliciting reports of and for recording and reporting an adverse event.
- Description of the statistical methods including the number of subjects planned to be enrolled and the reason for choice of sample size.
- Data handling and record keeping.
- Financing and insurance if not addressed in a separate agreement.
- Publication policy.

14.2.1   How to ask a clinical question that can be answered

The aim of a clinical trial is to answer a question. It is thus obvious that this question has to be formulated before the beginning of the trial, and that the design of the trial should be adapted to answer the question. There are two major types of questions concerning efficacy of a product: is it more efficient than another product or is it as efficient as another product? The design of the trial, the number of subjects to include and the statistical tests depend on the type of question. Another point is that the evidence that supports the validity (truthfulness) of the results of a study depends primarily on the studies methodology. The optimal methodology aims to avoid biases. These can be defined as the systematic tendency of any aspects of the design, conduct, analysis and interpretation of the results to make the estimate of a product effect deviate from its true value. David Sackett, a pioneer in study methodology, identified no less than 56 possible biases that may arise in clinical trials and stressed that more than half of them were related to the study design.[1,2] Consequently, one should always seek statistical help in planning a study, both on sample size and study design (in other words the statistician should be involved when planning the study, not only when it is done, and that the results need to be analysed). Study methodologies such as randomisation, blinding and accounting for all patients help ensure that the study results are not biased.

14.2.2   Selection of the major end-point and of the control group

The first critical step is to select the major end-point of the study. If the major end-point of a study is not clearly stated in an article (and when several end-points are used), the reader must always consider that positive results may be biased by the author opinion.[3] The choice of the major end-point should be based on its relevance. Knowledge of the results of the major end-point in the control group (mean and standard deviation) is needed to calculate the number of subjects (see below).

The choice of the control group is also very important. Four types of controls can be used: no treatment, placebo, different dose of the study product, or different active product. A placebo is a dummy product that must be as close as possible to the test product in terms of colour, texture, taste, etc., but that does not contain the active ingredient of the test product. Preparing such a placebo is often difficult with milk products, as their flavour and texture are subtle. The composition of the placebo should also be chosen in order to avoid nocebo effect. (The term nocebo or 'negative placebo' is used for conditions that reduce the therapeutic effectiveness of a product, causing deteriorating symptoms or even new ones.) For example, a lactose-rich product is not a good placebo for a study trying to prove that a fermented dairy product will help shortening diarrhoea in subjects with gastroenteritis (because lactose can induce diarrhoea).

The use of a 'no treatment' or placebo control often raises problems of ethics, acceptability and feasibility, which have to be carefully considered before the study. One of the possibilities is to plan for 'early escape' from the protocol in case of treatment failure or to use an add-on design. The term 'add-on study' is used for a placebo-controlled trial of a product conducted in subjects also receiving standard treatment. For example, Guslandi compared the efficacy of 3 g/d mesalazine to 2 g/d mesalazine plus 1 g of *Saccharomyces boulardii* in patients with Crohn's disease.[4] With such a design, all the patients received at least 2 g of the classical treatment mesalazine.

14.2.3   Randomisation and blinding

Randomisation and blinding are used to minimise bias due to the subject or the investigator opinion on the product.[3,5] Double-blind randomised studies provide a far stronger evidence for efficacy than non-blinded and open studies.[3] Indeed, randomisation is the best way to avoid a systematic difference between the groups at the beginning of the study ('baseline'). However, there is still a possibility that the two groups will differ by chance and this has to be tested when analysing the results. Subjects and investigators may be consciously or unconsciously influenced to report favourable effects with the product if they can differentiate it from the control. Blinding is intended to minimise this bias. In single-blind studies, only the patients ignore which product they receive, while in *double-blind* studies, both the patients and the investigators do not know which product is taken. The organisation of this 'blindness' needs external helps (usually by the statistician). The blindness code should not be broken before the end of the study except in the case of a severe adverse event and this has to be monitored.

14.3   Statistical analysis

One of the major aims of the statistical analysis of the results of a clinical trial is to evaluate the chance (or risk) that they could be due to chance and not to the effect of the product. Everyone will intuitively understand that this risk is lower when high numbers of subjects are studied, but how many subjects should there be?[6,7] There are two types of statistical errors. The type I error (alpha error) is to reject the hypothesis tested (i.e. the existence of a difference of efficacy between two products) when it is true. The type II error (beta error) is to reject the hypothesis tested when a given alternative hypothesis is true. The alpha and beta errors are linked and are also linked to the statistical power of a trial.

Most studies make the hypothesis that a product will be more efficient than the other; the power calculation can then be used to determine how many subjects are needed to have a reasonable (usually 80% or 90%) chance of detecting a clinically important difference between the two products (Table 14.2). The power of a test is the probability that the test will reject the hypothesis tested when a specific alternative hypothesis is true. To calculate the power of a given test, it is necessary to specify alpha and to specify a specific alternative hypothesis.[8]

Several statistical tests can be used in equivalence studies. The absence of difference between two groups using common statistical tests designed to test for differences does not allow us to establish the absence of a difference. The three

Table 14.2   Information needed to calculate the number of subjects for a clinical trial

1. What is the major study end-point (single result or score, percentage, mean of several measurements, survival time...)?
2. What values would be expected in the control group? (including standard deviation for measurements)? (Look at the literature*)
3. How large a difference between the tested product and the control would be considered clinically relevant?
4. Do you want to test for equivalence between the two groups or superiority of one product? (Equivalence testing needs more subjects)

*If the information does not exist, the study will have to be considered as a pilot study.

double-blind RCT studies that compared the probiotic mutaflor (which contains 25 × 10⁹ viable *E. coli* strain Nissle 1917 per 100 mg) to mesalazine to reduce the risk of recurrence of ulcerative colitis illustrate that perfectly. In the first study, Kruis *et al.*[9] included 120 patients in a 12-week double-blind, double-dummy study. Half of them received 1.5 g/d of mesalazine, and the other half received 200 mg/d of mutaflor. After 12 weeks, 11.3% of the subjects receiving mesalazine and 16% of those receiving the probiotic had relapsed. This difference was not significant. However, the statistical power of the study was low as the percentage of recurrence after 12 weeks was low in both groups. Although the authors were convinced that the two products were equivalent, their conclusion was not accepted by the medical community.[10] A second RCT was thus performed that included 116 patients followed for 1 year. Once again equivalence tests were not used and the power of the study was not calculated accordingly. Although this study suggested no significance difference between the two treatments, the finding could still not be accepted by statisticians and by gastroenterologists.[10] A third RCT has recently been presented in an abstract form in which statistical tests were planned to assess equivalence between treatments and the numbers of subjects were calculated accordingly.[12] Some 222 patients were treated either with mesalazine or with mutaflor for one year. The relapse rate was 36.4% in the probiotic *vs.* 33% in the mesalazine group and equivalence was statistically proven. Non-inferiority trials are designed to demonstrate that a product is not relevantly inferior to a standard. Equivalence trials are usually designed to demonstrate equivalence with respect to a rate and extent of drug absorption between a reference and a generic version (bioequivalence goal).

14.4   Ethical issues

Clinical trials should be conducted in accordance with ethical principles. Ethical principles have been formulated in the Declaration of Helsinki.[13] They are consistent with good clinical practice (GCP) and the applicable regulatory requirements.[14] Every clinical trial must first be evaluated by an independent ethics committee. The legal status, composition, function, operations and regulatory requirements pertaining to independent ethics committees may differ among countries, but should allow the committee to act in agreement with GCP as described in this guidance.[14] The members of such a committee are both medical/scientific professionals and non-medical/non-scientific persons; their responsibility is

> to ensure the protection of the rights, safety, and well-being of human subjects involved in the trial and to provide public assurance of that protection, by, among other things, reviewing and approving/providing favourable opinion on the trial protocol, the suitability of the investigator(s), facilities, and the methods and material to be used in obtaining and documenting informed consent of the trial subjects.[14]

Table 14.3   Main points of GCP for the investigators and the sponsor (adapted from reference 14)

Investigator
- Investigator's qualifications and agreements
- Adequate resources
- Medical care of the trial subjects
- Communication with IRB/IEC
- Compliance with protocol
- Investigational product(s)
- Randomisation procedures and unblinding
- Informed consent of the trial subjects
- Records and reports
- Progress reports
- Safety reporting
- Premature termination or suspension of a trial
- Final report(s) by investigator/institution

Sponsor
- Quality assurance and quality control
- Trial design
- Information about the product(s)
- Manufacturing, packaging, labelling and coding the product
- Supplying and handling the product(s)
- Trial management, data handling, record keeping
- Monitoring committee
- Investigator selection
- Allocation of duties and functions
- Compensation to subjects and investigators
- Financing
- Notification/submission to regulatory authority(ies)
- Record access
- Safety information
- Adverse drug reaction reporting
- Monitoring
- Audit
- Non-compliance
- Premature termination or suspension of a trial
- Clinical trial/study reports

The independent ethics committee must obtain the following documents: trial protocol(s)/amendment(s), written informed consent form(s) and consent form updates that the investigator proposes for use in the trial, written information to be provided to subjects, available safety information, information about payments and compensation available to subjects, the investigator's current curriculum vitae and/or other documentation of qualifications. The independent ethics committee reviews the proposed clinical trial within a reasonable time and documents its views (approval/favourable opinion; modifications required prior to the approval/favourable opinion; disapproval/negative opinion; or termination/suspension of any prior approval/favourable opinion).

14.5   Managing a clinical trial

GCP is an international ethical and scientific quality standard for designing, conducting, recording and reporting trials involving the participation of human subjects.[14] Compliance with this standard provides public assurance that the rights, safety and well-being of trial subjects are protected, consistent with the principles of guidelines such as the ICH GCP Guideline,[14] which have been written to provide a unified standard for the European Union, Japan and the United States to facilitate the mutual acceptance of clinical data by the regulatory authorities in these jurisdictions. It is not possible to describe here in detail all the rules that need to be followed to respect GCP; however, a list of the headings of the main points for the investigators and the sponsor is shown in Table 14.3.

Monitoring is defined as the act of overseeing the progress of a clinical trial, and of ensuring that it is conducted, recorded and reported in accordance with the protocol, standard operating procedures, GCP and the applicable regulatory requirement(s). Clinical trials have to follow the GCP rules and be submitted to the Independent Ethics Committee Review. Authorities will base their decisions for labelling products on the quality and relevance of the clinical studies that are analysed by expert committees.

14.6   Assessing the validity of a clinical trial

The main points to analyse in order to answer the question of the validity of a clinical study are summarised in Table 14.4. The assignment of patients to either group (product or control) must be done by a random allocation to help eliminate conscious or unconscious bias (see above). The importance of the blinding has been stressed above. The study should begin and end with the same number of patients. Patients lost to the study must be accounted for or risk making the conclusions invalid. Indeed, patients may drop out because of the adverse effects of the therapy being tested. If not accounted for, this can lead to conclusions that may be over-confident about the efficacy of the therapy. Good studies must have more than 80% follow-up for their patients. The patients who were lost for follow-

Table 14.4   Main points to be considered in order to analyse the validity of the results of a clinical study (in order to avoid biased interpretation)

1. Was the assignment of patients to treatment randomised?
2. Were all the patients who entered the trial properly accounted for at its conclusion?
3. Was the follow-up complete?
4. Were patients analysed in the groups to which they were (originally) randomised?
5. Were patients, their clinicians, and study personnel 'blind' to the treatment?
6. Were the groups similar at the start of the trial?
7. Aside from the experimental intervention, were the groups treated equally?

up should be assigned to the 'worst-case' outcomes. Patients must be analysed within their assigned group: this is called 'intention to treat' analysis. Patients who forget or refuse their treatment should also not be eliminated from the study analysis. Indeed, excluding non-compliant patients introduces a bias.

14.7   Sources of further information and advice

Information on clinical trial methodology, statistics, evidence-based medicine, GCP, ethics and regulatory process is available and updated on the Internet. Some of the main addresses, which I consulted to write this chapter, are indicated below:

Statistics
    http://www.animatedsoftware.com/statglos/statglos.htm
Good clinical practice
    http://www.fda.gov/cder/guidance/959fnl.pdf
    www.fda.gov/oc/gcp/default.htm
Declaration of Helsinki
    http://ohsr.od.nih.gov/helsinki.php3
Evidence-based medicine
    cebm.jr2.ox.ac.uk
    www.ebmny.org/

14.8   References

1. SACKETT D L (1979), 'Bias in analytic research', *J Chronic Dis,* **32**(1–2), 51–63.
2. ALTMAN D G (1980), 'Statistics and ethics in medical research: misuse of statistics is unethical', *BMJ*, **281**, 1182–4.
3. GUYATT G H, SACKETT D L and COOK D J (1993), 'Users' guides to the medical literature. II. How to use an article about therapy or prevention. A. Are the results of the study valid?', Evidence-Based Medicine Working Group, *JAMA*, **270**, 2598–601.
4. GUSLANDI M, MEZZI G, SORGHI M and TESTONI P A (2000), '*Saccharomyces boulardii* in maintenance treatment of Crohn's disease', *Dig Dis Sci*, **45**, 1462–4.
5. ALTMAN D G and BLAND J M (1999), 'Statistics notes. Treatment allocation in controlled trials: why randomise?', *BMJ*, **318**, 1209–11.
6. ALTMAN D G (1980), 'Statistics and ethics in medical research. III. How large a sample?', *BMJ*, **281**, 1336–8.

7. http://members.aol.com/johnp71/javastat.html
8. http://www.animatedsoftware.com/statglos/statglos.htm
9. KRUIS W, SCHÜTZ E, FRIC P *et al.* (1997), 'Double-blind comparison of an oral *Escherichia coli* preparation and mesalazine in maintaining remission of ulcerative colitis', *Alim Pharmacol Ther*, **11**, 853–8.
10. MARTEAU P R, DE VRESE M, CELLIER C J and SCHREZENMEIR J (2001), 'Protection from gastrointestinal diseases with the use of probiotics', *Am J Clin Nutr*, **73**(2 Suppl), 430S–436S.
11. REMBACKEN B J, SNELLING A M, HAWKEY P M *et al.* (1999), 'Non-pathogenic *Escherichia coli* versus mesalazine for the treatment of ulcerative colitis: a randomised trial', *Lancet*, **354**, 635–9.
12. KRUIS W, FRIC P, STOLTE M and THE MUTAFLOR STUDY GROUP (2001), 'Maintenance of remission in ulcerative colitis is equally effective with *Escherichia coli* Nissle 1917 and with standard mesalamine', *Gastroenterology*, **120**, A139.
13. World Medical Association Declaration of Helsinki Adopted by the 18th World Medical Assembly Helsinki, Finland, June 1964 and amended by the 29th World Medical Assembly Tokyo, Japan, October, 1975 35th World Medical Assembly Venice, Italy, October 1983 and the 41st World Medical Assembly Hong Kong, September 1989 (http://ohsr.od.nih.gov/helsinki.php3).
14. FDA, (1997), 'Good clinical practice', *Federal Register, **62**, 25691–25709.

15

Consumers and functional foods

L. Lähteenmäki, VTT Biotechnology, Finland

15.1 Functional foods and consumers

Functional food is regarded as an entity that lies somewhere between medicines and conventional food products, but there is no common shared definition of functional foods. The closeness of functional foods to either medicines or foods varies among different applications. Japan has its own legislation for 'Food for Specified Health Uses', called FOSHU, in which functional foods are clearly regarded as food products that are eaten as part of an ordinary diet. In Europe, the definition suggested by an EU-funded concerted action project has widely been used. According to this definition, functional foods are 'satisfactorily demonstrated to affect beneficially one or more target functions in the body, beyond adequate nutritional effects in a way that is relevant to either an improved state of health and well-being and/or reduction of risk of disease' (Diplock *et al.*, 1999). In several European countries there are self-administered systems for endorsing health-related claims in products. Although there is no unanimous definition or conformity on what functional foods are, these definitions and endorsing systems widely agree that functional foods are related to health and they should be considered more as food than medicine. In this chapter, these foods are called functional foods in a wide sense.

The lack of official definition has been claimed to be one of the restricting factors in the success of functional foods among consumers. However, consumers very rarely define foods in health-related terms or nutritional terms; instead, foods are seen as categories that respond to the belief systems of individual consumers. Within these categories, whether they relate to use purpose, degree of liking or type of the food, perceived healthiness is one factor influencing choices among other factors, such as quality, taste, convenience, naturalness and preferences of family members. In a pan-European study, trying to eat healthily was mentioned by 32% of the respondents as one of the three most important reasons behind food choices (Lappalainen *et al.*, 1998). Functional foods have a clear health connotation, and although they have distinct features from conventional products, consumers evaluate their perceived healthiness similarly to other food products. The first part of this chapter deals with the role of health in food choices in general, then the special characteristics of functional foods are considered from the consumer's point of view. After this, a few studies on consumer responses to functional foods are examined in more detail. The last part concentrates on issues relevant to future acceptance on functional foods.

15.2   The role of health in food choice

Health is one reason consumers mention when they are asked about factors that influence their food choices. Correspondingly, when asked about health, wholesome and varied diet, together with regular exercise, were mentioned by seven respondents out of ten as the factors having the greatest impact on health in Denmark, Finland and the USA (Bech-Larsen *et al.*, 2001). Health is mostly a credence characteristic in food products (Grunert *et al.*, 2000). It has to be assumed, as in most cases, the possible health effects cannot be directly observed or experienced. Taste and other sensory qualities can be directly sensed and they give an immediate feedback for the eater. The pleasure derived from eating a pleasant tasting food is therefore a strong motivator for repeated choices. Physiological effects of foods influencing our well-being can, however, also have directly felt effects. Caffeine, which is the physiologically effective component in coffee, has a refreshing effect and research has shown that people learn to like caffeine-containing drinks because they want to feel the refreshing effect or they try to avoid possible caffeine deprivation (Yeomans *et al.*, 2000).

Health can have an impact on food choices through several mechanisms (Fig. 15.1). Mostly healthiness is assessed through the nutritional quality of foods and absence of any possibly harmful substances in foods. These characteristics cannot be directly observed from a food product; instead, consumers need information-based knowledge on the nutrient content and possible other health effects. The information has to be acquired and trusted by the consumer before it can have an impact on their behaviour. Our attitudes, which are either favourable or negative learned pre-dispositions towards an object or behaviour, guide what information we notice and how readily we integrate new beliefs into our knowledge system.

Nutrition experts emphasise that the degree of healthy eating needs to be judged on the level of diet, whereas no single food product can be categorised as health-promoting or not. Although this is true from a nutrition point of view, consumers tend to have ideas about the health-value of single products. Furthermore, the information about fat and vitamin content of foods tend to be generalised, so that, for example, all fruits are categorised as good sources of vitamin C and having low fat content. Similarly, experience with certain types of foods creates expectations

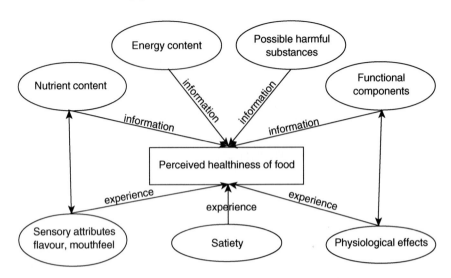

Fig. 15.1   Perceived healthiness in food products is a combination of knowledge-based beliefs and experienced attributes. Nutrient and energy content need to be conveyed through information, but sensory properties and perceived satiety may give feedback. The functional components very rarely have sensed physiological effects; instead they need to be verified through instrumental measures. With some effects the instrumental measure is not readily available and consumers must rely solely on the information provided.

on the sensory characteristics, e.g. low-fat products are assumed to have a different taste and mouth feel from regular products. Perceived healthiness can also depend on any possible harmful substances food is known to have. Some mushrooms, for example, have to be boiled in water before they are edible. In addition, for many consumers any artificial added components, such as food additives, are regarded as a potential risk for health. Functional components bring along promises of positive physiological effects, the value of which depends on consumers' health-related knowledge and viewpoints. Perceived healthiness depends on many factors and is based on the interplay between beliefs and experiences.

What is considered to be healthy eating varies among countries and depends on cultural conventions. As mentioned earlier, in a pan-European study carried out in all EU member states health was mentioned as one of the three most important reasons for food choices by 32% of the 14,331 respondents. The results varied among the countries from 24% in Luxembourg up to 50% in Austria. The lack of time and self-control were the most commonly mentioned barriers for healthy eating (Lappalainen *et al.*, 1998). However, when the interviewees were asked 'what is healthy eating?'- the results varied between countries. About half of the respondents considered low fat products and less fat in the diet as part of healthy eating. More fruit and vegetables and balanced and varied diet were mentioned by 42 and 41% of respondents, respectively (Margetts *et al.*, 1997). Cultural difference become apparent when frequencies of some responses are explored. Freshness and naturalness of foods are mentioned on average by 28% of the respondents, but among EU countries the range is from 6% in Denmark to 56% in Italy. Balanced diet is mentioned by 41% of EU citizens with a range of 11–74%. The results suggest that people are aware of nutritional guidelines in each country and these guidelines have an impact on the descriptions of healthy eating.

Health is clearly an essential criterion in food choices and it is largely defined through choices that are in accordance with nutritional education in each country. Roininen *et al.* (1999) have developed an attitude scale (General Health Interest) that measures an individual's health orientation in order to predict their willingness to follow nutritional recommendations. When scales were applied in a cross-cultural study, the health orientation was rather high in the three European countries studied (Finland, the Netherlands and the UK), with women showing clearly higher interest in health issues than men (Roininen *et al.*, 2001). In another UK study, 70 out of 100 respondents agreed with the statement that they are health conscious and 89 said that they tried to eat a balanced diet (Consumers' Association, 2000). The general health interest may not, however, correlate with consumers' willingness to use functional foods, as these food-related health behaviours have clear distinctions, as will be described in the following section.

15.3   Nutritional guidelines and health claims

Traditional nutrition education has emphasised the role of a health-promoting diet. Following nutritional guidelines will result in reduced risk to develop obesity and other chronic lifestyle-related diseases, such as cardiovascular diseases, diabetes and cancer. The message on the benefits of a healthy diet may be hard to convey to the consumer, as the end result is unsure and achieving the possible reward takes several years or even decades. The advice based on recommendations of dietary intakes of nutrients is usually given at a general level. Typically, guidelines suggest avoiding and favouring certain types of foods without giving any recommendations on any particular food products. Low-fat foods, low-salt foods or foods high in fibre are presented as categories rather than as single examples of products that should be consumed or not consumed. This categorisation may be confusing to the consumers as low-fat foods can be found in different food categories and their roles in the food system differ from each other.

Functional foods offer a new kind of positive health message for the consumers. Instead of avoiding certain types of foods, these food products promise positive physiological effects on bodily functions or even a reduction on risk level of diseases through eating a single product. By definition, the beneficial value of a functional food comes from producing a positive effect in the body, although sometimes the outcome is the avoidance of a negative effect, such as stomach upset or weakening of bones. As a reinforcement of desired behaviour, the positive reward is known to be more effective than the avoidance of punishment. Therefore, a health message in functional foods promising a positive health effect can be more appealing to consumers than an avoidance message, which is the basis of many health messages related to nutritional recommendations.

Functional messages may be easier for the consumer to understand since they contain results that can be clearly defined. Most of the promised effects of functional foods can be instrumentally measured, such as lowering the level of cholesterol in blood, decreasing the blood pressure or increasing the density of bone mass. However, some of the claimed effects are general and measuring their impact may be more difficult. In these cases the functional claims become purely credence characteristics that require consumers' trust on the claim. For example, products containing probiotics are largely credence products as most effects related to gut functions are difficult to verify by simple physiological measures. Consumers' own reported well-being cannot be used as an objective measure of any physiological effects, as many psychological studies have shown that the mere effect of paying attention to a phenomenon can always affect subjective responses.

One of the worries brought forward by consumer organisations and health professionals has been that health-related claims may confound the consumers and appear to offer easy options to promote health, so that consumers assume that a healthy diet is no longer essential. Instead the health effect can be gained through purchasing the particular products. Therefore, for example, in the UK Joint Health Claims Initiative (2002), one of the basic arguments has been that the health claims should only be combined with foods that have equal nutritional quality to those of their conventional counterparts. Quaker Oats tested how consumers understand health-related claims and found that those receiving an oatmeal or a fibre claim did not believe more often that one does not need to pay attention to the rest of the diet than those who received no claim in the package (Paul *et al.*, 1999). There seems to be no basis for believing that consumers would perceive the claims of functional foods as the salvation from having to follow a wholesome diet.

15.4   Consumers, claims and carrier products

Acceptance of any food product depends on the interaction between the consumer and the product (Fig. 15.2). Some people like one kind of flavour, whereas others dislike it. Similarly, price is more important to some individuals, while others put more emphasis on convenience. In so-called functional food products, the acceptability depends on the same product characteristics as in any conventional product, and these characteristics have to be as good as in any conventional product. Consumers are not ready to negotiate these basic requirements, but the health-related claim attached to the product can bring additional value to the product. Although health-related claims should boost the appeal of the product, they also bring an additional factor influencing acceptability. A claim carries a positive health-related message, but its weight depends on the individual consumer's motivation. Furthermore, the reception of the claim depends on the characteristics of the carrier product.

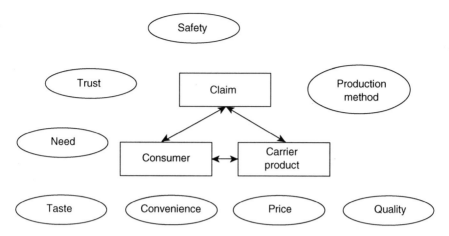

Fig. 15.2   Consumer responses to functional foods depend on the relationship between consumer, carrier product and the claim attached to the product. Having a health-related claim in a product requires trust in the claim, the effect has to be perceived safe and there must be a perceived need to use the product. As in any food product, overall quality (including sensory quality), price and convenience are also important. Furthermore, the method for producing the claimed effects may have an impact on the acceptability.

15.4.1   Claims

Health-related claims can vary in their nature. The claims can be broadly categorised into 'enhanced function' and 'reduction of risk of disease' claims, depending on the effect promised (Council of Europe, 2001). A majority of products marketed with health arguments belong to the enhanced function claim group. Products have been enriched with vitamins, mineral or trace elements, which are known to have physiologically active role in the body. Scientific proof for these claims relies on the nutritional literature and most nutritional effects are clear-cut causal effects. Enhanced function claims promise that food produces certain physiological responses, such as lowering of blood cholesterol. These claims leave open the link between the function and its final health effect. For consumers this means that they need to be informed about the link from sources other than the product itself. Therefore, marketing functional food products is a challenging task, as the health-related information needs to be gathered from various sources and a skilful marketing strategy should build up this jigsaw from pieces that support each other. In reduced risk claims the promised function is directly linked to the possible risk of disease or improvement of health. Such risks or improvements require more complex evidence, as they are based on probabilistic evidence between risk factors of diseases and likelihood of developing the illness or disorder.

Another distinction in claims is whether they are generic in their nature or product-specific. In generic claims the connection is made between the possible consequence and the active component or ingredient in product, whether the claim is for an enhanced function or a risk reduction. Most of the health-related claims allowed are generic in their nature and several countries have a list of accepted claims. For example, in Sweden the voluntary code 'Health claims in the labelling and marketing of food products' defined in 1990 eight generic claims, which link certain components with a physiological consequence (Asp and Trossing, 2001). These generic claims are based on the idea that a food product contains an ingredient or component that is known to have a lowering effect of disease risk and therefore the product is beneficial for health. The claim does not cover the product itself and can be used in all products that contain sufficient amount of the component. These claims are similar to traditional nutritional messages, which can be described as claims related to nutrient functions.

From the manufacturer's point of view, product-related claims can be more appealing as they link not only the compound or ingredient, but also the specific product with the promised benefit in the claim. The product-specific claim offers better protection against competitors to copy the products and means that the claim can be used effectively in marketing. In the revised Swedish code, these claims are called product-specific physiological claims (Asp & Trossing, 2001) and in UK they are called innovative claims (UK Joint Health Claims Initiative, 2002). The claims have to be based on adequate and sound scientific evidence, which means that the effect has to have been proven in independent clinical studies. The studies should give sufficient support for the evidence that products provide the promised benefits with portions that are eaten as a part of normal diet. However, translating the scientific probability-based information into consumer language is a tricky task as the way of thinking behind scientific and everyday beliefs differs dramatically.

Science provides the optimum knowledge on the phenomenon at any moment. Scientific knowledge increases all the time and new information challenges and replaces existing truths. Furthermore, the scientific evidence on health-related risks is based on probabilities. Consumers, on the other hand, base their food-related beliefs in approximate and rather crude rules of reasoning. Rather than using probabilities, consumers tend to make clear-cut assumptions on links between two statements or phenomena. If there is a reduced risk of a disease, for consumers this may mean that the risk has been removed. The equation where the product is effective in five cases out of ten is hard to comprehend; for the consumer the product should either be effective or not. An additional problem on probabilistic information is that the result cannot be promised to everyone. Although something is effective in most cases, the possibility of not personally gaining the benefit should be made understandable to the consumers. Consumers tend to be wary of linking food with reduction of disease, as this implies associating food that is commonly considered as a source of pleasure with something unpleasant, even if the promised effect would be positive. This conforms with another typical feature in everyday thinking, which tends to associate any two things that appear at the same time regardless whether they are causally connected or not. Food and health, in general, are both very sensitive topics and this adds to the challenge of communicating possible health-related claims to consumers in a comprehensible, usable and non-alarming manner.

15.4.2 Consumers

Consumers need to be aware of the health benefits offered by functional foods, but knowledge in itself is not a sufficient condition for willingness to use these products. The crucial factor in noticing and adopting messages related to the health effects of foods is the motivation of the respondent. The acceptability of new foods depends on the benefits these new foods provide for consumers (Frewer *et al.*, 1997). Wrick (1995) divided the potential consumers of functional foods into those who recognise themselves at risk for a disease and those who are health conscious. The motivational expectations of these two groups vary: the risk group wants measurable results, whereas the health-conscious group wants nutritional insurance and the ability to take care of oneself.

The most positive group towards functional foods have been women and middle-aged or elderly consumers (Poulsen, 1999; Bogue and Ryan, 2000). Age has also an impact on what kind of effects are interesting. With age, the reduction of risk levels becomes more attractive, while young respondents are more interested in increasing their energy levels. Therefore any generalisation of what age or sex is interested in functional foods is difficult to make, as the appeal depends on the benefits promised by products. Women seem to be more responsive to products that are associated with breast cancer (Arvola *et al.*, 2001) and men more responsive to products that lower the risk of prostate cancer (Hilliam, 2002).

In a cross-cultural study carried out in Denmark, Finland and the USA (Bech-Larsen *et al.*, 2001), the most attractive influences of functional foods were heart or cardiovascular diseases mentioned by 54–59%, prevention of stomach cancer mentioned by 34–48% depending on the country, and enhancing immune system mentioned by 36–39% of the respondents. In an Irish study, the two first were the same, but maintaining the health of teeth and bones rose as the third most appealing influence (Bogue and Ryan, 2000).

If a consumer feels a need to lower blood cholesterol level or reduce level of stress, then the products promising these effects are appealing. Wansink (2001) argues that the missing link between known health benefits of soy and self-relevant health consequences may hinder the success of soy products. The nutrition-related messages that were tailored in accordance with the receiver's own food habit and consumption patterns were more effective in adding their fibre consumption than messages that gave general advice on fibre-containing foods (Brinberg *et al.*, 2000). The tailored messages appeal to a certain group of people, whereas more general messages can be significant to a larger group, although their weight is not as strong. Producers of functional foods have to balance between the niche products that are highly relevant to a small group of consumers and messages that aim to produce general well-being in most consumers.

15.4.3 Carrier product effect

The existing product image has an impact on how the claims are received. The consumer perception of a health-related claim depends on its suitability to the product in question. In Denmark consumers were more positive about functional effects that were created through adding components that are naturally occurring ingredients or components in that food product (Poulsen, 1999; Bech-Larsen, 2001). Adding calcium into milk products may be more acceptable in consumers' minds than adding other minerals that do not originally belong to milk in any significant quantities. However, the health claim also needs to be in accordance with the previous health image of the product (Lähteenmäki, 2000). In a Finnish study, the possible positive health effects of conjugated linoleic acid were discredited in butter, because in Finland butter has such a strong negative health image. Butter is considered as a high source of saturated fatty acids and therefore any positive messages about butter and the possible positive health effects were considered to be artificial and irrelevant to the total image of butter.

Functional foods are defined as products that are frequently consumed so that the possible effects they claim to have can also be attained. Food and meal systems are culturally determined and the place of a carrier product in this system is well defined. The health-related claim needs to fit into this food system in order to be accepted. In some cases the functional food products have bypassed the food system by offering an additional product that is eaten or drank separately in small quantities. An example of this kind of product is the small bottles of probiotic drinks originally introduced by Yakult, which can be taken as a medicine or preventive action against possible stress. The acceptability of these products can be built in a different way as they do not require a place in the food system.

Functionality brings also a novel component to the food. Consumers are known to be suspicious about new foods (Pliner and Hobden, 1992), but this dread of novelty is mostly for foods that have unfamiliar sensory characteristics such as ethnic foods. The novelty in functional foods is, however, related to new components and the effects the foods have, although the food in itself looks very much the same as any conventional food product. This may also cause suspicion in consumers as the functionality is not a property that can be experienced. Also, the technology required to produce these functional foods may raise doubts in consumers' minds.

15.5   Consumer attitudes to functional foods

Very few studies have looked at consumer responses to so-called functional foods, although functional foods are a category of products that is seen to have a clear growth potential in the near future (Stanton et al., 2001). In Europe, gut health has attracted greatest interest, with dairy products as the main category providing probiotics and prebiotics to achieve this aim (Hilliam, 2002).

Consumers have become more aware of foods that have health benefits attached to them (Bogue and Ryan, 2000; Bech-Larsen et al., 2001). Some studies have shown rather strong reservations towards foods with health-related claims (Jonas and Beckmann, 1998; Consumers' Association, 2000), but others have revealed positive consumer attitudes towards claims (Bogue and Ryan, 2000; Bech-Larsen et al. 2001). In a UK study, consumers (n = 100) were asked opinions about three example products with health-related claims (Consumers' Association, 2000). The respondents were best aware of cholesterol-lowering spread (72%) and probiotic drink (58%), but the orange juice with bone benefits was recognised only by a few respondents (7%). The functional foods were perceived to be processed everyday foods, which are convenient to use. They were not considered to be medicines, but the orange juice and probiotic drink were regarded as supplements, whereas two-thirds of respondents did not think the cholesterol-lowering spread was a supplement. Although the benefits of functional foods were noticed, the opinion was that, with all example products, something else would have done the job as well. The claims were presented in two forms: improving health or preventing illness. The improvement claims received bigger approval than prevention claims. Allowing respondents to have a closer examination of the product had an impact on opinion. The benefits of yoghurt and spread became more agreeable, whereas fewer respondents agreed with the preventive power of orange juice after studying the package. The product-dependent differences were large, so that 82% of the interviewees agreed that cholesterol-lowering spread would improve specific health problems, whereas the agreement figure was 66% for the probiotic drink and 49% for orange juice.

Responses to a different kind of health-related claims were studied in three countries (Denmark, Finland, the USA; $n = 1533$) using the same method (Bech-Larsen et al., 2001). In addition to the no claim condition, two types of claims were used, namely physiological enhancement and reduction of risk of disease claims, which were both applied to three types of products: juice, yoghurt and spread. The components producing the effects of these claims were omega-3 fatty acids or oligosaccharides. Adding the claim in the products increased the perceived whole-someness of the products, regardless of the type of the claim. The effect was product-dependent, so that in juice and yoghurt the enriched products were rated as less wholesome, whereas in spread the situation was the other way around. In intentions to buy, the 'no enrichment' condition was favoured by the Danish and the American respondents, while in Finland the effect was product dependent. Juice and yoghurt were more highly rated when they were not enriched, but in spread the enrichment increased the intention to buy. Claim as such was a positive factor in all countries.

In Finland, an indirect approach was employed to study consumers' responses to functional foods. Instead of asking opinions about functional foods, consumers were asked for their impressions of buyers of functional foods. Respondents rated their impressions of a 40-year-old person who had either functional foods or conventional foods in his or her shopping list. The lists also contained filler items, either with positive or neutral health images, and respondents did not realise that functional foods were the target foods under examination. This approach was selected to avoid any moral and social issues that influence our health-related responses. Users of functional foods were considered more innovative than users of conventional foods and also more disciplined than those purchasing conven-tional foods with a neutral health image, but less disciplined than those choosing traditional options with a high health image (Saher et al., 2002). These results indicate that functional products present a new way of using food to improve health, which differs from the traditional way. Functional foods are regarded as a convenient way to acquire required nutrients and other beneficial compounds (Poulsen, 1999), but one that may not be required if the diet is already more varied and balanced (Bogue and Ryan, 2000).

Functional foods with health-related claims tend to be more expensive than their traditional counterparts. One of the questions raised is whether this higher price will be an obstacle for the success of functional foods. The foods are tailored for certain purposes and groups of consumers, and they are not meant as mass products. Within such a new product category, consumers' willingness to pay is hard to estimate. There seems to be a group of consumers who reported that they are ready to pay a premium for functional foods (Poulsen, 1999; Bogue and Ryan, 2000) or that the price had very little impact on intention to buy (Bech-Larsen, 2001).

15.6 Future trends

The ability to provide benefits to consumers is a crucial factor in the future success of functional foods. As the benefits are not directly sensed in the body, the credibility of health-related messages becomes the key factor. Functional dairy products have a immense potential to offer various health benefits for the consumer as described in other chapters of this book. Some of these may be easy for the consumers to accept, as they enhance the existing beliefs of the beneficial effects dairy products have, but some offer rather specialised high-technology solutions for targeted populations.

Substantiating the health effects and ensuring their presence in the final product requires scientifically sound research (Kwak & Jukes, 2001). The true challenge is to make this knowledge produced by science understandable and feasible for the consumer. So-called everyday thinking, which we apply in solving everyday problems, such as choices among food products, follows different rules from scientific thinking. The scientific information about the health benefits is likely to be probabilistic in its nature, which means that translating these messages for the consumer is important. Consumers need to have confidence in the source and content of the information. Producers of functional foods have to be careful in ensuring and maintaining this trust.

As approximation is one of the typical features in consumers' way of handling information, the health-related messages in products have to be simple and clear enough to be understood, but at the same time they need to be distinguishable from other messages. As an example, scientists and manufacturers invest a lot of effort in establishing the functional properties of different probiotic strains and products, but consumers may find it hard to make the difference between these strains and their outcomes. The targeted functions require also targeted messages to consumers of varying ages and cultural backgrounds. The issue of communicating health benefits to consumers effectively remains the ultimate challenge in the future success of functional foods. Only products that will be accepted and consumed can have a role in improving human health.

15.7   Sources of further information and advice

Further information of functional dairy products and their benefits can be obtained from manufacturers of the products.

More information about EU-funded research on consumer perceptions of functional foods can be obtained from two websites: Flair-Flow project disseminates information on all food-related research at the EU (www.flair-flow.com) and ProEuHealth-cluster consists of eight projects concentrating on gut heath (http://proeuhealth.vtt.fi).

15.8   References

ARVOLA A, URALA N and LÄHTEENMÄKI L (2001), 'Health claims as possible promoters of product acceptability', 4th Pangborn Sensory Science Symposium, Dijon, July 2001 (abstract).

ASP N-G and TROSSING M (2001), 'The Swedish code on health-related claims in action – extended to product-specific physiological claims', *Scand J Nutr*, **45**, 189.

BECH-LARSEN T, GRUNERT K G and POULSEN J B (2001), *The Acceptance of Functional Foods in Denmark, Finland and the United States*. Working Paper no. 73, MAPP, Aarhus, Denmark.

BOGUE J and RYAN M (2000*), Market-oriented New Product-development: Functional Foods and the Irish Consumer*. Agribusiness Discussion Paper no. 27, University College Cork.

BRINBERG D, AXELSON M L and PRICE S (2000), 'Changing food knowledge, food choice, and dietary fiber consumption by using tailored messages, *Appetite*, **35**, 35–43.

CONSUMERS' ASSOCIATION (2000), *Functional Food – Health or Hype?* London, Consumers' Association.

COUNCIL OF EUROPE (2001), Council of Europe's Policy Statements Concerning Nutrition, Food Safety and Consumer Health, *Guidelines Concerning Scientific Substantiation of Health-related Claims for Functional Food*. Technical Document, 2 July 2001 (http://www.coe.int/soc-sp).

DIPLOCK A T, AGGETT P J, ASHWELL M, BORNET F, FERN EB and ROBERTFROID MB (1999), 'Scientific concepts of functional foods in Europe: Consensus Document', *Brit J Nutr*, **81**(4), S1–S27.

FREWER L J, HOWARD C, HEDDERLEY D and SHEPHER R (1997) 'Consumer attitudes towards diferent food-processing technologies used in cheese production – the influence of consumer benefit', *Food Qual Pref*, **8**, 271–280.

GRUNERT K G, BECH-LARSEN T and BREDAHL L (2000), 'Three issues in consumer quality perception and acceptance of dairy products', *Int Dairy J*, **10**, 575–584.

HILLIAM M (2002), 'Functional food update', *The World of Food Ingredients*, **April/May**, 52–53.

JONAS M S and BECKMANN S C (1998), *Functional Foods: Consumer Perceptions in Denmark and England*. Working Paper no. 55, MAPP, Aarhus, Denmark.

KWAK N S and JUKES D (2001), 'Issues in the substantiation process of health claims', *Critical Rev Food Sci Nutr*, **41**, 465–479.

LÄHTEENMÄKI L (2000), 'Consumers' perspectives of functional foods'. In: Alander M and

Mattila-Sandholm T, *Functional Foods for EU Health in 2000*, VTT Symposium 198, Espoo, 19–21.

LAPPALAINEN R, KEARNEY J and GIBNEY M (1998), 'A Pan EU survey of consumer attitudes to food, nutrition and health: an overview', *Food Qual Pref*, **9**, 467–478.

MARGETTS B M, MARTINEZ J A, SABA A, HOLM L and KEARNEY M (1997), 'Definitions of "healthy" eating: a pan-EU survey of consumer attitudes to food, nutrition and health', *Eur J Clin Nutr*, **51** (Suppl. 2), S23–S29.

PAUL G L, INK S L and GEIGER C J (1999), 'The Quaker Oats health claim: a case study', *J Nutraceuticals, Funct and Med Foods*, **1**(4), 5–32.

PLINER P and HOBDEN K (1992), 'Development of a scale to measure the trait of food neophobia in humans', *Appetite*, **19**, 105–120.

POULSEN J B (1999), *Danish Consumers' Attitudes Towards Functional Foods.* Working paper no. 62, MAPP, Aarhus, Denmark.

ROININEN K, LÄHTEENMÄKI L and TUORILA H (1999), 'Quantification of consumer attitudes to health and hedonic characteristics of foods', *Appetite*, **33**, 71–88.

ROININEN K, TUORILA H, ZANDSTRA E H, DE GRAAF C, VEHKALAHTI K, STUBENITSKY K and MELA D J (2001), 'Differences in health and taste attitudes and reported behavior among Finnish, Dutch and British consumers: a cross-national validation of the health and taste attitude scales (HTAS)', *Appetite*, **37**, 33–45.

SAHER M, ARVOLA A, LINDEMAN M and LÄHTEENMÄKI L (2002), 'Impressions of functional food consumers', submitted to *Appetite*.

STANTON C, GARDINER G, MEEHAN H, COLLINS K, FITZGERALD G, LYNCH P B and ROSS P (2001), 'Market potential for probiotics', *Am J Clin Nutr*, **73** (Suppl), 476S–483S.

UK JOINT HEALTH CLAIMS INITIATIVE (2002), http://www.jhci.co.uk.

WANSINK B (2001). 'When does nutritional knowledge relate to the acceptance of functional food?', Research brief, http://www.consumerpsychology.com/insights/researching consumers.htm, Campaign, University of Illinois.

WRICK K L (1995), 'Consumer issues and expectations for functional foods', *Critical Rev Food Sci*, **35**, 167–173.

YEOMANS M R, JACKSON A, LEE MD, STEER B, TINLEY E, DURLACH P and ROGERS P J (2000), 'Acquisition and extinction of flavour preferences conditioned by caffeine in humans' *Appetite*, **35**, 131–141.

16

European research in probiotics and prebiotics: the PROEUHEALTH cluster

T. Mattila-Sandholm, L. Lähteenmäki and M. Saarela, VTT Biotechnology, Finland

16.1 Introduction: research projects within the PROEUHEALTH cluster

Health is one of the main reasons behind food choices, but other factors are also taken into consideration by the consumer (Sloan, 2000). As pleasantness and sensory quality can be experienced directly, they are known to be the essential factors in repeated choices. Even with foods aimed for health, the taste has to be good to secure repeated choice. Probiotics modifying the gut microbiota have physiological effects that cannot be experienced directly. The credibility of health-related messages is therefore critical to guarantee that the product has a reward value for the consumer (Bech-Larsen et al., 2001; Grunert et al., 2000). Consumers are selective in their attention so that messages that are relevant to us and congruent with our earlier beliefs are easily accepted. Therefore, food-related messages have to correspond with our existing beliefs and motivation to use a product. Preventing a possible stomach upset has little relevance to a healthy person, but can make a vast difference for someone suffering from inflammatory bowel disease (IBD) or someone who easily catches travellers' diarrhoea (Salminen et al., 1995, 1998a, b, c).

Keeping this in mind, microbes with a positive health impact are, and will remain, an important functional ingredient for years to come. New strains will be identified and foods will be developed to fulfil the needs of specific consumer groups. Increased understanding of interactions between gut microbiota, diet and the host will open up possibilities of producing novel ingredients for nutritionally optimised foods, which promote consumer health through microbial activities in the gut (Salminen *et al.*, 1998a, b, c; Naidu *et al.*, 1999).

The acceptance of probiotic, prebiotic and synbiotic products in the future will depend on the solid proof of the health benefits they promise at the moment. Therefore, research that provides the scientifically sound evidence to back up the health claims is needed. Making the knowledge produced by science comprehensible to the consumer is also a major challenge as consumers want clear-cut opinions about claims, whereas scientific thinking deals with probabilities and degrees of uncertainty. Consumers' trust in information depends on the source and content of a message. The critical point is that, with the right approach, producers of probiotic foods can gradually build and ensure consumer trust (Mattila-Sandholm and Saarela, 2000; Mattila-Sandholm *et al.*, 2000; Alander *et al.*, 2001; Crittenden *et al.*, 2001).

A number of steps are essential in the development of efficacious probiotic and prebiotic functional foods (Fig. 16.1). A prerequisite for mechanistic studies of probiotic action is an understanding of the composition and activity of the intestinal microbiota as well as interactions with the host in both healthy and diseased individuals. High-throughput molecular methods are required to examine the intestinal microbiota and to track the location and activity of probiotic strains in the intestinal tract. An understanding of the mechanisms by which probiotics exert beneficial effects on human health allows selection of strains with appropriate traits for hypothesis-driven clinical studies. The safety of new strains must be demonstrated before they are used in human clinical trials. An important area of research is the development of technologies to maximise the stability of functional traits of probiotics during manufacture, formulation, storage and in the intestinal tract. Additionally, the efficacy of products may be enhanced by exploiting synergistic interactions between functional ingredients as is potentially the case with synbiotics. Finally, an understanding of the most appropriate methods to communicate the benefits of the functional foods to consumers and the influence of health messages on consumer choice is essential to ensure that products are appropriately applied and targeted to benefit specific populations (Mercenier and Mattila-Sandholm 2001; Mattila-Sandholm *et al.*, 2002).

The Food, GI-tract Functionality and Human Health (PROEUHEALTH) Cluster brings together eight complementary, multicentre interdisciplinary research projects (Fig. 16.1). All have the common aim of improving the health and quality of life of European consumers. The collaboration involves 64 different research groups from 16 different European countries and is coordinated by leading scientists (Table 16.1). The research results from the cluster are disseminated through annual workshops and through the activities of three different platforms: a science, an industry and a consumer platform (Fig. 16.2 and website: http:// proeuhealth.vtt.fi). The cluster started in 2001 and will end in 2005.

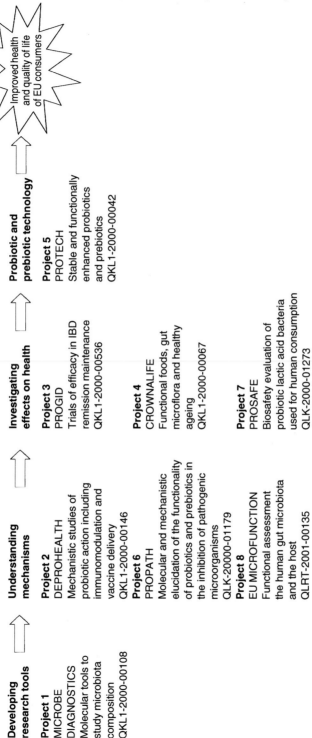

Developing research tools

Project 1
MICROBE DIAGNOSTICS
Molecular tools to study microbiota composition
QKL1-2000-00108

Understanding mechanisms

Project 2
DEPROHEALTH
Mechanistic studies of probiotic action including immunomodulation and vaccine delivery
QKL1-2000-00146

Project 6
PROPATH
Molecular and mechanistic elucidation of the functionality of probiotics and prebiotics in the inhibition of pathogenic microorganisms
QLK-20000-01179

Project 8
EU MICROFUNCTION
Functional assessment the human gut microbiota and the host
QLRT-2001-00135

Investigating effects on health

Project 3
PROGID
Trials of efficacy in IBD remission maintenance
QKL1-2000-00536

Project 4
CROWNALIFE
Functional foods, gut microflora and healthy ageing
QKL1-2000-00067

Project 7
PROSAFE
Biosafety evaluation of probiotic lactic acid bacteria used for human consumption
QLK-2000-01273

Probiotic and prebiotic technology

Project 5
PROTECH
Stable and functionally enhanced probiotics and prebiotics
QKL1-2000-00042

Improved health and quality of life of EU consumers

Fig. 16.1 The food, GI-tract functionality and human health cluster.

Table 16.1   Research groups in the PROEUHEALTH cluster

**Cluster Coordinator: Prof. Tiina
Mattila-Sandholm**
VTT Biotechnology
P.O. Box 1500
FIN-02044 VTT
Finland
Phone: +358 9 456 5200,
Mobile: +358 50 552 7243
Fax: +358 9 455 2103
E-mail: tiina.mattila-sandholm@vtt.fi

**EU Commission, Scientific Officer:
Dr Jürgen Lucas**
European Commission
Research DG, SDME 8-12
Rue de la Loi 200
B-1049 Brussels
Belgium
Phone: +32 2 296 4152
Fax: +32 2 296 4322
E-mail: jurgen.lucas@cec.eu.int

Consumer Platform: Dr Liisa Lähteenmäki
VTT Biotechnology
P.O. Box 1500
FIN-02044 VTT
Finland
Phone: +358 9 456 5965
Fax: +358 9 455 2103
E-mail: liisa.lahteenmaki@vtt.fi

Industry Platform: Prof. Charles Daly
University College Cork
Faculty of Food Science and Technology
Cork
Ireland
Phone: +353 21 902 007
Fax: +353 21 276 398
E-mail: dean.food@ucc.ie

Science Platform: Prof. Willem de Vos
Wageningen University
Laboratory of Microbiology
Department of Agrotechnology and Food
Sciences
Hesselink van Suchtelenweg 4
PO Box 8033
6703 CT Wageningen
The Netherlands

Phone: +31 317 483 100
Fax: +31 317 483 829
E-mail: willem.devos@algemeen.micr.wau.nl

Dr Annick Mercenier (DEPROHEALTH)
Bioscience Department
Nestlé Research Centre
Box 44, CH-1000
Lausanne 26
Switzerland
Phone: +41 21 785 8466
Fax: +41 21 785 8549
E-mail: annick.mercenier@rdls.nestle.com

Dr Joël Dore (CROWNALIFE)
INRA
Unité d'Ecologie et Physiologie du Système
Digestif
Domaine de Vilvert
78352 Jouy-en-Josas Cedex
France
Phone: +33 1 3465 2709
Fax: +33 1 3465 2492
E-mail: dore@jouy.inra.fr

Prof. Fergus Shanahan (PROGID)
National University of Ireland
University College Cork
Department of Microbiology and Medicine
Western Road
Cork
Ireland
Phone: +353 21 490 2843
Fax: +353 21 427 6318
E-mail: f.shanahan@ucc.ie

Prof. Luc De Vuyst (PROPATH)
Vrije Universiteit Brussel
Research Group of Industrial Microbiology
Fermentation Technology and Downstream
Processing
Pleinlaan 2
1050-Brussels
Belgium
Phone: +32 2 629 3245
Fax: +32 2 629 2720
E-mail: ldvuyst@vub.ac.be

Table 16.1 cont'd

Project coordinators: Prof. Michael Blaut
(MICROBEDIAGN0STICS)
Deutsches Institut für Ernährungsforschung
Potsdam-Rehbrücke
Gastrointestinale Mikrobiologie
Arthur-Scheunert-Allee 114-116
14558 Bergholz-Rehbrücke
Germany
Phone: +49 33 2008 8470
Fax: +49 33 2008 8407
E-mail: blaut@www.dife.de

Prof. Dietrich Knorr (PROTECH)
Technische Universität Berlin
Lebensmittelbiotechnologie und -prozesstechnik
Königin-Luise-Strasse 22
14195 Berlin
Germany
Phone: +49 30 3147 1250
Fax: +49 30 8327 662
E-mail: dietrich.knorr@tu-berlin.de

Prof. Herman Goossens (PROSAFE)
Universiteit Antwerpen
Universitair Ziekenhuis
Department of Medical Microbiology
Wilrijkstraat 10
2650 Edegem
Belgium
Phone: +32 3 821 3789
Fax: +32 3 825 4281
E-mail: herman.goossens@uza.uia.ac.be

Prof. Glenn Gibson
(EU&MICROFUNCTION)
University of Reading
School of Food Biosciences
Whiteknights
 Reading RG6 6AP
United Kingdom
Phone: +44 1189 357223
Fax: +44 1189 357222
E-mail: g.r.gibson@reading.ac.uk

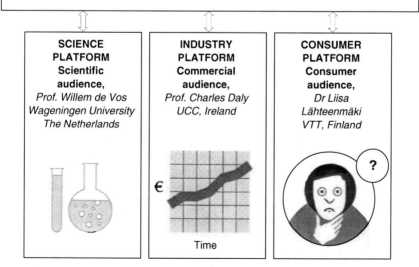

Fig. 16.2   Three platforms disseminating the aims and findings of the cluster to targeted audiences.

16.2   Developing research tools: MICROBE DIAGNOSTICS

16.2.1   Development and application of high throughput molecular methods for studying the human gut microbiota in relation to diet and health – MICROBE DIAGNOSTICS

The microbial gut ecosystem has a major impact on the health and well-being of humans. Since the make-up and activity of the gut microorganisms, collectively known as microbiota, are highly influenced by diet, the project is designed to better understand impact of microbiota on human health and to provide easy-to-use methods for monitoring its composition. Novel methods are developed to understand and exploit the nutrition-driven impact of the human gut microbiota on health. Since the presently used methods for analysing the intestinal microbiota are time-consuming and tedious, high-throughput methods for the automated detection of fluorescently labelled cells based on microscopic image analysis, flow cytometry and DNA arrays will be developed. *In situ* detection methods for monitoring bacterial gene expression at the cellular level will be developed to monitor the impact of dietary constituents on the transcription of bacterial genes. Samples from human populations will be analysed with the developed methods to identify important factors of microbiota development, composition and activity (Vaughan *et al.*, 1999; Satokari *et al.*, 2001; Simmering and Blaut 2001; de Vos, 2001; Heilig *et al.*, 2002).

Objectives

The project is aimed at developing, refining, validating and automating the most advanced molecular methods for monitoring the composition of the gut microbiota and bacterial gene expression in selected human populations in response to diet and life style. The specific objectives of the project are:

- to improve and facilitate gut microbiota monitoring with molecular methods;
- to understand antagonistic and synergistic interactions of the intestinal microbiota in response to nutrition; and
- to find links between major dysfunctions and the intestinal microbiota.

Future results and achievements

The project is expected to provide methods that allow the rapid detection of intestinal bacteria and their activities. The application of the developed methods will provide baseline data on the intestinal microbiota composition in response to origin and lifestyle. The project will also provide fundamental information on the biodiversity and phylogeny of the human gut microbiota. The obtained data will be used to develop a mechanistic concept for diet-induced microbiota development and to define the role of the intestinal microbiota in disease development. The project thereby contributes to investigate the role and impact of food on physiological function, the development of foods with particular benefits, links between diet and chronic diseases.

The project will provide technological solutions for the rapid detection of gut microorganisms with a high potential for a wide range of applications. The screening of gut microbiota composition will be applied in relevant biomedical research areas. It will facilitate the diagnosis of a disturbed microbiota and the development of rational therapies based on knowledge of disease mechanism. It will support the development of functional foods aimed at improved gastrointestinal function and other health benefits.

16.3   Understanding mechanisms of actions: DEPROHEALTH, PROPATH and EU MICROFUNCTION

16.3.1   Probiotic strains with designed health properties – DEPROHEALTH

Lactic acid bacteria (LAB) are well known for their extensive use in the production of fermented food products. The potential health benefits they may exert in humans were intensively investigated during the last century. However, the mechanisms underlying the health-promoting traits attributed to LAB, especially lactobacilli, remain mostly unknown and this has impaired the rational design of probiotic screening methods with accurate predictive value. This project aims at establishing a correlation between *in vitro* tests and mouse models mimicking important human intestinal disorders such as IBD and *Helicobacter pylori* and rotavirus infections. As these diseases correspond to major public health problems, a second generation of probiotic strains with enhanced prophylactic or therapeutic properties will also be designed during the project. These designed strains and the isogenic parental ones will be used to unravel mechanisms involved in the immunomodulation capacity of specific *Lactobacillus* strains (Steidler *et al.*, 2000; Grangette *et al.* 2001).

Objectives
The general aim of this project is to acquire knowledge about the molecular factors affecting the immunomodulation and/or immunogenicity of selected probiotic lactobacilli to allow for developing isolates with enhanced protective or therapeutic effect. The objectives of the proposal are to unravel mechanisms and identify key components of the immunomodulation capacity of probiotic lactobacilli and to design a so-called second generation of probiotic strains with enhanced properties against gastrointestinal disorders.

Two types of gastrointestinal diseases are targeted: inflammations such as IBD and infections such as those caused by *H. pylori* and rotavirus. For each of them, atherapeutic or prophylactic (i.e. vaccine) recombinant probiotic strains will be constructed and tested in relevant animal models (i.e. mimicking the human disease) to evaluate their capacity to induce or modulate the immune response in the proper way. This data will be correlated to *in vitro* testing of the immunomodulation capacity with an attempt to identify/develop screening methods that, in the future, will allow isolating efficient probiotic strains targeted to specific applications. Recombinant DNA technology will also be used to assess the importance of specific bacterial cell wall components and adhesion factors in immunomodulation.

Future results and achievements
Two types of modified probiotic strains will be constructed: mutant strains affected in their cell wall composition and adhesion proteins as well as recombinant strains with enhanced therapeutic or protective properties, focusing on gastrointestinal diseases of inflammatory or infectious origin. The final goal would be to prove that designed probiotic strains could be used as original therapeutic agents as, if successful, they would lead to novel anti-inflammatory treatments or oral vaccines against *H. pylori* and rotavirus. It is also expected that, by the end of the study, major progress will have been made towards the rational design of simplified probiotic screening methods with accurate predictive value.

16.3.2   Molecular analysis and mechanistic elucidation of the functionality of probiotics and prebiotics in the inhibition of pathogenic microorganisms to combat gastrointestinal disorders and to improve human health – PROPATH

Recently, much attention has been paid to the health-promoting properties of lactobacilli and bifidobacteria. Probiotics and prebiotics are the driving forces of the functional foods market. However, a major problem is that many of these health-promoting properties are still questioned. For instance, the fundamental basis of the inhibition of Gram-negative pathogenic bacteria – such as the enterovirulent diarrhoeagenic *Salmonellae*, and *H. pylori* causing gastritis and gastric ulcer disease – by probiotic LAB has not been elucidated. This project will focus on the identification of the responsible compounds. In addition, the mechanism of the inhibition of Gram-negative pathogens by probiotic lactobacilli and bifidobacteria will be studied using co-culture models (Coconnier *et al.*, 1998, 2000; Zamfir *et al.*, 1999; Avonts and De Vuyst, 2001).

Objectives
The aim is to obtain a selection of probiotic lactobacilli and bifidobacteria that display a clear inhibition of diarrhoeagenic Gram-negative pathogenic bacteria and *H. pylori*. Thereafter, the project focuses on identifying the metabolite(s) responsible for the inhibition and/or killing of Gram-negative pathogenic bacteria as well as in having the conditions and kinetics of the production of antimicrobials active towards Gram-negative pathogens and to predict their *in vivo* action. One of the aims will be to establish co-culture models (simulated gut fermentations, human cell lines, animal models) and to study the interaction between inhibitory strains of LAB and Gram-negative pathogens causing gastrointestinal disorders. Finally, the selected probiotic LAB will be tested in clinical studies.

Future results and achievements
The results of the molecular analysis and mechanistic elucidation of the functionality

of probiotics and prebiotics in the inhibition of pathogenic microorganisms will result in:

* a project culture collection of probiotic strains of lactobacilli and bifidobacteria, together with data on their characteristic and inhibitory spectrum;
* a molecular typing method for selected probiotic strains;
* identified compounds responsible for the inhibition of Gram-negative pathogenic bacteria;
* conditions of antimicrobial production in the gut environment; and
* co-culture models showing the inhibitory action by the probiotic strains.

16.3.3   Functional assessment of interactions between the human gut microbiota and the host – EU MICROFUNCTION

The aim of this project is to identify the effects of dietary modulation on the human gastrointestinal microbiota. The influence of probiotics, prebiotics and synbiotics will be ascertained. The main objectives are to clarify effects on the normal gut microbiota and on the host gastrointestinal function, as well as to determine mechanisms involved in pro-, pre- and synbiotic functionality. These objectives will be achieved through exploitation of model systems and state-of-the-art technology (Bengmark, 1998; Salminen *et al.*, 1998a, b, c; Collins and Gibson, 1999; Suau *et al.*, 1999, 2001; Steer *et al.*, 2000).

Objectives
The principal objectives are to assess the efficacy and safety of probiotics and prebiotics, to determine effective doses/combinations, to identify mechanisms of action and to investigate impacts on host function. An important aspect will be the development of new synbiotics and their use in a human trial. The work aims to identify the mechanisms of effect and to produce valuable information on the influence of dietary intervention on the activities of human gut microbiota. In addition, it provides essential means of validating probiotics and prebiotics, and will give information on the optimal combinational approach.

There is currently an imperative requirement to identify the realistic health outcomes associated with probiotic and prebiotic intake and, importantly, give rigorous attention towards determining their mechanisms of effect. This project will aim to do so through investigating probiotic and prebiotic influence on host functionality, including microbiological and physiological aspects.

Future results and achievements
The aim is to identify mechanistic interrelationships through fundamental scientific approaches. The following milestones will be achieved:

* efficient prebiotics and required dosage;
* active synbiotics;
* effects of functional foods on bacterial translocation;
* effects on host gene interactions;

- safety of functional foods determined;
- effects on selected health indices in humans.

16.4   Investigating effects on health: PROGID, CROWNALIFE and PROSAFE

16.4.1   Probiotics and gastrointestinal disorders: controlled trials of European Union patients – PROGID

This project will assess the efficacy of two previously selected probiotic micro-organisms, administered as dried fermented milk products, in alleviating the effects of IBD – Crohn's disease and ulcerative colitis. Specifically, two distinct long-term (12-month), large-scale, multicentred, randomised, double-blind, placebo-controlled feeding trials will be performed within a subset of the EU population suffering from these gastrointestinal disorders.

Background and objectives

In recent years, the International Organisation for the Study of Inflammatory Bowel Disease and others have published prominent articles highlighting the necessity for standards in the design of trials evaluating therapies in gastrointestinal disorders. In the assessment of probiotic products, however, most studies have been small, uncontrolled and poorly documented with imprecise definition of the end-points. In response, the European Commission has funded the PROGID project, which will evaluate two specific probiotic microorganisms in patients with ulcerative colitis or Crohn's disease from diverse geographical locations. The participating centres in these studies will assess the efficacy of *Bifidobacterium infantis* UCC35624 and *Lactobacillus salivarius* UCC118 in one year, randomised, double-blind, placebo-controlled trials for maintenance of remission. Both of the selected probiotic strains have a history of safety and efficacy in healthy adults and relapsed inflammatory bowel disease patients (Shanahan, 2000; Dunne, 2001; Dunne and Shanahan, 2002).

Future results and achievements

The anticipated results from the PROGID project include:

- the provision of qualitative and quantitative evidence that the evaluated probiotics may (or may not) have a role in maintaining remission of IBD;
- confirmation, or otherwise, of the involvement of specific members of the gastrointestinal microbiota as causative or contributory agents of IBD;
- generation of linear physiological and immunological data relevant to the disease and remission states of IBD;
- creation of a greater awareness among the EU population of functional foods and their potential benefits in maintaining healthy lifestyles; and
- the establishment of a repository of biological samples obtained from across the EU.

16.4.2   Functional food, gut microflora and healthy ageing – CROWNALIFE

Elderly people represent an increasing fraction of the European population. Their higher susceptibility to degenerative and infectious diseases leads to rising public health and social concerns. Appropriate preventive nutrition strategies can be applied to restore and maintain a balanced intestinal microbiota exerting protective functions against the above disorders. The project is based on the application of selected biomarkers in human studies and we will identify the structural and functional specificity of the elderly's intestinal microbiota across Europe. Using this baseline information, we will investigate functional food-based preventive nutrition strategies, aiming to beneficially affect the functional balance of the elderly's intestinal microbiota. Expected outcomes include nutritional recommendations as well as new concepts and prototype functional foods, specifically adapted for health benefits to the elderly population (Kruse *et al.*, 1999; Liévin *et al.*, 2000; Schwiertz *et al.*, 2000; Sghir *et al.*, 2000; Marteau *et al.*, 2001).

Objectives
The overall objective is to improve the quality of life of the elderly throughout the third age, with emphasis on the preservation of the period of independence recognised as the crown of life. The focus is on preventive nutrition and the application of functional food to derive health benefits for the ever-increasing European elderly population. Based on hypothesis-driven human studies, the specific objectives of the project are:

* to assess structural and functional alterations of the intestinal microbiota with ageing and across Europe; and
* to validate functional foods-based preventive nutrition strategies to restore and maintain a healthy intestinal microbiota in the elderly.

Implementations include nutritional recommendations as well as new concepts and prototype functional food specifically adapted for health benefits to the elderly population.

Future results and achievements
The results will establish six steps:

* the assessment of the gut microbiota diversity and composition in the European elderly, and identification of its alterations with ageing (baseline human study);
* the assessment of modulation of the intestinal microbiota (intervention human study with a functional food), and potential health benefits towards degenerative pathologies and infectious diseases;
* the improved health status of the ageing population via specific nutritional recommendations;
* the nutritional guidelines based on the complete assessment of a synbiotic product;
* the design and provision of adapted food supplements directed towards intestinal microbial function; and

- the validation of processes and rationale for the design of a new generation of functional foods to satisfy health benefits for the elderly, based on innovative technologies.

16.4.3   Biosafety evaluation of probiotic lactic acid bacteria used for human consumption – PROSAFE

Safety assessment is an essential phase in the development of any new pro- or prebiotic functional food. The safety record of probiotics is good, and lactobacilli and bifidobacteria have a long history of safe use. However, all probiotic strains must be evaluated for their safety before being used in human clinical studies and in functional food products. Conventional toxicology and safety evaluation alone is of limited value in the safety evaluation of probiotic bacteria. Instead, a multidisciplinary approach is necessary, involving contributions from pathologists, geneticists, toxicologists, immunologists, gastroenterologists and microbiologists (Salminen and von Wright, 1998; Salminen et al., 1998a, b, c; Sanders and Huis in't Veld, 1999; Crittenden et al., 2002).

Objectives
The safe use for human consumption of probiotic strains, selected LAB including lactococci, lactobacilli, pediococci and bifidobacteria, and other food-associated microorganisms such as enterococci, will be assured by proposing criteria, standards, guidelines and regulations. Furthermore, procedures and standardised methodologies of pre-marketing biosafety testing and post-marketing surveillance will be provided. The specific objectives of the project will include five stages:

- the taxonomic description of probiotics, LAB and other food-related microorganisms;
- the detection of resistance and horizontal transfer of antibiotic resistance genes;
- the careful analysis and definition of the nonpathogenic status of probiotic LAB;
- the immunological adverse effects of the studied bacteria in the experimental autoimmune encephalitis (EAE) mouse model; and
- the survival, colonisation and genetic stability of probiotic strains in the human gut.

Future results and achievements
Future results will finalise important goals relating to probiotic strains, selected LAB and other food-associated microorganisms and will consist of:

- establishing a culture collection and database;
- providing standardised methodologies to detect antibiotic resistance;
- investigating the potential virulence properties and their association with clinical disease and results obtained in rat endocarditis models;
- studying potential immunological adverse effects;
- analysing the genetic stability and colonisation of probiotic strains in the human gastrointestinal tract; and
- providing recommendations for biosafety evaluation of probiotic strains.

16.5   Probiotic and prebiotic technologies: PROTECH

16.5.1   Nutritional enhancement of probiotics and prebiotics: technology aspects on microbial viability, stability, functionality, and on prebiotic function – PROTECH

Maintaining the functional properties of probiotics during manufacture, formulation and storage are essential steps in delivering health benefits to consumers from these products. The overall objective of this project is to address and overcome specific scientific and technological hurdles that impact on the performance of functional foods based on probiotic–prebiotic interactions. Such hurdles include the lack of a strong knowledge on the primary factors responsible for probiotic viability, stability and performance. Limited information is available on the impact of processing and storage and of food matrices or food constituents on probiotic viability, stability and functionality. Furthermore, information about the interactions between probiotics and prebiotics in functional foods prior to consumption is also limited (Saarela *et al.*, 2000; Mattila-Sandholm *et al.*, 2001).

Objectives
This project has three general objectives:

- to explore effects of processing on the stability and functionality of probiotics and on the performance of prebiotics;
- to apply selected processing techniques for prebiotic modification to identify and optimise probiotic–prebiotic combinations; and
- to use the information generated as the basis for new process and product options.

As a result of achieving these objectives, selected prebiotics in combination with tailored manufacturing processes will not only contribute to probiotic functionality but also improve viability and stability of probiotic cultures within food matrices during processing and storage.

The project is divided into five specific objectives:

- **Effects on probiotic viability:** the aim is to consolidate quantitative data of processing effects on the physiology and viability of probiotic organisms. This will include effects during different fermentation processes, and freezing and drying strategies and also during passage through the GI tract. Effects on cell composition including stress responses, cell membrane composition and leakage, and cell viability will be measured. A viability model for probiotic cultures will be developed and validated.
- **Effects on probiotic stability:** experiments will be performed to investigate the influence of media composition and processing conditions on probiotic stability. This includes cryotolerance, resistance to dehydration, storage and genetic stability. Stability of the important probiotic characteristics of strains will also be investigated including acid and bile tolerance and adhesion to intestinal cells. New formulation techniques including micro-encapsulation will be assessed in real food matrices including dairy- and cereal-based foods.

- **Probiotic–prebiotic interaction:** the ultimate objective is to optimise probiotic–prebiotic interaction for maximum probiotic performance in culture, in food systems and in the GI tract. Special attention will be given to indigestible oligo- and polysaccharides with respect to fermentability and to the production of short chain fatty acids in animal models and *in vitro*.
- **Prebiotic function:** new prebiotics will be developed that are tailor-made for the stabilisation and optimum performance of probiotics. These will be examined for their physiological properties and their organoleptic and physicochemical properties will be technologically evaluated in real foods.
- **Probiotic functionality:** the effect of growth conditions and stress environments on the functionality of probiotic bacteria will be examined. Stresses tested will include high acid, bile, temperature, oxygen and osmotic tension. In addition, high hydrostatic pressure and controlled permeabilisation technologies will be evaluated for stress preconditioning of LAB, including bifidobacteria. Cross-protection against other stresses will be examined.

Future results and achievements

Expected achievements include the establishment of unique data sets that contain the identification of critical process parameters for probiotics and prebiotics and results from systematic studies suggesting means to overcome existing process and product limitations. The compilation of protocols for probiotic performance, prebiotic function, and probiotic–prebiotic interactions will also be provided. Further, probiotic viability models and functionality biomarkers will be established. In addition, optimisation of probiotic viability, stability in culture and in real food systems at a pilot scale, and the generation and modification of unique prebiotics, of probiotic interactions and of environmentally and processing induced functionality of probiotics will be attempted. Application of the expected results will lead to new process concepts for probiotics, for prebiotics and for probiotic–prebiotic combinations.

16.6   Consumers and the perceived health benefits of probiotics

Science-based knowledge on how probiotic bacteria can promote our well-being has increased in recent years. The success of probiotic products will be determined by consumers' willingness to buy and eat them. The perceived benefits in these new products are key factors for the consumer acceptance. Some of the benefits are relevant to a large group of people, but others are part of a well-defined target group. For ordinary consumers, preventing travellers' diarrhoea or helping with temporary gut disorders is a clear advantage, but still the motivation to use probiotics for patients suffering from irritable bowel syndrome (IBS) or IBD is likely to be higher if the probiotics can, at least partly, replace medication and reduce or even prevent symptoms (Poulsen, 1999; Roininen *et al.*, 1999).

Functional foods, i.e. products promising specified effects on physiological functions, represent a new kind of health message for the consumers. In traditional nutritional messages the emphasis is on diet and on avoiding or favouring certain types of foods rather than giving recommendations on particular products. Probiotic and other products containing the new kind of health message appear to better respond to the needs of different people than the idea of a nutritionally balanced diet (Roininen *et al.*, 1999; Sloan, 2000).

Although the most enthusiastic users of functional foods are believed to be middle-aged women who already follow healthy diets, in previous studies in Finland the general health interest, which measures people's willingness to comply with nutritional recommendations, has not been strongly linked with the willingness to use functional foods. The well-defined benefits promised by probiotic foods contain a positive message to consumers. The idea of gaining a benefit by using a product rather than avoiding a slow development of chronic diseases forms a different message. For Irish consumers the most interesting claims were related to heart diseases, cancer and bone health. Surprisingly, probiotics were not well recognised, even though yoghurt was among the most frequently used functional foods (Grunert *et al.*, 2000).

Developing probiotic foods also introduces a novelty aspect to the product. In general, consumers tend to be suspicious towards novelty in foods because it means uncertainty and thus threat. Therefore, the benefits promised by these new food products will be weighed against the perception of possible risks. Highly developed technologies are often involved in manufacturing the new probiotic foods. In several European countries healthiness has been associated with natural-ness. The need for highly advanced technology in the production of probiotic foods may lower the perceived naturalness, which may create distrust towards functional foods among consumers. Consumers seem to find functions that enforce the natural properties of a product more acceptable than those functions that are artificially added and in discordance with the earlier image of the product. Therefore, in addition to technological considerations, adding probiotics into food products requires understanding of the existing beliefs consumers have about these products (Lappalainen *et al.*, 1998; Poulsen, 1999; Bogue and Ryan, 2000; Lähteenmäki and Arvola, 2001).

Consumer responses to functional foods and their health-related messages require more research. Communicating gut health messages effectively to con-sumers of varying age and cultural backgrounds remains one of the key issues. Besides this, we need tremendous efforts towards communication and cooperation between medical doctors, nutritionists and pharmaceutics. As information is a vital factor, trust between all sides is required, so that messages will be given attention. The differences between functional properties of various strains are difficult to explain to consumers and tools for doing this should be created. Some consumers may find it hard to approve of food products working almost like medicines. Furthermore, food products need to have a place in the daily food system, whereas medicines are taken when instructed, and using food as a replacement for medi-cines requires adjustments in eating habits as well. Probiotic products can improve human health only if they are eaten frequently. To ensure this, the above issues

Table 16.2 The cluster projects

• Which bacterium is which?
• Second generation probiotics
• New therapeutics with potential against IBD
• A healthier retirement
• Keeping probiotics alive and healthy so that they can keep you healthy
• Probiotics can defend against bad bacteria
• Ensuring the safety of probiotic bacteria
• Why are probiotics effective?

must be addressed so that an appropriate set of products, which will be accepted and consumed, can be developed, such as the PROEUHEALTH cluster project topics in consumer language (Table 16.2).

16.7   Conclusions and future trends

Probiotics, prebiotics and synbiotics aimed at improving intestinal health currently represent the largest segment of the functional foods market in Europe, Japan and Australia. Evidence continues to emerge demonstrating that these ingredients have potential to improve human health in specific intestinal disorders. The European Commission, through its Fifth Framework Programme, is presently investing a substantial research effort in the intestinal microbiota, its interaction with its host, and methods to manipulate its composition and activity for the improvement of human health. Eight multicentre interdisciplinary research projects currently cover a variety of research topics required for the development of helpful probiotic foods, ranging from understanding probiotic mechanisms at a molecular level to developing technologies to ensure delivery of stable products and demonstrating safety and efficacy in specific disorders. This concerted research effort promises to provide an appreciable understanding of the human intestinal microbiota's role in health and disease, and new approaches and products to tackle a variety of intestinal afflictions.

The future research on food and microbes with a health impact will continue to develop, aimed specifically at:

• exploring the mechanisms of action of microbes and their health effects in the GI tract, especially in healthy individuals;
• developing sophisticated diagnostic tools for the gut microbiota and biomarkers for their assessing their functionality;
• examining the effects of food-derived bioactive compounds on GI diseases, GI infections and allergies;
• developing new therapeutic and prophylactic treatments for different patient and population groups;
• realising molecular understanding of immune modulation by bacteria with health effects;

- elucidating the role of colon microbiota in the conversion of bioactive compounds;
- analysing the effects of the metabolites to the colon epithelium or the effects after absorption;
- ensuring the stability of microbes with health effects and their bioactive compounds also in new types of food applications by developing feasible technologies; and finally
- providing the safety of the functional ingredients.

16.8   References

ALANDER M, MÄTTÖ J, KNEIFEL W, JOHANSSON M, CRITTENDEN R, MATTILA-SANDHOLM T and SAARELA M (2001), 'Effect of galacto-oligosaccharide supplementation on human faecal microflora and on survival and persistence of *Bifidobacterium lactis* Bb-12 in the gastrointestinal tract', *Int Dairy J*, **11**, 817–825.

AVONTS L and DE VUYST L (2001), 'Antimicrobial potential of probiotic lactic acid bacteria', *Mededelingen van de Faculteit Landbouwkundige en Toegepaste Biologische Wetenschappen Universiteit Gent*, **66/3b**, 543–550.

BECH-LARSEN T, GRUNERT K G and POULSEN J B (2001), *The Acceptance of Functional Foods in Denmark, Finland and the United States*, Working Paper no. 73, MAPP, Denmark.

BENGMARK S (1998), 'Ecological control of the gastrointestinal tract. The role of probiotic flora', *Gut*, **42**, 2–7.

BOGUE J and RYAN M (2000), *Market-oriented New Product Development: Functional Foods and the Irish Consumer*, Agribusiness Discussion Paper no. 27, Department of Food Economics, University College Cork, Ireland.

COCONNIER M F, LIÉVIN V, HEMERY E, LABOISSE C L and SERVIN A L (1998), 'Antagonistic activity of the human *Lactobacillus acidophilus* strain LB against *Helicobacter* infection *in vitro* and *in vivo*', *Appl Environ Microbiol*, **64**, 4573–4580.

COCONNIER M H, LIÉVIN V, LORROT M and SERVIN A L (2000), 'Antagonistic activity of *Lactobacillus acidophilus* LB against intracellular *Salmonella enterica* Serovar *Typhimurium* infecting human enterocyte-like Caco-2/TC-7 cells', *Appl Environ Microbiol*, **66**, 1152–1157.

COLLINS M D and GIBSON G R (1999), 'Probiotics, prebiotics and synbiotics: dietary approaches for the modulation of microbial ecology', *Am J Clin Nutr*, **69**, 1052–1057.

CRITTENDEN R, LAITILA A, FORSSELL P, MÄTTÖ J, SAARELA M, MATTILA-SANDHOLM T and MYLLÄRINEN P (2001), 'Adhesion of bifidobacteria to granular starch and implications in probiotic technologies', *Appl Environm Microbiol*, **67**(8), 3469–3475.

CRITTENDEN R, SAARELA M, OUWEHAND A, SALMINEN S, VAUGHAN E, DE VOS W, VON WRIGHT A and MATTILA-SANDHOLM T (2002), '*Lactobacillus paracasei* F19: survival, ecology and safety in the human intestinal tract', *Microb Ecol Health Dis*, **3**, 22–26.

DE VOS W M (2001), 'Advances in genomics for microbial food fermentations and safety', *Current Opin Biotechnol*, **12**(5), 493–498.

DUNNE C (2001), 'Adaptation of bacteria to the intestinal niche – probiotics and gut disorder', *Inflam Bowel Dis*, **7**(2), 136–145.

DUNNE C and SHANAHAN F (2002), 'The role of probiotics in the treatment of intestinal infections and inflammation', *Curr Opin Gastroenterol*, **18**(1), 40–45.

GRANGETTE C, MÜLLER-ALOUF H, GOUDERCOURT D, GEOFFROY M-C, TURNEER M and MERCENIER A (2001), 'Mucosal immune responses and protection against tetanus toxin after intranasal immunization with recombinant *Lactobacillus plantarum*', *Infect Immun*, **69**, 1547–1553.

GRUNERT K, BECH-LARSEN T and BREDAHL L (2000), 'Three issues in consumer quality perception and acceptance of dairy products', *Int Dairy J*, **10**, 575–584.

HEILIG H G, ZOETENDAL E G, VAUGHAN E E, MARTEAU P, AKKERMANS A D and DE VOS W M (2002), 'Molecular diversity of *Lactobacillus* spp. and other lactic acid bacteria in the human intestine as determined by specific amplification of 16S ribosomal DNA', *Appl Environ Microbiol*, **68**(1), 114–123.

KRUSE H P, KLEESSEN B and BLAUT M (1999), 'Effects of inulin on faecal bifidobacteria in human subjects', *Brit J Nutr*, **82**, 375–382.

LÄHTEENMÄKI L and ARVOLA A (2001), 'Food neophobia and variety seeking – consumer fear or demand for new food products', in Frewer L J, Risvik E, and Schifferstein H, *Food, People and Society. A European Perspective of Consumer Food Choices*, Springer-Verlag, Berlin, 161–175.

LAPPALAINEN R, KEARNEY J and GIBNEY M A (1998), 'Pan-European survey of consumers' attitudes to food, nutrition and health: an overview', *Food Qual Pref*, **9**, 467–478.

LIÉVIN V, PEIFFER I, HUDAULT S, ROCHAT F, BRASSART D, NEESER J-R and SERVIN A L (2000), '*Bifidobacterium* strains from resident infant human gastrointestinal microflora exert antimicrobial activity', *Gut*, **47**, 646–652.

MARTEAU P, POCHART P, DORÉ J, MAILLET C, BERNALIER A and CORTHIER G (2001), 'Comparative study of the human cecal and fecal flora', *Appl Environ Microbiol*, **67**, 4939–4942.

MATTILA-SANDHOLM T and SAARELA M (2000), 'Probiotic functional foods'. In: Gibson G R and Williams C M F, *Functional Foods – Concept to Product*, Woodhead Publishing Limited, 287–313.

MATTILA-SANDHOLM T, BLUM S, COLLINS J K, CRITTENDEN R, DE VOS W, DUNNE C, FONDEN R, GRENOV B, ISOLAURI E, KIELY B, MARTEU P, MORELLI L, OUWEHAND A, RENIERO R, SAARELA M, SALMINEN S, SAXELIN M, SCHIFFRIN E, SHANAHAN F, VAUGHAN E and VON WRIGHT A (2000), 'Probiotics: towards demonstrating efficacy', *Trends Food Sci Technol*, **10**, 393–399.

MATTILA-SANDHOLM T, MYLLÄRINEN P, CRITTENDEN R, MOGENSEN G, FONDEN R and SAARELA M (2001), 'Technological challenges for future probiotic foods', *Int Dairy J*, **12**, 1–10.

MATTILA-SANDHOLM T, BLAUT M, DALY C, DE VUYST L, DORÉ J, GIBSON G, GOOSSENS H, KNORR D, LUCAS J, LÄHTEENMÄKI L, MERCENIER A, SAARELA M, SHANAHAN F and DE VOS W M (2002), 'Food, GI-tract functionality and human health cluster: PROEUHEALTH', *Microb Ecol Health Dis*, **14**, 65–74.

MERCENIER A and MATTILA-SANDHOLM T (2001), 'The food, GI-tract functionality and human health European research cluster, PROEUHEALTH', *Nutr Metab Cardiovas Dis*, **11**(4), 1–5.

NAIDU A S, BIDLACK W R and CLEMENS R A (1999), 'Probiotic spectra of lactic acid bacteria (LAB)', *Crit Rev Food Sci Nutrit*, **38**(1), 13–126.

POULSEN J B (1999), *Danish Consumers' Attitudes Towards Functional Foods*, Working Paper no. 62, MAPP, Aarhus, Denmark.

ROININEN K, LÄHTEENMÄKI L and TUORILA H (1999), 'Quantification of consumer attitudes to health and hedonic characteristics of foods', *Appetite*, **33**, 71–88.

SAARELA M, MOGENSEN G, FONDEN G, MÄTTÖ J and MATTILA-SANDHOLM T (2000), 'Probiotic bacteria: safety, functional and technological properties', *J Biotechnol*, **84**, 197–215.

SALMINEN S and VON WRIGHT A (1998), 'Current probiotics – safety assured?', *Microb Ecol Health Dis*, **10**, 68–77.

SALMINEN S, ISOLAURI E and ONNELA T (1995), 'Gut flora in normal and disordered states', *Chemotherapy*, **41**, 5–15.

SALMINEN S, BOULEY C, BOUTRON-RUAULT M C, CUMMINGS J H, FRANCK A, GIBSON G R, ISOLAURI E, MOREAU M C, ROBERFROID M and ROWLAND I (1998a), 'Functional food science and gastrointestinal physiology and function', *British J Nutr*, **80** (Suppl), 147–

171.

SALMINEN S, OUWEHAND A C and ISOLAURI E (1998b), 'Clinical applications of probiotic bacteria', *Int Dairy J*, **8**, 563–572.

SALMINEN S, VON WRIGHT A, MORELLI L, MARTEAU P, BRASSART D, DE VOS W, FONDEN R, SAXELIN M, COLLINS K, MOGENSEN G, BIRKELAND S E and MATTILA-SANDHOLM T (1998c), 'Demonstration of safety of probiotics – a review', *Int J Food Microbiol*, **44**(1–2), 93–106.

SANDERS M E and HUIS IN'T VELD J H J (1999), 'Bringing a probiotic-containing functional food to the market: microbiological, product, regulatory and labeling issues', *Antonie van Leeuwenhoek*, **76**, 293–315.

SATOKARI R M, VAUGHAN E E, AKKERMANS A D, SAARELA M and DE VOS W M (2001), 'Polymerase chain reaction and denaturing gradient gel electroforesis monitoring of fecal bifidobacterium populations in a prebiotic and probiotic feeding trial', *Syst Appl Microbiol*, **24**(2), 227–231.

SCHWIERTZ A, LE BLAY G and BLAUT M (2000), 'Quantification of different *Eubacterium* spp. in human fecal samples with species-specific 16S rRNA-targeted oligonucleotide probes', *Appl Environ Microbiol*, **66**, 375–382.

SGHIR G, GRAMET A, SUAU V, ROCHET P, POCHART P and DORÉ J (2000), 'Quantification of bacterial groups within human fecal flora by oligonucleotide probe hybridization', *Appl Environ Microbiol*, **66**, 2263–2266.

SHANAHAN F (2000), 'Probiotics in inflammatory bowel disease: is there a scientific rationale?', *Inflam Bowel Dis*, **6**, 107–1115.

SIMMERING R and BLAUT M (2001), 'Pro- and prebiotics – the tasty guardian angels?', *Appl Microbiol Biotechnol*, **55**, 19–28.

SLOAN E (2000), 'The top ten functional food trends', *Food Technol*, **54**(4), 33–62.

STEER T, CARPENTER H, TUOHY K and GIBSON G R (2000), 'Perspectives on the role of the human gut microbiota in health and methods of study', *Nutr Res Rev*, **13**, 229–254.

STEIDLER L, HANS W, SCHOTTE L, NEIRYNCK S, OBERMEIER F, FALK W, FIERS W and REMAUT E (2000), 'Treatment of murine colitis by *Lactococcus lactis* secreting interleukin-10', *Science*, **289**, 1352–1355.

SUAU A, BONNET R, SUTREN M, GODON J J, GIBSON G R, COLLINS M D and DORÉ J (1999), 'Direct analysis of genes encoding 16S rRNA from complex communities reveals many novel molecular species within the human gut', *Appl Environ Microbiol*, **65**, 4799–4807.

SUAU A, ROCHET V, SGHIR A, GRAMET G, BREWAEYS S, SUTREN M, RIGOTTIER-GOIS L and DORÉ J (2001), '*Fusobacterium prausnitzii* and related species represent a dominant group within the human faecal flora', *Syst Appl Microbiol*, **24**, 139–145.

VAUGHAN E E, MOLLET B and DE VOS W M (1999), 'Functionality of probiotics and intestinal lactobacilli: light in the intestinal tract tunnel', *Curr Opin Biotechnol*, **10**(5), 505–510.

ZAMFIR M, CALLEWAERT R, CALINA CORNEA P, SAVU L, VATAFU I and DE VUYST L (1999), 'Purification and characterization of a bacteriocin produced by *Lactobacillus acidophilus* IBB 801', *J Appl Microbiol*, **87**, 923–931.

17

The market for functional dairy products: the case of the United States

L. Hoolihan, Dairy Council of California, USA

17.1 Introduction

It could be argued that the functional foods movement began decades ago with the fortification of milk with vitamins A and D, salt with iodine, and bread with B vitamins. These efforts and fortification of other 'staple' foods with specific nutrients were successful in eradicating diseases of the time such as rickets, goiter and pellagra. Similarly, the recent fortification of folic acid to cereals and grains was intended to help reduce the incidence of certain kinds of birth defects to the spinal cord. In the mid-1990s, however, a very distinct mindshift occurred in the thinking behind fortification of the food supply. Whereas previously foods were fortified with specific nutrients for public health reasons, the new goal was to fortify foods with various nutritive components for achievement of optimal health, vitality, longevity and improved physical performance and endurance. This mindshift marks the real introduction of the functional foods movement. Foods were no longer consumed just for their inherent nutrients but for health benefits beyond the traditional nutrients they contained. Within a period of a few years, a vast number and variety of functional foods were introduced to the market, including energy and sports bars, calcium-fortified orange juice, omega-3 enriched eggs, yoghurts with probiotics, cereals, snack foods, fortified juices and other beverages.

Foods classified as functional are generally associated with an individual's primary health concerns, which in the US include cardiovascular disease, high blood pressure, stroke, high cholesterol and cancer. Other major health concerns include weight, diabetes, and overall nutrition and diet (IFIC, 2002). Which foods the consumer perceives as 'functional' will vary depending on their particular health concerns and nutrition goals. Advertising messages and marketing campaigns will also influence how a consumer views a particular food product or type of food. One survey found that the top ten foods that consumers identify as providing health benefits beyond basic nutrition include broccoli, fish or fish oil, green leafy vegetables, oranges or orange juice, carrots, garlic, fiber, milk, oats/oat bran/oatmeal and tomatoes (IFIC, 2002).

17.2   Drivers of the functional foods market

Along with the general interest in achieving optimal health and well-being, there are a number of factors fueling interest in the functional foods market. Escalating health-care costs are causing people to look for means of preventing disease rather than treating it, and for alternatives to traditional medical care. Our aging population with the desire of living longer, healthier and more active lives prefers to achieve these goals through diet and lifestyle choices rather than through pharmaceutical means. Scientific evidence over the past several years has strengthened the link between diet and wellness, proving to even the most skeptical that diet plays an indisputable role in one's health. The current age of information and technology has also played into this movement. With a wealth of health-related information literally at one's fingertips through the Internet, consumers are educating themselves and increasingly taking charge of their own health. This 'self-health' movement has definitely contributed to the interest in functional foods. Finally, recent legislation (discussed in Section 17.4) has made it easier for food manufacturers to include health-related information on their products, providing an opportunity to educate the consumer at point-of-purchase.

Surveys show that more North Americans than ever before are seeking information on diet and health and taking action to improve their eating habits. The American Dietetic Association has conducted surveys every two years for the past decade to track the evolution of people's attitudes, knowledge and behaviors around diet and nutrition. In their most recent survey, 85% of American consumers say that diet and nutrition are important to them personally, 75% say they carefully select foods in order to achieve balanced nutrition and a healthful diet, and 58% say they actively seek information about nutrition and healthful eating (ADA, 2002). The strength of the perceived link between diet and nutrition has continued to increase from their previous surveys.

The International Food Information Council (IFIC) also tracks consumer perceptions of functional foods. Their most recent survey of over 1000 adults in Massachusetts, USA, found that almost all consumers (94%) believe that certain foods provide health benefits that go beyond basic nutrition and may reduce the risk of disease or other health concerns (IFIC, 2002). This level of interest has been consistently strong since their initial survey in 1998. In addition, a majority (68%) of consumers believe they have a 'great amount' of control over their own health and 78% believe that food and nutrition play a 'great role' in maintaining or improving overall health. Almost two-thirds (63%) of Americans say they are eating at least one food to receive a functional health benefit, up from 53% in 1998. Eight-five percent of Americans are 'very interested' or 'somewhat interested' in learning more about functional foods, a level of interest consistently strong since 1998.

A survey by HealthFocus found that 91% of shoppers use fortified foods and 27% 'always' or 'usually' purchase foods because they are fortified with additional vitamins and minerals. Thirty-nine percent strongly agree or agree that it is worth a small extra premium to buy foods fortified with added nutrients (HealthFocus, 2001a). Between 1992 and 2000, the percentage of US shoppers who believed that some foods contain active components that reduce the risk of disease and improve long-term health increased from 69% to 72%; the percentage who believed that foods can be used to reduce the use of drugs and other medical therapy increased from 44% to 52% over this same time period (HealthFocus, 2001b).

Although the consumer is unquestionably the driver behind the functional foods movement, the medical community is not far behind. The American Dietetic Association's position statement on functional foods states: 'Functional foods, including whole foods and fortified, enriched or enhanced foods, have a potentially beneficial effect on health when consumed as part of a diet on a regular basis, at effective levels' (ADA, 1999). Health professionals are starting to include recommendations of functional foods to their patients and clients on an individual basis. In a recent survey conducted by Market Enhancement Group, Inc., 74% of all health professionals including family practitioners, pediatricians, nurses, nurse practitioners and health educators, indicated that they recommend a diet that emphasizes functional foods to their clients or patients (DCC, 2001). The top reasons health professionals cited for recommending functional foods include improving health generally, preventing osteoporosis, improving heart health, preventing diabetes and increasing energy levels.

17.3   The growth of the functional foods market in the US

The marketing numbers substantiate the increasing interest in this relatively new functional foods arena. The US market accounted for $18.25 billion in total functional food sales in 2001, representing more than 38% of the total worldwide market share, and easily the largest (NBJ, 2001). Europe followed in second place at $15.4 billion, then Japan at $11.8 billion. During the six-year period between 1995 and 2001, the global functional foods market grew an incredible 59%, from $30 billion to an estimated $47.6 billion. The US market grew at a rate of 7.3% in 2001 and over the next five years is expected to remain at a strong growth rate of about 6–8% per year, after which it is predicted to level off to reach sales of $31 billion in 2010 (NBJ, 2002). Still, in terms of total food sales, the US functional foods proportion represents a mere 3.7% of the total $503 billion in food sales in 2001.

The beverage category currently is, and is expected to remain, the leading category of functional foods sales, topping $8.9 billion in the US in 2001.

Table 17.1   US functional food sales in 2001

	$ billion in sales	% of total sales
Beverages (juice, soft drinks, beer, tea, coffee)	$8.9	48
Breads and grains (bread, cereal, pasta, rice, baked goods)	$4.9	27
Packaged/prepared foods (soups, meals, deli, baby, jam)	$1.6	9
Dairy (milk, cheese, eggs, ice cream, yoghurt)	$1.1	6
Snack foods (candy, chips, bars)	$1.6	9
Condiments (oils, dressings, spreads, sauces, spices, sweeteners)	$0.1	1
Total	$18.25	

Source: NBJ (2002).

Beverages are followed in sales by breads and grains in sales at $4.9 billion, packaged/prepared foods at $1.6, dairy at $1.1, snack foods at $1.6 and condiments at $0.15 billion/year (NBJ, 2002) (see Table 17.1).

17.4   The regulatory context in the US

17.4.1   Terms used to describe functional foods
There are a variety of terms in the scientific as well as the lay literature used to describe the collective group of foods that provide some functionality or added health benefit. Functional foods, perhaps one of the more well-known and accepted terms, was first defined in 1994 as 'any modified food or food ingredient that may provide a health benefit beyond the traditional nutrients it contains' (Thomas & Earl, 1994). Other terms commonly used do not enjoy a universal definition but include nutraceuticals, pharmafood and designer foods. Nutraceuticals are considered any substance that may be considered a food or part of a food and provides medical or health benefits, including the prevention and treatment of disease. Pharmafoods are foods or nutrients that claim medical or health benefits, including the prevention and treatment of disease. Designer foods, which may involve genetic engineering, are processed foods that are supplemented with food ingredients naturally rich in disease-preventing substances.

In spite of the quest for optimal health and the drive towards consuming certain 'healthful' foods to attain it, consumers express little familiarity with these terms commonly used to describe the concept. Recent IFIC research found that the majority of consumers, 62%, prefer the term 'functional foods' over other terms in describing foods that provide health benefits beyond basic nutrition (IFIC, 2002). It is possible that, although the consumer is the driver of this movement, it has been the food industry, health professionals and health/nutrition organizations that have coined terms to express and promote the concept.

A point of confusion among experts has been that there is no clear distinction between a 'non-functional' and a 'functional' food. There is disagreement as to whether functional foods should be limited to those foods to which some adulteration has taken place to alter its nutrient content, or if the term should apply as well to unadulterated products that naturally provide health benefits beyond their inherent nutrients. *Nutrition Business Journal*, which has been following the business angle of functional foods since their inception, recently segmented functional foods into categories including 'substantially fortified' foods representing 58% of the total functional foods sales, 'inherently functional' foods representing 22%, and 'performance' foods representing 20% (NBJ, 2002). This categorization may help to bring the food industry and regulators to a working consensus over what constitutes a functional food.

17.4.2   Legislative acts influencing the functional foods movement

Promotion of the concept of functional foods from the regulatory perspective has occurred in stages since the mid-1980s through the enactment of three US legislative acts. The Nutrition Labeling and Education Act (NLEA) of 1990 mandated that the Food and Drug Administration (FDA) establish regulations requiring most foods to have a uniform nutrition label. The NLEA also established circumstances under which claims about content and disease prevention could be made about nutrients in foods (NLEA, 1990). Following this legislation, the Dietary Supplement Health and Education Act of 1994 provided a strict definition for dietary supplements and created a consistent means for dealing with safety issues, regulating of health claims, and labeling of dietary supplements (DSHEA, 1994). Finally, the FDA Modernization Act of 1997 in effect was an amendment of the federal Food, Drug, and Cosmetic Act by allowing health claims that are not preauthorized by the FDA if the claims are based on 'authoritative statements' of government agencies such as the National Institutes of Health (FDAMA, 1997). All three of these legislative acts have resulted in the food and supplement industries being held more accountable for claims made and labeling of products. This, in turn, has probably had a positive impact on consumer perceptions and confidence around claims and information provided on food labels.

The federal Food, Drug, and Cosmetic Act does not provide a statutory definition of functional foods, or any of the other terms mentioned above, and hence has no authority to establish a formal regulatory category for such foods. Because functional foods do not constitute a distinctly separate category of foods, the FDA regulates them in the same way it regulates all foods. The safety of the ingredients must be assured and all claims must be substantiated, truthful and non-misleading. In addition, functional foods, like all other foods, are regulated in part by their intended use, determined by label claims, the directions for use on the product label, and any other accompanying label information.

Scientific and regulatory agencies are considering the best way to regulate claims made for functional foods or components. The FDA currently regulates food products depending upon their intended use and the nature of claims made on the labels, with three types of statements allowed. Structure–function claims are used to describe an effect a food or food component might have on the normal functioning of the body but does not affect a disease state. Such claims do not need official approval by the FDA. However, in order to use a structure–function claim the manufacturer must be able to back up the claim with sound scientific data. Regulations regarding use of structure–function claims are not clearly defined for food because it is not known how far a marketer can go before triggering concern by the FDA that the claim is stepping into the more highly regulated health claim arena. Health claims are claims on the food label that describe a relationship between a food or food component and a disease or health-related condition. Health claims for food labels must be approved prior to use by the FDA and must be supported by significant scientific agreement. Nutrient content claims are terms that describe the level of a nutrient in a food. Table 17.2 provides descriptions of these three types of label claims and examples of each.

Table 17.2 Types of label claims

Type of claim	Requirements for use	Example
Health claim: claim on the food label that describes a relationship between a food or food component and a disease or health-related condition.	Prior FDA approval	• While many factors affect heart disease, diets low in saturated fat and cholesterol may reduce the risk of disease. • Diets low in sodium may reduce the risk of high blood pressure, a disease associated with many factors. • Diets containing foods that are good sources of potassium and low in sodium may reduce the risk of high blood pressure and stroke.
Structure–function claim: claim that refers to a nutrient that maintains the structure or function of the human body but does not affect a disease state	Ability to show that the claim is substantiated by credible research	• The calcium, vitamin D and phosphorus in low-fat milk help build strong bones. • The protein in yoghurt helps maintain strong muscles. • The vitamin A in fat-free milk helps promote normal vision.
Nutrient content claim: claim that describes the level of a nutrient in a food	Substantiation that claim is accurate and not misleading	• Excellent source of vitamin D • Low fat • Good source of vitamin A • Low in saturated fat • A low sodium food

17.5   The potential for functional dairy foods in the US

One of the largest areas of functional foods in Europe and Japan has been in the dairy market, accounting for about 65% of all functional food sales in Europe. This trend in the US has been considerably slower to catch on, with functional dairy products accounting for only 6.4% of the total functional foods sales in 2000 (NBJ, 2001).

However, there remains considerable potential for dairy products within the functional foods movement in the US. Over the past several years, consumption of dairy products has been linked to improved bone health (Miller *et al.*, 2000), reduced blood pressure (Bucher *et al.*, 1996; Appel *et al.*, 1997), an enhanced immune system (Kelley *et al.*, 2001; Erickson and Hubbard, 2000; Early *et al.*, 2001), reduced risk of colon (Holt *et al.*, 1998) and breast cancers (Hjartaker *et al.*, 2001). Newer research is showing that the calcium from dairy products may even play a role in improved body weight maintenance (Zemel & Morgan, 2002). Specific components under investigation for these effects include calcium, conjugated linoleic acid, probiotics, whey protein, lactoferrin, butyric acid and sphingolipids, among others.

Milk is in fact considered by some the 'original functional food', not just for the nutrients it inherently contains but also for its ability to deliver those nutrients to the body. Lactoferrin present in milk enhances iron absorption by ensuring that the iron remains in a stable, non-toxic form. Research on the bioavailability of other nutrients from milk show that vitamin E is absorbed twice as well from milk as it is from fortified orange juice or from supplements (Hayes *et al.*, 2001), and that the triglyceride structure in milk enhances the absorption of both long chain fatty acids and calcium (Kennedy *et al.*, 1999). In addition, the 'cluster' of nutrients in milk, including calcium, vitamin D, protein, phosphorus, magnesium, vitamin A, vitamin B6 and trace elements such as zinc – all important for bone health – are present in milk in levels appropriate to maximize calcium absorption (Anderson, 2000).

Other components not inherently concentrated in milk, but that are receiving attention as possible fortification opportunities, include beta-glucan for maintaining cardiovascular health, glucosamine for maintaining joint health, lactoferrin for health gastrointestinal microflora, lutein for healthy vision and general eye health, and magnesium for cardiovascular health, among others (FMSTI, 2001). Consumer reaction to, and acceptance of, the fortification of milk with these ingredients is currently being investigated so that processors will know which to offer.

One of the most successful strategies for positioning functional foods, identified in the 1998 IFIC survey, is to place functional foods in the most traditional context possible. This makes milk, one of the most traditional foods available, well positioned in the functional foods movement. Milk not only contains a myriad of important nutrients for growth, development and maintenance of optimal health but as discussed above is also being recognized as an ideal carrier for other nutrients that may not be highly concentrated in milk. This presents obvious opportunities for the industry to utilize milk as a fortification vehicle for nutrients for which consumption is marginal or inadequate, further enhancing the functional food status of dairy products.

17.5.1   Specific functional dairy foods in the market

Recognizing the potential for functional dairy products above and beyond traditional dairy products, various producers have introduced a number of products into the market. As examples of functional milks introduced in the US in July 2000, Parmalat USA unveiled three new functional milks, each claiming to expand the natural benefits of milk. *Milk-E* containing additional vitamin E and biotin was touted to promote healthy skin and faster metabolism. *Lactose Free Plus* was introduced as a lactose-free milk containing inulin, a source of dietary fiber, and *Lactobacillus acidophilus* and *Bifidobacterium*, well-known probiotics with substantial health benefits (See chapter 10). Finally, *Skim Plus* contains 34% more calcium and 37% more protein than traditional milk. Since then Parmalat has introduced Plus Omega-3, a UHT milk containing 80 mg omega-3 per 100 ml with vitamins E, C and B6 as well as calcium.

Danone's *Actimel* is another US example of a functional dairy product, currently in its third year of test marketing in Colorado. *Actimel* is a yoghurt drink fermented by *Lactobacillus bulgaricus* and *Streptococcus thermophilus* and containing added *Lactobacillus casei immunitass*. Stonyfield Farms is also test-marketing a drinkable yoghurt intended for adults, which is fortified with six live, active probiotic cultures. Suiza Foods manufactures *Kidsmilk*, described as 'containing 50% more of the 11 vitamins found in regular milk'. Campina produces a buttermilk and apple juice product with added vitamin D and calcium, as well as raftilose intended to increase calcium absorption.

There are considerable regional differences in the availability of these various products in the US, making broad-reaching marketing campaigns ineffective and inappropriate. As most of these products are either in the test-marketing stage or early in their life cycle, it remains to be seen how successful they will be and to what extent the dairy industry will continue to develop and market new functional dairy products to meet the needs of specific segments of the population.

17.5.2   Perceptions of functional dairy foods

Although fruits and vegetables initially received most of the spotlight in the functional foods arena for their healthful and disease-preventative attributes, dairy has become a strong player in the eyes of the health professional as well as the consumer. A recent survey of health professionals in California indicated that 91% of those surveyed considered dairy products as a functional food. Yoghurt, milk and cottage cheese drew the most attention in the dairy functional foods category, followed by cheese, dry milk added to foods, and whey (DCC, 2001). A similar survey in consumers found that 64% of consumers believe that dairy foods can help decrease the risk of disease and/or optimize the health and wellness of one or more members of their family (DCC, 2002). Enhancing immunity, improving the absorption and utilization of nutrients, and decreasing breast cancer risk were some of the benefits of dairy consumption cited by consumers.

17.6   Future trends

17.6.1   The future of functional diary products

Development and marketing of functional dairy products is promising in a number of areas. Fortifying dairy foods with iron is one area recently receiving attention. Because dairy foods represent a major food source for many cultures worldwide and they are inherently low in iron, dairy products may be a reasonable vehicle for iron fortification. Certain population subsegments such as children, women of child-bearing age and vegetarians whose iron consumption may be marginal would, in particular, benefit from fortification of dairy foods with iron. Research is currently underway to examine ways to get around the adverse effects added iron may have on the lipids, caseins and whey proteins in dairy products. Complexing added iron with lactoferrin may be one solution to fortify with iron without the color, flavor and quality issues generally associated with iron fortification. Remembering that texture, appearance and taste prevail over nutrient content and health benefits in purchasing decisions, successful fortification of a product must consider these factors. Convenience and pricing will also play a big role in the ultimate success of a product.

Prebiotics – inulin and oligofructose – is another area receiving attention for the ability to increase calcium absorption, improve bone mineral density, improve functioning of the gastrointestinal tract, and enhance immunity among other potential health benefits (Milner & Roberfroid, 1999). Stonyfield Farms' family of yoghurts currently contains inulin, along with their trademark six live active probiotic cultures. A variety of other food manufacturers are experimenting with the fortification of prebiotics to their products as well.

Other substances that may play a role in future functional dairy products include whey proteins for their immune-enhancing and cancer-fighting properties, sphingolipids for their ability to ward off some forms of cancer, conjugated linoleic acid for its body composition and cancer benefits, and bioactive peptides for their immune-enhancing effects. Research on these components is, at this point, preliminary, mostly taking place in laboratory and animal studies. Future research will undoubtedly focus on determining levels needed for consumers to reap the benefits, identifying the most appropriate dairy product to use as a fortification vehicle and determining stability and shelf-life of products incorporating these components, among other manufacturing concerns.

Thus dairy foods, long seen as ideal vehicles for fortification with added nutrients, are positioned strongly in the functional foods movement. Dairy foods are already considered to be natural, healthy and wholesome foods. Sensory studies consistently show that the consumer enjoys their taste and other sensory qualities. They are also moderately priced and represent a 'traditional' food category identified as having the most promise in successful marketing and sales of functional foods. It remains to be seen which nutrients and other ingredients will come out on top in terms of consumer perception and acceptance and, perhaps just as important, how quickly the dairy industry chooses to

act in developing and marketing functional products incorporating those ingredients.

17.6.2   Future trends in the overall functional foods market

In spite of the overall positive perceptions and attitudes towards functional foods
amongst the consumer and health professional, the continued growth of the market
will depend upon a number of factors. How consumers learn about diet and health
connections will be paramount, with many consumers skeptical when reading
marketing claims of food companies or on product labels. Consumer awareness
campaigns such as promotion via health professionals or other credible authorities
will be necessary in the successful marketing of functional food products. The
expected health benefits of consuming a particular food and the levels and
frequency of consumption necessary to make a difference are areas that will need
to be addressed in some form of consumer education. Knowledge about food–drug
interactions is also needed, as functional foods may either block or increase the
absorption of drugs, either reducing their effectiveness or increasing the risk of
toxicity.

The pricing of functional food products will also impact the long-term success
of this broad food category. Typically, functional foods attract a premium price,
and even in cases in which a food might be proven beneficial for a certain
condition, its relatively high price may discourage people from purchasing the
product. This will be particularly the case in the lower socio-economic group,
which paradoxically is often the group with the greatest need for good nutrition.

A strong regulatory framework will also be essential in the long-term growth
and success of the functional foods market. Over-aggressive marketing of foods
with added health benefits, misleading information on food labels and lack of a
strong regulatory system for overseeing the production and marketing of these
foods have led to a decline in consumer confidence. Finally, taste remains the
number one reason for food purchases, emphasizing the need to the food industry
to develop foods that, above all else, are palatable.

Citing the number of failed functional food products that have been pulled from
the marketplace for sluggish sales over the past few years, US food manufacturers
are hesitant to jump on the functional foods bandwagon that other countries have
been riding. Rather than researching, developing and marketing brand new products, the emphasis will continue to be on enhancing existing products by fortifying
with high-profile nutrients such as calcium, iron, and antioxidant vitamins E, C and
A. Traditional, familiar foods and food groups such as dairy will do well here as the
consumer already has considerable trust in these products. Given the growing
bottled water market and the strong functional food beverage category,
'aquaceuticals' – fortified bottled water – is also predicted to become mainstream
over the next few years. Probiotics for protection against infection and enhanced
immune system, fiber and potassium for blood pressure, and foods fortified with
calcium will likely be other emerging functional food categories.

The growth of the market will depend on these factors and others as the consumer, food industry and regulators alike learn what works and what does not. There will undoubtedly be more product failures than successes in this market as manufacturers juggle regulatory limitations with consumer demands and how to deliver credible information about their products. Regulatory agencies will need to put into place a framework that allows for new foods to be aggressively developed and marketed, yet which maintains safety standards and regulates information provided on product labels. In turn, the consumer will be overwhelmed with the number of food choices available as well as with the abundance of health- and nutrition-related information relevant to themselves and their families. Given the increasingly strong interest in and link between health and nutrition, however, the functional foods arena will undoubtedly move forward; the question is not if and when, but how quickly and aggressively.

17.7   Sources of further information and advice

Functional Foods for Health Program; Department of Food Science and Human Nutrition; University of Illinois at Urbana-Champaign, http://www.ag.uiuc.edu/~ffh/ffh.html
Institute of Food Technologists, www.ift.org
International Food Information Council Foundation, http://www.ific.org/food
American Dietetic Association, www.eatright.org
Food Science Program at Rutgers University, http://foodsci.rutgers.edu
USDA/ARS Food and Nutrition Information Center, http://www.nal.usda.gov/fnic

17.8   References

ADA (AMERICAN DIETETIC ASSOCIATION) (1999), 'Functional foods – Position of ADA', *J Am Diet Assoc*, **99**, 1278–1285.
ADA (AMERICAN DIETETIC ASSOCIATION) (2002), 'Tracking trends', *Dietetics in Practice*, 2(1), 1–4, http://www.eatright.org/pr/2002/trends2002.html.
ANDERSON J J B (2000), 'Role of calcium and vitamin D in bone modeling and remodeling in adolescent girls', IFT Annual Meeting abstract no. 28–2.
APPEL L J, MOORE T J, OBARZANEK E, VOLLMER W M, SVETKEY L P, SACKS F M, BRAY G A, VOGT T M, CUTLER J A, WINDHAUSER M M, LIN P H and KARANJA N (1997), 'A clinical trial of the effects of dietary patterns on blood pressure', *N Engl J Med,* **336**, 1117–24.
BUCHER H C, COOK R J, GUYATT G H, LANG, J D, COOK D J, HATALA R and HUNT D L (1996), 'Effects of dietary calcium supplementation on blood pressure: a meta-analysis of randomized controlled trials', *J Amer Med Assoc*, **275**, 1016–1022.
DCC (DAIRY COUNCIL OF CALIFORNIA) (2001), Functional Foods Study, June 2001; Market Enhancement Group, Inc.
DCC (DAIRY COUNCIL OF CALIFORNIA) (2002), DCC Functional Foods Study, September 2002; Market Enhancement Group, Inc.
DSHEA (DIETARY SUPPLEMENT HEALTH AND EDUCATION ACT) (1994), Public Law 103–417, section 13(a).
EARLY E M, HARDY H, FORDE T and KANE M (2001), 'Bactericidal effect of a whey protein concentrate with anti-*Helicobacter pylori* activity', *J Appl Microbiol*, **90**(5), 741–748.
ERICKSON K L and HUBBARD N E (2000), 'Probiotic immunomodulation in health and disease', *J Nutr*, **130**, 403–409.
FDAMA (FEDERAL DRUG ADMINISTRATION MODERNIZATION ACT) (1997), Public law, 105–115.

FMSTI (FLUID MILK STRATEGIC THINKING INITIATIVE) (2001), 'Fluid Milk's Role in the Functional Foods Movement: Milk's Unique Nutrient Profile and Functional Ingredient Opportunities', www.milkplan.org.

HAYES KC, PRONCZUK A and PERLMAN D (2001), 'Vitamin E in fortified cow milk uniquely enriches human plasma lipoproteins', *Am J Clin Nutr*, **74**, 211–218.

HEALTHFOCUS (2001a), 'Functional Foods and Nutraceuticals', Consumer Trend Report.

HEALTHFOCUS (2001b), 'National Study of Public Attitudes and Actions Toward Shopping and Eating', Consumer Trend Report.

HJARTAKER A, LAAKE P and LUND E (2001), 'Childhood and adult milk consumption and risk of premenopausal breast cancer in a cohort of 48,844 women – the Norwegian women and cancer study', *Int J Cancer*, **93**(6), 888–893.

HOLT PR, ATTILLASOY EO, GILMAN J, GUSS J, MOSS SF, NEWMARK H, FAN K, YANG K and LIPKIN M (1998), 'Modulation of abnormal colonic epithelial cell proliferation and differentiation by low-fat dairy foods', *J Amer Med Assoc*, **280**, 1074–1079.

IFIC (INTERNATIONAL FOOD INFORMATION COUNCIL) (2002), 'Functional foods attitudinal research: the consumer view on functional foods: yesterday and today', *Food Insight*, **May/June.**

KELLEY D S, SIMON V A, TAYLOR P C, RUDOPH I L, BENITO P, NELSON G J, MACKEY B E and ERICKSON K L (2001), 'Dietary supplementation with conjugated linoleic acid increased its concentration in human peripheral blood mononuclear cells, but did not alter their function', *Lipids*, **36**, 669–674.

KENNEDY K, FEWTRELL M S, MORLEY R, ABBOTT R, QUINLAN P T, WELLS J C K, BINDELS J G and LUCAS A (1999), 'Double-blind, randomized trial of a synthetic triacylglycerol in formula-fed infants: effects on stool biochemistry, stool characteristics, and bone mineralization', *Am J Clin Nutr*, **70**, 920–927.

MILLER G D, JARVIS J K and MCBEAN L D (2000), *Handbook of Dairy Foods and Nutrition.* 2nd ed. Boca Raton, London, New York, Washington, DC. CRC Press.

MILNER JA and ROBERFROID M (1999), 'Nutritional and health benefits of inulin and oligofructose; proceedings of a conference held May 18–19, 1998', *J Nutrition*, **129**(7S), 1395S–1502S.

NAS THOMAS P R and EARL R EDS (1994), *Opportunities in the nutrition and food sciences; research challenges and the next generation of investigators.* Institute of Medicine, National Academy of Sciences, Washington, DC, National Academy Press, 109.

NBJ (NUTRITION BUSINESS JOURNAL) (2001), 'Functional foods V', *Nutr Bus J*, **7**(10), 1–8, www.nutritionbusiness.com.

NBJ (NUTRITION BUSINESS JOURNAL) (2002), 'NBJ's annual overview of the nutrition industry VII', *Nutr Bus J*, **7**(5/6), 1–11, www.nutritionbusiness.com.

NLEA (NUTRITION LABELING and EDUCATION ACT) (1990), Code of Federal Regulations, **101**, 14(a)(1).

ZEMEL M B and MORGAN K (2002), 'Regulation of adiposity and obesity risk by dietary calcium: mechanisms and implications', *J Am Coll Nutr*, **21**, 146S–151S.

Index

Lightning Source UK Ltd.
Milton Keynes UK
178398UK00001B/12/P